Mathematical
Surveys
and
Monographs

Volume 147

Classifying Spaces of Sporadic Groups

David J. Benson
Stephen D. Smith

American Mathematical Society

2000 *Mathematics Subject Classification.* Primary 20D08, 20J06, 55R35; Secondary 20B25, 20J05, 55R40.

For additional information and updates on this book, visit
www.ams.org/bookpages/surv-147

Library of Congress Cataloging-in-Publication Data

Benson, D. J. (David J.), 1955–
Classifying spaces of sporadic groups / David J. Benson, Stephen D. Smith.
p. cm. — (Mathematical surveys and monographs, ISSN 0076-5376 ; v. 147)
Includes bibliographical references and index.
ISBN 978-0-8218-4474-8 (alk. paper)
1. Sporadic groups (Mathematics) 2. Classifying spaces. I. Smith, Stephen D., 1948– II. Title.

QA177.B465 2008
512′.2—dc22 2007060587

∞ The paper used in this book is acid-free and falls within the guidelines
established to ensure permanence and durability.
Visit the AMS home page at http://www.ams.org/

10 9 8 7 6 5 4 3 2 1 13 12 11 10 09 08

ABSTRACT. For each of the 26 sporadic finite simple groups, we construct a 2-completed classifying space, via a homotopy decomposition in terms of classifying spaces of suitable 2-local subgroups; this leads to an additive decomposition of the mod 2 group cohomology. We also summarize the current status of knowledge in the literature about the ring structure of the mod 2 cohomology of those groups.

Our decompositions arise via recent homotopy colimit theorems of various authors: in those results, the colimit is indexed by a collection of 2-subgroups, which is "ample" in the sense of affording the desired cohomology decomposition. Furthermore our decompositions are "sharp" in the sense of being simplified by the collapse of an underlying spectral sequence.

Among the various standard ample collections available in the topological literature, we make a suitably minimal choice for each group—and we further interpret that collection in terms of an equivalent "2-local geometry" from the group theory literature: namely as a simplicial complex determined by certain 2-local subgroups. In particular, we complete the verification that for each sporadic group, an appropriate 2-local geometry affords a small ample collection. One feature which emerges in each case is that the geometry has "flag-transitive" action by the group, so that the orbit complex (the quotient space modulo that action) is a simplex: and then the diagram of classifying spaces of subgroups which indexes the homotopy decomposition is a pushout n-cube of the relevant dimension n (though sometimes that orbit simplex diagram is further simplified via cancellations).

The work begins with a fairly extensive initial exposition, intended for non-experts, of background material on the relevant constructions from algebraic topology, and on local geometries from group theory. The subsequent chapters then use those structures to develop the main results on individual sporadic groups.

Contents

Introduction

The 26 sporadic simple groups have played an exceptional role in the theory and classification[1] of the finite simple groups, ever since the discovery of the first few such groups in the late 1800s. In the last decade or so, various authors have also observed exceptional behavior in the mod p group cohomology of some of the sporadic groups, especially at the prime $p = 2$; see for example Benson [**24, 25, 26**], Benson and Wilkerson [**29**], Milgram [**92, 93, 94**]. Determining (or at least describing) the cohomology of sporadic groups represents an area of considerable current activity; and such study combines methods of algebraic topology with information and techniques from finite group theory.

In this work, we provide for each sporadic group a description of a 2-completed classifying space, in terms of classifying spaces of suitable 2-local subgroups. In particular, this leads to an additive decomposition of the mod 2 group cohomology. Similar decompositions can be obtained for odd primes p; but the indexing set of p-local subgroups is usually smaller and often less interesting.

Although our focus here is on sporadic groups, we emphasize that this work should be also be viewed in relation to work on cohomology of the other simple groups. For example, our decompositions for sporadic groups are largely inspired by a standard decomposition (see Example 5.1.18) of the mod p cohomology for a group of Lie type of characteristic p. However, even in that model case for Lie type groups, it remains a major open problem to give an explicit description of the individual component terms and their cohomology. In contrast, the cohomology and completed classifying spaces of groups of Lie type in coprime characteristic are rather well understood in the light of the extraordinary work of Quillen [**103**] for the general linear groups, and subsequent work along the same lines by Fiedorowicz and Priddy [**60**] for the remaining classical groups and Kleinerman [**81**] for the groups of exceptional Lie type. Cohomology and classifying spaces of symmetric and alternating groups are understood to some extent (Nakaoka [**99**], Mui [**97**], Mann [**89, 90**], Gunawardena, Lannes and Zarati [**70**], Ha and Lesh [**71**], etc).

The context of this work. Naturally the structure of the p-local subgroups[2] in a finite group G should provide an important ingredient in the determination of its mod-p group cohomology. One p-local approach to cohomology, often used as a tool in modern work on individual sporadic groups, is provided by the topological literature on *cohomology decompositions*. In particular, we mention Dwyer's analysis [**51, 53**] of *ample* collections $\mathcal{C}$ of p-subgroups of G, namely collections which

[1] The reader who is uneasy with the classification of the finite simple groups may insert "known" before "sporadic simple groups" here. The classification is not used in our analysis in this work, which studies only the properties of those 26 individual sporadic groups.

[2] A *p-local subgroup* is the normalizer of a non-identity p-subgroup.

afford an additive decomposition of the mod p group cohomology of G, where the components are given by the cohomology of certain p-local subgroups of G (for example, the normalizers of the p-groups in $\mathcal{C}$). The origins of these decompositions go back to work of Brown, Quillen, and Webb in the 1970s and 1980s. More recently, treatments of certain ample collections (notably by Jackowski, McClure, Oliver, Dwyer, and Grodal) consider *homotopy decompositions*, namely corresponding decompositions not at the level of cohomology, but instead at the deeper topological level of the classifying space BG. Here the decomposition is in fact afforded by a homotopy colimit, in which the component terms are classifying spaces of suitable subgroups of G, and the terms are indexed by a category determined by the collection $\mathcal{C}$. For these purposes, there are various standard ample collections in the literature, such as $\mathcal{S}_p(G)$ defined by all nontrivial p-subgroups; but often the standard collections are too large to be useful for practical computation, so that it is also of interest to determine ample subcollections which are as *small* as possible.

Several years ago, the first author initiated the project that became this work: namely the construction, via a homotopy decomposition over subgroups, of a 2-completed classifying space $BG_2^\wedge$ for each sporadic group G. One motivation was the observation that for a number of the sporadic groups, the group theory literature (especially that on *2-local geometries*) already contained information sufficient to determine a small ample collection, and hence a homotopy decomposition—for example, collections corresponding to the 2-local geometries treated in Smith and Yoshiara [**116**], and the geometry used in Benson [**24**]. For the remaining sporadic groups, progress was initially slower, since the larger groups require substantially more complicated analysis. However, recent work of various authors on the 2-local structure of large sporadic groups has led for example to the completion in Yoshiara [**131**] of the determination of the 2-radical subgroups of the remaining sporadic groups. So, combining such results in group theory with the homotopy colimit theorems indicated above, in the present work we exhibit for each of the 26 sporadic groups a small ample collection related to a 2-local geometry, and the corresponding homotopy decomposition. In each case, since the group acts flag-transitively on the geometry, the relevant homotopy colimit is over an indexing category given by a *simplex*; so that the diagram of classifying spaces is that of a pushout n-cube in the relevant dimension n. (Occasionally cancellations simplify the calculation to an even smaller subdiagram.) We show furthermore that each decomposition is *sharp*, in Dwyer's sense of affording an alternating sum decomposition of group cohomology via the formula of Webb—this simplification arises from the collapse of an underlying spectral sequence. In particular, these results confirm a conjecture (see [**116**, Conj. 1, p. 376]) concerning such a connection between group theory and algebraic topology. That conjecture was necessarily somewhat vague, since the 2-local geometries in the literature were defined in a number of different ways; however, in this work we do indicate for each group a 2-local geometry suitable for our purposes.

Outline of the work. Our treatment is fairly lengthy, so we indicate some of its main features in overview.

Chapter 1 gives a brief summary of our main results.

Since we wish to make our account accessible both to group theorists and to homotopy theorists, we then present in Part 1 an extensive exposition for non-experts of some of the background material involved in each area.

Most of Part 1 is devoted to a review of selected topological material leading up to, and into, the modern literature[3] on decompositions of group cohomology: Chapter 2 recalls some basics of the group cohomology of a finite group G; including aspects of the topological approach via the classifying space BG, and of "approximating" BG—via the Borel construction on a suitable G-space. In the classical literature, these spaces of interest are viewed as topological spaces; but in Chapter 3, we will indicate the more modern viewpoint on spaces as simplicial sets—emphasizing the standard equivalence (due to Quillen) of those two viewpoints, in terms of the homotopy categories they determine. In Chapter 4, we then adopt the viewpoint of simplicial sets—since that is the appropriate general context for our description there of two important constructions due to Bousfield and Kan, namely completions and homotopy colimits. Chapter 5 then reviews some of the more specific literature on homotopy decompositions of the classifying space, given in Dwyer's viewpoint via a homotopy colimit indexed by a collection of p-subgroups; the chapter then examines some of the standard particular collections which are ample in the sense of affording such a decomposition.

Part 1 concludes with Chapter 6, which reviews various notions from the group theory literature, primarily relevant to 2-local geometries for sporadic groups; especially from the viewpoint of their observed connections with the ample collections of 2-subgroups and decompositions described in Chapter 5.

Part 2 then presents the main results of our work. Chapter 7 contains a section on each of the individual sporadic groups G: We indicate a homotopy decomposition for the 2-completed classifying space $BG^{\wedge}_2$, in terms of a small ample collection determined by a suitable 2-local geometry—with a corresponding additive decomposition of the mod 2 group cohomology of G. We also add, where available, further remarks on the status of knowledge in the literature about the full ring structure of the mod 2 cohomology. In some cases, our results are fairly immediate deductions from results already in the literature; but in the remaining cases where we require a more substantial argument, we sometimes postpone the detailed proofs to sections corresponding to the relevant groups in our final Chapter 8.

We close this Introduction with a brief mention of a topic which we will touch on repeatedly later, but for which we have *not* tried to develop our own exposition in this work:

Foreword on Lie type groups and buildings. Our main results in Chapter 7 are concerned with the 26 sporadic simple groups, viewed via their 2-local geometries; but at various points in our expository Chapters 4–6, it will also be natural to consider examples given instead by the simple groups of Lie type, together with their natural geometries given by Tits buildings. There are a number of reasons for this, among them:

- Buildings are simplicial complexes providing convenient and very natural spaces for a number of the topological constructions we will be examining.

[3]In particular we will frequently quote from a very useful recent exposition on decompositions and their background by Dwyer [**53**]—which we strongly urge readers to consult alongside our exposition.

• Various standard results on groups of Lie type and buildings provide a "model case" for homology decompositions and related research areas—often providing motivation for generalizations to other groups, or even to all finite groups.

• The 2-local geometries for sporadic groups were inspired by certain analogies with buildings.

In particular, our main results for sporadic groups involve a homotopy decomposition indexed by a simplex—and these can be regarded as analogues of the corresponding standard result for Lie type groups, see Example 5.1.18.

When we discuss examples from the theory of Lie type groups, our approach will be to quote the relevant properties from the literature, wherever they arise in our development, for example in Example 4.6.3—rather than attempting to provide a single exposition of that material beforehand; for nowadays a number of fuller introductions to the Lie theory are readily available. (But we do review a fair number of relevant properties at Example 5.1.1.)

In particular, an expository treatment of groups of Lie type, tailored to the general context of our work here, can be found in Section 6.8 of the book [**22**] of the first author. We also mention the expository discussion of buildings given in Chapter 2 of the book [**115**] now in preparation[4] by the second author. These two treatments also list various additional references: on groups of Lie type, we mention especially Carter [**43**] as well as Ch. 2–4 in the third volume [**64**] of the series of Gorenstein, Lyons, and Solomon; and on buildings, Brown [**36**] and Ronan [**105**].

Acknowledgments. We are grateful to a number of colleagues who have read earlier versions of this work, and made a number of corrections and very helpful comments and suggestions; including Jianbei An, Masato Sawabe, Satoshi Yoshiara, John Maginnis, Silvia Onofrei, Jesper Grodal, and Bob Oliver.

Benson is partly supported by NSF grant DMS-0242909 and by a Senior Scientist Prize from the Humboldt Foundation. Smith is partly supported by NSA grant MDA 903-03-1-0100 and subsequent grants.

[4]As of this writing (June 2007), revised Chapters 1–3 are visible on the Web at the URL http://www.math.uic.edu/~smiths/book/book.ps; further revisions will continue at least through later 2007.

Some general notation and conventions

We indicate here at the outset some fairly standard general notation, which we normally follow throughout this work. (Of course we will introduce more specific notation as needed later on in the course of the text.)

We write $A := B$ or $B =: A$ to indicate that A is *defined* to be equal to B; whereas we write $A = B$ to denote an equality between already defined objects.

Conventions for groups. We use G to denote some *finite* group.We emphasize that at least initially, G can be an arbitrary finite group; we will not restrict our focus to simple groups (and in particular, sporadic groups) until Chapter 6. We indicate below some common conventions for general finite groups; for further details, see standard treatments such as Gorenstein, Lyons, and Solomon [**64**].

For a prime p, we write $O_p(G)$ for the largest normal p-subgroup of G, and $O^p(G)$ for the smallest normal subgroup of G for which $G/O^p(G)$ is a p-group. Thus $O^p(G)$ is generated by the p'-elements of of G.

If H and K are subgroups of a group G, we write $[H, K]$ for the subgroup of G generated by the commutators $h^{-1}k^{-1}hk$ where $h \in H$ and $k \in K$. We write $L_n(G)$ for the nth term in the *lower central series* of G, defined by $L_0(G) := G$ and $L_{n+1}(G) := [L_n(G), G]$. We write $L_\infty(G)$ for the intersection of the $L_n(G)$, namely the final term in the lower central series. This is the smallest normal subgroup of G for which the quotient is nilpotent. Since a finite nilpotent group is the direct product of its Sylow p-subgroups, we have

$$L_\infty(G) = \bigcap_p O^p(G); \qquad G/L_\infty(G) \cong \prod_p G/O^p(G). \tag{0.0.1}$$

Suppose that $H \leq K \leq G$ are finite groups. We say that H is *weakly closed* in K with respect to G if the only G-conjugate of H in K is H itself. In particular, this implies that H is normal in K.

Atlas conventions, with some variations. For the names of simple groups, and for various notation related to their subgroup structure, we use the conventions of the Atlas [**47**]; see the Atlas for fuller details. We summarize some of these usages below, indicating several small modifications that we use in this work. Many of these conventions can be seen in subgroups of the sporadic groups given in the table in Chapter 1.

Group extensions: The notation $N{:}H$ indicates a split extension with normal subgroup N and quotient H; while $N{}^{\cdot}H$ indicates a nonsplit extension. The notation $N\,H$ makes no statement about whether the extension splits. For example, all three of these notations for extensions arise for subgroups in the entry for the Fischer group Fi_{24}' in the table in Chapter 1.

Structure of certain p-groups: If m and n are positive integers, we write m^n for an abelian group $(\mathbb{Z}/m)^n$, and in particular we use a number m to denote $\mathbb{Z}/m$, except where confusion might result from these abbreviations. The notation $p^{a+b+\cdots}$ denotes a p-group having an ascending central series with quotients of rank $a, b, \cdots$; square brackets $[a]$ indicate order p^a with no assertion about a series. In particular, p^{1+2n} denotes an extraspecial p-group of width n; with exponent p when p is odd. When $p = 2$, the two types of extraspecial 2-groups of width n are denoted by 2^{1+2n}_+ and 2^{1+2n}_-, where the sign indicates the type of the orthogonal form on the central quotient. In width $n = 1$, we have $2^{1+2}_+ = D_8$, the dihedral group of order 8; and $2^{1+2}_- = Q_8$, the quaternion group of order 8. More generally D_{2m} denotes the dihedral group of order $2m$.

Almost-simple groups: We write Σ_n for the symmetric group of degree n, and Alt_n for the alternating group of degree n (the Atlas uses S_n and A_n).[5] We write $\Omega^+_{2n}(2)$ and $\Omega^-_{2n}(2)$ for the simple subgroup (usually the commutator subgroup, of index two) in the orthogonal group over the field $\mathbb{F}_2$ of two elements (the Atlas uses $O^+_{2n}(2)$ and $O^-_{2n}(2)$, which elsewhere is often used instead for the full orthogonal group). Finally, we write $Sp_{2n}(2)$ for the symplectic group over $\mathbb{F}_2$ (the Atlas uses $S_{2n}(2)$). We follow the Atlas in writing $L_n(q)$ and $U_n(q)$ for the simple linear and unitary groups; and $G_2(2)$ for the group of exceptional Lie type G_2 over $\mathbb{F}_2$.

We write $G \wr H$ for the wreath product (copies of G permuted by H); there is really only one instance where we use this, namely in one of the 2-local subgroups of the Harada–Norton group HN in Section 7.19, where

$$\mathrm{Alt}_5 \wr 2 = (\mathrm{Alt}_5 \times \mathrm{Alt}_5){:}2.$$

Conventions for coefficient rings and homology. We let R denote a commutative ring with unit; typically R appears as the coefficient ring for homological algebra.

Often we will specialize R to rings of particular interest: we write $\mathbb{Z}$ for the ring of rational integers, and $\mathbb{F}_p$ for the Galois field of p elements. We write $\mathbb{Z}_{(p)}$ for the ring of p-local integers; namely the local subring of the rationals $\mathbb{Q}$ obtained from $\mathbb{Z}$ by inverting all primes other than p. The quotient ring $\mathbb{Z}_{(p)}/p\mathbb{Z}_{(p)}$ is isomorphic to the field $\mathbb{F}_p$. In Chapter 7 on the sporadic groups, we will specialize to $p = 2$ (though similar results can be obtained for odd p).

For homology and cohomology, we follow the convention that a comma used as a delimiter indicates *group* (co)homology, in expressions such as $H^*(G, R)$; while a semicolon as a delimiter, in expressions such as $H^*(X; R)$, indicates the (co)homology of a *space* X. See for example Proposition 2.2.7 (1) where both notations appear.

NOTATION 0.0.2 (Spectral sequences). Beginning at Section 2.6, we will be considering certain spectral sequences for homology and cohomology. For more detail on general spectral sequences, see for example Chapter 3 of [**22**].

We index homology spectral sequences $E^2_{s,t}$ and cohomology spectral sequences $E_2^{s,t}$ in such a way that s is the horizontal coordinate and t is the vertical coordinate.[6] Consequently the vertical axis is given by the terms with $s = 0$.

[5]We have chosen to use Alt_n, to avoid confusion with our later practice in Chapters 6–8 of using A_n to denote an elementary abelian 2-group of rank n.

[6]Warning for the non-expert: notice this is the convention for Cartesian coordinates; it is the opposite of the linear algebra convention for indexing the rows and columns of a matrix.

We say that the spectral sequence is *first quadrant* if the terms are zero for $s < 0$ and for $t < 0$. The spectral sequences we encounter will be first quadrant.

We recall, for example from [**22**, p. 99]: A (homology) spectral sequence is said to be *strongly convergent* if given s and t, there exists some finite value $n \geq 2$ such that $E^n_{s,t} = E^\infty_{s,t}$; and further for fixed m, the terms $E^\infty_{s,t}$ on the diagonal $s + t = m$ form a set of filtration quotients for the m-th homology group that the spectral sequence is designed to compute. For cohomology spectral sequences, one uses the corresponding notation $E_n^{s,t}$ and $E_\infty^{s,t}$ in the definition. *It is standard that first quadrant spectral sequences are strongly convergent, so this convergence holds for the sequences we will encounter.* ◊

NOTATION 0.0.3 (Limits and colimits). We shall use the terms *limit* and *colimit* (written $\varprojlim$ and $\varinjlim$) to refer to what are sometimes known as *inverse limit* and *direct limit* (see for example Spanier [**117**, p. 18]), or as *projective limit* and *inductive limit* respectively (depending on when and where you were educated).

We recall some features of these functors related to the standard setting in homological algebra of exactness, derived functors, and adjointness; for further reference, see for example Benson [**21**, p. 22] [**22**, Sec. 7.2] or Weibel [**125**, Sec. 2.6]. For modules, the functor $\varprojlim$ is left exact; and the right derived functors of $\varprojlim$ are written ${\varprojlim}^i$. The functor $\varinjlim$ is right exact; and its left derived functors are written ${\varinjlim}_i$. (See for example [**21**, Exer., p. 244]; and 2.6.9 in [**125**], as well as 2.6.4 and 2.6.1 there.) These left derived functors will be described in more detail when we need them later, in Section 4.8. ◊

Conventions for spaces. Through Chapter 2, the letter X will usually denote a topological space. For reference on classical algebraic topology, see standard texts such as Spanier [**117**].

In the chapters thereafter, much of the topological material will be developed instead in the language of simplicial sets (see for example May [**91**] for further reference); so starting just before Section 3.7, we will use the word "space" to mean simplicial set. Then X will usually denote such a space, and we will apply various topological notions to spaces in this new sense.

However, we will recall in Section 3.6 that the viewpoints of topological spaces and of simplicial sets are in fact equivalent for our present purposes. Hence even in our later chapters, the reader can typically still think of a space X informally just as a classical topological space.

We denote a space (in either of the above senses) which consists of a single point by a star $\star$; sometimes $\star$ will be contained as a basepoint in a larger space. We use a 5-pointed star, to avoid conflict with the use of an asterisk $*$ (6-pointed star) for several other purposes: namely the customary sequence of dimensions in homological expressions such as $H^*(G, R)$, and the join $X * Y$ of simplicial complexes or topological spaces.

If f and g are maps of spaces from X to Y, we write $f \simeq g$ if f and g are homotopic. We write $X \simeq Y$ if spaces X and Y are homotopy equivalent. (We use the symbol $\cong$ to denote an isomorphism, and we usually then also specify the category in which that map is a morphism.)

We follow the notational convention from topology that if X is a set or a space admitting action of the group G, then X^g denotes the fixed points of $g \in G$, and

X^G the fixed points of G. If X is a subset of G, we use $g^{-1}Xg$ to denote conjugation by g; this avoids the ambiguity of using the group theory convention of X^g for this conjugation.

We shall write $X_p^\wedge$ for the Bousfield–Kan $\mathbb{F}_p$-completion $(\mathbb{F}_p)_\infty X$ of X; see Notation 4.3.5.

Finally, we mention that our Index also serves as a reference for further notation and definitions.

CHAPTER 1

Overview of our main results

In this chapter, we briefly summarize our homotopy colimit decompositions for the 26 sporadic groups G. The detailed statements of those results, and a great deal of further information, will be presented in Chapter 7. In each case there, we express the 2-completed classifying space $(BG)^\wedge_2$ as the completion of the homotopy colimit of a diagram of classifying spaces of subgroup of G. That diagram is a simplex, determined by an "incident" set of 2-local subgroups H_i $(i \in I)$ and all their intersections $H_J := \bigcap_{i \in J} H_i$ for $\varnothing \neq J \subseteq I$; see Definition 4.6.2. The following table indicates, for each group G, the subgroups H_i $(i \in I)$ used for this decomposition.

After the table, we will further discuss some of these diagrams of subgroups.

Group	Subgroups
Mathieu group M_{11}	$Q_8 : \Sigma_3$ Σ_4
Mathieu group M_{12}	$2^{1+4}_+ : \Sigma_3$ $4^2 : D_{12}$
Mathieu group M_{22}	$2^4 : \mathrm{Alt}_6$ $2^3 : L_3(2)$ $2^4 : \Sigma_5$
Mathieu group M_{23}	$2^4 : \mathrm{Alt}_7$ $2^3 : L_3(2)$ $2^4 : (3 \times \mathrm{Alt}_5) : 2$
Mathieu group M_{24}	$2^4 : \mathrm{Alt}_8$ $2^6 : (L_3(2) \times \Sigma_3)$ $2^6 : 3 \cdot \Sigma_6$
Janko group J_1	$2^3 : 7 : 3$
Janko group J_2	$2^{1+4}_- : \mathrm{Alt}_5$ $2^{2+4} : (3 \times \Sigma_3)$
Janko group J_3	$2^{1+4}_- : \mathrm{Alt}_5$ $2^{2+4} : (3 \times \Sigma_3)$ $2^4 : (3 \times \mathrm{Alt}_5)$

Group	Subgroups
Janko group J_4	$2_+^{1+12}:3\cdot M_{22}:2$ $2^{3+12}\cdot\big(\Sigma_5 \times L_3(2)\big)$ $2^{10}:L_5(2)$ $2^{11}:M_{24}$
Higman–Sims group HS	$4\cdot 2^4:\Sigma_5$ $4^3:L_3(2)$ $2^4:\Sigma_6$
McLaughlin group McL	$2\cdot \mathrm{Alt}_8$ $2^4:\mathrm{Alt}_7$ $2^4:\mathrm{Alt}_7$
Suzuki group Suz	$2_-^{1+6}\cdot U_4(2)$ $2^{2+8}:(\mathrm{Alt}_5 \times \Sigma_3)$ $2^{4+6}:3\cdot \mathrm{Alt}_6$
Conway group Co_3	$2\cdot Sp_6(2)$ $2^{2+6}3^{1+2}2^2$ $2^4\cdot L_4(2)$
Conway group Co_2	$2_+^{1+8}:Sp_6(2)$ $2^{4+10}(\Sigma_3 \times \Sigma_5)$ $(2_+^{1+6} \times 2^4)L_4(2)$ $2^{10}:M_{22}:2$
Conway group Co_1	$2_+^{1+8}\cdot\Omega_8^+(2)$ $2^{2+12}:\big(\Sigma_3 \times L_4(2)\big)$ $2^{4+12}\cdot\big(3\cdot\Sigma_6 \times \Sigma_3\big)$ $2^{11}:M_{24}$
Fischer group Fi_{22}	$\big(2 \times 2_+^{1+8}:U_4(2)\big):2$ $2^{5+8}:(\Sigma_3 \times \mathrm{Alt}_6)$ $2^{10}:M_{22}$ $2^6:Sp_6(2)$
Fischer group Fi_{23}	$(2^2 \times 2_+^{1+8})\big(3 \times U_4(2)\big)2$ $2^{6+8}:(\Sigma_3 \times \mathrm{Alt}_7)$ $2^7:Sp_6(2)$ $2^{11}\cdot M_{23}$

Group	Subgroups
Fischer group Fi'_{24}	$2_+^{1+12} \cdot 3U_4(3)2$ $2^{3+12}\big(\mathrm{Alt}_6 \times L_3(2)\big)$ $2^{6+8}\big(\Sigma_3 \times L_4(2)\big)$ $2^8 : \Omega_8^-(2)$ $2^{11} \cdot M_{24}$
Harada–Norton group HN	$2_+^{1+8}(\mathrm{Alt}_5 \wr 2)$ $2^{3+2+6}\big(3 \times L_3(2)\big)$ $2^6 \cdot \Omega_6^-(2)$
Thompson group Th	$2_+^{1+8} \cdot \mathrm{Alt}_9$ $2^5 \cdot L_5(2)$
Baby Monster $B = F_2$	$2_+^{1+22} \cdot Co_2$ $2^{2+10+20}(M_{22}:2 \times \Sigma_3)$ $2^3 2^{[32]}\big(\Sigma_5 \times L_3(2)\big)$ $2^{5+5+10+10}L_5(2)$ $2^{9+16}Sp_8(2)$
Fischer–Griess Monster $M = F_1$	$2_+^{1+24} \cdot Co_1$ $2^{2+11+22}(M_{24} \times \Sigma_3)$ $2^{3+6+12+18}\big(3 \cdot \Sigma_6 \times L_3(2)\big)$ $2^{5+10+20}\big(\Sigma_3 \times L_5(2)\big)$ $2^{10+16} \cdot \Omega_{10}^+(2)$
Held group He	$2_+^{1+6}L_3(2)$ $2^6 : 3 \cdot \Sigma_6$ $2^6 : 3 \cdot \Sigma_6$
Rudvalis group Ru	$2 \cdot 2^{4+6} : \Sigma_5$ $2^{3+8} : L_3(2)$ $2^6 : G_2(2)$
O'Nan group $O'N$	$4 \cdot L_3(4) : 2$ $4^3 \cdot L_3(2)$
Lyons group Ly	$2 \cdot \mathrm{Alt}_{11}$ $\big((2^{2+4} : (3 \times 3) : 2) \times 3\big) : 2$ $2^3 \cdot L_3(2)$ $(2^4 \times 3)\mathrm{Alt}_7$

Here is an illustration of a diagram arising from the subgroups in the table: in the case $G = M_{11}$, we have a diagram of groups

$$\begin{array}{ccccc} Q_8 : \Sigma_3 & & & & \Sigma_4 \\ & \nwarrow & & \nearrow & \\ & & D_8 & & \end{array}$$

described by the two listed subgroups, and their (single) intersection. This gives us a diagram of classifying spaces

$$\begin{array}{ccccc} B(Q_8 : \Sigma_3) & & & & B\Sigma_4 \\ & \nwarrow & & \nearrow & \\ & & BD_8 & & \end{array}$$

and the map from the homotopy colimit of this diagram to BM_{11} is a mod 2 cohomology equivalence. This is equivalent to the statement that after Bousfield–Kan 2-completion, the map is a homotopy equivalence.

In each case, the diagram of groups is a simplex formed by taking the intersections of an "incident" (cf. Condition 6.2.7) set of representatives of the listed subgroups, and inclusions between these intersections. This simplex in fact arises as an orbit space, namely the quotient of a corresponding simplicial complex by a natural action of G; the result depends on the flag-transitivity (cf. Definition 6.4.2) of the group action, which is verified for each of the groups in turn.

In three of the 26 sporadic cases, we are able to reduce to a slightly smaller diagram than the simplex diagram. For J_3, we reduce the pushout cube to the following diagram:

$$\begin{array}{ccccccc} 2_-^{1+4} : \mathrm{Alt}_5 & & & 2^{2+4} : (3 \times \Sigma_3) & & & 2^4 : (3 \times \mathrm{Alt}_5) \\ & \nwarrow & \nearrow & & \nwarrow & \nearrow & \\ & 2^{2+4} : (3 \times 2) & & & & 2^{2+4} : 3^2 & \end{array}$$

For Fi'_{24} and Ly we reduce to more complicated diagrams described in the relevant sections of Chapter 7.

In these three cases, the diagram can be thought of as being indexed by the set of simplices in a simplicial complex consisting of two simplices of the same dimension which share a common face of codimension one. For J_3 this complex is 1-dimensional, for Ly it is 2-dimensional, and for Fi'_{24} it is 3-dimensional. We picture these complexes below:

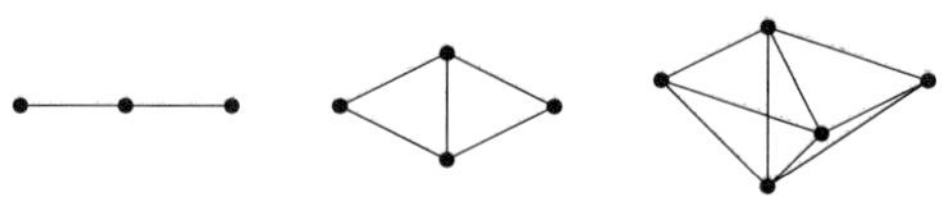

J_3 Ly Fi'_{24}

Part 1

Exposition of background material

CHAPTER 2

Review of selected aspects of group cohomology

One important motivation for our main results on sporadic groups in Chapter 7 is to contribute to a description of their group cohomology. So in this section, we begin our exposition by recalling certain features of group cohomology, which are used at various points in the remainder of our development. In particular we lay some foundations for our viewpoint on decompositions of cohomology.

There are of course many standard sources for group cohomology. We mention first that the material reviewed in this section corresponds roughly to that in Section 2 of Dwyer's exposition in [**53**]. However, Dwyer begins directly with the approach to group cohomology via the classifying space; whereas we felt that for readers whose background is in group theory, it might be better to begin by assuming less topological experience. In particular, we will first review group cohomology essentially in the classical way, via the homological algebra underlying the bar construction; so our coverage in the first section will primarily refer to the corresponding treatment in Benson [**21**, Sec. 3.4]. Our discussion in the later sections of the chapter will refer instead primarily to the exposition in Benson [**22**, Ch. 2]; and also to standard material in such sources as Adem and Milgram [**6**], Brown [**35**], and Evens [**58**]. As a further reference on group cohomology which is particularly well suited to our later viewpoint of decompositions, we recommend the treatment in [**6**], especially Chapter V there.

2.1. Some features of group cohomology via the algebraic approach

The main aim of this preliminary section is to recall, from the viewpoint of homological algebra, the standard primary decomposition for the $\mathbb{Z}$-cohomology of a finite group G (though much of the material applies more generally to discrete groups).

The approach using homological algebra. Group theorists may tend to be familiar with the definition of the group homology (and cohomology) of G primarily using the *bar construction*—that is, using the classical formula for the boundary map ∂_n, defined explicitly on $(n+1)$-tuples of elements of G, as in [**21**, p. 58].

In modern algebraic approaches, that construction arises naturally in the context of homological algebra, as the boundary map of a corresponding chain complex. We will now quote certain points from the treatment in [**21**, Sec. 3.4], to which the reader is referred for fuller definitions and details. For further background in homological algebra, see standard texts such as Cartan and Eilenberg [**42**], Hilton and Stammbach [**74**], and Weibel [**125**].

The $(n+1)$-tuples from G span a free left R-module F_n, on which the natural action of the group algebra RG, via multiplication by elements of G, is also free.

The corresponding chain complex F_* (with boundary map ∂_n in dimension n) is the *standard resolution* (also called the bar resolution) of R regarded as a trivial RG-module—this resolution is RG-free, and so in particular is an RG-projective resolution. (See for example [**21**, 2.4.1] for projective resolutions.)

While our focus in this work will be on group cohomology, in various later situations (for example (2.6.8)) we will need cohomology with coefficients in some more general RG-module M. One can define *homology* with coefficients in a module by using the chain complex given by a suitable tensor product (over RG) of a *right* RG-module M with the standard resolution F_*:

$$H_*(G,M) := \mathrm{Tor}_*^{RG}(M,R) = H_*(M \otimes_{RG} F_*)$$

When M is a left module, we use a right module version of the bar complex F_*. For *cohomology* with coefficients in a module, one uses an associated cochain complex defined using Hom:

$$H^*(G,M) := \mathrm{Ext}^*_{RG}(R,M) = H^*\big(\mathrm{Hom}_{RG}(F_*,M)\big).$$

We obtain *group homology* and *group cohomology* as the special case where the module M is just the trivial RG-module R:

$$H_*(G,R) := \mathrm{Tor}_*^{RG}(R,R) = H_*(R \otimes_{RG} F_*) \tag{2.1.1}$$

$$H^*(G,R) := \mathrm{Ext}^*_{RG}(R,R) = H^*\big(\mathrm{Hom}_{RG}(F_*,R)\big). \tag{2.1.2}$$

In fact many of our later statements such as the properties in Proposition 2.1.5 below will be more naturally expressed in the language of *reduced* homology, which we now briefly recall; see for example Benson [**22**, 6.1.4] or Brown [**35**, p. 211] for details. Since the final term F_0 of the standard resolution is just the group algebra RG, it follows from the definitions above (see for example [**35**, III.1.5, III.1.8]) that in dimension zero we have:

$$H_0(G,R) \cong R \cong H^0(G,R) \quad \text{as } R\text{-modules.} \tag{2.1.3}$$

It can be convenient to eliminate these "automatic" copies of R, by replacing the original complex F_* with the corresponding *augmented* complex $\tilde{F}_*$—which is equipped with a further copy of R, as the new final term $\tilde{F}_{-1}$ in formal dimension -1. The use of $\tilde{F}_*$ in place of F_* in (2.1.2) above gives the *reduced* cohomology groups, which satisfy:

$$\text{For } n>0,\ \tilde{H}^n(G,R) = H^n(G,R);\ \text{but } \tilde{H}^0(G,R) = 0;$$

with of course a similar statement for the reduced group homology $\tilde{H}_*(G,R)$, arising from the use of $\tilde{F}_*$ in (2.1.1).

The primary decomposition. The "usual" coefficient ring for group cohomology is $R = \mathbb{Z}$, so that $H^*(G,\mathbb{Z})$ is the usual group cohomology.[1] But the coefficient ring $R = \mathbb{F}_p$ for a prime p (giving the *mod-p cohomology*) is also a case of interest—particularly in view of the primary decomposition in Proposition 2.1.5 (2) below; see our later discussion in Section 2.3 concerning the relationship between these coefficient rings.

We first recall the standard notation, in the primary decomposition of a finite abelian group A, that A_p denotes the *p-primary* part of A (i.e., the unique Sylow

[1] In the literature, the $\mathbb{Z}$-cohomology is sometimes denoted just by $H^*(G)$; however we will always specify our coefficient ring R.

p-subgroup of A). We may view A as a $\mathbb{Z}$-module, and then tensoring with $\mathbb{Z}_{(p)}$ (in which primes other than p are inverted) annihilates the p'-torsion; so that the p-primary part appears as:

$$A_p = A \otimes_{\mathbb{Z}} \mathbb{Z}_{(p)}. \tag{2.1.4}$$

PROPOSITION 2.1.5. *Assume that G is a finite group, and p a prime. Then*

(1) (i) *The reduced cohomology groups $\tilde{H}^n(G,\mathbb{Z})$ and $\tilde{H}^n(G,\mathbb{F}_p)$ are finite abelian groups, of exponent dividing $|G|$.*

(ii) *In particular, both the p-primary part $\tilde{H}^n(G,\mathbb{Z})_p = \tilde{H}^n(G,\mathbb{Z}) \otimes_{\mathbb{Z}} \mathbb{Z}_{(p)}$ of integral cohomology and the mod p cohomology $\tilde{H}^n(G,\mathbb{F}_p)$ are finite p-groups; and if further p does not divide $|G|$, then in fact $\tilde{H}^*(G,\mathbb{Z})_p = 0 = \tilde{H}^*(G,\mathbb{F}_p)$.*

(2) *We have the primary decomposition*

$$\tilde{H}^*(G,\mathbb{Z}) = \bigoplus_{p \text{ divides } |G|} \left(\tilde{H}^*(G,\mathbb{Z}) \otimes_{\mathbb{Z}} \mathbb{Z}_{(p)}\right)$$

as abelian groups. Analogous statements hold for group homology.

PROOF. The statement (1) (i) is a consequence of the standard formula for restriction followed by transfer (also called corestriction). For the formula in the case of cohomology, see for example Brown [**35**, III.10.2], or Evens [**58**, 4.2.3], or Hilton and Stammbach [**74**, VI.16.5]. Then (1) (ii) follows from (1) (i), using (2.1.4) to describe the p-primary part.

The proof of (1) also goes through in the case of homology. For example, finite generation of cycles holds, just as for cocycles, using the standard resolution F_* in the definition (2.1.1) of homology. And finite exponent also holds: for example in [**35**], one can use III.9.5.ii there to apply the proof of III.10.2 to the case of homology.

Next using (1), the finite abelian groups $\tilde{H}^n(G,\mathbb{Z})$ have the primary decomposition given in (2). (Cf. the decomposition after [**35**, III.10.2]; the formula there implicitly assumes the continuation of the hypothesis $n > 0$ from III.10.2.) □

2.2. Some features of group cohomology via the classifying space

We now supplement the algebraic approach of the previous section with the topological approach via the classifying space. For more detail comparing these viewpoints on group cohomology, see the extensive overview in [**22**, Sec. 2.1].

Our main aims in this section will be to first recall the usual context of classical homotopy theory; and then to review some features of Milnor's model of the classifying space.

DEFINITION 2.2.1 (Homology and homotopy). The ordinary (i.e., *singular*) homology $H_*(X;R)$ of a space X with coefficients in R is computed from the standard chain complex $C_*(X;R)$: where $C_n(X;R)$ is the R-span of the n-chains on X, namely continuous maps from n-simplices into X. The cohomology $H^*(X;R)$ is based on the corresponding cochain complex determined by applying $\mathrm{Hom}_R(-,R)$ to $C_*(X;R)$. See for example [**22**, Sec 1.3] or Spanier [**117**, Sec. 4.4] for definitions and details; and for example [**22**, 6.1.4] or [**117**, Sec. 4.3] for reduced homology $\tilde{H}_*(X;R)$ and cohomology $\tilde{H}^*(X;R)$. Also see for example [**117**, p. 184] for the *relative* homology $H_*(X,Y;R)$ for a *pair* $Y \subseteq X$ of spaces related by inclusion.

The *homotopy groups* $\pi_n(X)$ $(n \geq 0)$ of X (relative to some particular basepoint in X) are given by the homotopy classes of basepoint preserving maps from the sphere S^n into X; see for example [**22**, Defn. 1.2.2].

LEMMA 2.2.2. *Maps which are homotopic induce the same map in homology and cohomology with any coefficient ring R. In particular, a homotopy equivalence* $f : X \to Y$ *of spaces induces an isomorphism in homology and cohomology*

$$f_* : H_*(X; R) \to H_*(Y; R)$$
$$f^* : H^*(Y; R) \to H^*(X; R),$$

as well as an isomorphism of homotopy groups $f_* : \pi_*(X) \to \pi_*(Y)$.

PROOF. See for example Spanier [**117**, 4.4.9]. □

The classifying space. We now turn to the topological space used for the group cohomology of G. There are various equivalent constructions of a classifying space BG of G. We shall stick to finite groups, but much of the following works for more general groups. For further details, see [**22**, Ch. 2].

DEFINITION 2.2.3 (Contractible free G-space EG and classifying space BG). Let G denote a finite (or discrete) group. Assume that EG is some contractible CW-complex; and that EG admits a free (right) G-action—in other words, only the identity element has fixed points on EG. Also assume admissible G-action as in later Definition 2.5.5: that group elements take cells to cells, and that the setwise stabilizer of a cell is the same as its pointwise stabilizer. Then standard results (cf. [**22**, 2.4.4]) show that the the space $BG := EG/G$ of G-orbits determines a homotopy type which is independent of the particular such space EG chosen; the quotient BG is called a *classifying space* for G. We will discuss the quotient process further at Definition 2.5.5; in particular, BG inherits a cellular structure giving a CW-complex.

WARNING 2.2.4. We emphasize that "the" classifying space BG for G is only well-defined *up to homotopy equivalence.*

DIGRESSION 2.2.5 (Overview of Milnor's model of EG). Early in Section 2.1, we recalled the classical bar construction defined via sequences of elements of G, and the chain complex F_* of free RG-modules F_n of such sequences. It should seem plausible that a corresponding topological space might be constructed from some suitable infinite product of copies of G. For this purpose, the appropriate notion is not the Cartesian product, but the *join* $*$ of topological spaces—see for example [**22**, Defn. 2.4.5].[2] Thus Milnor takes the infinite join

$$EG := G * G * \cdots$$

of copies of G.

In our case, the finite group G is just a set of points, with the discrete topology; this topological space G is in particular a CW-complex. Then just as for the topology of the ordinary Cartesian product of spaces, we endow the infinite join EG with the compactly generated product topology. With this convention, it is clear that the join EG of CW-complexes is again a CW-complex; indeed here, cells are given by finite joins of points from different copies of G.

[2]The reader may wish to think in terms of the more elementary special case given by the join of simplicial complexes [**117**, 3.1.7].

Furthermore the G-action on EG is via simultaneous right multiplication on each factor; and so the freeness of the action is immediate from that of the usual right regular representation of G on itself.

The contractibility of EG is perhaps not so immediately evident. We will give here an informal sketch,[3] roughly following [**22**, 2.4.7], which will partly rely on some material which we will not introduce until later. The contractibility emerges as a feature of the infinite join, when we consider the homotopy groups $\pi_*(EG)$: Suppose f is a continuous map of the n-sphere S^n into EG (relative to some basepoint in EG). Since the sphere is compact, and we are using the compactly generated product topology on EG, the map f factors through the subjoin in EG of a finite number of copies of G—say the first m copies. For any $g \in G$, we let g_{m+1} denote the point corresponding to g in the $(m+1)$-th copy of G; we see that the image $f(S^n)$ lies in a subspace given by the join $f(S^n) * g_{m+1}$. This latter structure is a cone, and in particular is contractible: indeed we can contract to the cone point g_{m+1} (this should be clear just from the special case of the join of simplicial complexes); hence we can also contract to our original fixed basepoint. So our map f is homotopic to the trivial map to the basepoint. We conclude that $\pi_n(G) = 1$, for any basepoint; and this holds for all n. In particular, EG has the same (trivial) homotopy groups as a single-point space $\star$. So in the later language of Definition 3.5.1, EG is weakly homotopy equivalent to a point (that is, weakly contractible). However Whitehead's Theorem (see Proposition 3.5.2) shows that a weak homotopy equivalence of CW-complexes is in fact a homotopy equivalence. That is, EG is in fact contractible. $\diamondsuit$

DIGRESSION 2.2.6 (Some other approaches to EG). In the following chapter, in the context of simplicial sets, we shall give at Example 3.1.6 a closely related model for EG, called the *simplicial bar construction*. The contractibility of that model will perhaps seem more immediate.

Another construction of EG is to embed G in some unitary group $U(n)$, and then let it act on the Stiefel variety of n-frames in $\mathbb{C}^\infty$; see for example [**22**, p. 44].

We mention also that, in language used perhaps more often in the geometric literature than the topological literature, the map $EG \to BG$ in Milnor's model is a universal principal G-bundle: in the sense that principal G-bundles over any CW-complex X are in one-to-one correspondence with homotopy classes of maps from X to BG, via pullback of bundles. See for example [**22**, Defn. 2.4.3]. $\diamondsuit$

We now return to our main treatment. We summarize below some important properties of BG—beginning with the fact that its cohomology gives the group cohomology:

PROPOSITION 2.2.7 (see for example 2.4.11 in [**22**], 2.1 in [**53**]). *Assume that G is a finite group, and let R be any commutative ring of coefficients. Then*

(1) *$H_*(BG;R) \cong H_*(G,R)$ as R-modules, and $H^*(BG;R) \cong H^*(G,R)$ as R-algebras.*

(2) *BG is an Eilenberg–Mac Lane space of type $K(G,1)$; that is,*

(i) *BG has the homotopy type of a connected CW-complex.*

(ii) *$\pi_1(BG) \cong G$ and $\pi_i(BG) = 0$ for $i \neq 1$.*

---

[3]In the case of more general topological groups G, Milnor's argument for contractibility of EG is somewhat more delicate and technical.

PROOF. See the discussion following [**22**, 2.4.11] for the isomorphism in (1) of the cohomology of the space BG with the algebraic definition (2.1.2) of group cohomology based on the bar construction. That discussion also indicates the CW-complex appearing in (2) (i), which we have sketched earlier at 2.2.5. Indeed (2) is just part (iii) of [**22**, 2.4.11]: note that parts (i) and (ii) of (2) simply review the definition of an *Eilenberg–Mac Lane space* $K(\pi, n)$ given at [**22**, 2.2.1]. □

2.3. The relation between the coefficient rings $\mathbb{Z}$ and $\mathbb{F}_p$

From the primary decomposition in Proposition 2.1.5 (2), we see that to investigate the group cohomology $H^*(G, \mathbb{Z})$, we may as well determine one p-primary part $H^*(G, \mathbb{Z}) \otimes_{\mathbb{Z}} \mathbb{Z}_{(p)}$ at a time, for each prime p dividing $|G|$.

We will connect $\mathbb{Z}$ and $\mathbb{F}_p$ via the intermediate ring $\mathbb{Z}_{(p)}$; for example, we will see in Corollary 2.3.3 that $\mathbb{Z}_{(p)}$-homology (sometimes called the p-local homology) is afforded by the p-primary part in the homology form of the decomposition 2.1.5 (2). This relation is a consequence of the more general universal coefficient theorem in homology.

THEOREM 2.3.1 (Universal coefficient theorem in homology). *Assume that* R *is a* PID, X *is a space, and* k *is an* R*-module. Then for* $n \geq 1$ *there is an exact sequence*

$$0 \to H_n(X; R) \otimes_R k \to H_n(X; k) \to \operatorname{Tor}_1^R(H_{n-1}(X; R), k) \to 0. \tag{2.3.2}$$

PROOF. Apply the Künneth formula, see for example Cartan and Eilenberg [**42**, Theorem VI.3.3], or Hilton and Stammbach [**74**, Theorem V.2.5], to the complex of chains on X—the standard complex $C_*(X; R)$ of Definition 2.2.1. The modules in that complex are free over the PID R, and hence torsion free; this gives the submodule flatness condition required in the Künneth formula. □

COROLLARY 2.3.3. *Let* X *be a space and* p *a prime. Then*

$$H_*(X; \mathbb{Z}_{(p)}) \cong H_*(X; \mathbb{Z}) \otimes_{\mathbb{Z}} \mathbb{Z}_{(p)}.$$

In particular if G *is a finite (or discrete) group, then*

$$H_*(G, \mathbb{Z}_{(p)}) \cong H_*(G, \mathbb{Z}) \otimes_{\mathbb{Z}} \mathbb{Z}_{(p)}.$$

PROOF. It is standard that $\mathbb{Z}_{(p)}$ is flat as a $\mathbb{Z}$-module: see for example [**125**, Theoremm 3.2.2] on localizations; or just note that $\mathbb{Z}_{(p)}$ is torsion-free as a $\mathbb{Z}$-module. Then by flatness, the right-hand term in (2.3.2) coming from Tor must vanish, reducing that expression to an isomorphism, which gives the first assertion. The second assertion then follows as the special case where we set $X = BG$ (using of course the isomorphism given in Proposition 2.2.7 (1)). □

COROLLARY 2.3.4. *Let* X *be a space and* p *a prime. Then for* $n \geq 1$ *there is an exact sequence*

$$0 \to H_n(X; \mathbb{Z}) \otimes_{\mathbb{Z}} \mathbb{F}_p \to H_n(X; \mathbb{F}_p) \to \operatorname{Tor}_1^{\mathbb{Z}}(H_{n-1}(X; \mathbb{Z}), \mathbb{F}_p) \to 0.$$

In particular if G *is a finite (or discrete) group, then there is an exact sequence*

$$0 \to H_n(G, \mathbb{Z}) \otimes_{\mathbb{Z}} \mathbb{F}_p \to H_n(G, \mathbb{F}_p) \to \operatorname{Tor}_1^{\mathbb{Z}}(H_{n-1}(G, \mathbb{Z}), \mathbb{F}_p) \to 0.$$

PROOF. The first part is just (2.3.2) with the indicated choices of R and k. The second part follows by again setting $X = BG$. □

PROPOSITION 2.3.5. *Suppose $\rho\colon X \to Y$ is a map of spaces, and p is a prime. Then we have the implication (1) $\Rightarrow$ (2) for the following conditions:*

(1) ρ induces an isomorphism $\rho_*\colon H_*(X;\mathbb{Z}_{(p)}) \to H_*(Y;\mathbb{Z}_{(p)})$.

(2) ρ induces an isomorphism $\rho_*\colon H_*(X;\mathbb{F}_p) \to H_*(Y;\mathbb{F}_p)$.

If in addition the homology groups $H_n(X;\mathbb{Z}_{(p)})$ and $H_n(Y;\mathbb{Z}_{(p)})$ are finitely generated $\mathbb{Z}_{(p)}$-modules, then conversely (2) implies (1). This finiteness hypothesis holds for the classifying space BG by Proposition 2.1.5 (1).

The corresponding statements for cohomology also hold, with suitable modifications for the fact that cohomology reverses arrows.

PROOF. Multiplication by p (which we denote by just "p" below) induces a short exact sequence of rings

$$0 \to \mathbb{Z}_{(p)} \xrightarrow{p} \mathbb{Z}_{(p)} \xrightarrow{\pi} \mathbb{F}_p \to 0.$$

Using these rings as coefficients, we obtain corresponding short exact sequences for the chain complexes on X and Y. These in turn lead to standard long exact sequences (see for example [**22**, Prop. 2.3.7]) for their homology, with connecting homomorphisms which we will denote by δ. The long exact sequences are related by the map ρ_* of sequences induced from $\rho\colon X \to Y$, in the following commutative diagram:

$$\begin{array}{ccccccccc}
\cdots \to & H_{n+1}(X;\mathbb{F}_p) & \xrightarrow{\delta} & H_n(X;\mathbb{Z}_{(p)}) & \xrightarrow{p} & H_n(X;\mathbb{Z}_{(p)}) & \xrightarrow{\pi_*} & H_n(X;\mathbb{F}_p) & \xrightarrow{\delta} \cdots \\
& \downarrow{\rho_*} & & \downarrow{\rho_*} & & \downarrow{\rho_*} & & \downarrow{\rho_*} & \\
\cdots \to & H_{n+1}(Y;\mathbb{F}_p) & \xrightarrow{\delta} & H_n(Y;\mathbb{Z}_{(p)}) & \xrightarrow{p} & H_n(Y;\mathbb{Z}_{(p)}) & \xrightarrow{\pi_*} & H_n(Y;\mathbb{F}_p) & \xrightarrow{\delta} \cdots
\end{array}$$

For the proof of the forward implication, we assume that condition (1) holds. This says that, in the pair of adjacent columns for $\mathbb{Z}_{(p)}$ in the middle of the picture, the vertical arrows are isomorphisms. So the vertical arrow in any column for $\mathbb{F}_p$ has such a pair of isomorphisms to both its left and right;[4] hence the standard five lemma from homological algebra (see for example [**117**, 4.5.11]) shows that this vertical arrow ρ_* for $\mathbb{F}_p$ is also an isomorphism. That is, (2) holds.

For the proof of the reverse implication, we now assume in addition that the homology groups $H_n(X;\mathbb{Z}_{(p)})$ and $H_n(Y;\mathbb{Z}_{(p)})$ are finitely generated $\mathbb{Z}_{(p)}$-modules. And of course we begin with the assumption that (2) holds: that is, this time the vertical arrows ρ_* in the columns for $\mathbb{F}_p$ are assumed to be isomorphisms.

We claim first that the horizontal arrow p in the top row is surjective on the kernel of the vertical arrow ρ_* in the corresponding columns for $\mathbb{Z}_{(p)}$. The proof is a straightforward diagram chase on the above diagram. Such arguments are typically omitted in the topological literature; but we will indicate the easy details, since we know of no exact reference.[5] So consider some $x \in H_n(X;\mathbb{Z}_{(p)})$ with $\rho_*(x) = 0$. Then $\pi_*\rho_*(x) = 0$, so also $\rho_*\pi_*(x) = 0$—where this time the maps are those going around the top side of the relevant commutative square. Applying (2) for the vertical map ρ_* at $\mathbb{F}_p$, we obtain $\pi_*(x) = 0$. So by exactness of the rows, $x = py$ for some $y \in H_n(X;\mathbb{Z}_{(p)})$. Next by commutativity $p\rho_*(y) = \rho_*(py) = \rho_*(x) = 0$; so by exactness, $\rho_*(y) = \delta(z)$ for some $z \in H_{n+1}(Y;\mathbb{F}_p)$. Now by (2), $z = \rho_*(w)$ for some $w \in H_{n+1}(X;\mathbb{F}_p)$; so $\rho_*(y) = \delta\rho_*(w) = \rho_*\delta(w)$. We have $x = py = p(y - \delta w)$

[4] We may extend the right end of the diagram, namely $n = 0$, with zero maps.

[5] But see the related result called the Barratt-Whitehead lemma at [**65**, 17.4].

since $p\delta(w) = 0$ by exactness; while $\rho_*(y - \delta w) = 0$ by construction. This proves that multiplication by p is indeed surjective on $K := \mathrm{Ker}(\rho_*) \subseteq H_n(X; \mathbb{Z}_{(p)})$; that is, $K = pK$.

Next recall that $\mathbb{Z}_{(p)}$ is a local ring with unique maximal ideal $p\mathbb{Z}_{(p)}$; in particular pK lies in the Jacobson radical of the $\mathbb{Z}_{(p)}$-module K. Hence using our assumption that $H_n(X; \mathbb{Z}_{(p)})$ is finitely generated over $\mathbb{Z}_{(p)}$, so is K; and then we may apply Nakayama's lemma (see for example [**21**, 1.2.3]) to conclude from $K = pK$ that $K = 0$.

With this further information available, another diagram chase will now show that p is surjective on the cokernel of ρ_*, namely on $H_n(Y; \mathbb{Z}_{(p)})/\rho_*\big(H_n(X; \mathbb{Z}_{(p)})\big)$. For consider any $u \in H_n(Y; \mathbb{Z}_{(p)})$. By (2) we have $\pi_*(u) = \rho_*(v)$ for some $v \in H_n(X; \mathbb{F}_p)$. Then by commutativity and exactness, $\rho_*\delta(v) = \delta\rho_*(v) = \delta\pi_*(u) = 0$. Now ρ_* is injective by our proof above that $K = 0$; so $\delta(v) = 0$. Hence by exactness $v = \pi_*(w)$ for some $w \in H_n(X; \mathbb{Z}_{(p)})$; and then $\pi_*\rho_*(w) = \rho_*\pi_*(w) = \rho_*(v)$. It follows that $\pi_*\big(u - \rho_*(w)\big) = 0$; and so by exactness $u - \rho_*(w)$ is in the image of multiplication by p. This element represents the same coset as u in the cokernel; so we have shown that multiplication by p is indeed surjective on the cokernel of $\rho_* \colon H_n(X; \mathbb{Z}_{(p)}) \to H_n(Y; \mathbb{Z}_{(p)})$.

Another application of Nakayama's lemma, this time using the finite generation of $H_n(Y; \mathbb{Z}_{(p)})$, shows that the cokernel is zero. And this along with earlier $K = 0$ completes the proof that ρ_* in each column for $\mathbb{Z}_{(p)}$ is an isomorphism, so that (1) holds. □

2.4. The duality relation between homology and cohomology

Even though our primary interest is group cohomology, we will often prefer (as in the previous section) to use the language of group homology. This choice is based on the standard relation of cohomology to homology, which is given by the universal coefficient theorem.

THEOREM 2.4.1 (Universal coefficient theorem in cohomology). *Let R be a* PID *and X be a space. Then for $n \geq 1$ there is an exact sequence*

$$0 \to \mathrm{Ext}^1_R\big(H_{n-1}(X; R), R\big) \to H^n(X; R) \to \mathrm{Hom}_R\big(H_n(X; R), R\big) \to 0. \quad (2.4.2)$$

PROOF. Apply Brown [**35**, p. 8], Cartan and Eilenberg [**42**, Theorem VI.3.3a], or Hilton and Stammbach [**74**, Theorem V.3.3], to the standard complex $C_*(X; R)$ of chains on X from Definition 2.2.1. □

REMARK 2.4.3 (Projectivity and Ext). A projective R-module can be characterized in a number of ways; for example, as a direct summand of a free module (cf. Weibel [**125**, 2.2.1]). If M is a projective R-module, then $\mathrm{Ext}^n_R(M, R) = 0$ for $n \geq 1$ (cf. [**125**, 2.2.4, 1.6.8]).

COROLLARY 2.4.4. *Suppose that X is a path-connected space; and that for $n \geq 1$, $H_n(X; \mathbb{Z})$ is a finite abelian group. Then $H^1(X; \mathbb{Z}) = 0$; and for $n \geq 1$, $H^n(X; \mathbb{Z})$ is the dual abelian group to $H_{n-1}(X; \mathbb{Z})$.*

In particular, if G is a finite group, then we have $H^1(G, \mathbb{Z}) = 0$; and for $n \geq 2$, the finite abelian groups $H^n(G, \mathbb{Z})$ and $H_{n-1}(G, \mathbb{Z})$ are in duality (and so in particular are non-canonically isomorphic).

PROOF. Apply (2.4.2) with $R = \mathbb{Z}$, and recall that $n \geq 1$ there.

By hypothesis, $H^n(X;\mathbb{Z})$ is a finite abelian group, whereas $\mathbb{Z}$ has no nonzero elements of finite order. So the term $\mathrm{Hom}_{\mathbb{Z}}(H_n(X;\mathbb{Z}),\mathbb{Z})$ on the right of (2.4.2) vanishes. This leaves an isomorphism

$$\mathrm{Ext}^1_{\mathbb{Z}}(H_{n-1}(X;\mathbb{Z}),\mathbb{Z}) \cong H^n(X;\mathbb{Z}).$$

First consider the case $n = 1$. The hypothesis that X is path-connected gives $H_0(X;\mathbb{Z}) \cong \mathbb{Z}$ (cf. [**125**, Ex. 3.6.7]). Since $\mathbb{Z}$ is a free $\mathbb{Z}$-module, in particular it is projective; so the left-hand term above involving Ext^1 vanishes (cf. Remark 2.4.3). The isomorphism now gives the first assertion that $H^1(X,\mathbb{Z}) = 0$.

So consider $n > 1$. Now for a finite abelian group A, the dual abelian group is defined as $\mathrm{Hom}_{\mathbb{Z}}(A,\mathbb{R}/\mathbb{Z})$; and in particular the dual is (non-canonically) isomorphic to A, since it has the same invariant factors. But it is standard that the dual may also be computed as $\mathrm{Ext}^1_{\mathbb{Z}}(A,\mathbb{Z})$—see for example Brown [**35**, Exer. 2, Sec. VI.7] for further details. We may apply this characterization to the finite abelian group $H_{n-1}(X;\mathbb{Z})$ in the left-hand term of the above isomorphism, to obtain the duality assertion.

In particular, if G is a finite group, then by Proposition 2.1.5 (1) in the case of homology, the groups $H_n(G,\mathbb{Z}) = H_n(BG;\mathbb{Z})$ are finite for $n \geq 1$; and BG is path-connected by Proposition 2.2.7 (2). So we get the final assertion as a special case of the preceding assertions. □

COROLLARY 2.4.5 (Primary decomposition of group (co)homology). *For a finite group G,*

$$\tilde{H}^*(G,\mathbb{Z}) = \bigoplus_{p \text{ divides } |G|} \tilde{H}^*(G,\mathbb{Z}_{(p)})$$

as abelian groups, and the analogous statement holds for group homology. □

COROLLARY 2.4.6. *For any space X, $H^n(X;\mathbb{F}_p)$ is the dual $\mathbb{F}_p$-vector space of $H_n(X;\mathbb{F}_p)$ for all $n \geq 0$.*

In particular for a finite group G, the finite-dimensional $\mathbb{F}_p$-spaces $H_n(G,\mathbb{F}_p)$ and $H^n(G,\mathbb{F}_p)$ are in duality for all $n \geq 0$, and so are (non-canonically) isomorphic for all $n \geq 0$.

PROOF. Set $R = \mathbb{F}_p$. For $n = 0$, the assertion follows just from (2.1.3).

Next apply (2.4.2) with $R = \mathbb{F}_p$, recalling that $n \geq 1$ there. This time, in contrast with the proof of Corollary 2.4.4, we get vanishing of the left-hand term $\mathrm{Ext}^1_{\mathbb{F}_p}(H_{n-1}(X;\mathbb{F}_p),\mathbb{F}_p)$—since $\mathbb{F}_p$ is a field, so that all $\mathbb{F}_p$-modules are projective (cf. Remark 2.4.3); equivalently, every short exact sequence of $\mathbb{F}_p$-vector spaces splits. This leaves the isomorphism

$$H^n(X;\mathbb{F}_p) \cong \mathrm{Hom}_{\mathbb{F}_p}(H_n(X;\mathbb{F}_p),\mathbb{F}_p)$$

which gives the first assertion.

In particular when $X = BG$, by Proposition 2.1.5 (1) the $\mathbb{F}_p$-vector spaces $H^n(G,\mathbb{F}_p) = H^n(BG;\mathbb{F}_p)$ are finite dimensional; so we get the second assertion as a special case of the first. □

DEFINITION 2.4.7 (R-homology equivalence). We first recall that a map $X \to Y$ of topological spaces is an *R-homology equivalence* if it induces an isomorphism of all reduced homology groups over R; with an analogous definition of *R-cohomology* *equivalence.* By Lemma 2.2.2, a homotopy equivalence induces an R-(co)homology equivalence.

THEOREM 2.4.8. *Let R be a* PID. *If a map $f: X \to Y$ is an R-homology equivalence of spaces, then it is also an R-cohomology equivalence.*

If $f: X \to Y$ is an R-cohomology equivalence, and in addition the homology groups $H_n(X;R)$ and $H_n(Y;R)$ are finitely generated R-modules for each $n \geq 0$, then f is also an R-homology equivalence. (This finite generation holds for $X = BG$ and $R = \mathbb{Z}$ or $\mathbb{F}_p$ by 2.1.5 (1).)

PROOF. From (2.4.2) applied to each of Y and X, we see that f induces the following diagram of R-modules for $n > 0$; where we have written α, β, and γ for the appropriate members of the induced map f^* of sequences:

$$\begin{array}{ccccccccc} 0 & \longrightarrow & \mathrm{Ext}^1_R(H_{n-1}(Y;R),R) & \longrightarrow & H^n(Y;R) & \longrightarrow & \mathrm{Hom}_R(H_n(Y;R),R) & \longrightarrow & 0 \\ & & \downarrow \alpha & & \downarrow \beta & & \downarrow \gamma & & \\ 0 & \longrightarrow & \mathrm{Ext}^1_R(H_{n-1}(X;R),R) & \longrightarrow & H^n(X;R) & \longrightarrow & \mathrm{Hom}_R(H_n(X;R),R) & \longrightarrow & 0 \end{array}$$

For the first assertion, assume that f is an R-homology equivalence. Then the vertical maps α and γ are isomorphisms. Hence the five lemma shows that the middle vertical map β is also an isomorphism, giving the first assertion.

Next, without any assumptions on f, we apply the snake lemma (see for example [**21**, 2.3.9]) to the above diagram, to obtain the exact sequence of R-modules:

$$0 \to \mathrm{Ker}(\alpha) \to \mathrm{Ker}(\beta) \to \mathrm{Ker}(\gamma) \to \mathrm{Coker}(\alpha) \to \mathrm{Coker}(\beta) \to \mathrm{Coker}(\gamma) \to 0.$$

For the second assertion, we now assume instead that $f: X \to Y$ is an R-cohomology equivalence. Then β is an isomorphism, so that $\mathrm{Ker}(\beta) = 0 = \mathrm{Coker}(\beta)$; hence from the snake sequence above we deduce that $\mathrm{Ker}(\alpha) = 0 = \mathrm{Coker}(\gamma)$, and that $\mathrm{Ker}(\gamma) \cong \mathrm{Coker}(\alpha)$. Since R is a domain, the Hom terms in the above diagram are torsion free. So $\mathrm{Ker}(\gamma)$ is a submodule of a torsion-free module, and hence also torsion-free; and therefore so is $\mathrm{Coker}(\alpha)$.

Next assume in addition that the homology groups of X and Y are finitely generated R-modules. Using the standard basis theory for finitely generated modules over a PID, such a module is a direct sum of a free module and a torsion module. For such a module M, say with torsion submodule M_t, $\mathrm{Ext}^1_R(M,R)$ reduces to $\mathrm{Ext}^1_R(M_t,R)$, which is also a torsion R-module (cf. [**125**, 3.3.7]). So the Ext terms in the above diagram are all torsion. In particular, $\mathrm{Coker}(\alpha)$ is a quotient of torsion modules and hence torsion; while from the previous paragraph, $\mathrm{Coker}(\alpha)$ is torsion-free. We conclude then that $\mathrm{Coker}(\alpha) = 0 = \mathrm{Ker}(\gamma)$. Since we saw in the previous paragraph that $\mathrm{Ker}(\alpha) = 0 = \mathrm{Coker}(\gamma)$, this completes the proof that α and γ are isomorphisms. Now the second assertion follows. □

2.5. The Borel construction and G-equivariant homology

DEFINITION 2.5.1 (Equivariant homotopy).

- A topological space X with a G-action by continuous maps is called a *G-space*.
- A map of G-spaces which commutes with the G-actions is called *G-equivariant*, or just a *G-map*.
- A homotopy which is G-equivariant is called a *G-homotopy*.

• A homotopy equivalence of G-spaces X and Y consisting of G-equivariant maps $f\colon X \to Y$ and $g\colon Y \to X$ such that the compositions gf and fg are G-homotopic (rather than just homotopic) with Id_X and Id_Y is called a *G-homotopy equivalence.* ♢

WARNING 2.5.2. We emphasize that the above requirements for a G-homotopy equivalence are strictly stronger than those for a G-equivariant map f which induces an ordinary homotopy equivalence. For this latter condition, we will normally avoid using the term "G-equivariant homotopy equivalence", because it invites confusion with "G-homotopy equivalence". *Instead we will normally use the phrasing "homotopy equivalence via a G-map".*

EXAMPLE 2.5.3. Since the free G-space EG in Definition 2.2.3 is contractible, the map $EG \to \star$ to a point is a homotopy equivalence; and it is necessarily a G-map since the G-action on $\star$ is trivial. But it is not a G-homotopy equivalence, for any nontrivial group G. For by the free action of G on EG, there is no G-fixed point of EG; so that any candidate for a homotopy inverse to $EG \to \star$ (which must take $\star$ to some point in EG) could not be G-equivariant.

See also Warning 5.4.5 for another example of this distinction. ♢

REMARK 2.5.4 (The orbit space and its topology). For a general G-space X, the *orbit space* X/G is the set of G-orbits in X, endowed with the quotient topology—in which a collection of orbits is open in X/G, whenever the union is open when regarded as a subset of X. So directly from the definition, the natural projection $\pi\colon X \to X/G$ is continuous.

DEFINITION 2.5.5 (Admissible action). We say that a cellular action of a finite (or discrete) group G on a CW-complex X is *admissible* if the setwise stabilizer of each cell is in fact equal to its pointwise stabilizer.

Under this assumption on the action, it is then standard that the G-orbits on cells of X afford cells in a CW-structure on X/G. (See for example Lück [**85**, Example 1.7, p. 8], or [**10**, Example (1.1.1)(5), p. 7]; admissibility is also mentioned as a consequence of the the condition (*) in Bredon [**32**, p. I-1].)

If X is even a simplicial complex, so that the cells in the CW-complex are just the simplices, a simplex stabilizer then also fixes all the vertices of the simplex. ♢

WARNING 2.5.6. In the case where X is a simplicial complex, the quotient X/G by an admissible action is not necessarily a *simplicial* complex: an easy example is given by the cyclic group of order three, acting by rotating the boundary of a triangle, which we regard as a simplicial complex with three 0-simplices and three 1-simplices.

DIGRESSION 2.5.7. We also mention (see for example [**10**, Example (1.1.1)(5), p. 7]) that if our original G-action on some CW-complex X were not admissible, then we could pass to the *barycentric subdivision* $\mathsf{Sd}\,(X)$ of X, and there obtain admissibility of the induced action. (Cf. our later discussion in Definition 2.7.7 of the subdivision for simplicial complexes.) ♢

PROPOSITION 2.5.8. *Let X and Y be CW-complexes on which a finite group G acts admissibly.*

(i) *A G-map $f\colon X \to Y$ is a G-homotopy equivalence if and only if, for every subgroup $H \leq G$, the induced map of fixed points $X^H \to Y^H$ is a homotopy equivalence.*

(ii) *If the G-actions on X and Y are free, then any G-map $f\colon X \to Y$ which is an ordinary homotopy equivalence is in fact a G-homotopy equivalence; and then f also induces an ordinary homotopy equivalence of the orbit CW-complexes X/G and Y/G.*

PROOF. Part (i) is the main theorem of James and Segal [**78**]. And (ii) is a corollary, because the hypothesis of an ordinary homotopy equivalence $f\colon X \to Y$ gives that condition for the subspaces fixed by the trivial subgroup $H = 1$; while for a subgroup $H \neq 1$, the fixed subspaces X^H and Y^H are both empty by the hypothesis of free G-action. Thus from (i) we conclude that f is a G-homotopy equivalence of X and Y. Finally it is clear (using for example the fact that the projection maps to X/G and Y/G in Remark 2.5.4 are by definition open and hence continuous) that the map induced by f on the quotients is an ordinary homotopy equivalence of X/G with Y/G. □

DEFINITION 2.5.9 (The Borel construction). For a G-space X, the *Borel construction* on X is the space

$$EG \underset{G}{\times} X := (EG \times X)/G.$$

More precisely, this is the orbit space of the Cartesian product $EG \times X$, taken modulo the diagonal G-action given by $(e, x)g = (eg, g^{-1}x)$; in particular, we have identified $(eg, x) = (e, gx)$ in that quotient.

Ordinarily our space X will be a CW-complex (indeed a simplicial complex) with an admissible G-action. Then since EG is also a CW-complex with an admissible G-action, the Cartesian product $EG \times X$ has these properties as well; and thus as in our discussion in Definition 2.5.5, it follows that the quotient inherits a cellular structure:

LEMMA 2.5.10. *For a CW-complex X with an admissible G-action, the Borel construction $EG \underset{G}{\times} X$ is also a CW-complex.* □

One can show (cf. [**22**, 2.2.4]) that different choices for EG are homotopy equivalent via a G-map. So since their Cartesian products with X also have free action by G, it follows using Proposition 2.5.8 (ii) that the G-orbit spaces giving the corresponding Borel constructions for X are homotopy equivalent. That is, the Borel construction is well-defined.

Another term found in the literature for the Borel construction $EG \underset{G}{\times} X$ is the *homotopy orbit space*, with notation X_{hG}. Indeed we will see later in Remark 2.5.20 that the Borel construction has a decomposition indexed by G-orbits.

EXAMPLE 2.5.11. If X is a point $\star$, with trivial G-action, it is immediate from the definition that

$$EG \underset{G}{\times} \star \cong EG/G = BG. \tag{2.5.12}$$

Thus the classifying space BG can be regarded as the easiest special case of the Borel construction.

COROLLARY 2.5.13. *Let X and Y be CW-complexes on which a finite group G acts, and $f\colon X \to Y$ a G-map which is a homotopy equivalence (but not necessarily a G-homotopy equivalence). Then f induces a homotopy equivalence of Borel constructions:*

$$\mathrm{Id} \underset{G}{\times} f \; : \; EG \underset{G}{\times} X \;\to\; EG \underset{G}{\times} Y$$

PROOF. This is just a variant of the deduction we made after Lemma 2.5.10, that the Borel construction is well-defined. Here, the Cartesian products of EG with X and Y also have free G-action; so we see by Proposition 2.5.8 (ii) that their G-quotients giving the Borel constructions are homotopy equivalent. □

DEFINITION 2.5.14 (Equivariant homology). The ordinary (co)homology of the Borel construction is usually referred to as the *G-equivariant R-(co)homology* of the original G-space X, and denoted by:

$$H^*_G(X;R) := H^*(EG \underset{G}{\times} X; R)$$
$$H^G_*(X;R) := H_*(EG \underset{G}{\times} X; R).$$

EXAMPLE 2.5.15. If X is a point $\star$, then from (2.5.12) we have:

$$H^*_G(\star;R) = H^*(BG;R) = H^*(G,R) \tag{2.5.16}$$

PROPOSITION 2.5.17. *Let X and Y be CW-complexes on which a finite group G acts, and let $f\colon X \to Y$ be a G-equivariant map which is a homotopy equivalence (but not necessarily G-homotopy equivalence). Then*

$$f_*\colon H^G_*(X;R) \to H^G_*(Y;R)$$
$$f^*\colon H^*_G(Y;R) \to H^*_G(X;R)$$

are isomorphisms.

PROOF. This is immediate from Corollary 2.5.13 and Definition 2.5.14. □

We can consider the Borel construction on a G-space X from the viewpoint of the *stabilizer subgroups* or *isotropy subgroups* G_x of cells x. The Borel construction is particularly simple in the special case where X is a single transitive G-set. Here we may identify X with the coset space G/H, where $H = G_x$ as above. Then a term in the definition 2.5.9 of the Borel construction $EG \underset{G}{\times} G/H$ has the form (e, gH); and we may naturally identify this with the orbit $e(gH)$ in the quotient space EG/H.

Furthermore since EG is contractible with free G-action, in particular it also has free H-action; so from our earlier discussion in Definition 2.2.3 of EG, we may regard EG also as a suitable free H-space EH, so that

$$EG/H \text{ is a model for the classifying space } EH/H = BH. \tag{2.5.18}$$

The two remarks above lead to the following relation:

REMARK 2.5.19 (Borel construction as classifying space of point stabilizer). For a transitive G-set $X = G/G_x$ with $x \in X$,

$$EG \underset{G}{\times} (G/G_x) \text{ is homotopy equivalent to } EG_x/G_x = BG_x.$$

Of course the earlier expression (2.5.12) is just the special case of this fact where the G-action is trivial and hence $G_x = G$.

REMARK 2.5.20 (Expressions indexed by the orbit space X/G). If X is a CW-complex with an admissible action of a finite group G, then for each orbit Y of n-cells, with cell stabilizer H, we saw in Remark 2.5.19 that the space $EG \underset{G}{\times} Y$ is homotopy equivalent to $BH \times n$-cell.

So the Borel construction $EG \underset{G}{\times} X$ has a decomposition in terms of spaces of the form $BH \times n$-cell. The decomposition is indexed by G-orbits Y of cells in the space X; that is, by members of the orbit space X/G. ♢

2.6. Spectral sequences for G-equivariant homology

The equivariant homology of a G-space X described in the previous section can in fact be computed by means of its naturally associated Serre spectral sequence, which we will briefly describe below. We will then indicate a different spectral sequence, namely the isotropy spectral sequence, which also converges to the equivariant homology of X. Indeed we will be encountering still other spectral sequences later, for example in Section 4.8; and one recurring theme in our development (largely following Dwyer's analysis of decompositions) will be the interrelation of these various sequences.

Both the Serre spectral sequence and the isotropy spectral sequence are special cases of the Leray spectral sequence of a map, so we will actually begin by describing the Leray sequence.

For further background on spectral sequences in general, see Notation 0.0.2, and the references indicated there. We mention that our discussion below primarily emphasizes the context of *filtered complexes*, as in Benson [**22**, Sec. 3.2]. But we will also briefly mention the viewpoint of *double complexes*, as in [**22**, Sec. 3.4], since we will be returning to that viewpoint in later chapters.

The Leray spectral sequence. Every map of CW-complexes $f\colon X \to B$ gives rise to a Leray spectral sequence converging to the (co)homology of X, and we now describe that sequence. In the special case where f is a fibration, the E^2 page of the spectral sequence will be easier to describe than the general construction below, and that subcase is called the Serre spectral sequence.

DEFINITION 2.6.1 (Leray spectral sequence). Assume that we are given a map $f\colon X \to B$ of CW-complexes, where the map f is not necessarily a fibration. Write $\mathsf{Sk}^n B$, $n \geq 0$, for the *n-skeleton* (cf. Spanier [**117**, 3.1.5]) afforded by the subcomplex of cells of dimension at most n. For $n < 0$, we set $\mathsf{Sk}^n B = \varnothing$.

Using the inverse image of this skeletal filtration of B, we obtain a corresponding filtration on X:

$$F_n X := f^{-1}(\mathsf{Sk}^n B).$$

Notice that $F_n X = \varnothing$ for $n < 0$. Now each successive inclusion pair $F_{s-1}X \subseteq F_s X$ determines (cf. [**117**, p. 184]) a long exact sequence involving the relative homology of that pair. And the sequences for distinct pairs can then be suitably intertwined[6]

[6]Cf. [**22**, p. 107], which gives the corresponding construction for cohomology rather than homology

to form a homology spectral sequence whose E^1-page is given by

$$E^1_{s,t} := H_{s+t}(F_sX, F_{s-1}X; R). \tag{2.6.2}$$

We will not explicitly discuss here the derivation of the E^2-page; but later in Remark 4.4.12, we will describe it in the special case of the isotropy spectral sequence.

Since $F_nX = \varnothing$ for $n < 0$, we have $E^1_{s,t} = 0$ for $s < 0$. Since $F_nX \supseteq \mathsf{Sk}^nX$, from the long exact sequence for relative homology it follows (cf. Weibel [**125**, 5.4.3] and Dwyer [**53**, 9.1]) that $H_n(F_sX, F_{s-1}X; R) = 0$ for $n < s$, so that $E^1_{s,t} = 0$ for $t < 0$. So the spectral sequence is first quadrant, and hence strongly convergent as mentioned in Notation 0.0.2. The spectral sequence (2.6.2) is called the *Leray spectral sequence* for f, and it converges to $H_*(X; R)$.

There is also a corresponding Leray spectral sequence for cohomology:

$$E_1^{s,t} := H^{s+t}(F_sX, F_{s-1}X; R) \Rightarrow H^*(X; R) \tag{2.6.3}$$

which is obtained in an analogous way. ♢

WARNING 2.6.4. The Leray spectral sequence is not at all *homotopy invariant.* In other words, given two maps as in Definition 2.6.1, whose spaces are related by homotopy equivalences in a commutative diagram

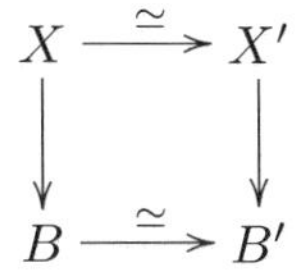

it turns out that the Leray spectral sequences (and in particular the E_2-pages) for $X \to B$ and $X' \to B'$ can be quite different—although both sequences do converge to $H_*(X; R)$. So we can *not*, for example, replace a map with an equivalent fibration, and then regard the Leray spectral sequence of the map as the Serre spectral sequence of the fibration, which is our next topic of discussion.

Fibrations. In preparation for the Serre spectral sequence, we provide here a brief review of fibrations. For further details, see for example [**22**, Sec. 1.6] or [**117**, Sec. 2.2].

DEFINITION 2.6.5 (Fibrations). The version of a surjection which is typically appropriate for topological spaces is that of a *covering map* $p \colon E \to B$ of a base space B, by a *total space* E. For many purposes in algebraic topology, coverings can be usefully generalized to appropriate kinds of *fibrations*, essentially by considering *homotopy lifting* properties: namely by asking for which kinds of spaces X can a homotopy into the base B, which also includes a lift of one end of the homotopy into E, then be extended into a lifting of the entire homotopy on $X \times [0, 1]$ into E?

We express this in a standard picture as follows. Assume we are given some fixed map $p \colon E \to B$. Then for some varying space X, we also assume we are given the homotopy h and the partial lifting f below:

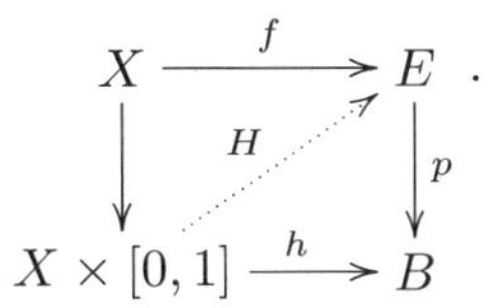

Then we can ask for which spaces X a full lifting H exists.

If the lifting H exists for all topological spaces X, then p is a *Hurewicz fibration* (or just a fibration). The preimage $p^{-1}(b)$ of a point b in the base space B is called a *fiber* of the map. Provided B is path-connected, this is independent of the chosen point. If instead we only require that the lifted homotopy H should exist for all simplicial complexes X, then p is a *Serre fibration.*

We close by just mentioning (and for further reference, indicating for example [**117**, Sec. 1.4]) that *cofibrations* arise in an essentially dual way to the above. If we consider injections instead of surjections, we can replace homotopy lifting by *homotopy extension*; and then *Hurewicz cofibrations* are obtained from a diagram suitably dual to the above, when the analogous extension map corresponding to H exists for all spaces X. Under mild conditions, the image of a cofibration is closed. It is often appropriate to add this as an extra condition, and this is the definition of a *closed* Hurewicz cofibration, see for example [**55**, 3.5]. ◇

The Serre spectral sequence. Now assume that $f\colon X \to B$ is a Serre fibration, with fiber F. Then the Leray spectral sequence of f, defined in (2.6.2) above via its E^1-page, is called the Serre spectral sequence of the fibration f. The E^2 page in this case can in fact be expressed in terms of the base and the fiber, rather than of the total space X:

$$E^2_{s,t} \cong H_s\big(B; H_t(F;R)\big). \tag{2.6.6}$$

Here, the coefficients $H_t(F;R)$ are to be regarded as a local system of coefficients on B; if B is path-connected, this amounts to an action of $\pi_1(B)$ on $H_t(F;R)$.

The Serre spectral sequence of equivariant cohomology. For further background on the Serre spectral sequence associated to the Borel construction, see for example Section 3.10 of [**22**].

Let X be a CW-complex with an admissible G-action. From the natural surjection $X \to \star$, we obtain maps

$$X \to EG \underset{G}{\times} X \to EG \underset{G}{\times} \star\,,$$

where the latter term is just BG using the expression (2.5.12). The second map, from the Borel construction of X to BG, is in fact a Serre fibration f with fiber X. Then there is a Serre spectral sequence corresponding to this fibration; and this sequence converges to the G-equivariant cohomology $H^*_G(X;R)$ in Definition 2.5.14. The roles of "X, B, F" in (2.6.6) are now played by $EG \underset{G}{\times} X$, BG, X, so that we have:

DEFINITION 2.6.7 (The spectral sequence of the Borel construction). For a G-space X, the *spectral sequence of the equivariant (co)homology* of X, that is, of the Borel construction of X, is

$$E_2^{s,t} = H^s\big(BG; H^t(X,R)\big) \Rightarrow H^{s+t}_G(X;R) \tag{2.6.8}$$

$$E^2_{s,t} = H_s\big(BG; H_t(X,R)\big) \Rightarrow H^G_{s+t}(X;R). \tag{2.6.9}$$

Notice that since the (co)homology groups are defined only for $s,t \geq 0$, the sequence is first quadrant—and hence strongly convergent as we mentioned in Notation 0.0.2.

The isotropy spectral sequence. As before, we let X be a CW-complex with an admissible G-action.

DEFINITION 2.6.10. To get the *isotropy spectral sequence* of X, instead of applying the Serre spectral sequence to the fibration $EG \underset{G}{\times} X \to BG$ as above, we apply the Leray spectral sequence of Definition 2.6.1 to the map $EG \underset{G}{\times} X \to X/G$, which is usually not a fibration.

2.7. Review of some aspects of simplicial complexes

We had mentioned after Definition 2.5.5 that our Borel constructions approximating BG will often be applied to simplicial complexes with a G-action. In fact simplicial complexes—and even the special case of a single simplex—are prominent in various other parts of our later development.

So we conclude the chapter with a review of some of the fundamental material related to simplicial complexes. This is probably not needed for most readers; but we include it for two reasons:

- the group geometries in Chapter 6 are defined as simplicial complexes; and
- certain features motivate generalizations to simplicial sets in Chapter 3.

Classically a topological space X is often replaced by a *triangulation*; that is, the space X is approximated up to homeomorphism by a more elementary structure, namely that of a simplicial complex Δ.

A complex Δ is in turn a union of individual simplices. Intuitively, a simplex is just the convex hull of a suitable set of vertices in a Euclidean space; and we will recall this "continuous" viewpoint in a moment in Definition 2.7.2. However, we will first indicate the "discrete" viewpoint, which is usually more convenient for our purposes: namely a simplex is identified with the *set* of its vertices. This discrete viewpoint is still more elementary, in that it is based just on the combinatorics of finite sets; and this focus on finite sets is continued in the material on simplicial sets in the subsequent chapter.

DEFINITION 2.7.1 (The abstract set-theoretic simplex). Fix some natural number n. We will use the standard notation Δ^n to denote the classical *n-simplex*, defined by a set of $n+1$ vertices.

More generally, we will often use Δ^I to denote the simplex defined by some finite nonempty index set I, with $|I| - 1$ giving the dimension of Δ^I. Thus we regard Δ^n as the special case where $|I| = n+1$; we will then often denote that vertex set by $\{0, 1, \ldots, n\}$.

We emphasize that

a simplex is determined by its set of vertices.

In particular, a simplex cannot have repeated vertices.

Of course each nonempty subset $J \subseteq I$ determines a *face* Δ^J of Δ^I, where each Δ^J is also a simplex in its own right. Sometimes the term face is used more specifically for a *maximal* sub-simplex; for example the usual boundary map on a simplex is the alternating sum of the maps to the faces in this maximal sense (see also Definition 3.1.1.)

In fact we will often regard a simplex Δ^I as the easiest case of a simplicial complex; in doing so we are abusing notation somewhat, in that we are then identifying Δ^I with the collection of all its faces Δ^J for $\varnothing \neq J \subseteq I$. Observe for Δ^I

regarded as a simplicial complex that the poset $\mathcal{Q}(\Delta^I)$ of faces and their inclusions in Δ^I (often called the *face poset* of Δ^I in the combinatorial literature) is isomorphic to the poset of nonempty subsets of I. Indeed we sometimes informally use the latter subposet $2^I \setminus \{\varnothing\}$ of the power set of I as shorthand for the simplicial complex Δ^I. $\diamondsuit$

We saw above that a simplex cannot have repeated vertices; but we mention that essentially allowing such repetitions is a part of the formalism of simplicial sets in the next chapter—and that many of the pleasant properties of simplicial complexes in this section are preserved under that more general formalism.

To a set-theoretic simplex, we associate a standard topological space:

DEFINITION 2.7.2 (The classical geometric realization of a simplex). One standard *geometric realization* $|\Delta^I|$ of an abstract simplex Δ^I is given by the convex hull in a Euclidean space of standard basis vectors indexed by I. (There are other possible constructions of the realization, all linearly homeomorphic.)

Thus for the abstract n-simplex Δ^n defined by $I = \{0, 1, \ldots, n\}$, the geometric realization $|\Delta^n|$ is the standard *Euclidean n-simplex*, namely the subset of $\mathbb{R}^{n+1}$ given by the coordinatization:

$$\{(u_0, \ldots, u_n) \in \mathbb{R}^{n+1} \mid u_i \geq 0 \text{ for } 0 \leq i \leq n, \text{ and } \sum_{i=0}^{n} u_i = 1\}.$$

In particular we may regard the vertices of the set Δ^n as being naturally embedded in the subspace $|\Delta^n|$ of Euclidean space. We will use this coordinatization later in Definition 3.3.2, in defining topological spaces corresponding to simplicial sets.

We also mention an alternative coordinatization of $|\Delta^n|$, now in $\mathbb{R}^n$ rather than $\mathbb{R}^{n+1}$: namely as the set of points with coordinates $(\tilde{u}_1, \ldots, \tilde{u}_n)$ satisfying

$$0 \leq \tilde{u}_1 \leq \cdots \leq \tilde{u}_n \leq 1.$$

The connection between this and the previous coordinatization is given by the formulas $\tilde{u}_i := u_0 + \cdots + u_{i-1}$ and $u_i = \tilde{u}_{i+1} - \tilde{u}_i$ (with the conventions $\tilde{u}_0 := 0$ and $\tilde{u}_{n+1} := 1$). This alternative coordinatization of $|\Delta^n|$ will be particularly useful later in Theorem 3.4.3, when we look at products of simplicial sets, from the viewpoint of their corresponding notion of geometric realization. $\diamondsuit$

DEFINITION 2.7.3 (Abstract simplicial complexes). We now return to the set-theoretic viewpoint on simplices in Definition 2.7.1. In that context, a finite *simplicial complex* Δ is determined from a finite set V of vertices, by taking a collection of subsets of V, where these subsets are called the *simplices* of Δ; and requiring that the collection should be closed under taking nonempty subsets of each subset in the collection. That is, the defining condition is:

$$\text{If a set } I \text{ of vertices is a simplex, then so is any nonempty subset } J \subseteq I. \quad (2.7.4)$$

Although the abstract simplicial complex Δ is not itself a topological space, note that the construction in Definition 2.7.2 of the geometric realization of a simplex extends directly to define the geometric realization $|\Delta|$ as a topological subspace of a suitable Euclidean space. As a result we follow the usual convention that

We may apply topological concepts to Δ, implicitly meaning to $|\Delta|$.

Thus for example we will later refer to the notion of homotopy equivalence between abstract simplicial complexes Δ and Δ'. ◊

As an example of an abstract simplicial complex, we already mentioned in Definition 2.7.1 the simplex Δ^I; indeed we observed there that the faces Δ^J satisfy the defining condition on subsets $J \subseteq I$ in (2.7.4).

We will also often make use of simplicial complexes obtained from posets, in the following standard way:

DEFINITION 2.7.5 (The complex of a poset). Assume that $\mathcal{Q}$ is a poset, say with order relation denoted $<$. Let $\Delta(\mathcal{Q})$ denote the collection of nonempty $<$-chains in $\mathcal{Q}$ (where the individual elements of $\mathcal{Q}$ are regarded as chains of length zero). Since a nonempty subchain of a chain is itself a chain, $\Delta(\mathcal{Q})$ satisfies the condition of closure under subsets in (2.7.4), and so defines a simplicial complex.

In the combinatorial literature, $\Delta(\mathcal{Q})$ is usually called the *order complex* of the poset $\mathcal{Q}$. However in the wider context of this work, it seems more natural to emphasize the poset itself rather than its order relation; so we will instead typically refer to $\Delta(\mathcal{Q})$ as the *poset complex* of $\mathcal{Q}$, or more informally just the complex of chains in the poset. We mention also (see the discussion at later Definition 3.3.7) that topologists often use the language of the *nerve* of the poset.

We emphasize that the strict order relation $<$ is used in $\Delta(\mathcal{Q})$; we could not use $\leq$ to get a simplicial complex, since a simplex in Definition 2.7.1 is not allowed to have repeated vertices. ◊

This construction allows us to extend to posets various of our earlier considerations for simplicial complexes:

REMARKS 2.7.6 (Some further aspects of poset complexes). Usually our poset $\mathcal{Q}$ will admit an action of G by *poset maps*, i.e., by permutations of the set $\mathcal{Q}$ preserving the order relation $<$; so that $\mathcal{Q}$ is a *G-poset*. It is then immediate from the definitions that the induced action of G on the poset complex $\Delta(\mathcal{Q})$ is simplicial; thus $\Delta(\mathcal{Q})$ is a *G-(simplicial) complex*.

It is also immediate that the stabilizer in G of a chain must stabilize each term of the chain; hence we see from Definition 2.5.5 that

For a G-poset $\mathcal{Q}$, the G-action on the poset complex $\Delta(\mathcal{Q})$ is admissible ...

(where we regard the simplicial complex in fact as a CW-complex). Hence from our discussion just after Definition 2.5.5 of the admissibility of the G-action on the CW-complex EG, we have the consequence:

... and so the Borel construction $EG \underset{G}{\times} |\Delta(\mathcal{Q})|$ is a CW-complex.

Borel constructions arising in this way from posets are prominent in our later development. Indeed typically our poset $\mathcal{Q}$ will be a subposet of the poset of all subgroups of G, often consisting of p-subgroups; with inclusion as the order relation, and G-conjugation as the action.

Our notation distinguishes between the abstract simplicial complex $\Delta(\mathcal{Q})$ given by the poset complex, and the topological space $|\Delta(\mathcal{Q})|$ given by its geometric realization. We may sometimes write $|\mathcal{Q}|$ as an abbreviated notation for the latter, leaving the intermediate step of taking the poset complex implicit. Similarly we may extend to posets the convention in Definition 2.7.3:

We apply topological concepts to a poset $\mathcal{Q}$, implicitly via $\Delta(\mathcal{Q})$ and $|\Delta(\mathcal{Q})|$.

For example, we will say that posets $\mathcal{Q}$ and $\mathcal{R}$ are *homotopy equivalent* when their poset complexes $\Delta(\mathcal{Q})$ and $\Delta(\mathcal{R})$ are homotopy equivalent. $\diamondsuit$

The poset viewpoint is also relevant to a standard construction for complexes:

DEFINITION 2.7.7 (The barycentric subdivision of a complex or poset). If we start with some abstract simplicial complex Δ, we can form the associated poset $\mathcal{Q}(\Delta)$ of simplices and their inclusions as in Definition 2.7.1. Then the corresponding poset complex $\Delta\big(\mathcal{Q}(\Delta)\big)$ is just the collection of all inclusion-chains of simplices of Δ; and this gives the standard abstract *barycentric subdivision* $\mathsf{Sd}\,(\Delta)$ of the original simplicial complex Δ. It is standard (see for example [**117**, 3.3.9 and (c), p. 121]) that the topological spaces $|\Delta|$ and $|\mathsf{Sd}\,(\Delta)|$ are homeomorphic; so in particular the complexes Δ and $\mathsf{Sd}\,(\Delta)$ are homotopy equivalent.

If instead we start with a poset $\mathcal{Q}$, then we can similarly define the *poset barycentric subdivision* $\mathsf{Sd}\,(\mathcal{Q})$ of $\mathcal{Q}$, via $\mathcal{Q}\big(\Delta(\mathcal{Q})\big)$: namely the poset of simplices in the complex $\Delta(\mathcal{Q})$ of $<$-chains from $\mathcal{Q}$, which just reflects the obvious poset structure on those chains. Again $\mathsf{Sd}\,(\mathcal{Q})$ is homotopy equivalent to $\mathcal{Q}$, from the viewpoint of posets. $\diamondsuit$

CHAPTER 3

Simplicial sets and their equivalence with topological spaces

In Chapter 4, we will review two standard general constructions due to Bousfield and Kan in [**31**], namely completions and homotopy colimits. But first in this chapter we recall some background on simplicial sets, which is the context in which those constructions are developed.

Our viewpoint on these topics for general spaces X will emphasize applications to the special case of the space $X = BG$ and group cohomology; and when relevant, we will focus on the aspect of decompositions mentioned in the previous chapter.

The material on completions and homotopy colimits corresponds roughly with Sections 3 and 4 in Dwyer's exposition [**53**]. We will sometimes refer to Dwyer for further details; but we also will also indicate some features outside Dwyer's treatment.

3.1. The context of simplicial sets

Many of the constructions of modern homotopy theory, including those of Bousfield and Kan in [**31**], are performed in the category of simplicial sets—rather than in the category of topological spaces (or CW-complexes, or even just simplicial complexes). So in this section, we will provide a brief discussion of the context of simplicial sets.

For a more detailed exposition on simplicial sets, we refer the reader especially to Section 3 of Dwyer [**53**]; other expository discussions are given in Benson [**22**, Sec. 1.8], May [**91**], and in [**31**, Secs. VIII.3 and VIII.4].

DEFINITION 3.1.1 (Simplicial sets). Let $\mathbf{\Delta}$ be the category whose objects are the finite nonempty totally ordered sets and whose arrows are the monotonic maps f—i.e., order preserving: $a \leq b$ implies $f(a) \leq f(b)$. Then a simplicial set is a functor from the opposite category $\mathbf{\Delta}^{\mathrm{op}}$ to the category $\mathfrak{Set}$ of sets (i.e., a contravariant functor from $\mathbf{\Delta}$ to $\mathfrak{Set}$). A map of simplicial sets is a natural transformation of functors. We write $\mathfrak{Sp}$ for the category of simplicial sets.

This abstract definition is usually applied a little more concretely in practice, as follows: We write $[n]$ for the totally ordered set

$$0 < 1 < \cdots < n.$$

The objects $[n]$ for $n \geq 0$ are representatives of the isomorphism classes in $\mathbf{\Delta}$. Indeed, the full subcategory $\mathbf{\Delta}^{-}$ consisting of these objects is a *skeletal subcategory*, meaning that the inclusion is an equivalence of categories. So to give a simplicial set X is to give, for each n, the image of $[n]$ in $\mathfrak{Set}$ under the functor X—the image may be thought of as the set X_n of "n-simplices in X". Furthermore for each map $f\colon [m] \to [n]$, we need to give a map $f^*\colon X_n \to X_m$ (generalizing the notion of face relationships); in Section 3.2 we give a presentation for the category $\mathbf{\Delta}^-$, and we'll see that as a consequence, to specify all these maps f^*, we only need give certain *face maps* and *degeneracy maps* satisfying appropriate relations. Finally a map $\phi\colon X \to Y$ of simplicial sets consists of maps $\phi_n\colon X_n \to Y_n$ for all $n \geq 0$ which commute with the face and degeneracy maps.

When an n-simplex (in the above sense of a member of X_n) can be obtained from an m-simplex for smaller $m < n$ (that is, via such a map f^* from $f\colon [n] \to [m]$), it is said to be *degenerate*, and should be thought of as a simplex which has repeated vertices. A simplex which is not degenerate is *nondegenerate*. A *face* of an n-simplex is an $n-1$-simplex obtained by applying f^* for one of the $n+1$ injective maps f from $[n-1]$ to $[n]$. ◊

EXAMPLE 3.1.2 (Discrete simplicial sets). If X is a set, we can regard X as a *discrete* simplicial set by using the constant functor—taking each object $[n]$ to the set X, and each map to the identity map. In a discrete simplicial set, the nondegenerate simplices are just the 0-simplices, one for each element of the set X.

EXAMPLE 3.1.3 (Simplicial complexes). If Y is a simplicial complex, we may obtain a corresponding simplicial set X as follows. We convert Y to an ordered simplicial complex by choosing a total order on the vertices of Y; and then X is the functor taking $[n]$ to the set of order preserving maps f from $[n]$ to the set of vertices of Y, where each f has the property that its image is the set of vertices of a simplex. The degenerate simplices are then the non-injective maps. Not all simplicial sets arise from simplicial complexes this way.

In particular we now see how the simplicial set structure generalizes the ordered simplicial complex: for a face map takes a simplex to a maximal subset essentially just as in Definition 2.7.1; while a degeneracy map acts by just repeating some vertex. ◊

EXAMPLE 3.1.4. The easiest example of a simplicial complex is an n-simplex. We write Δ^n for the "simplicial n-simplex" obtained from such a classical simplex by applying the construction of Example 3.1.3. This simplicial set can be very succinctly given as the representable functor $\mathrm{Hom}_{\mathbf{\Delta}}(-,[n])\colon \mathbf{\Delta} \to \mathfrak{Set}$.

In particular Δ^0 gives the *simplicial point* $\star$. For example we often choose a simplicial subset $\star$ for a basepoint in a space. ◊

EXAMPLE 3.1.5 (Skeleton). If X is a simplicial set, its *n-skeleton* $\mathsf{Sk}^n X$ is defined to be the simplicial subset generated by the simplices of dimension at most n in X. In other words, an element $x \in X_m$ is in $\mathsf{Sk}^n X$ precisely when there is a map $f\colon [m] \to [n]$ such that x is in the image of $f^*\colon X_n \to X_m$.

EXAMPLE 3.1.6 (The simplicial bar construction). If G is a group, we denote by EG the simplicial set $[n] \mapsto \mathrm{Hom}_{\mathfrak{Set}}([n], G)$. The group G acts on the right on EG via its action by right multiplication on the second factor. This action is free, since no n-simplex can be fixed by a non-identity element of G. We write BG for the quotient by the action; an n-simplex is an orbit of G on n-simplices of EG.

More informally, BG has just one 0-simplex. There is a 1-simplex for each element of G, with both ends equal to the 0-simplex. Then for each pair of elements g, h in G, there is a 2-simplex whose edges are the 1-simplices corresponding to g, h and gh. For each triple of elements there is a 3-simplex, and so on. ◊

REMARK 3.1.7. We have already defined EG and BG in Definition 2.2.3. The relationship of the example above with that earlier definition is via Example 3.4.7 below. Namely, we shall describe a construction called the geometric realization; when applied to the simplicial bar construction, it gives topological spaces for which we temporarily use the same notation EG and BG, such that EG is contractible with free right G-action, and $BG = EG/G$. Keep in mind, though, that BG is only well defined up to homotopy, and the simplicial bar construction is only a convenient functorial way of constructing it.

DEFINITION 3.1.8 (The nerve of a category). Suppose that $\mathfrak{C}$ is a small category (i.e., the collection of objects is small enough to be a set). Then we form a simplicial set $\mathcal{N}\mathfrak{C}$, called the *nerve* of $\mathfrak{C}$ as follows. Let $\mathbf{n}$ be the category corresponding to the totally ordered set $[n]$; so $\mathbf{n}$ has $n+1$ objects $0, \ldots, n$, and one map $i \to j$ whenever $i \leq j$, with the obvious composition. A map $[m] \to [n]$ in $\mathbf{\Delta}$ gives rise in an obvious way to a functor $\mathbf{m} \to \mathbf{n}$. The n-simplices in the nerve $\mathcal{N}\mathfrak{C}$ are the functors $\mathbf{n} \to \mathfrak{C}$, and maps in the simplicial set come from the functors between the categories $\mathbf{n}$. In other words, an n-simplex in $\mathcal{N}\mathfrak{C}$ is given by a list of n composable arrows in $\mathfrak{C}$.

An obvious example is given by the *trivial category*, with just one object and one map; the nerve is then just the simplicial point $\star$.

A functor $\mathfrak{C} \to \mathfrak{D}$ gives rise to a map of simplicial sets $\mathcal{N}\mathfrak{C} \to \mathcal{N}\mathfrak{D}$. $\diamondsuit$

EXAMPLE 3.1.9 (The simplicial bar construction as a nerve). The *Cayley category* $\mathfrak{C}_G$ of a group G has one object for each element $g \in G$, and arrows given by the left multiplications by elements of G. So G acts on $\mathfrak{C}_G$ by right multiplication. The nerve $\mathcal{N}\mathfrak{C}_G$ is the simplicial bar construction EG described above, and inherits a right G-action.

The quotient of $\mathfrak{C}_G$ by the right action of G is a category $\mathfrak{B}_G$ with just one object and the group G as its arrows, with composition given by the group multiplication. Then the nerve $\mathcal{N}\mathfrak{B}_G$ is equal to the quotient $\mathcal{N}\mathfrak{C}_G/G$, which is therefore equal to the simplicial bar construction BG of Example 3.1.6. $\diamondsuit$

EXAMPLE 3.1.10 (The nerve of a poset). Similarly, if $\mathcal{Q}$ is a poset, we can regard $\mathcal{Q}$ as a category with one object for each element of $\mathcal{Q}$, and one arrow $x \to y$ for each inequality $x \leq y$ in $\mathcal{Q}$. In this way, we get the *nerve* of a poset $\mathcal{N}\mathcal{Q}$.

3.2. A presentation for the category $\mathbf{\Delta}$ of finite ordered sets

In this section we give a presentation for the category $\mathbf{\Delta}$, or rather for the skeletal subcategory $\mathbf{\Delta}^-$ consisting of the objects $[n]$ for $n \geq 1$. This is often useful for seeing what information is necessary in order to give a simplicial set; for example, in Definition 4.2.1 we shall use this presentation in constructing a cosimplicial space.

Let $d^i\colon [n] \to [n+1]$ $(0 \leq i \leq n+1)$ be the injection missing the element i, and $s^i\colon [n] \to [n-1]$ $(0 \leq i \leq n-1)$ be the surjection sending i and $i+1$ to i. These maps generate $\mathbf{\Delta}^-$, and all relations are a consequence of the following ones.

$$d^i d^j = d^{j+1} d^i \qquad i \leq j$$

$$s^i d^j = \begin{cases} d^{j-1} s^i & i < j-1 \\ \mathrm{Id} & i = j-1, j \\ d^j s^{i-1} & i > j \end{cases} \tag{3.2.1}$$

$$s^i s^j = s^j s^{i+1} \qquad i \geq j$$

This means that to give a simplicial set is to give sets X_n $(n \geq 0)$ together with *face maps* $d_i = (d^i)^*\colon X_{n+1} \to X_n$ $(0 \leq i \leq n)$ and *degeneracy maps* $s_i = (s^i)^*\colon X_{n-1} \to X_n$ $(0 \leq i \leq n-1)$ satisfying the dual of the relations (3.2.1)—namely the order of composition is reversed in the form using subscripts rather than superscripts.

The proof that the generators and relations above give a presentation for $\mathbf{\Delta}^-$ is straightforward. Namely, using the relations (3.2.1), every word in the given generators can be written in the form

$$d^{j_1} \dots d^{j_m} s^{i_1} \dots s^{i_n}$$

with $i_1 < \cdots < i_n$ and $j_1 > \cdots > j_m$. It is straightforward to check that every map in $\mathbf{\Delta}^-$ can be written in this form; and that the expression is unique.

REMARK 3.2.2. Every degenerate simplex $x \in X_n$ in a simplicial set X can be expressed uniquely using the maps $s_i = (s^i)^*$ as $x = s_{i_j} \cdots s_{i_1} y$ where $y \in X_{n-j}$ is nondegenerate and $i_j > \cdots > i_1 \geq 0$. Then x is in the image of s_i if and only if $i \in \{i_1, \dots, i_j\}$. The maps $d_i = (d^i)^*$ take nondegenerate simplices to nondegenerate simplices.

3.3. Singular simplices and geometric realization

In this section, we describe adjoint functors connecting topological spaces with simplicial sets. This will lead to an equivalence of homotopy categories.

We recall from Definition 2.7.2 that the standard Euclidean n-simplex in $\mathbb{R}^{n+1}$ is denoted by $|\Delta^n|$, with the coordinatization

$$\{(u_0, \dots, u_n) \in \mathbb{R}^{n+1} \mid u_i \geq 0 \text{ for } 0 \leq i \leq n, \text{ and } \sum_{i=0}^{n} u_i = 1\}.$$

Then a monotonic map $f\colon [m] \to [n]$ gives rise to a linear map $f_\sharp\colon |\Delta^m| \to |\Delta^n|$ defined by $f_\sharp(u_0, \dots, u_m) = (u'_0, \dots, u'_n)$ where $u'_i = \sum_{f(j)=i} u_j$.

REMARK 3.3.1. Every point in $|\Delta^n|$ can be written uniquely as $d_\sharp^{j_k} \cdots d_\sharp^{j_1} u$, where u is in the interior of $|\Delta^{n-k}|$, and $j_1 > \cdots > j_k \geq 0$. The maps $s_\sharp^i$ take interior points to interior points.

DEFINITION 3.3.2 (Singular simplices). For any topological space Y, the *singular simplices* form a simplicial set: We recall that a singular n-simplex of Y is a continuous map $|\Delta^n| \to Y$. These can then be used in defining the functor

$$\mathsf{Sing}(Y)\colon \mathbf{\Delta} \to \mathfrak{Set}$$

sending $[n]$ to the set of singular n-simplices of Y. A map $f\colon [m] \to [n]$ is sent to composition with $f_\sharp$.

A continuous map of topological spaces gives rise in an obvious way to a map of singular simplices, so that Sing is a functor from the category $\mathfrak{T}$ of topological spaces to the category $\mathfrak{Sp}$ of simplicial sets. ◇

Next we consider a functor in the other direction, namely from $\mathfrak{Sp}$ to $\mathfrak{T}$.

DEFINITION 3.3.3 (Geometric realization). Given a simplicial set X, one can form a CW-complex called the *geometric realization* $|X|$ by taking one copy of $|\Delta^n|$ for each nondegenerate n-simplex in X, and gluing the copies together—using the monotonic maps in the category $\mathbf{\Delta}$ and the induced linear maps $f_\sharp$ as above. More precisely, we define

$$|X| := \coprod_n (X_n \times |\Delta^n|)/\sim$$

where each X_n is regarded as a discrete set, and $\sim$ is the equivalence relation generated by $(f^*x, u) \sim (x, f_\sharp u)$. Sometimes faces of a given n-simplex get identified this way, so the resulting space is not necessarily a simplicial complex, but it is always a CW-complex.

EXAMPLE 3.3.4. It is easy to check that the geometric realization $|\Delta^n|$ of the simplicial n-simplex Δ^n is the standard Euclidean n-simplex $|\Delta^n|$; so that it is in fact consistent to use the same notation for both.

DEFINITION 3.3.5. A representative $(x, u) \in X_n \times |\Delta^n|$ of an element of $|X|$ is said to be *nondegenerate* if x is a nondegenerate n-simplex and u is in the interior of $|\Delta^n|$.

LEMMA 3.3.6. *Every point in $|X|$ has a unique nondegenerate representative.*

PROOF. Set $W := \coprod_n (X_n \times |\Delta^n|)$ and let $(x, u) \in W$. Recall from the discussion after (3.2.1) that we abbreviate $(s^i)^*$ by s_i, and $(d^i)^*$ by d_i. Using Remark 3.2.2, we write $x = s_{i_j} \cdots s_{i_1} y$ with $y \in X_{n-j}$ nondegenerate and $i_j > \cdots > i_1 \geq 0$; and using Remark 3.3.1 we write $u = d_\sharp^{j_k} \cdots d_\sharp^{j_1} v$ with v an interior point in $|\Delta_{n-k}|$ and $j_1 > \cdots j_k \geq 0$. Define maps $\alpha, \beta \colon W \to W$ via $\alpha(x, u) = (y, s_\sharp^{i_j} \cdots s_\sharp^{i_1} u)$ and $\beta(x, u) = (d_{j_k} \cdots d_{j_1} x, v)$. Then each point (x, u) is equivalent to $\beta(x, u)$, and hence to $\alpha\beta(x, u)$; and $\alpha\beta(x, u)$ is nondegenerate since we saw at 3.3.1 that the maps $s_\sharp^i$ preserve the interior property for v. Furthermore, if (x, u) is nondegenerate then $(x, u) = \alpha(x, u) = \beta(x, u) = \alpha\beta(x, u)$, which proves uniqueness. □

The following special case of geometric realization will be important to us later.

DEFINITION 3.3.7 (Realization of the nerve, and poset complex). If $\mathfrak{C}$ is a small category, we usually write $|\mathfrak{C}|$ as an abbreviation for the geometric realization $|\mathcal{N}\mathfrak{C}|$ of its nerve.

For a poset $\mathcal{Q}$ regarded as a small category, the nerve $\mathcal{N}\mathcal{Q}$ is described in Example 3.1.10; and then the geometric realization $|\mathcal{Q}| = |\mathcal{N}\mathcal{Q}|$ turns out to give not just a CW-complex but a Euclidean simplicial complex—in fact, it just reproduces the classical geometric realization $|\Delta(\mathcal{Q})|$, in the sense discussed after Definition 2.7.2, of the poset complex $\Delta(\mathcal{Q})$ in Definition 2.7.5. Recall we also abbreviated that sense of $|\Delta(\mathcal{Q})|$ with the expression $|\mathcal{Q}|$ in earlier Remarks 2.7.6.

Thus the nerve $\mathcal{N}\mathcal{Q}$ provides an alternative approach, via simplicial sets, to our observations about poset topology in Remarks 2.7.6—such as the language of homotopy equivalence for posets, for example in the case of the barycentric subdivision in Definition 2.7.7. ◇

We conclude the section by stating the standard relation: the singular simplices functor is right adjoint to the geometric realization functor

$$\mathfrak{T} \underset{|\ |}{\overset{\mathsf{Sing}}{\rightleftarrows}} \mathfrak{Sp} \tag{3.3.8}$$

In other words, if X is a simplicial set and Y is a topological space then

$$\mathrm{Hom}_{\mathfrak{T}}(|X|, Y) \cong \mathrm{Hom}_{\mathfrak{Sp}}(X, \mathsf{Sing}(Y)).$$

3.4. Products and function spaces

One advantage of simplicial sets over simplicial complexes is that products and function spaces are easier to treat.

We recall that if X and Y are CW-complexes, then to make $X \times Y$ into a CW-complex, we must not take the product topology, because this does not have enough open subsets. So we increase the number of open subsets by making a set open if and only if its intersection with each compact subset in the product topology is open. Then $X \times Y$ is a CW-complex.

DEFINITION 3.4.1. Define the product of two simplicial sets X and Y by

$$(X \times Y)_n = X_n \times Y_n.$$

In other words, as a functor applied to each $[n]$, the product is taken "pointwise".

At first sight, this definition may look wrong, because it doesn't work like that for simplicial complexes. But this is one place where the degenerate simplices are helpful. A product of a degenerate simplex in X with a simplex in Y which is degenerate in a different way may be a nondegenerate simplex in $X \times Y$.

EXAMPLE 3.4.2. The simplicial set Δ^1 has geometric realization $|\Delta^1| = [0,1]$, the unit interval. Using the informal viewpoint of repeated vertices mentioned just before Example 3.1.2, the nondegenerate 1-simplex 01 determines via s_0 and s_1 two degenerate 2-simplices 001 and 011; and one sees in the product $\Delta^1 \times \Delta^1$, that $(001, 011)$ and $(011, 001)$ are not in the image of either $s_0 \times s_0$ or $s_1 \times s_1$, and hence are nondegenerate 2-simplices. Indeed the realization $|\Delta^1 \times \Delta^1|$ is a unit square, cut diagonally into the two triangles given by these 2-simplices, which share the edge $(01, 01)$ given by the images under $d_1 \times d_1$.

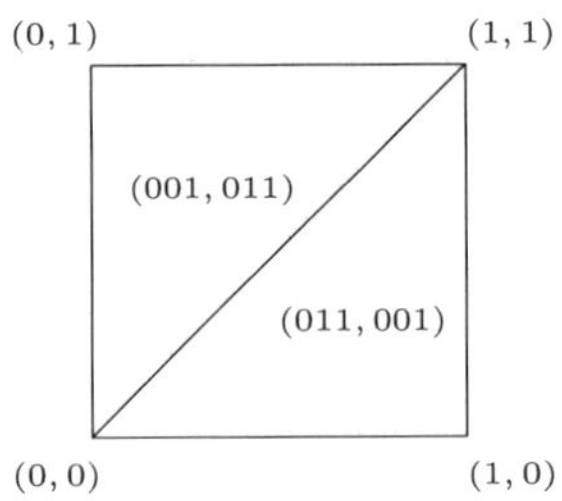

THEOREM 3.4.3. *If X and Y are simplicial sets, then there is a natural map of CW-complexes*

$$|X \times Y| \to |X| \times |Y|$$

which is a homotopy equivalence (provided that the product on the right is given the compactly generated topology so that it is a CW-complex).

PROOF. Let $\pi_1\colon X \times Y \to X$ and $\pi_2\colon X \times Y \to Y$ be the projections. We claim that

$$|\pi_1| \times |\pi_2|\colon |X \times Y| \to |X| \times |Y|$$

is a homeomorphism. We define an inverse map

$$\phi\colon |X| \times |Y| \to |X \times Y|$$

as follows. Let (x, u) be a point in $|X|$; by Lemma 3.3.6, we can take x to be a nondegenerate simplex in X; and $u = (\tilde{u}_1, \ldots, \tilde{u}_m)$ to be in the interior of $|\Delta^m|$—where now we use the alternative coordinatization of Δ^m given in Definition 2.7.2. Similarly, let (y, v) be a point in $|Y|$ with y nondegenerate and $v = (\tilde{v}_1, \ldots, \tilde{v}_n)$ in the interior of $|\Delta^n|$. We define a sequence of real numbers

$$0 < \tilde{w}_1 < \cdots < \tilde{w}_r < 1$$

by putting the *distinct* quantities among the $\tilde{u}_i$ and $\tilde{v}_j$ in order. Thus the $\tilde{w}_k$ define a point w in the interior of $|\Delta^r|$, and r is some integer at least as big as m and n and at most as big as $m + n$. Define $i_1 < \cdots < i_{r-m}$ to be those values of k for which $\tilde{w}_k$ is not among the $\tilde{u}_i$; and $j_1 < \cdots < j_{r-n}$ to be those values of k for which $\tilde{w}_k$ is not among the $\tilde{v}_j$. The map ϕ is then defined by

$$\phi\big((x, u), (y, v)\big) = \big((s_{i_{r-m}} \cdots s_{i_1} x, s_{j_{r-n}} \cdots s_{j_1} y), w\big).$$

Since $\{i_1, \ldots, i_{r-m}\} \cap \{j_1, \ldots, j_{r-n}\} = \varnothing$, using Remark 3.2.2 we see that the term $(\ s_{i_{r-m}} \cdots s_{i_1} x\ ,\ s_{j_{r-n}} \cdots s_{j_1} y\ )$ is not in the image of a degeneracy map of $X \times Y$. So as w is in the interior of $|\Delta^r|$, we conclude that $\phi\big((x, u), (y, v)\big)$ is a nondegenerate representative in $|X \times Y|$. It is not hard to check that ϕ is inverse to $|\pi_1| \times |\pi_2|$. Finally, one checks that ϕ is continuous on each cell. Since the compactly generated topology on the product is the same as the weak topology with respect to the cells (i.e., a set is open if and only if its intersection with each cell is open), ϕ is continuous. □

THEOREM 3.4.4. *If $\mathfrak{C}$ and $\mathfrak{D}$ are small categories then*

$$\mathcal{N}\mathfrak{C} \times \mathcal{N}\mathfrak{D} \;\cong\; \mathcal{N}(\mathfrak{C} \times \mathfrak{D}).$$

PROOF. An n-simplex on either the left or the right is given by a pair consisting of a sequence of maps $X_0 \to \ldots \to X_n$ in $\mathfrak{C}$ and a sequence of maps $Y_0 \to \ldots \to Y_n$ in $\mathfrak{D}$. □

EXAMPLE 3.4.5. We saw in Definition 3.1.8 that a functor $F\colon \mathfrak{C} \to \mathfrak{D}$ gives rise to a map of simplicial sets $\mathcal{N}F\colon \mathcal{N}\mathfrak{C} \to \mathcal{N}\mathfrak{D}$; so we get a map of CW-complexes $|F|\colon |\mathfrak{C}| \to |\mathfrak{D}|$. A natural transformation of functors from F to F' can be viewed as a functor from $\mathfrak{C} \times (\cdot \to \cdot)$ to $\mathfrak{D}$, where $(\cdot \to \cdot)$ is a category with two objects and one non-identity arrow, whose nerve is just Δ^1. Taking the geometric realization of the nerve, and using Theorems 3.4.4 and 3.4.3, this gives rise to a map $|\mathfrak{C}| \times [0, 1] \to |\mathfrak{D}|$ which is a homotopy from $|F|$ to $|F'|$. In particular,

An equivalence of categories gives a homotopy equivalence of nerves. (3.4.6)

EXAMPLE 3.4.7. The Cayley category $\mathfrak{E}_G$ of Example 3.1.9 is equivalent to the trivial category with one object and one arrow, whose geometric realization is a point. Namely, there is a natural transformation on $\mathfrak{E}_G$ sending each object g to the unique arrow from 1 to g, showing that $\mathfrak{E}_1 \to \mathfrak{E}_G$ is an equivalence of categories. So using (3.4.6), we see that the geometric realization $|EG|$ (of the simplicial bar construction EG viewed as $\mathcal{N}\mathfrak{E}_G$) is a contractible CW-complex with free G action, and with quotient $|BG|$. This means that Definition 2.2.3 is satisfied for these spaces.

Next, we discuss function spaces. We recall that for topological spaces X and Y, we form the space of continuous maps $\mathsf{Map}(X,Y)$. The usual topology for such a function space is the compact-open topology. But as with products, because we are interested in CW-complexes, we then increase the number of open subsets by making a set open if and only if its intersection with each compact subset in the compact-open topology is open. With this definition, it is a theorem of Milnor [**95**] that for CW-complexes X and Y, $\mathsf{Map}(X,Y)$ has the homotopy type of a CW-complex. Furthermore, there is a homeomorphism

$$\mathsf{Map}(Y,\mathsf{Map}(X,Z)) \cong \mathsf{Map}(X\times Y,Z), \tag{3.4.8}$$

which has the general form of the familiar adjointness statement for set mappings. Below we will want to obtain analogous adjointness statements for suitable spaces of functions between simplicial sets.

DEFINITION 3.4.9 (Function spaces). If X and Y are simplicial sets, we define a simplicial set $\mathsf{Map}(X,Y)$ as follows. The set $\mathsf{Map}(X,Y)_n$ of n-simplices is defined to be the set $\mathrm{Hom}_{\mathfrak{Sp}}(X\times\Delta^n,Y)$ of maps of simplicial sets from $X\times\Delta^n$ to Y. The face and degeneracy maps among these simplices for various dimensions n come from the maps of simplicial sets among the various Δ^n.

PROPOSITION 3.4.10. *Function spaces and products are adjoint, in the sense that if X, Y and Z are simplicial sets, then there is a natural isomorphism of sets*

$$\mathrm{Hom}_{\mathfrak{Sp}}(Y,\mathsf{Map}(X,Z)) \cong \mathrm{Hom}_{\mathfrak{Sp}}(X\times Y,Z).$$

PROOF. Giving a morphism in $\mathfrak{Sp}$ on the left-hand side means giving maps of sets $Y_n\to\mathsf{Map}(X,Z)_n$ which are are compatible with the maps in $\mathbf{\Delta}$ for n. But then we may similarly decompose $\mathsf{Map}(X,Z)_n=\mathrm{Hom}_{\mathfrak{Sp}}(X\times\Delta^n,Y)$ into maps of sets in corresponding dimensions m, so that we are giving maps of sets

$$Y_n\to\mathrm{Hom}_{\mathfrak{Set}}(X_m\times\Delta^n_m,Z_m)$$

compatible now for both n and m. By the usual adjointness for sets, this is the same as giving maps of sets

$$X_m\times\Delta^n_m\times Y_n\to Z_m$$

subject to the same compatibility. We will see below that this is equivalent to giving maps of sets

$$X_m\times Y_m\to Z_m$$

compatible with the maps in $\mathbf{\Delta}$ for m, namely a morphism in $\mathfrak{Sp}$ taking $X\times Y\to Z$.

For recall from Example 3.1.4 that Δ^n_m is just $\mathrm{Hom}_{\mathbf{\Delta}}([m],[n])$; and in particular, $\mathrm{Id}_{[m]}\in\Delta^m_m$. Now if $\phi_{m,n}\colon X_m\times\Delta^n_m\times Y_n\to Z_m$ are compatible maps as above, we define corresponding compatible maps $\psi_m\colon X_m\times Y_m\to Z_m$ by $\psi_m(x,y)=\phi_{m,m}(x,\mathrm{Id}_{[m]},y)$. Conversely, if we are given ψ_m, then for $f\in\Delta^n_m=\mathrm{Hom}_{\mathbf{\Delta}}([m],[n])$ we define $\phi_{m,n}(x,f,y)=\psi_m(x,f^*y)$. Note that by compatibility in n, we have

$$\phi_{m,n}(x,f,y)=\phi_{m,m}(x,\mathrm{Id}_{[m]},f^*y)$$

so that these processes are mutually inverse. □

REMARK 3.4.11. The proof of this proposition can be interpreted as saying that the symbol Δ^n_m acts like a Kronecker delta, in the sense of translating between off-diagonal terms and diagonal terms. The process can also be viewed as a manifestation of Yoneda's lemma (see for example [**21**, 2.1.4]). The same Kronecker viewpoint will reappear in later proofs: for example, the proof of the following proposition is very similar. But see also Definition 4.4.4, and Lemmas 4.1.7 and 4.1.9.

PROPOSITION 3.4.12. *Let X, Y and Z be simplicial sets. Then there is a natural isomorphism of simplicial sets*

$$\mathsf{Map}(Y, \mathsf{Map}(X, Z)) \cong \mathsf{Map}(X \times Y, Z). \qquad \square$$

3.5. Categorical settings for homotopy theory

As we had mentioned earlier, the classical setting for homotopy theory is the *homotopy category of CW-complexes:* this category has objects given by CW-complexes, and morphisms given by homotopy classes of maps (as opposed to just individual maps). In particular, the isomorphisms in this homotopy category are the homotopy equivalences.

The homotopy category of topological spaces needs to be set up more carefully. There are two competing notions, and we need to be careful to distinguish between them. In the first definition, which we shall *not* use, we just take the topological spaces and homotopy classes of maps. The second definition, which we *shall* use, puts more emphasis on CW-complexes; this is the category which appears naturally in algebraic topology. It is obtained from the category of topological spaces as the category of fractions, where *weak homotopy equivalences* are formally inverted in a suitable sense.

DEFINITION 3.5.1 (Weak homotopy equivalence). A weak homotopy equivalence is defined to be a map which induces an isomorphism of all homotopy groups with respect to all choices of basepoint (see for example [**22**, 1.5.7]).

Since the homotopy groups are defined by homotopy classes of maps from spheres, a homotopy equivalence is always a weak homotopy equivalence; but not every weak homotopy equivalence is a homotopy equivalence. However, the two notions do coincide for CW-complexes:

PROPOSITION 3.5.2 (Whitehead; see for example 7.5.9 in Spanier [**117**]).

(i) *A weak homotopy equivalence between CW-complexes is a homotopy equivalence.*

(ii) *A weak homotopy equivalence induces an isomorphism in (co)homology with any coefficients.*

(iii) *Every topological space is weakly homotopy equivalent to a CW-complex, namely the geometric realization of its singular simplices.* $\square$

Since classifying spaces are CW-complexes from Definition 2.2.3, we have the following corollary:

COROLLARY 3.5.3. *A weak homotopy equivalence between classifying spaces of discrete groups is a homotopy equivalence.*

THEOREM 3.5.4. *The inclusion of the CW-complexes in the category of all topological spaces induces an equivalence of homotopy categories.*

PROOF. This follows immediately from Proposition 3.5.2 (i) and (iii). □

The homotopy category of simplicial sets is set up in a similar way. There is a combinatorial description of the homotopy groups of a simplicial set, which is rather complicated (see for example Kan [**79**] or May [**91**, Corollary 26.7]), and a weak homotopy equivalence is a map which induces an isomorphism on homotopy groups with any choice of basepoint. These maps are then formally inverted in the homotopy category.

After some preparations in the following section, we shall see in Theorem 3.6.10 that the homotopy categories of simplicial sets, CW-complexes, and topological spaces are equivalent; so that in effect, we can just choose which of those contexts to use. For many purposes, CW-complexes form a good category for doing homotopy theory. But for some purposes, especially for analyzing homotopy colimits and p-completions (which will be our main focus), the technical advantages of simplicial sets are important.

3.6. Model categories, Kan complexes and Quillen's equivalence

Quillen [**100**] set up a general framework for discussing equivalences of homotopy categories, called *model categories.*[1] We shall not give the full set of axioms for a model category, but an essential point is that there are three distinguished classes of maps, called *fibrations*, *cofibrations*, and *weak equivalences.* Any two of these classes determine the third in a categorical manner. The homotopy category $\mathsf{Ho}(\mathfrak{C})$ of a model category $\mathfrak{C}$ is formed by suitably inverting the weak equivalences, just as for the categories indicated in the previous section.

An object in a model category $\mathfrak{C}$ is said to be *fibrant* if the map to the terminal object is a fibration; and *cofibrant* if the map from the initial object is a cofibration. We write $\mathfrak{C}_{cf}$ for the full subcategory of $\mathfrak{C}$ whose objects are those which are both cofibrant and fibrant. For general objects in $\mathfrak{C}$, there are two competing notions of homotopy of maps, called *left homotopy* and *right homotopy*; but for objects in $\mathfrak{C}_{cf}$, these notions coincide, and give a well behaved notion of a homotopy between maps. The following is proved in [**100**] (compare with the Whitehead theorem 3.5.2):

PROPOSITION 3.6.1. *Let $\mathfrak{C}$ be a model category.*

(i) *Every object in $\mathfrak{C}$ is weakly equivalent to an object in $\mathfrak{C}_{cf}$.*

(ii) *A weak equivalence between objects in $\mathfrak{C}_{cf}$ is a homotopy equivalence.*

(iii) *Let $\pi\mathfrak{C}_{cf}$ be the category whose objects are the objects in $\mathfrak{C}_{cf}$, and whose arrows are homotopy classes of maps. Then the inclusion $\mathfrak{C}_{cf} \to \mathfrak{C}$ induces an equivalence of categories $\pi\mathfrak{C}_{cf} \to \mathsf{Ho}(\mathfrak{C})$.*

The category of topological spaces has at least two naturally occurring model category structures. In the first, the fibrations are the Hurewicz fibrations (recall 2.6.5), the cofibrations are the closed Hurewicz cofibrations, and the weak equivalences are the homotopy equivalences. This is not the model category structure normally of interest in algebraic topology, and not the one we shall use. Rather, we use the second—which we will only briefly sketch in the remainder of this paragraph. Here the fibrations are the Serre fibrations, the weak equivalences are the

[1]Quillen used the term *closed* model category; but later authors have not found much use for model categories that are not closed, so the adjective "closed" is usually implicitly assumed.

weak homotopy equivalences, and the cofibrations are the maps defined by these two classes using the model category axioms. These turn out to be the retracts $X \to Y$ of maps $X \to Y'$ where Y' is obtained from X by attaching cells (i.e., there is a commutative diagram

$$\begin{array}{ccc} X & \longrightarrow & Y \\ \| & & \uparrow\downarrow \\ X & \longrightarrow & Y' \end{array}$$

where the composite $Y \to Y' \to Y$ is the identity). In this model category, all objects are fibrant, and the cofibrant objects are the spaces which are homotopy equivalent to a CW-complex; hence the CW-complexes afford the subcategory $\mathfrak{C}_{cf}$.

EXAMPLE 3.6.2 (The homotopy category of $\mathfrak{Sp}$). We put a model category structure on the category of simplicial sets, in which the weak equivalences are the weak homotopy equivalences, and the cofibrations are the maps $f\colon X \to Y$ such that each $f_n\colon X_n \to Y_n$ is a monomorphism. Finally the fibrations are the *Kan fibrations*, defined as follows.

DEFINITION 3.6.3 (Kan fibrations). Let $\Delta^{n,i}$ denote the subcomplex of the simplex Δ^n generated by all the faces except the ith face. A *Kan fibration* is a map of simplicial sets $X \to Y$ such that for all $n \geq 1$ and $0 \leq i \leq n$, given any map from $\Delta^{n,i}$ to X and a compatible map from Δ^n to Y, then there is a lift to a map $\Delta^n \to X$, making both triangles in the following diagram commute.

$$\begin{array}{ccc} \Delta^{n,i} & \longrightarrow & X \\ \downarrow & \nearrow & \downarrow \\ \Delta^n & \longrightarrow & Y \end{array}$$

In the model category of simplicial sets, all objects are cofibrant, and the fibrant objects are called *Kan complexes*. These are the objects X such that every map from $\Delta^{n,i}$ to X extends to a map from Δ^n to X. In particular, the Kan complexes afford the subcategory $\mathfrak{C}_{cf}$.

WARNING 3.6.4. The simplicial sets constructed from simplicial complexes in Example 3.1.3 are usually not Kan complexes.

Consider first the abstract simplicial complex defined by two edges with one vertex in common. We may also view this configuration as the simplicial subset Δ of the simplicial 2-simplex Δ^2 which (again using shorthand notation of potentially-repeated vertices) is generated by nondegenerate 1-simplices 01 and 12. For the unique nondegenerate 2-simplex $y_2 = 012$ of Δ^2, these are the faces $d_2(y_2)$ and $d_0(y_2)$; so Δ is in fact just the configuration denoted by $\Delta^{2,1}$ in Definition 3.6.3. Now Δ is not a Kan complex, since the simplicial subset $\Delta^{2,1}$ has no 2-simplex such as 012 to satisfy the Kan extension condition from itself.

Even the standard simplicial n-simplex Δ^n is not a Kan complex for $n \geq 2$; cf. [**63**, Sec. I.5]. $\diamondsuit$

Fortunately the failure of the Kan property in standard simplices and other examples in Warning 3.6.4 does not turn out to be a problem, either in the theory or in our applications. In particular, we will see shortly in Example 3.6.7 that from those examples, or indeed from an arbitrary simplicial set X, we can in fact obtain a related Kan complex, by passing to the simplicial set given by $\mathsf{Sing}(|X|)$; that is, by composing the two basic constructions of Section 3.3.

It is elementary to check from the definition in 3.6.2 that the product of Kan fibrations is again a Kan fibration (see for example Gabriel and Zisman [**62**, p. 66]); it then follows (cf. [**63**, p. 40]) that

$$\text{The Cartesian product of Kan complexes is a Kan complex.} \tag{3.6.5}$$

EXAMPLE 3.6.6 (Fundamental groupoid and group). If X is a simplicial set, we define its *fundamental groupoid* to be the category with one object for each 0-simplex in X. The arrows are generated by the 1-simplices, and are defined in both possible directions: Namely if x is a 1-simplex, then it is regarded as an arrow from $d_1 x$ to $d_0 x$; and it also defines an inverse arrow from $d_0 x$ to $d_1 x$. Relations are given by the 2-simplices; if y is a 2-simplex then the composition $d_0 y \circ d_2 y$ is set equal to $d_1 y$. (In particular the result is indeed a *groupoid*, namely a small category in which all arrows have inverses.)

If X is a Kan complex, then the Kan condition for $\Delta^{2,1}$ shows that there always exists such a 2-simplex for any pair of composable arrows, so that the arrows in the fundamental groupoid are just equivalence classes of 1-simplices, as opposed to longer compositions. In a sense, the Kan condition has done the composing for us.

For a choice of basepoint $\star$ in X, the *fundamental group* $\pi_1(X, \star)$ is defined to be the automorphisms of the object $\star$ in the fundamental groupoid. So for a Kan complex, this consists of equivalence classes of 1-simplices beginning and ending with $\star$.

In exactly the same way, the higher homotopy groups of a Kan complex are much easier to define than for a general simplicial set. If $\star$ is a basepoint in X, then $\pi_n(X, \star)$ consists of equivalence classes of n-simplices with the property that all the faces are equal to the degeneracy $s_0^{n-1}\star$ of the basepoint. The equivalence relation is that $x \sim y$ if there exists an $n+1$-simplex z with $d_0 z = x$, $d_1 z = y$ and $d_i z = s_0^n \star$ for $i > 1$. ♢

EXAMPLE 3.6.7. It is easy to check that the simplicial set of Definition 3.3.2, namely the singular simplices on a topological space, is in fact always a Kan complex. This is because there is a continuous map $|\Delta^n| \to |\Delta^{n,i}|$ which is left inverse to the inclusion in the opposite direction.

So we see as promised after Warning 3.6.4 that a simplicial set X determines a related Kan complex $\mathsf{Sing}(|X|)$. Furthermore we see in the adjunction (3.3.8) that the functor Sing has its image in the subcategory of Kan complexes; recall we had already seen in Definition 3.3.3 that the realization functor $|\quad|$ in the other direction has its image in the subcategory of CW-complexes. ♢

PROPOSITION 3.6.8. *For a groupoid $\mathfrak{C}$, the nerve $\mathcal{N}\mathfrak{C}$ is a Kan complex. In particular, the simplicial sets EG and BG described in Example 3.1.6 are Kan complexes, for any discrete group G.*

PROOF. See for example [**63**, I.3.5]; as the proof is elementary, we will sketch it—primarily in order to indicate the subtlety which requires the hypothesis of groupoid, as opposed to just any small category.

We recall from the informal viewpoint on the nerve in Definition 3.1.8 that an n-simplex σ in $\mathcal{N}\mathfrak{C}$ is given by a list of n composable maps $f_1, \ldots, f_n$ in $\mathfrak{C}$, connecting $n+1$ endpoints—which we will now number just as 0 to n:

$$\sigma := (0 \xrightarrow{f_1} 1 \cdots\cdots i-1 \xrightarrow{f_i} i \xrightarrow{f_{i+1}} i+1 \cdots\cdots n-1 \xrightarrow{f_n} n)$$

We saw there that the ith face $d_i(\sigma)$ of the simplex is obtained by omitting endpoint i. We emphasize particularly that for $i = 0$ or n, that face actually consists of a *sub*list of the original list of maps, namely the sublist omitting either f_1 or f_n. However when $i \neq 0, n$, that face is a list of $n-1$ maps, with the composition $f_{i+1} \circ f_i$ at position i; while the remaining maps are uncomposed members from the original list of n.

We now assume that we are given an embedding of $\Delta^{n,i}$ in $\mathcal{N}\mathfrak{C}$. This corresponds to omitting the ith face as described above: it is a set of n lists, each consisting of $n-1$ maps, with the consequent overlap and compatibility properties. It then remains to reconstruct from them a list of n composable maps corresponding to Δ^n.

In case $n = 1$, we are given an embedding of $\Delta^{1,i}$, for $i = 0$ or 1: namely one list, consisting of no maps; that is, just a single object $\star$ of $\mathfrak{C}$, representing either the beginning or the endpoint of a single arrow in some extension corresponding to Δ^1. So for that extension, we only need a map either from or to $\star$—there is no restriction on the object of $\mathfrak{C}$ at the other end of that arrow, so we may as well choose it also to be $\star$. And then the identity map $\mathrm{Id}_\star$ works for either case $i = 0$ or 1.

Now assume that $n \geq 2$. Then i cannot be both 0 and n; say $i \neq 0$—the argument below is easily adjusted for the case $i \neq n$. So $\Delta^{n,i}$ in particular contains the 0th face of an n-simplex; hence from our discussion above, the first of our n given lists has the form $(f_2, f_3, \ldots, f_n)$ for a suitable extension—that is, we are missing only the initial map f_1 for an extension. Furthermore that earlier discussion showed that the list for the 1st face of an n-simplex starts with $f_2 f_1$, while the lists for the remaining jth faces with $j \geq 2$ start with f_1 itself. So assume that we are even in the case $n \geq 3$. Then the omitted i can be at most one of the values of j between 2 and n; so $\Delta^{n,i}$ does contain at least one of the jth faces for $j \geq 2$, and thus we do recover f_1 to complete our extension.

But this argument breaks down in the remaining case $n = 2$, with $i = 0$ or 2—say $i = 2$: Here an embedding of $\Delta^{2,2}$ gives us 2 lists of length 1, of the form f_2 and $f_2 f_1$. So we need invertibility of f_2 in order to recover f_1; and this is why we require the groupoid hypothesis.

For the final statement about EG and BG, we see from the description of either of those simplicial sets at Example 3.1.6 that the morphism represented by a group element g has an inverse represented by g^{-1}; so the corresponding categories $\mathfrak{E}_G$ and $\mathfrak{B}_G$ in Example 3.1.9 are indeed groupoids. □

Notice in particular that we now obtain the analogue of Corollary 3.5.3 for the simplicial set version of BG–this time, we use Propositions 3.6.8 and 3.6.1 (ii), in place of the Whitehead theorem 3.5.2 for topological spaces.

REMARK 3.6.9 (The simplicial Borel construction and Kan complexes). We had mentioned in Chapter 2 that our later decompositions can be regarded as beginning with an abstract simplicial complex Δ, and then proceeding to its Borel construction—which we had obtained there in its topological space form as $(EG \times$

$|\Delta|)/G$. We saw in Lemma 2.5.10 that this is a CW-complex, for which weak equivalences are homotopy equivalences by the Whitehead theorem 3.5.2 (i).

Now later we will wish to work in the context of simplicial sets, rather than topological spaces; and in particular, we will use the simplicial Borel construction $(EG \times \Delta)/G$ defined using the simplicial set version of EG in 3.1.6. However, we saw in Warning 3.6.4 that an abstract simplicial complex Δ, when regarded as a simplicial set, need not be a Kan complex. Consequently the simplicial Borel construction $EG \underset{G}{\times} \Delta$ need not be a Kan complex. The remedy is to replace Δ by a weakly equivalent Kan complex before applying the Borel construction. The quotient of a Kan complex by a free G-action is again a Kan complex (the argument is given in [**63**, V.3.7]). ♢

We now return from our digression on Kan complexes, to state Quillen's general equivalence result:

THEOREM 3.6.10. *The adjunction* (3.3.8) *passes down to an equivalence between the homotopy categories of topological spaces and of simplicial sets; and restricts to an equivalence between the homotopy categories of CW-complexes and of simplicial sets.*

PROOF. The proof can be found in Quillen [**100**] and in May [**91**]; and further discussion may be found in Bousfield and Kan [**31**, VIII.3 and VIII.4]. □

The effect of Theorem 3.6.10 is that we can replace the category of topological spaces (or CW-complexes) with the category of simplicial sets. So we make the following convention.

CONVENTION 3.6.11. From now on, the word "space" means simplicial set.

3.7. Simplicial R-modules and homology

Recall from Definition 3.1.1 that $\mathbf{\Delta}$ is the category of finite nonempty totally ordered sets, and that a simplicial set is a functor from $\mathbf{\Delta}^{\mathrm{op}}$ to sets.

DEFINITION 3.7.1. A *simplicial object* in a category $\mathfrak{C}$ is a functor from $\mathbf{\Delta}^{\mathrm{op}}$ to $\mathfrak{C}$. So for example a simplicial set is a simplicial object in the category of sets. Similarly, we have simplicial groups, simplicial schemes, and so on. The simplicial objects in a category $\mathfrak{C}$ are often denoted by a subscript dot indicating the contravariance, as in $X_{\bullet}$; and together these objects form a category, correspondingly denoted $\mathfrak{C}_{\bullet}$.

In this section, we are interested in simplicial R-modules. The following proposition shows that these are always Kan complexes.

PROPOSITION 3.7.2. *A simplicial group, when regarded as a simplicial set by forgetting the group structure, is always a Kan complex.*

PROOF. In fact, a stronger statement is true. A simplicial set X with the property that each X_n is a group and such that the face maps are group homomorphisms is called a *grouplike* simplicial set. A map $f\colon X \to Y$ of grouplike simplicial sets is said to be a *homomorphism* if each $f_n\colon X_n \to Y_n$ is a group homomorphism. Every surjective homomorphism of grouplike simplicial sets is a Kan fibration, and in particular, every grouplike simplicial set is a Kan complex. The proof is essentially due to Moore [**96**], see also May [**91**]. □

DEFINITION 3.7.3 (Simplicial R-modules and homology). Let R be a coefficient ring. There is an equivalence of categories, called the Dold–Kan correspondence, between simplicial R-modules and chain complexes of R-modules indexed by the nonnegative integers. In particular, the case $R = \mathbb{Z}$ gives an equivalence between simplicial abelian groups and chain complexes of abelian groups. The same construction can be applied to simplicial objects in any abelian category.

If $A.$ is a simplicial R-module, the *normalized chain complex* NA_* is defined so that NA_n is equal to the intersection of the kernels of $d_1, \ldots, d_n$. The remaining face map d_0 is used as the boundary map d. From the dual version of the relations (3.2.1), we obtain the identity $d_0 d_0 = d_0 d_1$—which shows here that $d^2 = 0$ since NA_n is in the kernel of d_1, so that we do get a chain complex. Conversely, given a chain complex C_*, we let $A.$ be the simplicial R-module whose nondegenerate simplices are the elements of C_*, and with face maps given on the nondegenerate simplices by $d_0 = d$ and $d_i = 0$ for $i > 0$. More formally, A_n is equal to the sum $\bigoplus_{[n]\twoheadrightarrow[k]} C_k$ defined using only surjective mappings; and the face and degeneracy maps can be worked out using the dual version of the relations (3.2.1) from the effect on nondegenerate simplices given above. From the definition of homotopy of a Kan complex, it is not hard to see that the homology of the chain complex $H_n(C_*)$ is the same as the homotopy $\pi_n A..$

DEFINITION 3.7.4. The homology of a space X with coefficients in R can be described as follows. We define $R \otimes X$ to be the simplicial set whose n-simplices are the finite formal linear combinations $\sum_i r_i x_i$ where $x_i \in X_n$. This is a simplicial R-module, and the homology of the corresponding normalized chain complex of R-modules is $H_*(X; R)$:

$$H_n(X; R) = \pi_n(R \otimes X).$$

A related construction leads to reduced homology. We define RX to be the simplicial subset of $R \otimes X$ consisting of the linear combinations satisfying $\sum_i r_i = 1$. A choice of basepoint $\star$ in X makes each $(RX)_n$ into a free R-module on the basis given by the subset $X_n \setminus s_0^n(\star)$, because the coefficient of $s_0^n(\star)$ in such a linear combination on X_n is determined by the other coefficients. So RX is a simplicial R-module, and $\tilde{H}_*(X; R)$ is the homology of the corresponding normalized chain complex of R-modules:

$$\tilde{H}_n(X; R) = \pi_n(RX).$$

If Y is a subspace of X, then we define the *relative homology* $H_n(X, Y; R)$ of the pair X, Y to be the homology of the quotient of the normalized chain complex of $R \otimes X$ by the normalized chain complex of $R \otimes Y$. So the short exact sequence of chain complexes given by this quotient gives rise to a long exact sequence in homology

$$\cdots \to H_n(Y; R) \to H_n(X; R) \to H_n(X, Y; R) \to H_{n-1}(Y; R) \to \cdots$$

The dual notions for cohomology of a space with coefficients in R are defined by taking the normalized chain complexes as above, which are chain complexes of free R-modules, and applying $\mathrm{Hom}_R(-, R)$ to obtain cochain complexes.

CHAPTER 4

Bousfield-Kan completions and homotopy colimits

4.1. Cosimplicial objects

In this section, we discuss cosimplicial spaces and the total space construction. These constructions will be used in our discussion of Bousfield–Kan completions.

We start with the concept naturally dual to Definition 3.7.1:

DEFINITION 4.1.1. A *cosimplicial object* in a category $\mathfrak{C}$ is a (covariant) functor from $\mathbf{\Delta}$ to $\mathfrak{C}$. The cosimplicial objects in a category $\mathfrak{C}$ are often denoted by a superscript dot, as in $X^{\cdot}$; together these objects form a category, correspondingly denoted $\mathfrak{C}^{\cdot}$.

In particular, a *cosimplicial space* can be regarded either as a functor from $\mathbf{\Delta}$ to $\mathfrak{Sp}$—or as a functor from $\mathbf{\Delta}^{\mathrm{op}} \times \mathbf{\Delta}$ to $\mathfrak{Set}$. The cosimplicial spaces give the category $\mathfrak{Sp}^{\cdot}$.

If $X^{\cdot}$ is a cosimplicial space and k and n are integers, we write X^n_k for the value of $X^{\cdot}$ on $([k],[n])$. So $X^{\cdot}$ can be regarded as a series of spaces X^n for $n \geq 0$ with a map $X^n \to X^m$ associated to each monotonic function $[n] \to [m]$. This amounts to giving *coface maps* $d^i \colon X^n \to X^{n+1}$ $(0 \leq i \leq n+1)$ and *codegeneracy maps* $s^i \colon X^n \to X^{n-1}$ $(0 \leq i \leq n-1)$ satisfying the relations (3.2.1). ♢

EXAMPLE 4.1.2 (The cosimplicial point). The *cosimplicial point* $\star^{\cdot}$ is the constant functor sending each $[n]$ to the simplicial point $\star$. This is a terminal object in the category $\mathfrak{Sp}^{\cdot}$ of cosimplicial spaces.

EXAMPLE 4.1.3 (The cosimplicial simplex). The cosimplicial simplex $\Delta^{\cdot}$ has as its nth space the simplicial n-simplex Δ^n, with the obvious maps. More formally, this is the same as saying that $\Delta^{\cdot}$ is the Hom functor $\Delta^n_k = \mathrm{Hom}_{\mathbf{\Delta}}([k],[n])$.

DEFINITION 4.1.4 (Cosimplicial mapping space). If X and Y are spaces then $\mathsf{Map}^{\cdot}(X,Y)$ is the cosimplicial space with

$$\mathsf{Map}^{\cdot}(X,Y)^n_k = \mathrm{Hom}_{\mathfrak{Set}}(X_n, Y_k).$$

The face and degeneracy maps come from Y, and the coface and codegeneracy maps come from X.

DEFINITION 4.1.5. If $X^{\cdot}$ is a cosimplicial space and Y is a space, then we make the product $X^{\cdot} \times Y$ into a cosimplicial space in a way analogous to Definition 3.4.1 of the product for spaces. Namely,

$$(X^{\cdot} \times Y)^n_k = X^n_k \times Y_k.$$

DEFINITION 4.1.6 (Function spaces). If $X^{\cdot}$ and $Y^{\cdot}$ are both cosimplicial spaces, we define a mapping space: in a way analogous to Definition 3.4.9 for spaces, but now instead using the product in Definition 4.1.5. Namely $\mathsf{Map}(X^{\bullet}, Y^{\bullet})$ has n-simplices $\mathsf{Map}(X^{\bullet}, Y^{\bullet})_n$ given by the morphisms in $\mathfrak{Sp}^{\bullet}$ of the cosimplicial spaces $X^{\bullet} \times \Delta^n \to Y^{\bullet}$, with face and degeneracy maps coming from those among the Δ^n.

There is a version involving cosimplicial spaces of the adjointness result for spaces in Proposition 3.4.10:

LEMMA 4.1.7. *If $X^{\bullet}$ and $Z^{\bullet}$ are cosimplicial spaces and Y is a space, then there is a natural isomorphism of sets*

$$\mathrm{Hom}_{\mathfrak{Sp}}\big(Y, \mathsf{Map}(X^{\bullet}, Z^{\bullet})\big) \cong \mathrm{Hom}_{\mathfrak{Sp}^{\bullet}}(X^{\bullet} \times Y, Z^{\bullet}) \ .$$

PROOF. As in the proof of 3.4.10, we express a morphism in $\mathfrak{Sp}$ on the left-hand side in terms of compatible maps of sets on n-simplices; but this time the term $\mathsf{Map}(X^{\bullet}, Z^{\bullet})_n = \mathrm{Hom}_{\mathfrak{Sp}^{\bullet}}(X^{\bullet} \times \Delta^n, Z^{\bullet})$ then decomposes into maps of sets for both indices l, m of the cosimplical-space structure, so that we have compatible maps of sets

$$Y_n \to \mathrm{Hom}_{\mathfrak{Set}}(X^l_m \times \Delta^n_m, Z^l_m)$$

which by the usual adjointness for sets translates to compatible maps

$$X^l_m \times \Delta^n_m \times Y_n \to Z^l_m.$$

And as in the proof of 3.4.10, this is equivalent to giving compatible maps

$$X^l_m \times Y_m \to Z^l_m,$$

namely a morphism in $\mathfrak{Sp}^{\bullet}$ from $X^{\bullet} \times Y$ to $Z^{\bullet}$. This is another example of the Kronecker delta interpretation of Δ^n_m, see Remark 3.4.11. □

DEFINITION 4.1.8 (Total space of a cosimplicial space). If $X^{\bullet}$ is a cosimplicial space, we define its *total space* to be

$$\mathsf{Tot}\, X^{\bullet} = \mathsf{Map}(\Delta^{\bullet}, X^{\bullet}).$$

For example, we shall see in Proposition 4.1.10 that the total space of the cosimplicial mapping space is the ordinary mapping space.

LEMMA 4.1.9. *If X, Y and Z are spaces then there is an isomorphism of sets*

$$\mathrm{Hom}_{\mathfrak{Sp}^{\bullet}}\big(\Delta^{\bullet} \times Z, \mathsf{Map}^{\bullet}(X, Y)\big) \cong \mathrm{Hom}_{\mathfrak{Sp}}(X \times Z, Y).$$

PROOF. A morphism in $\mathfrak{Sp}^{\bullet}$ from $\Delta^{\bullet} \times Z$ to $\mathsf{Map}^{\bullet}(X, Y)$ consists of compatible maps

$$\Delta^n_m \times Z_m \to \mathrm{Hom}_{\mathfrak{Set}}(X_n, Y_m),$$

or via the usual adjointness for sets

$$X_n \times \Delta^n_m \times Z_m \to Y_m.$$

As before (see Remark 3.4.11) this is equivalent to

$$X_m \times Z_m \to Y_m,$$

namely a morphism in $\mathfrak{Sp}$ from $X \times Z$ to Y. □

PROPOSITION 4.1.10. *If X and Y are spaces, then the space $\mathsf{Map}(X, Y)$ of Definition 3.4.9 is related to the cosimplicial space $\mathsf{Map}^{\bullet}(X, Y)$ of Definition 4.1.4 by*

$$\mathsf{Map}(X, Y) = \mathsf{Tot}\, \mathsf{Map}^{\bullet}(X, Y).$$

PROOF. By Yoneda's lemma (see for example [**21**, 2.1.4]), it suffices to show that these two objects give rise to isomorphic representable functors. And indeed evaluating on an arbitrary object Z in the category $\mathfrak{Sp}$ of spaces, we have isomorphisms of sets

$$\begin{aligned}
\mathrm{Hom}_{\mathfrak{Sp}}\big(Z, \mathsf{Tot}\,\mathsf{Map}^{\cdot}(X,Y)\big) &= \mathrm{Hom}_{\mathfrak{Sp}}\big(Z, \mathsf{Map}(\Delta^{\cdot}, \mathsf{Map}^{\cdot}(X,Y)\big) && \text{by Defn. 4.1.8} \\
&\cong \mathrm{Hom}_{\mathfrak{Sp}^{\cdot}}\big(\Delta^{\cdot} \times Z, \mathsf{Map}^{\cdot}(X,Y)\big) && \text{by Lemma 4.1.7} \\
&\cong \mathrm{Hom}_{\mathfrak{Sp}}(X \times Z, Y) && \text{by Lemma 4.1.9} \\
&\cong \mathrm{Hom}_{\mathfrak{Sp}}\big(Z, \mathsf{Map}(X,Y)\big) && \text{by Prop. 3.4.10.} \quad \square
\end{aligned}$$

4.2. Bousfield–Kan completions

One topological construction which we will use in our later main results is the p-completion. This is a functorial construction which turns mod p homology equivalences into homotopy equivalences. We could have avoided mentioning this construction by always talking about mod p homology equivalences instead of homotopy equivalences; this was the route taken by Dwyer [**51, 52, 53**].

As a source for this material, the literature refers primarily to the original treatment of Bousfield and Kan in [**31**]; and we know of no more expository development for non-experts. So in this section, we will review some basic properties of more general completions; this will also allow us as usual to emphasize applications to group cohomology. The work of [**31**] is developed in the context of simplicial sets, as discussed in Section 3.1.

We mention that one motivation for the study of completions is given by Bousfield and Kan in the discussion of "fracturing" in their Introduction in [**31**, p. 2]): for example, their result [**31**, VII.4.2] gives a notion of "primary decomposition" for suitable completed spaces. So as indicated earlier, we will see in Theorem 4.3.1 (iv) that for group cohomology, this gives a homotopy decomposition at the level of classifying spaces, which underlies the primary decomposition of cohomology given in Proposition 2.1.5 (2).

It is also worth mentioning that for nilpotent spaces X (i.e., for any choice of basepoint, $\pi_1 X$ is nilpotent, and acts nilpotently on $\pi_i X$ for $i > 1$) there is a more classical topologically defined R-completion which agrees with the Bousfield–Kan completion. However, most of the spaces we shall be considering are not nilpotent.

Completions and their properties. Let R be any coefficient ring; keeping in mind that for the case of group cohomology, we will be particularly interested in the rings $R = \mathbb{Z}$ and $\mathbb{F}_p$. The following is the definition of the Bousfield–Kan R-completion.

DEFINITION 4.2.1 (Bousfield–Kan [**31**, I.4]). Let X be a space, and let RX be defined as in Definition 3.7.3. There are maps $\phi\colon X \to RX$ given by $\phi(x) = 1x$, and $\psi\colon RRX \to RX$ given by multiplication in R:

$$\psi\Big(\sum_i r_i\big(\sum_j \alpha_{ij} x_{ij}\big)\Big) = \sum_{i,j} (r_i \alpha_{ij}) x_{ij}.$$

These maps are used to make a cosimplicial space $R^{\cdot}X$ as follows. We set $(R^{\cdot}X)^n = R^{n+1}X$, with coface maps $d^i = 1^i \phi 1^{n+1-i}\colon (R^{\cdot}X)^n \to (R^{\cdot}X)^{n+1}$ and codegeneracy maps $s^i = 1^i \psi 1^{n-1-i} \colon (R^\bullet X)^n \to (R^\bullet X)^{n-1}$. We have the picture

$$RX \;\substack{\xrightarrow{\;d^0\;}\\ \xleftarrow{\;s^0\;}\\ \xrightarrow{\;d^1\;}}\; RRX \;\substack{\xrightarrow{\;d^0\;}\\ \xleftarrow{\;s^0\;}\\ \xrightarrow{\;d^1\;}\\ \xleftarrow{\;s^1\;}\\ \xrightarrow{\;d^2\;}}\; RRRX \qquad \cdots$$

The cosimplicial space $R^\bullet X$ is a sort of free resolution of X over R, and its total space

$$R_\infty X = \mathsf{Tot}\, R^\bullet X$$

is defined to be the *R-completion* of X. ◇

EXAMPLE 4.2.2. If $X = \star$, then $R^\bullet \star$ is a cosimplicial space which has each space $(R^\bullet \star)^n$ also equal to $\star$; that is, $R^\bullet \star$ is the cosimplicial point $\star^\bullet$ of Example 4.1.2. Now $\star^\bullet$ is a terminal object in $\mathfrak{Sp}^\bullet$, i.e., there is a unique mapping to it; so following through Definition 4.1.8 for the total space of $R^\bullet \star$ via mappings, this implies that $R_\infty \star = \star$.

The construction of the completion is functorial, so that a map of spaces $f \colon X \to Y$ induces a map of cosimplicial spaces $R^\bullet f \colon R^\bullet X \to R^\bullet Y$ and a hence a map of completions $R_\infty f \colon R_\infty X \to R_\infty Y$.

There is also a natural map $X \to R_\infty X$ making the following diagram commute.

$$\begin{array}{ccc} X & \xrightarrow{\;f\;} & Y \\ \downarrow & & \downarrow \\ R_\infty X & \xrightarrow{\;R_\infty f\;} & R_\infty Y \end{array}$$

PROPOSITION 4.2.3. [**31**, I.4.2] *For any space X, the R-completion $R_\infty X$ is a Kan complex.* □

COROLLARY 4.2.4. *If $f \colon R_\infty X \to R_\infty Y$ is a weak equivalence, then f is in fact a homotopy equivalence.*

PROOF. This follows from Propositions 4.2.3 and 3.6.1. □

Bousfield and Kan also show that the third of our notions of equivalence of spaces, namely at the level of homology, is actually strengthened to a homotopy equivalence when we pass to completions.

PROPOSITION 4.2.5. [**31**, I.5.5] *A map of spaces $f \colon X \to Y$ is an R-homology equivalence if and only if $R_\infty f \colon R_\infty X \to R_\infty Y$ is a homotopy equivalence of completions.* □

This rather general property demonstrates one of the important technical advantages of working with the context of simplicial sets, in which completions are defined. In particular, in our standard application given by group cohomology, we see that the study of maps preserving the R-homology of BG amounts to the study of maps preserving the homotopy type of the completion $R_\infty BG$.

We next discuss another property which interrelates notions of equivalence; focusing now on the particular map of X to its completion $R_\infty X$. Bousfield and Kan in [**31**, I.5.1] first establish some terminology for the strength of the equivalence which this map might possibly exhibit.

DEFINITION 4.2.6 (Goodness and completeness). A space X is
(1) *R-complete* if $X \to R_\infty X$ is a weak homotopy equivalence;
(2) *R-good* if $X \to R_\infty X$ is an R-homology equivalence; and
(3) *R-bad* if it is not R-good.

As this terminology suggests, it should seem desirable for a completion of X to be complete; but unfortunately that property is not automatic. However, a fundamental result of Bousfield and Kan shows in particular that R-goodness of X is sufficient (and indeed necessary) for that property; they establish the following relations among the conditions in Definition 4.2.6 for X and $R_\infty X$:

PROPOSITION 4.2.7. [**31**, I.5.2] *For a space X, the following are equivalent.*
(i) *X is R-good.*
(ii) *$R_\infty X$ is R-complete.*
(iii) *$R_\infty X$ is R-good.* □

In particular, observe in the situation of this result that completeness and goodness then propagate to further iterations $(R_\infty)^i X$ of the completion process. Thus an R-bad space never becomes R-good, no matter how many times it is completed.

COROLLARY 4.2.8. [**31**, I.5.3] *If X is R-good, then all the spaces $(R_\infty)^i X$ $(i \geq 1)$ are homotopy equivalent.*

PROOF. This follows from Proposition 4.2.7 and Corollary 4.2.4. □

PROPOSITION 4.2.9. *If $f \colon X \to Y$ is an R-homology equivalence, then X is R-good if and only if Y is R-good.*

PROOF. This follows from Proposition 4.2.5. □

The following results of Bousfield and Kan demonstrate that the desirable property of R-goodness indeed holds for many spaces and rings.

DEFINITION 4.2.10. We say that a group G is *R-perfect* if $R \otimes_{\mathbb{Z}} (G/[G,G]) = 0$.

PROPOSITION 4.2.11. [**31**, VII.3.2] *Suppose that X is connected and that $\pi_1(X)$ is R-perfect (i.e., $H_1(X;R) = 0$), with $R \subseteq \mathbb{Q}$ or $R = \mathbb{F}_p$. Then X is R-good and $R_\infty X$ is simply connected.* □

We shall apply the following to BG for G finite.

PROPOSITION 4.2.12. [**31**, VII.4.3.iii] *Suppose that X is connected and that $\pi_n(X)$ is finite for all $n \geq 1$. Then X is R-good for every subring R of $\mathbb{Q}$.* □

PROPOSITION 4.2.13. [**31**, VII.5] *Suppose that X is connected and that $\pi_1(X)$ is finite. Then X is $\mathbb{F}_p$-good.*[1] □

EXAMPLE 4.2.14. The real projective plane $\mathbb{R}P^2$ is not $\mathbb{Z}$-good, see [**31**, VII.5.2]. So having π_1 finite together with finitely generated homotopy or homology groups is not sufficient for a space to be $\mathbb{Z}$-good.

---

[1]Note that our $\mathbb{F}_p$ is denoted Z_p in [**31**]—this should not be confused with the standard notations of $\mathbb{Z}_{(p)}$ for the p-local integers and $\mathbb{Z}_p$ for the p-complete integers.

PROPOSITION 4.2.15. *Let X be a connected space such that $\pi_n(X)$ is finite for all $n \geq 1$.*

(i) *The natural map $\mathbb{Z}_\infty X \to \prod_p (\mathbb{F}_p)_\infty X$ is a homotopy equivalence.*

(ii) *The natural map $(\mathbb{Z}_{(p)})_\infty X \to (\mathbb{F}_p)_\infty X$ is a homotopy equivalence.*

(iii) $\pi_1\big((\mathbb{F}_p)_\infty X\big) \cong \pi_1 X / O^p(\pi_1 X)$.

(iv) $\pi_1\big(\mathbb{Z}_\infty X\big) = \pi_1 X / L_\infty(\pi_1 X)$.

(v) *If $L_\infty(\pi_1 X)$ is perfect (i.e., equal to its own commutator subgroup) then $\mathbb{Z}_\infty X$ is homotopy equivalent to X^+, the Quillen plus construction on X with respect to $L_\infty(\pi_1 X)$.*

PROOF. Parts (i) and (ii) are special cases of [**31**, VII.4.1]. Part (iii) can be seen by examining the proof of [**31**, VII.4.3]. Part (iv) follows from (i) and (iii), using (0.0.1). For part (v), it follows from (iv) and Proposition 4.2.12 that $\mathbb{Z}_\infty X$ satisfies the defining properties (see for example [**22**, Sec. 2.10]) of the Quillen plus construction. □

4.3. Completion of the classifying space BG

We are now ready to assemble a theorem giving the effects of completion on classifying spaces of finite groups.

THEOREM 4.3.1. *Let G be a finite group.*

(i) *The classifying space BG is $\mathbb{Z}$-good. So the map in integral homology*

$$\tilde{H}_*(BG;\mathbb{Z}) \to \tilde{H}_*(\mathbb{Z}_\infty BG;\mathbb{Z})$$

is an isomorphism, and $\mathbb{Z}_\infty BG$ is $\mathbb{Z}$-complete.

(ii) *The map in integral cohomology*

$$\tilde{H}^*(\mathbb{Z}_\infty BG;\mathbb{Z}) \to \tilde{H}^*(BG;\mathbb{Z})$$

is an isomorphism.

(iii) *The fundamental group of $\mathbb{Z}_\infty BG$ is given by*

$$\pi_1 \mathbb{Z}_\infty BG \cong G/L_\infty G.$$

(iv) *The natural map $\mathbb{Z}_\infty BG \to \prod_p (\mathbb{F}_p)_\infty BG$ is a homotopy equivalence, where the product is over primes dividing $|G|$.*

(v) *The classifying space BG is $\mathbb{Z}_{(p)}$-good and $\mathbb{F}_p$-good, and*

$$(\mathbb{Z}_{(p)})_\infty BG \to (\mathbb{F}_p)_\infty BG$$

is a homotopy equivalence. So the maps in p-local and mod p homology

$$H_*(BG;\mathbb{Z}_{(p)}) \to H_*\big((\mathbb{F}_p)_\infty BG;\mathbb{Z}_{(p)}\big)$$
$$H_*(BG;\mathbb{F}_p) \to H_*\big((\mathbb{F}_p)_\infty BG;\mathbb{F}_p\big)$$

are isomorphisms.

(vi) *The maps in p-local and mod p cohomology*

$$H^*\big((\mathbb{F}_p)_\infty BG;\mathbb{Z}_{(p)}\big) \to H^*(BG;\mathbb{Z}_{(p)})$$
$$H^*\big((\mathbb{F}_p)_\infty BG;\mathbb{F}_p\big) \to H^*(BG;\mathbb{F}_p)$$

are isomorphisms.

(vii) *The maps*

$$\tilde{H}_*\big((\mathbb{F}_p)_\infty BG;\mathbb{Z}\big) \to \tilde{H}_*\big((\mathbb{F}_p)_\infty BG;\mathbb{Z}_{(p)}\big)$$
$$\tilde{H}^*\big((\mathbb{F}_p)_\infty BG;\mathbb{Z}\big) \to \tilde{H}^*\big((\mathbb{F}_p)_\infty BG;\mathbb{Z}_{(p)}\big)$$

are isomorphisms.

(viii) *The fundamental group of* $(\mathbb{F}_p)_\infty BG$ *is given by*

$$\pi_1\big((\mathbb{F}_p)_\infty BG\big) \;\cong\; G/O^p(G).$$

PROOF. By Proposition 2.2.7, BG has finite homotopy groups and is connected. So part (i) follows from Proposition 4.2.12 and Proposition 4.2.7. Part (ii) follows from (i) using Theorem 2.4.8. Part (iii) follows from Proposition 4.2.15 (iv) using Proposition 2.2.7 (2)(ii). Part (iv) follows from Proposition 4.2.15 (i): To see that the product only involves the primes dividing $|G|$, notice that if p does not divide $|G|$, then by Proposition 2.1.5 (1) (ii) we have $\tilde{H}^*(BG;\mathbb{F}_p) = 0$. So the map $BG \to \star$ is a mod p homology equivalence, and hence by Proposition 4.2.5 a homotopy equivalence after applying $(\mathbb{F}_p)_\infty$. But by Example 4.2.2, $(\mathbb{F}_p)_\infty\star$ is just $\star$, which makes no difference in the product. Part (v) follows from Propositions 4.2.12, 4.2.13 and 4.2.15 (ii). Part (vi) follows from (v) using Theorem 2.4.8. Part (vii) follows from [**31**, VI.5.6]. Part (viii) follows from Proposition 4.2.15 (iii) again using Proposition 2.2.7 (2)(ii). □

REMARK 4.3.2. Theorem 4.3.1 can be regarded as an explanation, at the level of spaces, of the primary decomposition of group cohomology in Proposition 2.1.5 (2). Namely, we apply $\tilde{H}^*(-,\mathbb{Z})$ to the decomposition of (iv); and then use part (ii) to identify the left side with $\tilde{H}^*(G,\mathbb{Z})$, and use parts (vii) and (vi) to identify the terms on the right hand side via Corollary 2.3.3 with $\tilde{H}^*(G,\mathbb{Z}) \otimes_{\mathbb{Z}} \mathbb{Z}_{(p)}$.

REMARK 4.3.3. It follows from part (iii) of Theorem 4.3.1 that BG is only $\mathbb{Z}$-complete in case G is nilpotent. If G is not nilpotent, then the completion $\mathbb{Z}_\infty BG$ usually has many nontrivial homotopy groups, so it is not an Eilenberg–Mac Lane space. So, although passing to the $\mathbb{Z}$-completion of BG has no effect on $\mathbb{Z}$-cohomology, it can profoundly change the homotopy type.

EXAMPLE 4.3.4. If $G = SL_2(5)$, then the loop space $\Omega(\mathbb{Z}_\infty BG)$ is homotopy equivalent to the homotopy fiber of a degree 120 self-map on a 3-sphere (Cohen and Levi [**46**]). So $\pi_1(\mathbb{Z}_\infty BG)$ is trivial, and $\pi_i(\mathbb{Z}_\infty BG) \neq 0$ for infinitely many values of i; by contrast of course $\pi_1(BG) \cong G$ and $\pi_i(BG) = 0$ for $i > 1$.

NOTATION 4.3.5 (The p-completion). It follows from Theorem 4.3.1 (iv) that it makes sense to concentrate on the $\mathbb{F}_p$-completion of BG for p dividing $|G|$. When $R = \mathbb{F}_p$ for a prime p, the completion $(\mathbb{F}_p)_\infty X$ is called the *p-completion* of X, which we shall denote by $X^\wedge_p$.

By Theorem 4.3.1 (v) and Proposition 2.2.7 (1), we have

$$H^*(G,\mathbb{F}_p) \cong H^*(BG;\mathbb{F}_p) \cong H^*(BG^\wedge_p;\mathbb{F}_p). \tag{4.3.6}$$

For the rest of this work, we shall be focusing on homotopy decompositions of $BG^\wedge_p$, and the resulting information on the mod p group cohomology $H^*(G,\mathbb{F}_p)$. ◊

4.4. Simplicial spaces

In this section, we define simplicial spaces, and examine the diagonal construction. This construction will be used when we discuss homotopy colimits.

DEFINITION 4.4.1 (Simplicial spaces). A *simplicial space* is defined to be a simplicial object (see Definition 3.7.1) in the category of spaces. A simplicial space can be regarded as a *bisimplicial set*, because a simplicial space can be thought of as a functor from $\mathbf{\Delta}^{\mathrm{op}} \times \mathbf{\Delta}^{\mathrm{op}}$ to $\mathfrak{Set}$ (or a contravariant functor from $\mathbf{\Delta} \times \mathbf{\Delta}$ to $\mathfrak{Set}$). Indeed, there are two different ways of regarding a bisimplicial set as a simplicial space, taking $([m], [n])$ either to the m-simplices in the nth *horizontal* space $X_{-,n}$, or to the n-simplices in the mth *vertical* space $X_{m,-}$. We may call such a term $X_{m,n}$ an (m, n)*-bisimplex* of X.

EXAMPLE 4.4.2 (Bisimplicial R-modules). Since chain complexes of R-modules form an abelian category, we can repeat the Dold–Kan construction of Definition 3.7.3 to get an equivalence of categories between bisimplicial R-modules and double complexes of R-modules.

DEFINITION 4.4.3 (External product). If X and Y are spaces, then there is an *external product* $X \times_{\bullet} Y$ which is a simplicial space with

$$(X \times_{\bullet} Y)_{m,n} = X_m \times Y_n.$$

DEFINITION 4.4.4 (Diagonal space). For $X_{\bullet}$ a simplicial space, the *diagonal space* of $X_{\bullet}$ is defined by taking as simplices the diagonal elements from $X_{\bullet}$:

$$\mathsf{Diag}\,(X_{\bullet})_n = X_{n,n};$$

and for face and degeneracy maps, taking the maps from $X_{\bullet}$ with the diagonal form

$$d_i = (d^i, d^i)^* \qquad s_i = (s^i, s^i)^*$$

in terms of the usual maps d^i, s^i on $\mathbf{\Delta}$ in (3.2.1).

Here is a useful alternative way of looking at the diagonal. We form the disjoint union

$$\coprod_m (X_m \times \Delta^m), \tag{4.4.5}$$

where $X_m = X_{m,-}$ is the mth vertical space of $X_{\bullet}$, and Δ^m is the simplicial m-simplex (viewed as $\mathrm{Hom}_{\mathbf{\Delta}}(-, [m])$ as in Example 3.1.4); and then we pass to a quotient of this union, in which the representatives are given by the terms of form $X_{m,m} \times \Delta^m_m$, which in turn play the role of the sets $X_{m,m}$ in the diagonal. Namely we make identifications much as in Definition 3.3.3. This time we use certain non-diagonal maps from $X_{\bullet}$; the relevant definitions amount to $\big((f,1)^*x, u\big) \sim (x, f \circ u)$. In particular, each general term (x, f) in $X_{m,n} \times \Delta^m_n$ is identified with a diagonal term $\big((f,1)^*x, \mathrm{Id}\big)$ in $X_{n,n} \times \Delta^n_n$, as promised. The symbol Δ^m_n here is behaving like a Kronecker delta, just as we saw in Remark 3.4.11.

In this latter viewpoint, we evaluate the maps $(f, f)^*$ of the diagonal by breaking up into horizontal and vertical directions in $X_{\bullet}$. First the map $(1, f)^*$ on $X_{m,m}$ just amounts to applying on $X_{m,m} \times \Delta^m_m$ the map determined by f in the product space structure of $X_{m,-} \times \Delta^m$. In particular this uses the vertical space $X_{m,-}$, and it has image in $X_{m,n} \times \Delta^m_n$. Next applying $(f, 1)^*$ in the equivalence as above uses the horizontal space $X_{-,n}$, to bring the image back into the diagonal term $X_{n,n} \times \Delta^n_n$.

EXAMPLE 4.4.6. The product $X \times Y$ of spaces, as in Definition 3.4.1, is equal to $\mathsf{Diag}\,(X \times_{\bullet} Y)$.

EXAMPLE 4.4.7 (Spectral sequence of a bisimplicial R-module). If $A_{\bullet}$ is a bisimplicial R-module then one form of the Eilenberg–Zilber theorem (see for example Weibel [**125**, 8.5.1]) gives a homology equivalence between the chain complex C_* of R-modules corresponding as in Definition 3.7.3 to the simplicial R-module $\mathsf{Diag}\,A_{\bullet}$ and the total complex of the double complex of R-modules corresponding as in Example 4.4.2 to $A_{\bullet}$. This is analogous to Theorem 3.4.3, and is another place where the degeneracies are important. It follows that the spectral sequence of this double complex takes the form

$$\pi_{s,-}\pi_{-,t}A_{\bullet} \Rightarrow \pi_{s+t}\mathsf{Diag}\,A_{\bullet} \tag{4.4.8}$$

(recall from Definition 3.7.3 that π_* of a simplicial R-module is H_* of the corresponding chain complex). We can also obtain a *second* spectral sequence by reversing the order of the two coordinates of the bisimplicial R-module. This has a different E_2 page, but converges to the same thing, namely the homotopy of the diagonal (homology of the total complex). When both are being considered, the spectral sequence (4.4.8) is called the *first* spectral sequence of the bisimplicial R-module.

In particular, if $X_{\bullet}$ is a simplicial space then we can apply to each space in X the constructions determined by R in Definition 3.7.4, to obtain bisimplicial R-modules $R \otimes X_{\bullet}$ and $RX_{\bullet}$; then taking homotopy in the first coordinate, $\pi_{*,-}(R \otimes X_{\bullet}) = H_{*,-}(X_{\bullet}; R)$ and $\pi_{*,-}(RX_{\bullet}) = \tilde{H}_{*,-}(X_{\bullet}; R)$ are simplicial R-modules. Identifying $R \otimes \mathsf{Diag}\,X_{\bullet}$ with $\mathsf{Diag}\,(R \otimes X_{\bullet})$, and $R\,\mathsf{Diag}\,X_{\bullet}$ with $\mathsf{Diag}\,RX_{\bullet}$, the spectral sequence (4.4.8) gives the homology spectral sequence of a simplicial space

$$\pi_{s,-}H_{-,t}(X_{\bullet}; R) \Rightarrow H_{s+t}(\mathsf{Diag}\,X_{\bullet}; R), \tag{4.4.9}$$

$$\pi_{s,-}\tilde{H}_{-,t}(X_{\bullet}; R) \Rightarrow \tilde{H}_{s+t}(\mathsf{Diag}\,X_{\bullet}; R). \tag{4.4.10}$$

Again, there are second spectral sequences obtained by reversing the order of the coordinates of the simplicial space.

There is another way to view the spectral sequence of a simplicial space, which is as follows. We take the increasing filtration of $\mathsf{Diag}\,X_{\bullet}$

$$F_0\,\mathsf{Diag}\,X_{\bullet} \subseteq F_1\,\mathsf{Diag}\,X_{\bullet} \subseteq \cdots$$

given by letting $F_s\,\mathsf{Diag}\,X_{\bullet}$ be the image of $\coprod_{m \le s}(X_{m,-} \times \Delta^m)$ in (4.4.5) under the appropriate identifications; in other words, the skeletal filtration with respect to the first coordinate s. This gives rise to an increasing filtration of each of the simplicial R-modules $R \otimes \mathsf{Diag}\,X_{\bullet}$ and $R\,\mathsf{Diag}\,X_{\bullet}$, and hence of the corresponding chain complexes of R-modules. Then (4.4.9) and (4.4.10) are the spectral sequences of these filtered complexes, from the E^2 page onwards.

REMARK 4.4.11 (Equivariant homology spectral sequence and isotropy spectral sequence). From (2.6.9) and Definition 2.6.10, the equivariant homology spectral sequence and the isotropy spectral sequence for a G-space X are the Leray spectral sequences of the maps $EG \underset{G}{\times} X \to BG$ and $EG \underset{G}{\times} X \to X/G$. We can view these as the two homology spectral sequences of a simplicial space as follows. First, form the external product $EG \times_{\bullet} X$. The group G acts on this, and we write $EG \underset{G}{\times}_{\bullet} X$

for the quotient. So an (s,t)-bisimplex in $EG \underset{G}{\times_{\bullet}} X$ is a G-orbit of (s,t)-bisimplices in $EG \times_{\bullet} X$, and

$$\mathsf{Diag}\,(EG \underset{G}{\times_{\bullet}} X) \cong EG \underset{G}{\times} X \quad .$$

The first spectral sequence (4.4.9) of the simplicial space $EG \underset{G}{\times_{\bullet}} X$ is equal to the equivariant homology spectral sequence (2.6.9) because both arise from the skeletal filtration in the first coordinate s, corresponding to dimensions in EG. The second spectral sequence of the same simplicial space is the isotropy spectral sequence of Definition 2.6.10, corresponding to filtration by dimensions in X.

REMARK 4.4.12 (Isotropy spectral sequences and isotropy subgroups). This point of view enables us to form an understanding of the E^1 and E^2 pages of the isotropy spectral sequence. Namely, since this is usually regarded as a first spectral sequence, we can reverse the roles of s and t in the Example above, so that the coordinate s now refers to X, and t refers to EG. We regard X_s as a set, and the E^1 page is its equivariant homology $H_*(EG \underset{G}{\times} X_s; R)$. The horizontal map d^1 on the E^1 page of the spectral sequence is induced by the alternating sum of the face maps from X_s to X_{s-1}, and E^2 is the homology with respect to d^1. The nondegenerate simplices give rise to a subcomplex of E^1 with the same homology (the quotient has no homology), and so we can replace the E^1 page with a chain complex whose sth term is the homology of the Borel construction on the set of nondegenerate s-simplices in X. We can then decompose into orbits, and use Remark 2.5.19 to express this as a sum of homologies of stabilizers of simplices, namely homologies of isotropy subgroups of G. If X has a finite number of nondegenerate simplices, then the spectral sequence has a finite number of nonzero columns.

4.5. Diagrams of spaces and homotopy colimits

We now turn to a way of assembling component spaces called the homotopy colimit, in the sense of Bousfield and Kan. This material corresponds roughly to Section 4 of Dwyer [**53**]; the reader is referred to that source for fuller details.

DEFINITION 4.5.1. Let $\mathfrak{C}$ be a small category, called the *indexing category.* A $\mathfrak{C}$-*diagram of spaces* is a functor from $\mathfrak{C}$ to the category $\mathfrak{Sp}$ of spaces. The $\mathfrak{C}$-diagrams of spaces form a category $\mathfrak{Sp}^{\mathfrak{C}}$ in which the morphisms are the natural transformations.

So a $\mathfrak{C}$-diagram of spaces consists of a space X_i for each object i in $\mathfrak{C}$, and a morphism $X_i \to X_j$ for each arrow $i \to j$ in $\mathfrak{C}$. The homotopy colimit is a functor from diagrams of spaces to spaces

$$\underset{\overrightarrow{\mathfrak{C}}}{\underline{\mathrm{Hocolim}}}\colon \mathfrak{Sp}^{\mathfrak{C}} \to \mathfrak{Sp},$$

which will be defined in Definition 4.5.12, after we have given an extensive discussion of the special case of the homotopy pushout.

The homotopy colimit construction in effect patches together the component spaces, subject to suitable relations determined by the maps in the diagram; but in a homotopy invariant way. In other words, given a map from one diagram of spaces to another, which is a weak equivalence on each of the spaces in the diagram, the induced map of homotopy colimits is a weak equivalence.

To see what is wrong with the ordinary colimit in this respect, consider a diagram of the form

$$\begin{array}{ccc} X & \longrightarrow & \star \\ \downarrow & & \\ \star & & \end{array}$$

for a space X. The colimit of this diagram in the category of spaces is $\star$. On the other hand, consider the diagram

$$\begin{array}{ccc} X & \longrightarrow & CX \\ \downarrow & & \\ CX & & \end{array}$$

where CX is the cone on X. This time the colimit is the suspension of X, which is usually not weakly equivalent to $\star$, in spite of the fact that there is a map from the second diagram to the first which is a weak equivalence at each space.

Our main application of the construction will be to express the p-completed classifying space $BG^{\wedge}_p$ as the p-completion of the homotopy colimit of BH for suitable subgroups H of G. Furthermore the application of the Bousfield–Kan spectral sequence to that homotopy colimit will give a corresponding decomposition of homology in terms of that of the component spaces.

REMARK 4.5.2. In contrast with the situation for completions in the previous section, the homotopy colimit need *not* be a Kan complex—even when the component spaces X_d are Kan complexes (as will usually be our case by Proposition 3.6.8 since we will have component spaces BH for subgroups H of G). However, we shall pass to the p-completion of the homotopy colimit, and so obtain a Kan complex. But we also add one further caution: the operations of homotopy colimit and p-completion in general do not commute. So we emphasize that we will always perform any relevant p-completions (as in Definition 5.3.3) *after* taking any relevant homotopy colimits.

The homotopy pushout. Before describing the homotopy colimit in the general case, we will first give a fairly extensive discussion of the construction in what is usually regarded the simplest nontrivial case—a case which we will in fact encounter repeatedly later. We consider the diagram of spaces given by:

$$\begin{array}{ccc} X & \longrightarrow & Y \\ \downarrow & & \\ Z & & \end{array}$$

The category underlying this diagram we will denote by:

$$\begin{array}{ccc} \{1,2\} & \longrightarrow & \{1\} \\ \downarrow & & \\ \{2\} & & \end{array}$$

This notation is in fact the case of the 1-simplex category $\mathfrak{D}_2$ in our later general Definition 4.6.1—it arises from the poset *opposite* to (or the *dual* poset to) the poset of faces of the abstract 1-simplex on the vertex set $I_2 := \{1, 2\}$; the morphisms are given by containment (i.e., reverse inclusion $\supseteq$). Thus in the notation of Definition 4.5.1, we have $X_{12} := X$, $X_1 := Y$, and $X_2 := Z$ (where for brevity we have suppressed the set notation of braces and commas in the subscripts).

EXAMPLE 4.5.3 (The pushout as an ordinary colimit). Before describing the homotopy colimit in this case, for background we first recall the usual colimit of spaces (Spanier [**117**, p. 18]) over this 1-simplex category. In fact our diagram of spaces above is the *pushout square*; cf. [**22**, p. 244] or Dwyer [**53**, after 3.9]. The resulting colimit is the *pushout*, namely a space W which we can add to the diagram above, with arrows from Y and Z, completing that diagram to a commutative square in a universal way. To achieve this, we simply identify the image in Y of each point X with the corresponding image in Z of that point. We represent the component spaces as disks below, and obtain the following picture.

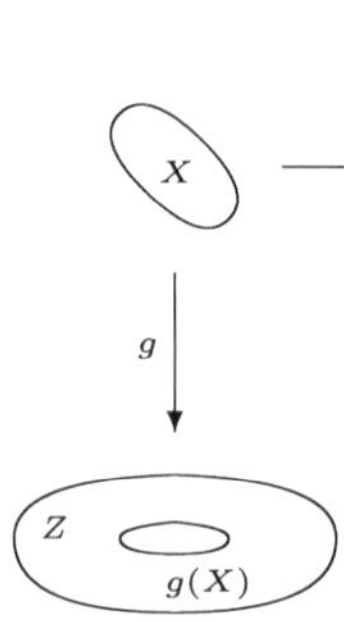

Now for the homotopy colimit, we will want to add some further flexibility to the process of identifications via maps among our component spaces. For each map, we "thicken" the relevant component spaces into cylinders, to reflect the freedom of working up to homotopy equivalence. More precisely, in this case of the pushout square (cf. [**53**, 4.18]), the *homotopy colimit* is realized as a double *mapping cylinder*—see for example [**22**, 1.4.1]. It is formed by taking two copies of $X \times [0, 1]$ joined along the left ends, and attaching their right ends to Y and Z, again using the maps from the original diagram of spaces.

EXAMPLE 4.5.4 (A sketch of the homotopy colimit of the pushout square). We replace the diagram above of the colimit by the diagram below for the homotopy colimit:

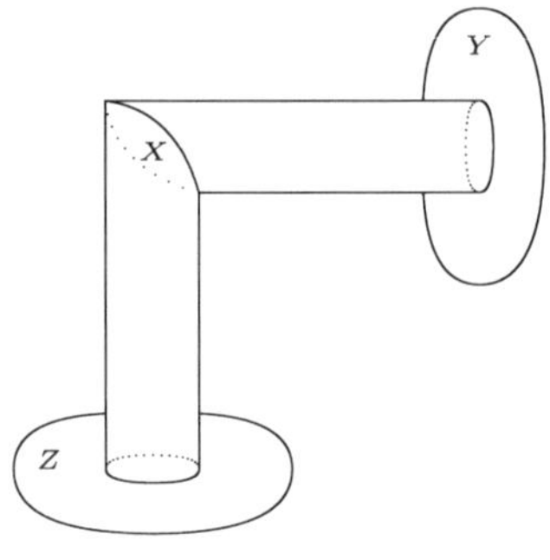

This picture may in fact be a little deceptive: it might suggest that we can always just homotopically contract down both cylinders toward X, to get a homotopy equivalence with the usual colimit. That this is not the case is seen from the example at the beginning of this section, with two cones and a suspension. However, Bousfield and Kan do indicate in [**31**, XII.3.1.iv] that in this pushout square setup, the homotopy colimit and the usual colimit will in fact be weakly equivalent, *if* at least one of the maps from X is a cofibration. But we also emphasize that for the more general indexing categories considered later, the natural map from the homotopy colimit to the ordinary colimit need not be a weak equivalence; see for example [**31**, XII.2.5].

In our applications of the pushout square, there will typically be a covariant functor from $\mathfrak{D}_2$ to finite groups, in such a way that the two maps are inclusions of a common subgroup $H_{12} \leq H_1 \cap H_2$ into two subgroups H_1 and H_2 of a finite group G.

$$\begin{array}{ccc} H_{12} & \longrightarrow & H_1 \\ \downarrow & & \\ H_2 & & \end{array}$$

We denote this situation via the rank two case $\mathcal{H}_2$ of our general Definition 4.6.2 below, giving a "1-simplex of subgroups of G".

EXAMPLE 4.5.5 (1-simplices of subgroups in some simple groups). Ordinarily we will obtain the family $\mathcal{H}_2$ as a 1-simplex of subgroups of G defined by a pair H_1 and H_2 of subgroups which generate G, and which further satisfy $H_{12} = H_1 \cap H_2$. We mention that an important example of such a 1-simplex of subgroups arises in a group of Lie type of Lie rank two, from the two maximal parabolic subgroups[2] H_1 and H_2 containing a fixed Borel subgroup H_{12}. But also in Chapter 7, we will work with an analogous 1-simplex defined by a suitable H_1 and H_2, in a total of five out of the 26 sporadic groups G, namely:

$$M_{11}, M_{12}, J_2, Th, O'N.$$

At the present stage of our exposition, we need not assume that the groups H_J described above (for nonempty $J \subseteq \{1,2\}$) are subgroups of any fixed finite group G: for we can consider instead the abstract group defined by their free product with amalgamation, namely $H := H_1 *_{H_{12}} H_2$. When the H_J are indeed subgroups of some G (as in Example 4.5.5), then there is a natural map from the free product with amalgamation to G, whose image is the subgroup generated by H_1 and H_2.

To obtain an expression relevant to the group cohomology of H, we simply take the classifying spaces BH_J of the H_J, so that our original pushout square of spaces takes the particular form:

$$\begin{array}{ccc} BH_{12} & \longrightarrow & BH_1 \\ \downarrow & & \\ BH_2 & & \end{array} \tag{4.5.6}$$

THEOREM 4.5.7. *Suppose that the maps $H_{12} \to H_1$ and $H_{12} \to H_2$ are injective. Then the homotopy colimit of the diagram* (4.5.6) *is weakly equivalent to the classifying space $B(H_1 *_{H_{12}} H_2)$ of the amalgamated product.*

---

[2]For details on these subgroups in a Lie type group, see Example 5.1.1 and the references there, such as [**22**, p. 235], [**43**, 8.3.4].

PROOF. See Theorem 1B.11 and Example 1B.12 in the book of Hatcher [**73**]. Note that the identification of $H_1 *_{H_{12}} H_2$ with π_1 of the homotopy colimit is just the Van Kampen theorem (see for example Spanier [**117**, p. 151]). The rest of the proof is involved in showing that the universal cover of the homotopy colimit is contractible. □

REMARK 4.5.8. The amalgamated product H is infinite whenever the inclusions $H_{12} \to H_1$ and $H_{12} \to H_2$ are proper, in contrast to our usual focus in this work on finite groups.

REMARK 4.5.9. The homotopy pushout above is never a Kan complex, even if H_{12}, H_1 and H_2 are all the trivial group. So the weak equivalence of Theorem 4.5.7 is never a homotopy equivalence. This is at least one reason why we $\mathbb{F}_p$-complete after taking homotopy colimits, rather than before. See for example Lemma 4.5.10 (iv), and Definition 5.3.3.

The weak equivalence indicated above is especially useful under certain further assumptions on the H_J; so we provide here a brief preliminary discussion on why those conditions usually hold in our later applications: As mentioned in Example 4.5.5, it is natural from the viewpoint of group theory to assume that H_1 and H_2 are subgroups generating a finite group G, and further that $H_{12} = H_1 \cap H_2$. And since we typically wish to compute mod p (co)homology for some prime p, it is natural to also assume that H_{12} contains a Sylow p-subgroup of G: for then it is standard that the injections $BH_{12} \to BH_i$ induce injections in cohomology and surjections in homology. And indeed this further Sylow hypothesis holds for the 1-simplex $\mathcal{H}_2$ of parabolics in Example 4.5.5; and also usually in the sporadic examples indicated there. However, the Sylow condition fails for one of the H_i in the case of M_{11}.

Under these further hypotheses, an optimal result might be for the natural surjection $H_1 *_{H_{12}} H_2 \to G$ to induce an $\mathbb{F}_p$-cohomology equivalence. The following lemma of Benson and Wilkerson gives some equivalent characterizations of this desirable situation.

LEMMA 4.5.10. [**29**, 3.1] *Suppose that H_1 and H_2 are subgroups which generate a finite group G, and that $H_{12} = H_1 \cap H_2$ contains a Sylow p-subgroup of G. Write F for the kernel of the map $H_1 *_{H_{12}} H_2 \to G$; observe that F is a finitely generated free group. Then the following are equivalent.*

(i) $H^*(G, H^1(F, \mathbb{F}_p)) = 0$.

(ii) *The (signed) restriction maps give a short exact sequence of $\mathbb{F}_p$-vector spaces:*

$$0 \to H^*(G, \mathbb{F}_p) \to H^*(H_1, \mathbb{F}_p) \oplus H^*(H_2, \mathbb{F}_p) \to H^*(H_{12}, \mathbb{F}_p) \to 0.$$

(iii) *The map $H_1 *_{H_{12}} H_2 \to G$ is a mod p cohomology isomorphism (i.e., an $\mathbb{F}_p$-cohomology equivalence).*

(iv) *The map of p-completed classifying spaces induced by the map in* (iii), *namely*

$$B(H_1 *_{H_{12}} H_2)^{\wedge}_p \to BG^{\wedge}_p,$$

is a homotopy equivalence. Note that by Theorem 4.5.7, $B(H_1 *_{H_{12}} H_2)$ is weakly equivalent to the homotopy colimit of the pushout square (4.5.6). □

We indicate some ways in which the result above involves a number of main themes in our exposition so far, and in our exposition still to come. For example we can apply, to the map in part (iii), the classifying space construction B regarded as a functor; and we saw earlier that the resulting left side $B(H_1 *_{H_{12}} H_2)$ is weakly equivalent to the homotopy colimit of the pushout square of the classifying spaces for the subgroups H_J. Application of cohomology functors to the homotopy decomposition given by this square then "explains" the cohomology decomposition in (ii). Indeed the condition (iii) corresponds essentially to the property that the collection given by the G-conjugates of the subgroup H_J is "ample" in the sense of Definition 5.3.3. Part (ii) says that the corresponding homology decomposition is "sharp" in the sense of Definition 5.8.1.

There is a slightly more complicated case which works the same way. Write $\mathfrak{D}_{2+2}$ for the category

$$
\begin{array}{ccccc}
\cdot & & \cdot & & \cdot \\
& \nwarrow \quad \nearrow & & \nwarrow \quad \nearrow & \\
& \cdot & & \cdot &
\end{array}
$$

This is the opposite of the category associated to the poset of simplices in the simplicial complex which is obtained by identifying two copies of Δ^1 along a common vertex. A $\mathfrak{D}_{2+2}$ diagram of classifying spaces is of the form

$$
\begin{array}{ccccc}
BH_1 & & BH_2 & & BH_3 \\
& \nwarrow \quad \nearrow & & \nwarrow \quad \nearrow & \\
& BH_{12} & & BH_{2,3} &
\end{array}
\tag{4.5.11}
$$

The homotopy colimit is formed from two double mapping cones with the middle space BH_2 identified. This is the classifying space of the amalgamated free product

$$H_1 *_{H_{12}} H_2 *_{H_{2,3}} H_3.$$

In the case of the group J_3 we shall eventually reduce to a diagram of this form.

Both the pushout square and the slightly more complicated diagram (4.5.11) are special cases of the classifying space of a *graph product of groups*, sometimes called the fundamental group of a graph of groups. For trees, one ends up with an amalgamated free product of groups; more generally, one chooses a spanning tree and forms the amalgamated free product, and then cycles in the graph are realized with HNN extensions. For more details see Appendix 1B of Hatcher [**73**], and in particular, Theorem 1B.11. In some sense, this is the one-dimensional case of the more general homotopy colimit construction, which we describe next; we shall not give a general description of the fundamental group of a homotopy colimit, but it can be found in Dror [**50**].

General homotopy colimits. We now turn to more general indexing categories. We begin with the technical definition of homotopy colimit, and then give a more digestible description. For further details, see Sec. XII of [**31**] or [**53**, 4.13].

DEFINITION 4.5.12 (Homotopy colimit). Let $\mathfrak{C}$ be an indexing category, and let F be a $\mathfrak{C}$-diagram in $\mathfrak{Sp}$. We define a simplicial space whose vertical spaces come from the "component" spaces $F(i)$ for objects $i \in \mathfrak{C}$; and whose horizontal maps come from the morphisms in $\mathfrak{C}$.

We implement this as follows. The simplicial space $\amalg.F$ has nth vertical space $(\amalg.F)_{n,-}$ given by

$$\coprod_{\sigma\in(\mathcal{N}\mathfrak{C})_n} F(\sigma_0)$$

where the disjoint union is over all n-simplices $\sigma = (\sigma_0 \to \cdots \to \sigma_n)$ in the nerve $\mathcal{N}\mathfrak{C}$, and we are just using the *initial* space $F(\sigma_0)$ associated to σ. The degeneracy map s_i in the horizontal direction takes the copy of $F(\sigma_0)$ indexed by σ isomorphically to the copy indexed by $s_i(\sigma)$. The face map d_i in the horizontal direction takes it isomorphically to the copy indexed by $d_i(\sigma)$—except when $i = 0$, in which case it takes it via the map $F(\sigma_0 \to \sigma_1)$ in the diagram to the copy of $F(\sigma_1)$ indexed by the face $d_0(\sigma) = (\sigma_1 \to \cdots \to \sigma_n)$.

The *homotopy colimit* is defined by

$$\underset{\mathfrak{C}}{\underrightarrow{\mathrm{Hocolim}}}\, F = \mathsf{Diag}\,\amalg.F,$$

namely the diagonal space of $\amalg.F$ in the sense of Definition 4.4.4. When we wish to emphasize the objects in $\mathfrak{C}$, we use the notation

$$\underset{i\in\mathfrak{C}}{\underrightarrow{\mathrm{Hocolim}}}\, F(i)$$

and leave the arrows implicit. ◇

EXAMPLE 4.5.13. Consider the $\mathfrak{C}$-diagram F of spaces where each space $F(\sigma_0)$ is just the simplicial point $\star$. The homotopy colimit of this diagram just reproduces the nerve $\mathcal{N}\mathfrak{C}$ of the indexing category. This is because the set of n-simplices of the diagonal space in the definition has for each $\sigma \in (\mathcal{N}\mathfrak{C})_n$ just a singleton set, namely the n-simplex set $(\star)_n$ of the simplicial point $\star$; and the face and degeneracy maps in the diagonal which connect these singletons are then naturally identified with the corresponding maps in the nerve among the simplices σ themselves.

According to the discussion of the diagonal following Definition 4.4.4, we can view $\underset{\mathfrak{C}}{\underrightarrow{\mathrm{Hocolim}}}\, F$ as the quotient of

$$\coprod_n \coprod_{\sigma\in(\mathcal{N}\mathfrak{C})_n} (F(\sigma_0) \times \Delta^n) \tag{4.5.14}$$

by the equivalence relation from maps $(f, 1)^*$, which use the horizontal space structure of the simplicial space $\amalg.F$; namely the horizontal face and degeneracy maps described in Definition 4.5.12. Now the degeneracy maps just give isomorphisms of terms $F(\sigma_0)$, so we can essentially restrict our attention to the effect of the face maps; and similarly, even to cases where a face map does not necessarily give an isomorphism; that is, to cases where the simplex σ consists of non-identity arrows from $\mathfrak{C}$.

DESCRIPTION 4.5.15 (Construction of homotopy colimits). To spell this out more graphically, set $X_i := F(i)$. We begin with the disjoint union of one copy of each X_i for i in $\mathfrak{C}$, each copy regarded as the product of X_i with a zero-simplex $\Delta^0 = \star$. Then for each non-identity morphism $i \to j$ in $\mathfrak{C}$, we take a copy of $X_i \times \Delta^1$ (where Δ^1 is just a simplicial unit interval). We glue this into the previous space as follows. We use the map $X_i \to X_j$ to identify points in one end $X_i \times d_0(\Delta^1)$ with their image in $X_j \times \star$, and we use the identity map on X_i to identify the other end $X_i \times d_1(\Delta^1)$ with $X_i \times \star$.

Next, for each pair of composable non-identity morphisms $i \to j \to k$ in $\mathfrak{C}$, we take a copy of $X_i \times \Delta^2$. The faces are glued into the previously constructed space as follows. We identify $X_i \times d_0(\Delta^2)$ inside the copy of $X_j \times \Delta^1$ indexed by $j \to k$, via $X_i \to X_j$. We identify $X_i \times d_1(\Delta^2)$ with the copy of $X_i \times \Delta^1$ indexed by $i \to k$, and the $X_i \times d_2(\Delta^2)$ with the copy indexed by $i \to j$.

We continue this way with sequences of composable non-identity maps in $\mathfrak{C}$ of arbitrary length; when we are done, we have constructed the homotopy colimit. ◊

WARNING 4.5.16. To form a homotopy colimit, it is not sufficient to be given a diagram in the *homotopy* category of spaces. We must be given spaces and actual maps. A diagram in the homotopy category need not lift to spaces; and if it does lift, it may lift in several ways with different homotopy colimits. See Example 4.7.5 for a way of constructing diagrams of classifying spaces with many lifts, or with no lifts at all.

PROPOSITION 4.5.17. (i) *If F is a $\mathfrak{C}$-diagram of spaces, then there is a natural map*

$$\underrightarrow{\mathrm{Hocolim}}_{\mathfrak{C}} F \to \underrightarrow{\mathrm{colim}}_{\mathfrak{C}} F$$

which is usually not a weak equivalence.

(ii) *Let $f\colon F \to F'$ be a map of $\mathfrak{C}$-diagrams of spaces. Suppose that, for each object i in $\mathfrak{C}$, the map $f(i)\colon F(i) \to F'(i)$ is a weak equivalence. Then the map*

$$\underrightarrow{\mathrm{Hocolim}}_{\mathfrak{C}} F \to \underrightarrow{\mathrm{Hocolim}}_{\mathfrak{C}} F'$$

is also a weak equivalence.

PROOF. For (i), we project each $F(\sigma_0) \times \Delta^n$ in (4.5.14) onto the first factor, and then map to the colimit. These maps are easily seen to be compatible, and hence define a map from the homotopy colimit.

Part (ii) is proved in [**31**, XII.4.2]. □

EXAMPLE 4.5.18 (Homotopy colimit of discrete spaces as a nerve). A construction of Grothendieck (see for example Dwyer [**51**, Defn. 2.8] or [**53**, Sec. 5]) allows us to describe the homotopy colimit of a diagram of discrete spaces (cf. Example 3.1.2) as the nerve of a category. Namely, given a $\mathfrak{C}$-diagram of sets F, regarded as a $\mathfrak{C}$-diagram of discrete spaces, we form the *transport category* or *Grothendieck construction* on F, denoted $\mathsf{Gr}(F)$. This is a small category whose objects are the pairs (i,x) with i is an object in $\mathfrak{C}$ and $x \in F(i)$. A map $(i,x) \to (i',x')$ consists of a map $f\colon i \to i'$ in $\mathfrak{C}$ such that $F(f)(x) = x'$. An n-simplex in $\mathcal{N}\mathsf{Gr}(F)$ is given by an n-simplex $i_0 \to \cdots \to i_n$ in the nerve $\mathcal{N}\mathfrak{C}$ of $\mathfrak{C}$ together with an element $x \in F(i_0)$. This is the same as the description of an n-simplex in the homotopy colimit, and it is easy to see that the face and degeneracy maps are the same, and so we have the following isomorphism.

$$\underrightarrow{\mathrm{Hocolim}}_{\mathfrak{C}} F \cong \mathcal{N}\mathsf{Gr}(F). \tag{4.5.19}$$

We shall need to use the fact that homotopy colimits commute with each other. In particular, we shall interpret the Borel construction as a homotopy colimit in Section 4.7, and we shall make use of the fact that homotopy colimits commute with the Borel construction.

Let $\mathfrak{C}$ and $\mathfrak{C}'$ be small categories. Then $\mathfrak{C} \times \mathfrak{C}'$-diagrams in $\mathfrak{Sp}$ are equivalent to $\mathfrak{C}$-diagrams in $\mathfrak{Sp}^{\mathfrak{C}'}$, and also to $\mathfrak{C}'$-diagrams in $\mathfrak{Sp}^{\mathfrak{C}}$.

THEOREM 4.5.20. *Let $F\colon \mathfrak{C} \times \mathfrak{C}' \to \mathfrak{Sp}$ be a $\mathfrak{C} \times \mathfrak{C}'$-diagram of spaces. Then*

$$\underset{\mathfrak{C}}{\underrightarrow{\mathrm{Hocolim}}}(\underset{\mathfrak{C}'}{\underrightarrow{\mathrm{Hocolim}}}\, F) \;\cong\; \underset{\mathfrak{C}\times\mathfrak{C}'}{\underrightarrow{\mathrm{Hocolim}}}\, F \;\cong\; \underset{\mathfrak{C}'}{\underrightarrow{\mathrm{Hocolim}}}(\underset{\mathfrak{C}}{\underrightarrow{\mathrm{Hocolim}}}\, F).$$

PROOF. An n-simplex in each of these three expressions corresponds to a triple consisting of an n-simplex $\sigma = (\sigma_0 \to \cdots \to \sigma_n)$ in $\mathcal{N}\mathfrak{C}$, an n-simplex $\sigma' = (\sigma'_0 \to \cdots \to \sigma'_n)$ in $\mathcal{N}\mathfrak{C}'$, and an n-simplex in $F(\sigma_0, \sigma'_0)$. Face and degeneracy maps act the same way in the three cases. □

We shall also make use of the Cofinality Theorem, which provides a method for changing categories for homotopy colimits.

DEFINITION 4.5.21 (Cofinal functors). Let $f\colon \mathfrak{C} \to \mathfrak{D}$ be a functor between small categories, and let Y be an object in $\mathfrak{D}$. We write $Y{\downarrow}f$ for the category whose objects are pairs (X, ρ), where X is an object in $\mathfrak{C}$ and $\rho\colon Y \to f(X)$ is an arrow in $\mathfrak{D}$. An arrow $(X, \rho) \to (X', \rho')$ in $Y{\downarrow}f$ consists of an arrow $\gamma\colon X \to X'$ in $\mathfrak{C}$ such that $f(\gamma) \circ \rho = \rho'$.

A functor $f\colon \mathfrak{C} \to \mathfrak{D}$ between small categories is *right cofinal* if for each object Y in $\mathfrak{D}$ the nerve $\mathcal{N}(Y{\downarrow}f)$ is weakly contractible.

THEOREM 4.5.22 (Cofinality Theorem, see Dwyer and Kan [**54**, 9.4]). *Let $f\colon \mathfrak{C} \to \mathfrak{D}$ be a right cofinal functor between small categories, and let $F\colon \mathfrak{D} \to \mathfrak{Sp}$ be a $\mathfrak{D}$-diagram of spaces. Then the natural map*

$$\underset{\mathfrak{C}}{\underrightarrow{\mathrm{Hocolim}}}\,(F \circ f) \to \underset{\mathfrak{D}}{\underrightarrow{\mathrm{Hocolim}}}\, F$$

is a weak equivalence.

PROOF. Dwyer and Kan simply refer to the proof of the corresponding dual theorem for homotopy limits given in Bousfield and Kan [**31**, XII.9.2]. □

REMARK 4.5.23. If, in the Cofinality Theorem above, f is equivariant with respect to G-actions on $\mathfrak{C}$ and $\mathfrak{D}$, then the weak equivalence of homotopy colimits is also a G-map. It is not necessarily a G-homotopy equivalence. But if f is G-cofinal in the sense that for each object Y in $\mathfrak{D}$, the nerve $\mathcal{N}(Y{\downarrow}f)$ is weakly G_Y-contractible, then it is a G-homotopy equivalence.

EXAMPLE 4.5.24 (Subdivision). If $\mathfrak{C}$ is a small category, we form a new category $s\mathfrak{C}$ (called the *subdivision* of $\mathfrak{C}$ because $|s\mathfrak{C}|$ is the barycentric subdivision of $|\mathfrak{C}|$) whose objects are the nonempty finite chains of arrows

$$\sigma = (X_0 \to \cdots \to X_n)$$

in $\mathfrak{C}$. The arrows $\sigma \to \sigma'$ in $s\mathfrak{C}$ are the containments of chains $\sigma \supseteq \sigma'$; so that σ' has been obtained by deleting objects in σ and composing arrows in $\mathfrak{C}$ as appropriate. There is a functor $f\colon s\mathfrak{C} \to \mathfrak{C}$ which sends a chain $(X_0 \to \cdots \to X_n)$ to X_0, and a containment of chains to the corresponding composite of maps between initial elements. If X is an object in $\mathfrak{C}$ then the objects in $X{\downarrow}f$ are the chains of maps $X \to X_0 \to \cdots \to X_n$ in $\mathfrak{C}$. An arrow in $X{\downarrow}f$ is given by a containment of chains as before, but where the initial object X may not be deleted. There are natural transformations

$$(X \to X_0 \to \cdots \to X_n) \longleftarrow (X \to X \to X_0 \to \cdots \to X_n) \longrightarrow (X \to X)$$

connecting the identity functor to the constant functor at $X \to X$, and these give rise to a contracting homotopy on the nerve $\mathcal{N}(X{\downarrow}f)$. Thus the functor f is right cofinal. It follows from Theorem 4.5.22 that given a $\mathfrak{C}$-diagram of spaces, the $s\mathfrak{C}$-diagram of spaces obtained by composing with f has weakly equivalent homotopy colimit. In particular, using Example 4.5.13 we see that the induced map of nerves $\mathcal{N}(s\mathfrak{C}) \to \mathcal{N}\mathfrak{C}$ is a weak homotopy equivalence.

4.6. The homotopy colimit over a simplex category

It turns out that for our eventual applications in Chapter 7, we will not need to consider arbitrary finite indexing categories $\mathfrak{C}$. One striking feature of our analysis of the sporadic groups is that in each case (for reasons connected with flag-transitive group actions, see Remark 5.6.11), we can in fact obtain the desired homotopy colimit over a *simplex diagram*, namely a diagram indexed over a category arising from a simplex—indeed a simplex of dimension between zero and four; and then the homotopy colimit is over a diagram of spaces given by a pushout n-cube, for $1 \leq n \leq 5$. In a few cases, namely J_3, Fi'_{24}, Ly, there are "cancellations" which reduce the computation to a suitable subdiagram of the full pushout diagram, using the Cofinality Theorem 4.5.22.

Consequently for our remaining exposition in this section, we will consider only examples given by simplex categories and pushout n-cube diagrams. So we now establish some corresponding notation:

DEFINITION 4.6.1 (Simplex category). Let I denote a finite index set; often we will use $I_n = \{1, \ldots, n\}$. Consider the abstract simplex Δ^I on the set I of vertices, having the nonempty subsets $J \subseteq I$ as its faces. The poset $\mathcal{Q}(\Delta^I)$ of faces of this simplex as in Definition 2.7.1 is given by the nonempty members $2^I \setminus \{\varnothing\}$ of the power set of I, with their inclusions. We define the corresponding *simplex category* $\mathfrak{D}_I$ as the category arising from the poset opposite to that poset of faces. So the objects of $\mathfrak{D}_I$ are the nonempty subsets $J \subseteq I$, and there is an arrow $J \to J'$ whenever $J \supseteq J'$. The use of the opposite poset causes the arrows in the category to go in the same direction as the natural face maps on the simplex, which is the reverse of the natural inclusions of the maximal faces in the simplex.

If $I = I_n$, we write $\mathfrak{D}_n$ for $\mathfrak{D}_{I_n}$. Note that the numbering differs by one from the dimension of the corresponding simplex, so that $\mathfrak{D}_n$ corresponds to a simplex Δ^{n-1}. In using the subscript n we are following the convention from group theory, where n corresponds to the *rank* of a geometry or amalgam associated to a diagram of n subgroups, given by a functor from $\mathfrak{D}_n$ to the category of subgroups of a group G. ♢

We already saw in our earlier discussion of the pushout square of spaces that $\mathfrak{D}_2$ naturally describes that diagram. Now observe more generally that $\mathfrak{D}_n$ is the natural category underlying the diagram of spaces given by a pushout n-cube: namely the missing subset given by the empty set $\varnothing$ then corresponds to the position that would be occupied by the usual colimit of the spaces over the diagram.

Now the spaces in our diagrams are usually classifying spaces of subgroups of a group G. So it is also natural to define notation for the configuration of subgroups which realizes the maps in the simplex category $\mathfrak{D}_I$—so that on applying the classifying space construction B regarded as a functor, we will obtain the relevant diagram of spaces.

DEFINITION 4.6.2 (Simplex of subgroups). Let I be a finite index set, as in Definition 4.6.1. Suppose that for each $i \in I$, we are given a subgroup H_i of G. Then the *simplex $\mathcal{H}_I$ of subgroups of G*, determined by the H_i as initial data, is the $\mathfrak{D}_I$-diagram of subgroups of G which sends a nonempty subset $J \subseteq I$ to

$$H_J := \bigcap_{i \in J} H_i.$$

We will assume this definition of H_J via intersections, whenever we later encounter a simplex $\mathcal{H}_I$ of subgroups.

For I_n as in Definition 4.6.1, we abbreviate $\mathcal{H}_{I_n}$ by $\mathcal{H}_n$. We follow the group theory literature on amalgams in saying that $|I|$ is the *rank* of the simplex $\mathcal{H}_I$. $\diamondsuit$

EXAMPLE 4.6.3 (Subgroup simplices from simple groups, continued). As we had already mentioned in earlier Example 4.5.5 for the case $n = 2$, a standard example of a simplex $\mathcal{H}_n$ of subgroups arises in a group G of Lie type and Lie rank n, given by the parabolic subgroups H_J over a fixed Borel subgroup H_I (cf. Carter [**43**, 8.3.4]); here the initial data H_i ($i \in I_n$) giving the vertices are the maximal parabolics containing the given Borel subgroup. Similarly for sporadic groups in Chapters 6 and 7, we will encounter such simplices $\mathcal{H}_I$ of subgroups for suitable index sets I of rank from one through five; the initial data H_i will be maximal 2-local subgroups which are "incident" in a suitable sense.

Our use of the term "simplex" of subgroups corresponds roughly to the usage of the term *amalgam* in the group theory literature. But the only case of the amalgamated product from that literature which we use in this work is the case $n = 2$ as in Lemma 4.5.10.

We also mention that for a simplex $\mathcal{H}_I$ in our later applications, distinct index subsets J and J' from I will usually determine distinct subgroups H_J and $H_{J'}$ of G. However, this distinctness is certainly not required. When coincidences of subgroups for distinct index subsets do occur, they will lead to cancellations in our homology decompositions—and so will have the effect (mentioned earlier for a few sporadic groups) of reducing the indexing of decompositions. Thus a smaller category than $\mathfrak{D}_I$ may be used, where identifications are made between some of the objects.

In the usual case where the H_J are distinct, we see that the poset $\mathcal{H}_n$ actually is the opposite poset to the poset of faces of the simplex Δ^{n-1}, which is of course the motivation for the terminology of subgroup simplex.

We now revisit the homotopy colimit construction of Description 4.5.15; we begin by establishing some conventions for the context above of a simplex category as indexing category:

NOTATION 4.6.4 (The homotopy colimit over a simplex category). Given a simplex $\mathcal{H}_I$ of subgroups of G, regarded as a functor from $\mathfrak{D}_I$ to subgroups of G as in Definition 4.6.2, we obtain a $\mathfrak{D}_I$-diagram of spaces $B\mathcal{H}_I$ by first applying $\mathcal{H}_I$, and then applying the functor B, which takes those subgroups to their classifying spaces. Finally, we take the homotopy colimit

$$\underset{\mathfrak{D}_I}{\underrightarrow{\mathrm{Hocolim}}}\, B\mathcal{H}_I = \underset{\varnothing \neq J \subseteq I}{\underrightarrow{\mathrm{Hocolim}}}\, BH_J .$$

The diagram of classifying spaces is a pushout n-cube, where $n = |I|$ is the rank of $\mathcal{H}_I$; its terms are of form BH_J for nonempty $J \subseteq I$.

Here the ordinary colimit over the diagram would be the pushout—namely it would afford the missing term corresponding to the empty set, completing the diagram to an actual n-cube. Correspondingly our aim will be for the homotopy colimit over the diagram to afford a homotopy decomposition for the classifying space BG; so that we could visualize BG as occupying that empty position.

To further illustrate the construction of homotopy colimits beyond our earlier example of $\mathfrak{D}_2$, we will now more briefly discuss the larger simplex categories $\mathfrak{D}_3$, $\mathfrak{D}_4$, and $\mathfrak{D}_5$, which we will need in Chapter 7.

So suppose first that we have the simplex category $\mathfrak{D}_3$, realized by a simplex $\mathcal{H}_3$ of subgroups of G: thus we start with subgroups H_1, H_2, and H_3, and consider their intersections $H_J = \bigcap_{i \in J} H_i$ for nonempty $J \subseteq I_3 = \{1,2,3\}$. Again for brevity we will omit set braces and commas in our notation for subscripts; and also now we will use the abbreviation $B_J := BH_J$ for the classifying spaces. Then the corresponding diagram of spaces is:

$$\begin{array}{ccccc} & & B_{12} & & \\ & \swarrow & \uparrow & \searrow & \\ B_1 & & & & B_2 \\ \uparrow & & B_{123} & & \uparrow \\ & \swarrow & & \searrow & \\ B_{13} & & & & B_{23} \\ & \searrow & & \swarrow & \\ & & B_3 & & \end{array} \tag{4.6.5}$$

Here the diagram of spaces is the *pushout cube*—where the ordinary colimit giving the pushout would appear at a further vertex opposite to B_{123}, with maps from B_1, B_2, and B_3.

We now examine for this case the homotopy colimit construction given in Description 4.5.15: notice in the diagram of spaces above that there are twelve non-identity maps. Three are not explicitly drawn, but arise as the compositions from the six composable pairs among the nine explicitly drawn maps. So for the homotopy colimit, we use the following: at each of the seven vertices, the initial spaces (crossed with a point); for each of the twelve arrows, a 1-simplex crossed with the corresponding initial space; and finally for each of the six composable pairs of arrows, a 2-simplex crossed with the space B_{123} which is the starting point of each of those compositions.

We will provide below a sketch of this homotopy colimit, in the spirit of Example 4.5.4. However for dimensional reasons, we will represent the space B_{123} as a point rather than a disk.

EXAMPLE 4.6.6 (A sketch of the homotopy colimit of the pushout cube). To the picture (4.6.5) of the pushout cube, we add an edge from B_{123} to each B_i, corresponding to the compositions of the sequences of length 2; then the various 2-simplices crossed with the point B_{123} form a hexagon centered at that vertex, as pictured below. We represent the spaces at the other vertices as disks; each pair of arrows emanating from a B_{ij} determines a copy of our earlier sketch in Example 4.5.4. So we see that the various spaces are glued together in $\underset{\varnothing\neq J\subseteq I_3}{\underrightarrow{\operatorname{Hocolim}}}\ BH_J$ in the manner indicated by the following diagram:

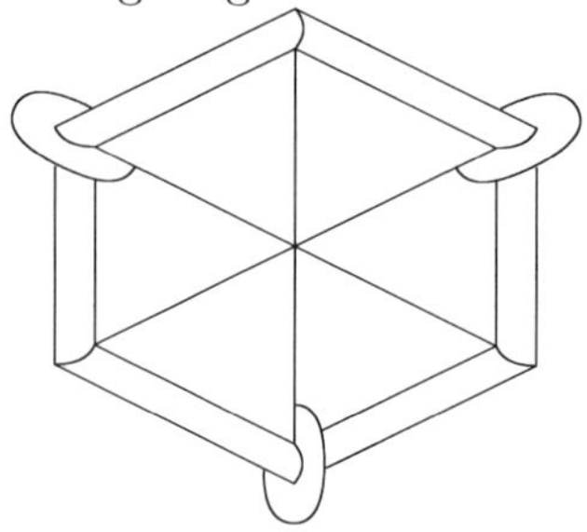

The pushout cube diagram (4.6.5) occurs in our analysis of 11 of the sporadic groups; in one of those cases (namely J_3) we reduce further to a diagram of the form (4.5.11), so this really only leaves ten cases. They are

$$M_{22},\ M_{23},\ M_{24},\ HS,\ McL,\ Suz,\ Co_3,\ HN,\ He,\ Ru.$$

We shall also need the analogous setup for the simplex category $\mathfrak{D}_4$, realized by $\mathcal{H}_4$ based on subgroups H_1, H_2, H_3, H_4. In Diagram (4.6.7), we have drawn the corresponding diagram of classifying spaces B_J, namely the *pushout hypercube* (4-dimensional cube). Notice that in this case, we have also added dotted lines from the vertices B_i to the usual "missing vertex" opposite to B_{1234}, corresponding to the missing subset given by the empty set.

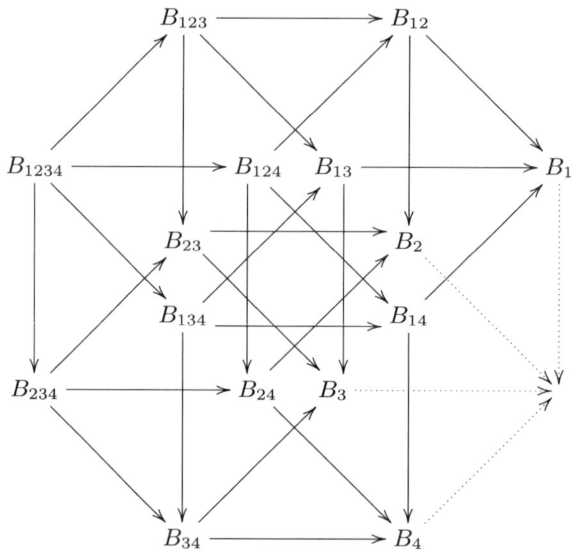

(4.6.7)

As the diagram (4.6.7) is already sufficiently complicated, we will not attempt to draw a sketch in the spirit of Example 4.6.6 for $\underset{\varnothing\neq J\subseteq I_4}{\underrightarrow{\operatorname{Hocolim}}}\ BH_J$. This diagram occurs in our analysis of six of the sporadic groups. They are:

$$J_4,\ Co_2,\ Co_1,\ Fi_{22},\ Fi_{23},\ Ly.$$

EXAMPLE 4.6.8. We shall also need the simplex category $\mathfrak{D}_5$ and the corresponding simplex $\mathcal{H}_5$ of subgroups. This time it is too complicated to try to draw even the corresponding pushout diagram (5-dimensional cube) of classifying spaces; much less, to attempt to sketch $\underset{\varnothing \neq J \subseteq I_5}{\underrightarrow{\text{Hocolim}}}\ BH_J$. This diagram occurs in our analysis of three of the sporadic groups, namely Fi'_{24}, B and M.

EXAMPLE 4.6.9. Finally, in one case we shall need the degenerate example of a homotopy colimit over the trivial category with one object and one arrow. This is the simplex category $\mathfrak{D}_1$. In this case, the homotopy colimit is just the single space involved in the diagram.

In our analysis, the point of this degenerate case is to capture the situation where the inclusion of a single p-local subgroup H into G is a mod p homology equivalence. This occurs, for example, when G has abelian Sylow p-subgroups, by Swan's theorem, see Theorem 5.7.4. We have the abelian Sylow situation for just one sporadic group when $p = 2$, namely the Janko group J_1. The degenerate case also occurs when the distinct Sylow p-subgroups have trivial intersections. This does not occur in a sporadic simple group with $p = 2$, but it does occur for finite groups of Lie type and Lie rank one at their defining characteristic. The simple examples with $p = 2$ are $L_2(2^n)$, $U_3(2^n)$, and the non-sporadic Suzuki groups $Sz(2^{2n+1}) = {}^2B_2(2^{2n+1})$. ◇

4.7. The Borel construction as a homotopy colimit

We turn now to another standard application of the homotopy colimit, which can be used to give an alternative theoretical approach to various constructions related to group cohomology. Namely, we examine what happens when we consider diagrams of spaces using the category $\mathfrak{B}_G$ of Example 3.1.9. We recall that this category has just one object, with the elements of the group G giving the arrows; and the nerve of $\mathfrak{B}_G$ gives the simplicial bar construction BG of Example 3.1.6.

DESCRIPTION 4.7.1 ($\mathfrak{B}_G$-diagrams). Now consider what information is encoded in a $\mathfrak{B}_G$-diagram of spaces; i.e., a functor $\mathfrak{B}_G \to \mathfrak{Sp}$. Since $\mathfrak{B}_G$ has only one object, such a functor names a single space X. To each arrow in $\mathfrak{B}_G$, namely to each element $g \in G$, is associated a map from X to itself; and composition is preserved. In other words, specifying a $\mathfrak{B}_G$-diagram of spaces is the same as specifying a space X together with an action of G on X.

The following theorem relates homotopy colimits with the Borel construction.

THEOREM 4.7.2. *If $F\colon \mathfrak{B}_G \to \mathfrak{Sp}$ is a $\mathfrak{B}_G$-diagram of spaces, corresponding to a G-space X, then*

$$\underset{\mathfrak{B}_G}{\underrightarrow{\text{Hocolim}}}\, F = EG \underset{G}{\times} X.$$

PROOF. We begin by considering functors $\rho\colon \mathfrak{C}_G \to \mathfrak{Sp}$ from the Cayley category $\mathfrak{C}_G$ of Example 3.1.9 to spaces. Such a functor specifies a space X_g for each $g \in G$, and compatible isomorphisms $X_g \to X_h$ for each pair g, h in G. Using the isomorphisms $X_1 \to X_g$, the simplicial space $\amalg.\rho$ of Definition 4.5.12 can be identified with the external product of the nerve of $\mathfrak{C}_G$ with X_1, namely $EG \times. X_1$.

Now if ρ is obtained by composing the quotient functor $\mathfrak{C}_G \to \mathfrak{B}_G$ with a $\mathfrak{B}_G$-diagram F of spaces, then the identifications used to identify $\amalg.\rho$ with $EG \times. X$

are not the identity map on X, they are given by the action of G. So if we let G act diagonally on $EG \times_\bullet X$, then the action is free on bisimplices, and the bisimplices of $\amalg_\bullet F$ are the G-orbits of bisimplices in $\amalg_\bullet \rho$. Taking the diagonal, we see using Example 4.4.6 that $EG \times_G X$ is equal to $\underrightarrow{\mathrm{Hocolim}}_{\mathfrak{B}_G} F$. □

REMARK 4.7.3. If X is a single point, leading to the nerve as in Example 4.5.13, then the case of $\mathfrak{B}_G$ takes the following form in view of Example 3.1.9:

$$\underrightarrow{\mathrm{Hocolim}}_{\mathfrak{B}_G} \star = \mathcal{N}\mathfrak{B}_G = BG = EG \times_G \star.$$

We can compose the equivalence of Theorem 4.7.2 with the earlier connection in Remark 2.5.19 of the Borel construction with isotropy subgroups:

LEMMA 4.7.4. *Let $X = G/H$ be a transitive G-set, regarded as a discrete G-space using Example 3.1.2. Then $\underrightarrow{\mathrm{Hocolim}}_{\mathfrak{B}_G} X = EG \times_G X$, which in turn is homotopy equivalent to BH.* □

EXAMPLE 4.7.5. Here are some examples to illustrate earlier Warning 4.5.16. Suppose that we are given an action of a group H on an abelian group N. This gives rise in an obvious way to a $\mathfrak{B}_H$-diagram in the homotopy category of spaces, where the object is taken to BN, and the arrows are taken to the homotopy classes of self-maps corresponding to the action.

Given any extension

$$1 \to N \to G \to H \to 1$$

where the conjugation action of H on N is the given one, we use EG as a model for EN. Then EG/N is BN, and this admits an action of $G/N \cong H$. So we obtain a $\mathfrak{B}_H$-diagram of spaces lifting the above diagram in the homotopy category. The homotopy colimit is $(EG/N)/(G/N)$, that is, $EG/G = BG$. In this example, the homotopy classes of lifts of this diagram from the homotopy category to the category of spaces exactly correspond to the equivalence classes of group extensions, namely $H^2(H, N)$.

On the other hand, if N is not abelian, then there are nontrivial inner automorphisms of N. The group of inner automorphisms is given by $\mathrm{Inn}(N) \cong N/Z(N)$. The action of an inner automorphism on BN is homotopic to the identity map, so a $\mathfrak{B}_H$-diagram in the homotopy category is given by a map from H to $\mathrm{Out}(N) = \mathrm{Aut}(N)/\mathrm{Inn}(N)$, the group of *outer automorphisms* of N. If we could lift to a diagram of spaces, so that H is acting on BN, then π_1 of the homotopy colimit would be a group G having N as a normal subgroup, with $G/N \cong H$, and where the corresponding map from H to $\mathrm{Out}(N)$ is the given one.

The following example then shows that lifting to the category of spaces is not always possible. Let $N = 2^{\cdot}\mathrm{Alt}_6$ and $H = \mathbb{Z}/2$, with the map from H to $\mathrm{Out}(N)$ giving the automorphism corresponding to $M_{10} = \mathrm{Alt}_6{}^{\cdot}2$. This gives a $\mathfrak{B}_{\mathbb{Z}/2}$-diagram in the homotopy category where the object goes to BN. If there were a lift to the category of spaces, then the homotopy colimit would be a classifying space for a group G with N a normal subgroup of G such that $G/Z(N) \cong M_{10}$. But there is no such extension; see for example the statement that there is no group "$2.G.2_3$" at the bottom of page 5 of the Atlas [**47**], together with the underlying discussion of "isoclinism" starting on page xxiii and especially page xxiv of that work. If g is an element of order eight in G, then that discussion shows that g would have to conjugate g^2 to g^{-2}, which is clearly absurd. It follows that there is no lift of the given diagram from the homotopy category to the category of spaces. $\diamondsuit$

Finally, in our discussion of homology decompositions in Section 5.3, we shall make a great deal of use of the fact that the Borel construction commutes with homotopy colimits; this is a key feature of Dwyer's viewpoint on decompositions.

PROPOSITION 4.7.6. *Let $X\colon \mathfrak{C} \to G\text{-}\mathfrak{Sp}$ be a $\mathfrak{C}$-diagram of G-spaces. Then*

$$EG \underset{G}{\times} (\underset{\overrightarrow{\mathfrak{C}}}{\mathrm{Hocolim}}\, X) \;\cong\; \underset{\overrightarrow{\mathfrak{C}}}{\mathrm{Hocolim}}\, (EG \underset{G}{\times} X).$$

PROOF. This follows from Theorems 4.5.20 and 4.7.2. $\square$

4.8. The Bousfield–Kan homology spectral sequence

We close this chapter with a discussion of the homology of a homotopy colimit, via the *Bousfield–Kan homology spectral sequence.*

Let $\mathfrak{C}$ be a (small) indexing category. We begin with the left derived functors of the right exact functor $\underrightarrow{\mathrm{colim}}$ applied to the category $R\text{-}\mathfrak{Mod}^{\mathfrak{C}}$ of $\mathfrak{C}$-diagrams of R-modules. The general recipe for describing left derived functors of a functor is to take a projective resolution, apply the functor, and take homology. So we need to begin by constructing enough projectives. For each object i in $\mathfrak{C}$ and R-module M, we can construct a functor $P(i, M)\colon \mathfrak{C} \to R\text{-}\mathfrak{Mod}$ via

$$P(i, M)(j) \;=\; \bigoplus_{\alpha\colon i \to j} M_\alpha,$$

where the sum is over all arrows $\alpha\colon i \to j$, and M_α is a copy of M indexed by the arrow $\alpha\colon i \to j$. The maps in this diagram are given as follows. If $j \xrightarrow{\beta} k$ is an arrow in $\mathfrak{C}$ then the map $P(i, M)(\beta)\colon P(i, M)(j) \to P(i, M)(k)$ takes M_α isomorphically to $M_{\beta\circ\alpha}$. We shall see in Corollary 4.8.2 that if M is a projective R-module then $P(i, M)$ is a projective diagram of R-modules.

For example, if $\mathfrak{C}$ is the pushout category

$$\begin{array}{ccc} a & \longrightarrow & b \\ \downarrow & & \\ c & & \end{array}$$

then the diagrams $P(a, M)$, $P(b, M)$ and $P(c, M)$ are as follows:

$$\begin{array}{ccc} M & \xrightarrow{\mathrm{Id}} & M \\ {\scriptstyle\mathrm{Id}}\downarrow & & \\ M & & \end{array} \qquad \begin{array}{ccc} 0 & \longrightarrow & M \\ \downarrow & & \\ 0 & & \end{array} \qquad \begin{array}{ccc} 0 & \longrightarrow & 0\,. \\ \downarrow & & \\ M & & \end{array}$$

Another example is given by $\mathfrak{C} = \mathfrak{B}_G$, the category discussed in Definition 3.1.9 and Description 4.7.1, with one object and a group G of arrows. In this case, $R\text{-}\mathfrak{Mod}^{\mathfrak{C}}$ is the category of RG-modules, and if x is the unique object in $\mathfrak{C}$, then $P(x, R)$ is the free RG-module of rank one.

PROPOSITION 4.8.1. *There is a natural isomorphism of sets*

$$\mathrm{Hom}_{R\text{-}\mathfrak{Mod}^{\mathfrak{C}}}\big(P(i, M), F\big) \cong \mathrm{Hom}_R\big(M, F(i)\big).$$

PROOF. Given a natural transformation from $P(i,M)$ to F, we obtain an R-module homomorphism $M \to F(i)$ by looking at the copy of M corresponding to the identity arrow $i \to i$.

Conversely, given an R-module homomorphism $\rho\colon M \to F(i)$, there is a unique homomorphism in $R\text{-}\mathfrak{Mod}^{\mathfrak{C}}$ extending it; on M_α for $\alpha\colon i \to j$ it is

$$F(\alpha)\circ\rho\colon M_\alpha \to F(j). \qquad \square$$

COROLLARY 4.8.2. *If M is a projective R-module and i is an object in $\mathfrak{C}$, then $P(i,M)$ is a projective $\mathfrak{C}$-diagram of R-modules.*

PROOF. Since M is projective, $\mathrm{Hom}_R(M,-)$ is right exact. It then follows using Proposition 4.8.1 that $\mathrm{Hom}_{R\text{-}\mathfrak{Mod}^{\mathfrak{C}}}\big(P(i,M),-\big)$ is also right exact, so that $P(i,M)$ is projective. $\square$

The following corollary shows that there are enough projective $\mathfrak{C}$-diagrams of R-modules.

COROLLARY 4.8.3. *Every $\mathfrak{C}$-diagram of R-modules is a quotient of a direct sum of ones of the form $P(i,M)$ where i is an object in $\mathfrak{C}$ and the M is a projective R-module.*

PROOF. If $F\colon \mathfrak{C} \to R\text{-}\mathfrak{Mod}$ is a $\mathfrak{C}$-diagram of R-modules, let $M(i)$ be a projective R-module surjecting onto $F(i)$. Then $\bigoplus_i P\big(i, M(i)\big)$ is a projective $\mathfrak{C}$-diagram of R-modules surjecting onto F, where the direct sum is over the objects i of $\mathfrak{C}$. $\square$

COROLLARY 4.8.4. *If i is an object in $\mathfrak{C}$ and M is an R-module then*

$$\underrightarrow{\mathrm{colim}}\, P(i,M) \cong M.$$

PROOF. If N is another R-module, let F_N be the constant functor with value N. Then the colimit is determined by

$$\mathrm{Hom}_{R\text{-}\mathfrak{Mod}^{\mathfrak{C}}}\big(P(i,M),F_N\big) \cong \mathrm{Hom}_R\big(\underrightarrow{\mathrm{colim}}\, P(i,M),N\big).$$

Now apply Proposition 4.8.1 to the left side to obtain $\mathrm{Hom}_R(M,N)$, and the Corollary follows by Yoneda's lemma. $\square$

If $F\colon \mathfrak{C} \to R\text{-}\mathfrak{Mod}$ is a $\mathfrak{C}$-diagram of R-modules, we let $\amalg_{\bullet}F$ be the simplicial R-module of Definition 4.5.12. Recall from Definition 3.7.3 that the homotopy π_n of a simplicial R-module $A_{\bullet}$ is the same as the homology H_n of the corresponding normalized chain complex $NA_{\bullet}$ of R-modules.

LEMMA 4.8.5. *If i is an object in $\mathfrak{C}$ and M is an R-module then*

$$\pi_n \amalg_{\bullet} P(i,M) = H_n N \amalg_{\bullet} P(i,M)$$

is equal to zero if $n>0$, and is isomorphic to $\underrightarrow{\mathrm{colim}}\, P(i,M) \cong M$ if $n=0$.

PROOF. The simplicial R-module $\amalg_{\bullet}P(i,M)$ has as its n-simplices a direct sum of copies $M_{i\to j_0\to\cdots\to j_n}$ of M indexed by chains of maps $i \to j_0 \to \cdots \to j_n$ in $\mathfrak{C}$. In the corresponding chain complex of R-modules, we only use non-identity maps. The boundary map in this complex is the obvious alternating sum obtained by deleting elements of this chain and composing if necessary.

We construct a contracting homotopy h on the corresponding chain complex, by sending $M_{i\to j_0\to\cdots\to j_n}$ via the identity map to $M_{i\to i\to j_0\to\cdots\to j_n}$. And it is easy to check that $hd+dh$ is equal to the identity map in positive degrees.

The 0-simplices in $\amalg.P(i, M)$ form a direct sum of copies $M_{i\to j}$ of M, one for each arrow $i \to j$ in $\mathfrak{C}$. The relations coming from the boundaries of the 1-simplices identify all of these with the copy of M corresponding to the identity arrow at i, in the same way as it happens in the colimit. □

THEOREM 4.8.6. *If $F\colon \mathfrak{C} \to R\text{-}\mathfrak{Mod}$ is a $\mathfrak{C}$-diagram of R-modules, then*

$$\underset{\mathfrak{C}}{\underrightarrow{\operatorname{colim}}}{}_n F \cong \pi_n \amalg.F.$$

Here, the homotopy π_n is the same as the homology H_n of the corresponding chain complex of R-modules as in Definition 3.7.3.

PROOF. Choose a projective resolution

$$\cdots \to P_2 \to P_1 \to P_0 \to F \to 0$$

of F by direct sums of projective diagrams as in Corollary 4.8.3. Then take the normalized chain complex $N\amalg.P_n$. This gives us (with suitable signs) a double complex, and hence two spectral sequences, depending on the order in which we apply the differentials. If we apply the (horizontal) differentials coming from the normalized chain complexes first, then by Lemma 4.8.5, the E^1 page just consists of the modules $\underset{\mathfrak{C}}{\underrightarrow{\operatorname{colim}}}\, P_n$ on the zero column of the double complex, and zero elsewhere. So the E^2 page is the derived functors of colim on this column, and there are no further differentials; thus the spectral sequence converges to $\underset{\mathfrak{C}}{\underrightarrow{\operatorname{colim}}}{}_* F$.

On the other hand, let us examine what happens if we apply the vertical differential coming from the resolution first. The nth column decomposes as a direct sum indexed by the sequences of non-identity arrows $\sigma_0 \to \cdots \to \sigma_n$. The terms indexed by this sequence form a copy of $P_*(\sigma_0)$, a projective resolution of $F(\sigma_0)$. So the column is a projective resolution of $N\amalg_n F$. It follows that the E^1 page is $N\amalg.F$ along the zero row of the double complex, and zero elsewhere. So the E^2 page consists of the modules $\pi_n\amalg.F$ on this row; thus the spectral sequence converges to $\pi_*\amalg.F$ as well. □

THEOREM 4.8.7. *Let $F\colon \mathfrak{C} \to \mathfrak{Sp}$ be a diagram of spaces. Consider the bisimplicial R-modules $\amalg.R \otimes F$ and $\amalg.RF$ obtained by applying the construction of Definition 3.7.4 to each space of $\amalg.F$. The spectral sequence* (4.4.8) *of these bisimplicial R-modules takes the form*

$$E^2_{s,t} = \underset{i\in\mathfrak{C}}{\underrightarrow{\operatorname{colim}}}{}_s\, H_t\big(F(i); R\big) \Rightarrow H_{s+t}\big(\underset{i\in\mathfrak{C}}{\underrightarrow{\operatorname{Hocolim}}}\, F(i); R\big). \tag{4.8.8}$$

$$E^2_{s,t} = \underset{i\in\mathfrak{C}}{\underrightarrow{\operatorname{colim}}}{}_s\, \tilde{H}_t\big(F(i); R\big) \Rightarrow \tilde{H}_{s+t}\big(\underset{i\in\mathfrak{C}}{\underrightarrow{\operatorname{Hocolim}}}\, F(i); R\big) \tag{4.8.9}$$

PROOF. By Theorem 4.8.6, the left hand side of (4.4.8) is

$$\pi_{s,-}\pi_{-,t}\amalg.(R\otimes F) \cong \pi_{s,-}\amalg.H_t(F;R) \cong \underset{i\in\mathfrak{C}}{\underrightarrow{\operatorname{colim}}}{}_s\, H_t\big(F(i); R\big).$$

The right hand side of (4.4.8) is equal to $H_{s+t}\big(\underset{i\in\mathfrak{C}}{\underrightarrow{\operatorname{Hocolim}}}\, F(i); R\big)$ using the definition 4.5.12 of the homotopy colimit in terms of the diagonal. The same argument applies in the case of reduced homology. □

REMARK 4.8.10. The spectral sequence (4.8.9) is called the Bousfield–Kan homology spectral sequence of the homotopy colimit. It is described in Bousfield and Kan [**31**, XII.5.7] (to reconcile the language here with theirs, note that reduced homology is a connected theory which "comes from" an Eilenberg–Mac Lane spectrum) and in Dwyer [**53**, Sec. 8].

There are also analogous Bousfield–Kan cohomology spectral sequences [**31**, XII.4.5, XII.5.8]

$$E_2^{s,t} = \underset{i\in\mathfrak{C}}{\varprojlim}{}^s\, H^t\big(F(i);R\big) \Rightarrow H^{s+t}\big(\underset{i\in\mathfrak{C}}{\underrightarrow{\mathrm{Hocolim}}}\, F(i);R\big) \tag{4.8.11}$$

$$E_2^{s,t} = \underset{i\in\mathfrak{C}}{\varprojlim}{}^s\, \tilde{H}^t\big(F(i);R\big) \Rightarrow \tilde{H}^{s+t}\big(\underset{i\in\mathfrak{C}}{\underrightarrow{\mathrm{Hocolim}}}\, F(i);R\big), \tag{4.8.12}$$

constructed in a similar way.

REMARK 4.8.13. We can also view the spectral sequences (4.8.8) through (4.8.12) as Leray spectral sequences. Namely, if we write $\star$ for the constant diagram with value a single point $\star$, then there is a map of diagrams $F \to \star$, and hence a map

$$f\colon \underset{\mathfrak{C}}{\underrightarrow{\mathrm{Hocolim}}}\, F \to \underset{\mathfrak{C}}{\underrightarrow{\mathrm{Hocolim}}}\, \star = \mathcal{N}\mathfrak{C},$$

see Example 4.5.13. Looking back at Example 4.4.7, we see that the dimensions $(\mathsf{Diag}\,\amalg.F)_n$ afford a filtration F_n of the diagonal which agrees with the inverse image of the skeletal filtration on the nerve

$$F_n(\mathsf{Diag}\,\amalg.F) = f^{-1}(\mathsf{Sk}^n\mathcal{N}\mathfrak{C}),$$

giving rise to a Leray spectral sequence as in (2.6.2). So from the E^2 page onwards, the spectral sequence (4.8.8) is the Leray spectral sequence of the map f.

In the special case of a homotopy colimit of a diagram of *transitive* G-sets, this Remark enables us to identify the Bousfield–Kan homology spectral sequence with the isotropy spectral sequence of Definition 2.6.10. The following proposition will be the basis for our discussion of Dwyer's viewpoint on sharpness in Chapter 5.

PROPOSITION 4.8.14 (Dwyer [**53**, 9.10]). *Let $F\colon \mathfrak{C} \to G\text{-}\mathfrak{Set}$ be a $\mathfrak{C}$-diagram of transitive G-sets. Then the Borel construction of F, namely*

$$\underset{i\in\mathfrak{C}}{\underrightarrow{\mathrm{Hocolim}}}\big(EG \underset{G}{\times} F(i)\big) \cong EG \underset{G}{\times} \big(\underset{\mathfrak{C}}{\underrightarrow{\mathrm{Hocolim}}}\, F\big)$$

(see Proposition 4.7.6) has the Bousfield–Kan homology spectral sequence

$$\underset{i\in\mathfrak{C}}{\underrightarrow{\mathrm{colim}}}{}_s\, H_t^G\big(F(i);R\big) \Rightarrow H_{s+t}^G\big(\underset{\mathfrak{C}}{\underrightarrow{\mathrm{Hocolim}}}\, F;R\big),$$

which can in turn be identified with the isotropy spectral sequence in homology (Definition 2.6.10) for the action of G on $\underset{\mathfrak{C}}{\underrightarrow{\mathrm{Hocolim}}}\, F$. Similarly, the Bousfield–Kan cohomology spectral sequence

$$\underset{i\in\mathfrak{C}}{\varprojlim}{}^s\, H_G^t\big(F(i);R\big) \Rightarrow H_G^{s+t}\big(\underset{\mathfrak{C}}{\underrightarrow{\mathrm{Hocolim}}}\, F;R\big)$$

can be identified with the isotropy spectral sequence in cohomology for this action.

PROOF. An n-simplex in $\underrightarrow{\operatorname{Hocolim}}_{\mathfrak{C}} F$ is given by a sequence $i_0 \to \cdots \to i_n$ of arrows in $\mathfrak{C}$ together with an element $x \in F(i_0)$ (regarded as a degenerate n-simplex in this discrete space). Since the G-action on $F(i_0)$ is transitive, it follows that an orbit simplex in $(\underrightarrow{\operatorname{Hocolim}}_{\mathfrak{C}} F)/G$ corresponds to a simplex in

$$\underrightarrow{\operatorname{Hocolim}}_{\mathfrak{C}}(F/G) = \underrightarrow{\operatorname{Hocolim}}_{\mathfrak{C}} \star.$$

So we get the following diagram, in which the vertical arrows are isomorphisms by Proposition 4.7.6:

$$\begin{array}{ccc} \underrightarrow{\operatorname{Hocolim}}_{\mathfrak{C}}(EG \underset{G}{\times} F) & \longrightarrow & \underrightarrow{\operatorname{Hocolim}}_{\mathfrak{C}} \star \\ \downarrow \cong & & \downarrow \cong \\ EG \underset{G}{\times} (\underrightarrow{\operatorname{Hocolim}}_{\mathfrak{C}} F) & \longrightarrow & (\underrightarrow{\operatorname{Hocolim}}_{\mathfrak{C}} F)/G. \end{array}$$

From the E^2, respectively E_2 page onwards, the Leray spectral sequence of the upper horizontal map is by Remark 4.8.13 the Bousfield–Kan homology spectral sequence for the homotopy colimit of the Borel construction $EG \underset{G}{\times} F$. The Leray spectral sequence for the lower horizontal map is just the isotropy spectral sequence as in Definition 2.6.10 for the action of G on $\underrightarrow{\operatorname{Hocolim}}_{\mathfrak{C}} F$. $\square$

In our later applications to sporadic groups in Chapter 7, the relevant spectral sequence will always turn out to collapse at its E^2-page onto the vertical axis, resulting in a much simpler homology decomposition—indeed in the form of an alternating sum, as in Webb's celebrated formula, which we will discuss in the following section as Theorem 5.1.16. This collapse is the property of "sharpness" discussed in Dwyer's more general context in Definition 5.8.1.

CHAPTER 5

Decompositions and ample collections of p-subgroups

In the previous several chapters, we have considered various topological constructions for general spaces X—usually then indicating their application in the case where X is the classifying space BG of G. From this chapter onward, we will focus more exclusively on BG, and on decompositions at the level of its homotopy and homology.

We emphasize that G can still be any finite group; we do not further specialize to the case of sporadic simple groups until Chapter 7.

So in this chapter, we will discuss some aspects of the modern literature on homotopy and homology decompositions for BG. In particular, we will lead up to Dwyer's notion of "homology decomposition" as given at 7.1 in [**53**]. Indeed that notion of homology decomposition is the principal theme of Dwyer's exposition; and our coverage in this chapter corresponds largely with the material in Dwyer's Sections 7 and 8—on collections of subgroups of G which are "ample" and "sharp". We also give an overview of some material from Dwyer's sections after those two.

Some further general references for the topic of decompositions via collections of subgroups include Webb's expository article [**123**], as well as Section V.3 of the book [**6**] of Adem and Milgram, and Section X.7 of Brown [**35**].

Overview of some high points in the chapter. Here is an outline of our treatment in this chapter:

The first section reviews some of the history of the subject of decompositions. Our aim is to summarize how results of Brown and of Quillen, making various uses of collections of p-subgroups of a group G, led Webb to a more specific study of decompositions of group cohomology; later developments include more recent work on decompositions at the level of the classifying space BG, via a homotopy colimit over a suitable collection of subgroups, by such authors as Jackowski, McClure, Oliver, Dwyer, and Grodal.

The following seven sections then examine various aspects of that work on decompositions—using Dwyer's overall analysis, which views all that work inside a unified setting, namely of the Borel construction of a suitable diagram of transitive G-sets. The first key notion is that of

- an *ample* collection of subgroups in Definition 5.3.3;

this is essentially one form of Dwyer's notion of a homology decomposition of BG, expressed as a homotopy colimit. Ampleness is then examined in individual sections on each of the three standard decompositions from the literature—namely the normalizer, centralizer, and subgroup decompositions. (The normalizer decomposition is the form which we will in fact use in our main results in Chapter 7; we will also use an obvious extension of the notion of ampleness from a collection to a more general simplicial complex, in Definition 5.6.6.) The other key notion is

- a *sharp* decomposition in Definition 5.8.1,

which gives a comparatively simple Webb-type formula for computing the decomposition, based on the collapse of the relevant associated spectral sequence.

The final three sections discuss various particular cases of ample collections used in the literature: Of these, the first considers the G-homotopy type of Brown's original ample collection $\mathcal{S}_p(G)$ of all nontrivial p-subgroups—which includes certain other standard collections such as $\mathcal{A}_p(G)$ and $\mathcal{B}_p(G)$; we shall see that the inclusions of these subcollections are G-homotopy equivalences. The remaining two sections then consider suitable ample subcollections of $\mathcal{A}_p(G)$ and of $\mathcal{B}_p(G)$ respectively—subcollections which typically are not homotopy equivalent to those originals. In our main results in Chapter 7 on the individual sporadic groups, our treatment will be based on one or more of the collections described in these three sections.

5.1. Some history of homology and homotopy decompositions

We now survey some of the literature related to decompositions of group cohomology. Much of the relevant research has been in the area of algebraic topology, though related themes have also been pursued in combinatorics and in finite group theory.

The reader who wishes to skip this historical section may safely go straight to Section 5.2, where the definitions and basic properties of the decompositions are given.

The history is primarily intended to provide some more general context, and to show how a number of ostensibly separate strands of research fit together in the bigger picture. In particular it will introduce several different viewpoints on decompositions, which were subsequently unified in Dwyer's overall analysis.

Some prehistory: groups of Lie type and their buildings. Our history will mainly focus on research developments from algebraic topology, beginning around 1975, and related to posets of p-subgroups; but it is natural to first include some mention of the influence of the earlier work of Tits [**119**] on buildings for simple groups of Lie type, which can also be approached from the viewpoint of p-subgroups. So we briefly review just a few relevant aspects of Tits' theory.

EXAMPLE 5.1.1 (Parabolic subgroups and the building). See for example Carter [**43**, Sec. 8.3] or Gorenstein, Lyons and Solomon [**64**, Ch. 3] for more detailed reference on the following material.

Let G denote a finite group of Lie type.[1] We recall that G has an associated root system and *Weyl group* W; and these are in turn classified by the corresponding *Dynkin diagrams*, of types A_n, B_n, C_n, D_n, G_2, F_4, E_6, E_7, E_8. The subscript denotes number of generators of W as a reflection group, giving the *Lie rank* of G.

We will let $\mathcal{P}$ denote the poset of *parabolic* subgroups of G (namely the overgroups of the *Borel subgroups* of G). We recall from Definition 2.7.5 that the inclusion-chains in the poset $\mathcal{P}$ define the associated simplicial complex $\Delta(\mathcal{P})$; and

[1]For simplicity we restrict the discussion to untwisted groups.

this complex in fact gives one standard viewpoint[2] on the Tits *building* for G, cf. [**43**, p. 293]. We let $\mathcal{U}$ denote the poset of *unipotent radicals* of the parabolics; this in fact affords the opposite poset to $\mathcal{P}$, and hence the corresponding complex $\Delta(\mathcal{U})$ also affords the building.

We now single out certain features, for which we will see analogies later in the subgroup structure of sporadic simple groups.

The group G of Lie type is in fact a suitable linear group over a finite field of some prime characteristic p (see for example [**22**, Sec. 6.8]). The unipotent radicals $\mathcal{U}$ are p-groups for this p, so that

- $\mathcal{U}$ is a poset of p-subgroups of G.

The parabolics $\mathcal{P}$ are the normalizers of the unipotent radicals (cf. [**43**, 8.5.2]), so that

- $\mathcal{P}$ is a poset of p-local subgroups of G.

Indeed the unipotent radical U of a parabolic P is the largest normal p-subgroup $O_p(P)$; so as the full normalizer $N_G(U)$ of U is given by P, it follows that also $N_G(P) = P$. That is,

- The parabolics $\mathcal{P}$ are *self-normalizing.*

In particular, this means that the permutation representation on the cosets of members of $\mathcal{P}$ is the same as that on the conjugates. The isomorphism of $\mathcal{P}$ with $\mathcal{U}$ mentioned above holds essentially because

- The unipotent radicals $\mathcal{U}$ are *weakly closed*

in a Sylow p-subgroup of G; see for example Grodal [**66**, 4.2] for a proof of this standard fact.

For the study of G of Lie type, the topology of the building is fundamental in a number ways; for example:

- The homology group of the building in the highest dimension (given by the Lie rank) affords the projective *Steinberg module* of G. (This is the Solomon–Tits Theorem, see for example [**22**, pp. 236–237].)

We particularly mention the viewpoint of group cohomology. The parabolics and the building are highly relevant to the important open problem of determining the mod p cohomology of G in the natural characteristic p; this problem is often expressed from the viewpoint of the p-groups given by the unipotent radicals $\mathcal{U}$. Indeed from the further theory of groups of Lie type, the collection $\mathcal{U}$ controls various aspects of the p-local structure of G—notably the p-fusion (see for example Gorenstein, Lyons and Solomon [**64**, 3.1.6]). It follows from control of fusion (cf. [**21**, Sec. 3.8]) that $\mathcal{U}$ similarly controls the mod p group cohomology of G, in the sense of *stability*—see for example Cline, Parshall and Scott [**45**, Cor. 6.1]. Furthermore, for the more specific viewpoint of decompositions in this work, we will see in Example 5.1.18 that

- $\Delta(\mathcal{U})$ affords a decomposition for the mod p group cohomology. $\diamondsuit$

Posets of p-subgroups in the work of Brown and Quillen. For the history of collections of p-subgroups as a subject in its own right, we turn to the appearance of such collections in the literature in algebraic topology.

[2]In fact our focus here on posets of p-subgroups will entirely avoid the more usual approach via the actual axioms for a building—which describe it using the properties of the substructures called *apartments*. We indicate one reason for this later, shortly after Theorem 6.1.2.

In influential work in the 1970s such as [**33, 34**], Brown considered groups G which are not necessarily finite, but are of virtually finite cohomological dimension—this means that there exists a subgroup of G of finite index and finite *cohomological dimension*; cf. [**35**, Sec. VIII.2]. Brown studied for example the effect of elements of finite p-power order on aspects of cohomology such as the p-part of the Euler characteristic. He was led to the idea of viewing all this p-torsion together, in the form of the following G-poset—where *from now on, we further restrict to our usual situation of finite G, and we assume that p is a prime dividing $|G|$*:[3]

DEFINITION 5.1.2. $\mathcal{S}_p(G)$ denotes the poset of nontrivial p-subgroups of G.

The corresponding simplicial complex of inclusion-chains is the poset complex $\Delta(\mathcal{S}_p(G))$ of Definition 2.7.5. We also recall our convention in 2.7.6 that we may apply topological constructions directly to the poset $\mathcal{S}_p(G)$—implicitly this means we are applying them to the abstract simplicial complex $\Delta(\mathcal{S}_p(G))$, or more formally to the Euclidean topological space given by its classical geometric realization $|\Delta(\mathcal{S}_p(G))|$. When we require that geometric realization, we may write just $|\mathcal{S}_p(G)|$, leaving the poset complex construction $\Delta(-)$ implicit. (For this initial part of our historical discussion, it is not necessary to restrict the meaning of "spaces" to simplicial sets, as in our usual Convention 3.6.11.)

Many of the results of Brown and Quillen related to $\mathcal{S}_p(G)$ can be traced back to the following fundamental technical observation, namely the contractibility of certain fixed-point subcomplexes; the statement appears for example in the proof of [**33**, Sec. 3, Theorem 2]:

THEOREM 5.1.3. *For any nontrivial p-subgroup P of G, the P-fixed subcomplex $\Delta(\mathcal{S}_p(G))^P$ and the P-singular set*[4] $\bigcup_{1\neq Q\leq P} \Delta(\mathcal{S}_p(G))^Q$ *are contractible.*

One consequence of this contractibility result is Brown's striking congruence [**33**, Sec. 3, Cor. 2] on the Euler characteristic of the complex:

$$\chi(\Delta(\mathcal{S}_p(G))) \equiv 1 \pmod{|G|_p} \tag{5.1.4}$$

where $|G|_p$ is the highest power of p dividing the group order. This is often considered a homological version of the third Sylow theorem, which shows that the number of Sylow p-subgroups is congruent to 1 modulo p.

Brown's results, especially the congruence (5.1.4), soon inspired various further developments on the topology of posets and complexes of p-subgroups. In particular, we will next describe certain parts of Quillen's extremely influential paper [**104**]. (Although Quillen's paper did not address group cohomology explicitly, many of the results and ideas still played a role in the historical development of homology decompositions).

Quillen further developed applications of contractibility results like Theorem 5.1.3—especially as a tool for showing *homotopy equivalences* among various natural subgroup complexes. For example, he focused primarily not on the full G-poset $\mathcal{S}_p(G)$, but rather on the following G-subposet:

---

[3] The assumption that p divides $|G|$ ensures that posets of nontrivial p-subgroups such as $\mathcal{S}_p(G)$ are nonempty. Various later statements either have no content or are false without this assumption.

[4] See for example [**6**, p. 174] for the classical *singular set* of the action of P. In the present context, it is of particular interest when P is a Sylow p-subgroup of G.

DEFINITION 5.1.5. $\mathcal{A}_p(G)$ denotes the poset of nontrivial elementary abelian p-subgroups of G.

In particular, Quillen showed homotopy equivalence of this subposet with the full poset $\mathcal{S}_p(G)$. For convenience of later reference, we have stated this result below (as well as the next several results on equivalence), in the later extension to a G-homotopy equivalence given by Thévenaz and Webb (cf. Remark 5.1.12).

THEOREM 5.1.6 (Quillen [**104**, 2.1]). *The inclusion $\mathcal{A}_p(G) \subseteq \mathcal{S}_p(G)$ induces a G-homotopy equivalence.* □

The next result is another consequence of the contractibility in Theorem 5.1.3; see 15.4 in Brown [**34**] (and also [**35**, X.7.3]). It uses the Borel construction to obtain a connection with the group cohomology of G. The statement given in [**35**, X.7.3] uses $\mathcal{A}_p(G)$ rather than $\mathcal{S}_p(G)$; but in view of the equivalence in Theorem 5.1.6, we see by Corollary 2.5.13 that the corresponding Borel constructions are homotopy equivalent. This argument establishes the following result:[5]

THEOREM 5.1.7. *The map of Borel constructions*

$$EG \underset{G}{\times} \Delta\big(\mathcal{S}_p(G)\big) \to EG \underset{G}{\times} \star = BG$$

(determined by $\Delta(\mathcal{S}_p(G)) \to \star$) induces an $\mathbb{F}_p$-cohomology equivalence; that is, we obtain isomorphisms of $\mathbb{F}_p$-chain complexes:

$$\tilde{H}^*_G\big(\Delta(\mathcal{S}_p(G)); \mathbb{F}_p\big) \to \tilde{H}^*_G(\star; \mathbb{F}_p) \cong \tilde{H}^*(BG; \mathbb{F}_p) \cong \tilde{H}^*(G, \mathbb{F}_p).$$

This result can be regarded as the beginning of the modern literature on homotopy and homology decompositions: Note that it relates the classifying space BG, at the level of its cohomology, with the space given by the Borel construction $EG \underset{G}{\times} \Delta\big(\mathcal{S}_p(G)\big)$ arising from the p-subgroups complex. Indeed in view of the relation with the isotropy spectral sequence given in Remark 4.4.12, we get an expression in terms of classifying spaces BG_σ indexed (cf. 2.5.20) by σ in the quotient CW-complex $\Delta\big(\mathcal{S}_p(G)\big)/G$, so that the corresponding terms at the level of cohomology are of form $H^*\big(G_\sigma, \mathbb{F}_p\big)$.

This more specific modern notion of decompositions was further developed in later work of Webb; and in fact it would be possible for us to focus more exclusively on that theme, if we were to now skip directly to Webb's second approach to decompositions which appeared in [**124**]. But for a broader view of decompositions, it will also be relevant to instead continue our discussion of the further more general history—especially parts related to Webb's first approach to decompositions which appeared in [**122**].

Not long after Quillen's paper [**104**], Bouc expanded on an idea in 6.1 of that paper, considering a different subposet of $\mathcal{S}_p(G)$:

DEFINITION 5.1.8. $\mathcal{B}_p(G)$ denotes the poset of nontrivial *p-radical subgroups*, namely nontrivial p-subgroups P satisfying $P = O_p\big(N_G(P)\big)$.

Bouc established the equivalence analogous to Quillen's result Theorem 5.1.6 on the elementary subposet $\mathcal{A}_p(G)$:

---

[5] But for an alternative deduction of the result from the analogue of Theorem 5.1.3, see Adem and Milgram [**6**, V.3.1] (they also state the result for $\mathcal{A}_p(G)$).

THEOREM 5.1.9 (Bouc [**30**, Cor., p. 50]). *The inclusion $\mathcal{B}_p(G) \subseteq \mathcal{S}_p(G)$ induces a G-homotopy equivalence.*

Another equivalence established by Quillen strikingly emphasized a connection of these abstract collections of p-subgroups (defined for any group G)—with the theory of buildings (for the special case where G is of Lie type):

THEOREM 5.1.10 (Quillen [**104**, 3.1]). *For G a finite group of Lie type in characteristic p, $\mathcal{A}_p(G)$ (and hence also $\mathcal{S}_p(G)$) is G-homotopy equivalent to the building of G.*

REMARK 5.1.11 (Unipotent radicals as $\mathcal{B}_p(G)$). It was also later observed (see for example Thévenaz and Webb [**118**, 2.2.iv]) that for G of Lie type in characteristic p, $\mathcal{B}_p(G)$ is just the collection $\mathcal{U}$ of unipotent radicals indicated in Example 5.1.1. We saw there that associated poset complex $\Delta(\mathcal{U})$ also affords the building; so this gives a direct proof of the analogue of the equivalence in Theorem 5.1.10, where now $\mathcal{B}_p(G)$ plays the role of "$\mathcal{A}_p(G)$" there.

In view of these results, one could regard $\mathcal{S}_p(G)$ (and other posets homotopy equivalent to it via a G-map) as partial analogues—for an arbitrary group G—of the Tits building.

REMARK 5.1.12 (Extension to G-homotopy equivalences). To complete our present discussion of equivalences, we now more formally indicate the extensions which we had already mentioned in earlier equivalence statements: Thévenaz and Webb in [**118**, Theorem 2 and Rk. 2.2.iv] showed that the homotopy equivalences listed above are in fact G-homotopy equivalences.

Notice that by Corollary 2.5.13, it suffices to have just the weaker equivalence property given by a G-equivariant map which is a homotopy equivalence—in order to induce a homotopy equivalence at the level of Borel constructions. So it follows, now additionally using Theorem 2.5.17, that the subposets $\mathcal{A}_p(G)$ and $\mathcal{B}_p(G)$ (and when G is of Lie type in characteristic p, the building) also satisfy the decomposition in the conclusion of Theorem 5.1.7. ◊

Another direction of Quillen's development, already present in unpublished work of Brown, considers applications to the p-modular representation theory of G. The contractibility in Theorem 5.1.3 implies vanishing of reduced homology groups; so for a nontrivial cyclic p-group $P = \langle x \rangle$, we obtain the vanishing of the reduced Euler characteristic of the x-fixed subcomplex:

$$\tilde{\chi}\left(\Delta\left(\mathcal{S}_p(G)\right)^x\right) = 0.$$

By the standard Lefschetz fixed-point formula (see for example [**22**, p. 223]), this value gives the trace of x on the *reduced Lefschetz module* of $\mathcal{S}_p(G)$, namely the virtual $\mathbb{F}_pG$-module defined by the alternating sum of the reduced chain groups of $\Delta(\mathcal{S}_p(G))$:

$$\tilde{L}_G\left(\Delta\left(\mathcal{S}_p(G)\right); \mathbb{F}_p\right) := \sum_i (-1)^i \tilde{C}_i\left(\Delta\left(\mathcal{S}_p(G)\right); \mathbb{F}_p\right) \tag{5.1.13}$$

By the Hopf trace formula (see for example Spanier [**117**, p. 195]), the character value of x on the alternating sum of reduced chain groups is the same as the value on the alternating sum of the reduced homology groups; and we saw above that the latter value is zero. Now the above vanishing of the character on nontrivial elements of p-power order immediately suggests a standard property (see for example Feit

[**59**, IV.2.5]) which is possessed by any G-module in characteristic zero determining a *projective* $\mathbb{F}_pG$-module. Quillen (following Brown) gave a proof in [**104**, Cor. 4.3] of a result in the direction suggested by this observation: namely that the character of $\tilde{L}_G(\Delta(\mathcal{S}_p(G)); \mathbb{F}_p)$ is indeed the same as the character of some virtual projective module (that is, a formal difference of projective modules); and later Webb in [**122**, Theorem A′] (which we apply here just to $\mathcal{S}_p(G)$) strengthened that conclusion from the character to the module itself, showing the following:

THEOREM 5.1.14. *$\tilde{L}_G(\Delta(\mathcal{S}_p(G)); \mathbb{F}_p)$ is a virtual projective $\mathbb{F}_pG$-module.*

This result is often regarded as a generalization, to an arbitrary finite group G, of the projective property of the Steinberg module for G a group of Lie type in characteristic p, which we had mentioned in 5.1.1: For notice in the Lie type case that it follows from the G-homotopy equivalence in Theorem 5.1.10 (using the characterization of G-homotopy equivalence in [**22**, 6.4.3.ii], and the contractibility in Theorem 5.1.3) that we obtain the projectivity in Theorem 5.1.14 for the reduced Lefschetz module of the building. And this module also affords the homological construction of the projective Steinberg module in the Solomon–Tits Theorem as indicated in 5.1.1.

Cohomology decompositions in the work of Webb. We will begin by summarizing the approach to cohomology decompositions that Webb originally followed in establishing [**122**, Thm A]; that approach proceeded via the viewpoint of the representation-theoretic results just indicated.

Instead of treating particular complexes such as the complex $\Delta(\mathcal{S}_p(G))$ of chains in the poset $\mathcal{S}_p(G)$, Webb considered suitable sufficient conditions for the action of G on some more general simplicial complex Δ. We first recall one necessary technical condition: we assume that the action of G on Δ is admissible—see Definition 2.5.5. For example we observed in Remarks 2.7.6 that admissibility holds when Δ is a poset complex like $\Delta(\mathcal{S}_p(G))$.

Hypothesis (b) of [**122**, Thm A] is actually a slight weakening of the contractibility requirement in Theorem 5.1.3: namely it requires the fixed subcomplex Δ^P only to be *acyclic*—that is, to have vanishing reduced homology groups over $\mathbb{Z}$; and it requires this not for all members of $\mathcal{S}_p(G)$, but just for the minimal members, namely the subgroups P of order exactly p in G. Webb shows in [**122**, Thm. A′] that this weaker hypothesis suffices to obtain the projectivity in Theorem 5.1.14 for the reduced Lefschetz module $\tilde{L}_G(\Delta; \mathbb{F}_p)$.

In fact Theorem A of [**122**] considers cohomology for G with coefficients in some module M. Since we are interested in the mod p group cohomology, here for M we will take the trivial $\mathbb{F}_pG$-module $\mathbb{F}_p$; and then we can express the group cohomology in the form $\mathrm{Ext}^*_{\mathbb{F}_pG}(\mathbb{F}_p, \mathbb{F}_p)$ of (2.1.2). We now summarize Webb's approach to this expression. We begin by applying $\mathrm{Ext}^i_{\mathbb{F}_pG}(-, \mathbb{F}_p)$ to $\tilde{L}_G(\Delta; \mathbb{F}_p)$. By the projectivity of $\tilde{L}_G(\Delta; \mathbb{F}_p)$ in the previous paragraph, this vanishes for $i > 0$ (cf. Remark 2.4.3). Furthermore for $i = 0$,[6] the term is just $\mathrm{Hom}_{\mathbb{F}_pG}(\tilde{L}_G(\Delta; \mathbb{F}_p), \mathbb{F}_p)$; and this term also vanishes, in view of the multiplicity formula 5.3 in [**123**]—indeed the discussion before 5.3 there mentions that the projective cover of the trivial module never

[6] The case $i = 0$ does not seem clear from the discussion in [**122**], so we supply a relevant remark from the discussion in the corresponding expository paper [**123**].

appears in $\tilde{L}_G(\Delta; \mathbb{F}_p)$ when p divides the order of G. That is, we have

$$\operatorname{Ext}^*_{\mathbb{F}_p G}\big(\tilde{L}_G(\Delta; \mathbb{F}_p), \mathbb{F}_p\big) = 0. \tag{5.1.15}$$

(Here Ext really means an alternating sum of Ext-groups). Next, the chain spaces in the definition (5.1.13) of $\tilde{L}_G\big(\Delta(\mathcal{S}_p(G)); \mathbb{F}_p\big)$ break up into permutation-module constituents: namely the induced modules $\operatorname{Ind}^G_{G_\sigma}(\mathbb{F}_p)$, spanned by the cosets in the various quotient spaces G/G_σ. Observe in particular that the indexing for these induced modules is not by simplices σ in Δ itself, but rather by the corresponding orbits in the quotient Δ/G—we will let $\bar{\sigma}$ denote the G-orbit of σ. Then using the Eckmann–Shapiro lemma (see for example [**21**, 2.8.4 (i)]—a cohomological analogue of Frobenius reciprocity in representation theory), along with the definition (2.1.2) of group cohomology, we convert the term for each such permutation module into the form

$$\operatorname{Ext}^*_{\mathbb{F}_p G_\sigma}(\mathbb{F}_p, \mathbb{F}_p) = H^*(G_\sigma, \mathbb{F}_p) = H^*(BG_\sigma; \mathbb{F}_p).$$

Finally we may separate out the formal term in dimension -1, which corresponds to the stabilizer $G_\sigma = G$ and hence just gives $H^*(G_\sigma, \mathbb{F}_p)$ in the above form; and move it to the other side of the equation in (5.1.15). This delivers Webb's result (in our special case where Webb's M is taken to be the trivial module $\mathbb{F}_p$), which is as follows:

THEOREM 5.1.16. [**122**, Thm. A] *Assume, for an admissible G-complex Δ, that Δ^P is acyclic for all subgroups P of order p in G. Then*

$$H^*(BG; \mathbb{F}_p) = \bigoplus_{\bar{\sigma} \in \Delta/G} (-1)^{\dim \sigma}\ H^*(BG_\sigma; \mathbb{F}_p)$$

where σ denotes a representative of the G-orbit $\bar{\sigma}$.

This alternating sum is often called *Webb's formula*; it gives a cohomology decomposition for BG at p of a particularly simple form.

We turn next to Webb's later approach to the decomposition in 5.1.16—which proceeds instead via the Borel construction of Δ. This route may be suggested to the reader by Remark 2.5.19, namely the interpretation of the Borel construction of G/G_σ as the classifying space BG_σ of the stabilizer—as a possible alternative explanation of the BG_σ in Theorem 5.1.16.

Webb shows, now strengthening the hypothesis of Theorem 5.1.16 from acyclicity to contractibility (cf. Theorem 5.1.3), that the cohomology equivalence of spaces in Theorem 5.1.7 then holds for Δ playing the role of "$\Delta\big(\mathcal{S}_p(G)\big)$".

PROPOSITION 5.1.17 (2.5.1 in [**124**]). *Let Δ be a simplicial complex with admissible G-action. Assume that Δ^P is contractible for all nontrivial p-subgroups P of G. Then the map of Borel constructions*

$$EG \underset{G}{\times} |\Delta| \to EG \underset{G}{\times} \star = BG$$

(determined by $\Delta \to \star$) induces an $\mathbb{F}_p$-cohomology equivalence.

Webb also establishes in [**124**, 2.5.1] that the corresponding isotropy spectral sequence (see Definition 2.6.10) for the G-equivariant $\mathbb{F}_p$-cohomology $H^*_G(\Delta; \mathbb{F}_p)$ collapses at the E_2-page to the vertical axis; that is, that terms for $s > 0$ in the E_2-page, obtained from the cohomology version of the E_1-page of the spectral sequence in (2.6.2), must vanish. From our discussion of these pages at Remark 4.4.12 (or see

Dwyer [**53**, 9.9]), this means that $H^j(BG;\mathbb{F}_p)$ is expressed as the alternating sum of the terms in the jth row of the E_1-page in the alternative form we indicated there—namely as the jth cohomology groups of the subgroups G_σ stabilizing simplices. Thus from Theorem 5.1.17 we also recover the decomposition in Theorem 5.1.16; and the indexing is over Δ/G as in Remark 2.5.20. Notice that this reason for the indexing by orbits is different in flavor from the reason for that indexing in Webb's original approach via projectivity of the reduced Lefschetz module (namely the Eckmann–Shapiro lemma).

Webb in fact deduces 5.1.17 from a main result in [**124**] which is stated in the context of the homological algebra of *Mackey functors*. For a version of the proof of 5.1.16 which is based more directly on the G-equivariant cohomology of Δ, and which emphasizes the role of the contractibility of the singular set of Δ appearing in Theorem 5.1.3, see V.3.2 and V.3.3 of Adem and Milgram [**6**].

As an important example of the decomposition in Webb's formula 5.1.16, we mention the model case of groups of Lie type in their defining characteristic:

EXAMPLE 5.1.18 (Cohomology decomposition over the building). For G a group of Lie type in characteristic p, the G-homotopy equivalence in Theorem 5.1.10 of the building Δ with $\mathcal{S}_p(G)$ gives by Corollary 2.5.13 a homotopy equivalence of the corresponding Borel constructions. Thus the above approach via the Borel construction in 5.1.17, again using the characterization of G-homotopy equivalence in [**22**, 6.4.3 (ii)], and the contractibility in Theorem 5.1.3, shows that we obtain the decomposition of Theorem 5.1.16 over the quotient complex Δ/G of the building Δ (the quotient is in fact a simplex—see later Proposition 6.4.4).

Homotopy colimit theorems and the unifying analysis of Dwyer. We have just seen how Webb's viewpoint on the Borel construction in Proposition 5.1.17 leads to homology decompositions such as the formula in Theorem 5.1.16. We now turn to some subsequent work in the literature on homotopy colimits—this work was done by topologists for application usually to compact Lie groups, rather than just to finite groups G. However, when we restrict those results to finite G, they also provide two further results on homotopy decompositions, which "explain" homology decompositions similar to Webb's formula.

Before describing those two further decompositions, we mention that the homotopy colimit form of those results led Dwyer to also re-interpret Webb's Borel construction approach in the language of homotopy colimits, using the viewpoint in Theorem 4.7.2. Indeed since the stabilizers G_σ in Webb's formula are typically normalizers of members of a suitable collection of subgroups, and since the complex of inclusion-chains from the collection gives Δ, the formula will correspond to the *normalizer decomposition* in the language of Dwyer, which we will see later in Theorem 5.6.10.

We now go on to describe the original development via homotopy colimits of two other decompositions: the *centralizer decomposition* and the *subgroup decomposition*.

We begin with the restriction to finite G of the work of Jackowski and McClure in [**76**]. They work essentially with the collection $\mathcal{A}_p(G)$ of nontrivial elementary abelian p-subgroups—but they use a category which is more complicated than just the poset complex $\Delta\big(\mathcal{A}_p(G)\big)$ of inclusion-chains. First, they consider the opposite poset $\mathcal{A}_p(G)^{\mathrm{op}}$—this will be relevant for the centralizers to be used below, since if

$E' \geq E$ is a containment in $\mathcal{A}_p(G)^{\text{op}}$, then we get the inclusion $C_G(E') \leq C_G(E)$ of centralizers. Second, they consider further morphisms beyond just inclusions, namely those given by group conjugations followed by inclusions. So in order to distinguish this *conjugation category* notationally from the simpler poset-category for $\mathcal{A}_p(G)^{\text{op}}$, we will (temporarily until Definition 5.3.1) write $\mathfrak{A}_p(G)$ for their category. Next as the functor for their eventual homotopy colimit, which we will denote here by α_{JM}, they define a composite map: The first functor maps $E \in \mathcal{A}_p(G)$ to the G-orbit given by $G/C_G(E)$. The second part of the functor then maps the G-space $G/C_G(E)$ to its Borel construction $EG \underset{G}{\times} \big(G/C_G(E)\big)$. With these definitions, their main result shows the following:

THEOREM 5.1.19. *$\underset{\mathfrak{A}_p(G)}{\underrightarrow{\text{Hocolim}}}\, \alpha_{JM} \to BG$ is an $\mathbb{F}_p$-homology equivalence.*

PROOF. See Jackowski and McClure [**76**, 1.3] □

In view of the definition of α_{JM} using the Borel construction on transitive G-sets such as $G/C_G(E)$, we can use the standard relation given in Remark 2.5.19 (this corresponds to the "natural equivalence of functors" mentioned in [**76**, Sec. 2]), to regard the image of E as the classifying space $BC_G(E)$—since $C_G(E)$ is the stabilizer of the object E in the natural conjugation action of G on the groups appearing in the objects of the category $\mathfrak{A}_p(G)$. This motivates the terminology for Theorem 5.1.19 of the *centralizer decomposition* in the language of Dwyer, which we will give in Theorem 5.4.4. Furthermore, [**76**, 2.1] shows that the Bousfield–Kan spectral sequence of this homotopy colimit collapses at the E^2-page to the vertical axis—in our notation in (4.8.8), the terms for $s > 0$ vanish. (This will turn out to be the appropriate analogue of the collapse of the isotropy spectral sequence in Webb's work.) Consequently the homotopy decomposition produces an analogue for $\mathfrak{A}_p(G)$ of the homology decomposition in Theorem 5.1.16: the component terms are now of the form $H^*(C_G(E), \mathbb{F}_p)$; and they are indexed now by $E \in \mathcal{A}_p(G)/G$—in view of Remark 2.5.20 for the Borel construction in the definition of α_{JM}. But the corresponding alternating-sum decomposition is more complicated to write down than Theorem 5.1.16, because of the more complicated morphisms in the indexing category.

Next we consider the restriction to finite G of the work of Jackowski, McClure, and Oliver [**77**]. They work with the collection $\mathcal{B}_p(G)^e$ essentially consisting of the p-radical subgroups $\mathcal{B}_p(G)$—but when $O_p(G) = 1$ they also include the case of the identity subgroup, and they call the collection the *p-stubborn* subgroups. They also use an associated category which is more complicated than just the poset complex $\Delta\big(\mathcal{B}_p(G)^e\big)$ of inclusion-chains. Their overall context is the *G-orbit category* $\mathcal{O}(G)$ of G, whose objects are G-sets G/H, and whose morphisms are G-equivariant maps between such G-sets. Their particular indexing category, denoted there by $\mathcal{R}_p(G)$, is the full subcategory of $\mathcal{O}(G)$ consisting of the orbits G/H with $H \in \mathcal{B}_p(G)^e$. We will use the notation β_{JMO} to denote the functor defined in the introductory section of [**77**], which takes G/H for $H \in \mathcal{B}_p(G)^e$ to the quotient EG/H. They show:

THEOREM 5.1.20. *$\underset{\mathcal{R}_p(G)}{\underrightarrow{\text{Hocolim}}}\, \beta_{JMO} \to BG$ is a $\mathbb{Z}_{(p)}$-homology equivalence, and also an $\mathbb{F}_p$-homology equivalence.*

PROOF. See Theorem 4 (and also 1.4) of [**77**]. We mention that the analogue for $\mathcal{B}_p(G)$ rather than $\mathcal{B}_p(G)^e$ was established by Dwyer in [**52**, 10.4] (and see also Grodal [**66**, p. 448]). □

We recall from (2.5.18) that the image EG/H of the object G/H in $\mathcal{R}_p(G)$ can be identified with with the classifying space BH; and hence also by Remark 2.5.19 with the Borel construction $EG \underset{G}{\times} (G/H)$—since the subgroup H itself is the stabilizer of the transitive G-set G/H in the natural translation action by G. This motivates the terminology for Theorem 5.1.20 of the *subgroup decomposition* in the language of Dwyer, which we will give in Theorem 5.5.4. In this case also, there is an analogue of Webb's formula in Theorem 5.1.16: the components are of form $H^*(H, \mathbb{F}_p)$ coming from terms BH; and they are indexed by $H \in \mathcal{B}_p(G)^e/G$—in view of Remark 2.5.20, using the discussion above of the identification the image EG/H of G/H with the classifying space BH. This decomposition was essentially described in [**77**, 2.14, see also 2.2]—but in language somewhat different from that of the Webb formula; in fact the language in [**77**] is more closely related to the discussion of "acyclicity" in Dwyer [**53**, Sec. 11], which underlies part of our later discussion in Section 5.8 of his analysis of Webb-type formulas.

As we have already mentioned at several earlier points, subsequent work of Dwyer illuminated some common threads among these three decompositions (normalizer, centralizer, subgroup). We will end our historical overview of this section with just a brief further summary, since we will be discussing some basic concepts from Dwyer's analysis in more detail in the next few sections.

In [**51**], Dwyer introduced the notion of an *ample* collection, using essentially the form of the conclusion of Proposition 5.1.17; and he showed that ampleness is equivalent to the property that the relevant homotopy colimit in each of the three indicated decompositions should determine a mod p homology equivalence with BG. Subsequently in [**52**], Dwyer developed the notion of a *sharp* homology decomposition; namely the collapse at the E_2-page of the Bousfield–Kan spectral sequence for homology of the homotopy colimit; which leads to alternating sum expressions analogous to Webb's formula in Theorem 5.1.16 for those decompositions. We will be considering ampleness and sharpness in more detail in the remainder of the chapter.

We mention also that Grodal in [**66**, esp. Sec. 2.8] gave a further viewpoint on decompositions; and in particular in [**66**, Sec. 9], he improved on some of Dwyer's results on sharpness.

5.2. Homology decompositions defined by homotopy colimits

Before launching into the details of ampleness, and the three specific decompositions associated to it, we begin with a motivating discussion of Dwyer's more general notion of a homology decomposition arising from a homotopy colimit; we will essentially be following Section 7 of Dwyer's exposition [**53**].

Let G be a finite group. We might naturally want to define a homology decomposition of BG as arising from a $\mathfrak{C}$-diagram F of spaces (for $\mathfrak{C}$ a small indexing category), when the homotopy colimit gives a mod p homology equivalence:

$$\underset{\overrightarrow{\mathfrak{C}}}{\text{Hocolim}}\ F \ \to\ BG.$$

Furthermore we might typically want to require that the components in the homotopy colimit should satisfy a suitable condition connecting them to classifying spaces for subgroups of G, such as the following: For each object i in $\mathfrak{C}$, there should be given a weak equivalence $F(i) \simeq BH_i$ of spaces for some $H_i \leq G$; and this map is compatible with the above map, in the sense that the obvious diagram of spaces relating them

$$\begin{array}{ccc} F(i) & \xrightarrow{\ \simeq\ } & BH_i \\ \Big\downarrow & & \Big\downarrow{\scriptstyle B\,\circ\,\mathrm{incl}} \\ \underrightarrow{\mathrm{Hocolim}}_{\mathfrak{C}}\, F & \longrightarrow & BG \end{array} \tag{5.2.1}$$

commutes up to homotopy (that is, it is a commutative diagram when read in the homotopy category of spaces). We shall make an appropriate modification to this condition in Definition 5.2.4 below—which will be motivated by the analysis of a related alternative viewpoint given in the next few paragraphs:

One method for producing a $\mathfrak{C}$-diagram F whose components are essentially classifying spaces BH_i as in (5.2.1) would be to start with a $\mathfrak{C}$-diagram of transitive G-sets, say $X\colon \mathfrak{C} \to G\text{-}\mathfrak{Set}$—and then form the Borel construction to obtain classifying spaces for the stabilizers, in the spirit of Remark 2.5.19. Indeed we saw in Lemma 4.7.4 how to regard the Borel construction as a suitable homotopy colimit, so that the Borel construction commutes as in Proposition 4.7.6 with the homotopy colimit for our diagram X; hence those results give us:

$$EG \underset{G}{\times} (\underrightarrow{\mathrm{Hocolim}}_{\mathfrak{C}}\, X) \simeq \underrightarrow{\mathrm{Hocolim}}_{\mathfrak{C}}(EG \underset{G}{\times} X(-)) = \text{“}\underrightarrow{\mathrm{Hocolim}}_{i\in\mathfrak{C}}\, BH_i\text{”} \tag{5.2.2}$$

where H_i is the stabilizer of a point in the transitive G-set $X(i)$. The inverted commas here are because of Warning 4.5.16; the given expression involving the classifying space is ostensibly only a diagram in the homotopy category; so the expression in the middle of (5.2.2) is preferable when we need to be more precise.

To recognize the requirements of (5.2.1): We are taking “F” to be given by the composition of the Borel construction with the diagram $X(-)$. Then the component term at some particular i in (5.2.2) corresponds to the top row in (5.2.1). Furthermore if we apply the Borel construction $(EG \underset{G}{\times} -)$ to the map $\underrightarrow{\mathrm{Hocolim}}_{\mathfrak{C}}\, X \to \star$, and apply the commuting of the Borel construction with homotopy colimits in Proposition 4.7.6, then we obtain a map

$$\underrightarrow{\mathrm{Hocolim}}_{\mathfrak{C}}(EG \underset{G}{\times} X(-)) \to BG. \tag{5.2.3}$$

for the left-hand term of (5.2.2) corresponding to the bottom row of (5.2.1). Thus this choice of F does produce a decomposition as in (5.2.1).

We will now outline an argument showing that the method indicated, of producing F from the Borel construction of X, is in effect the general method of producing decompositions:

For suppose conversely that we are given some homology decomposition F in the sense of (5.2.1). There is a natural surjection of EG on $EG/G = BG$, which together with the composite map $F(i) \to BG$ from (5.2.1) gives a *pullback* square—essentially dual to the pushout square indicated at Example 4.5.3, see for example

[**22**, p. 244]. So we form the pullback $Y(i)$ as the inverse limit (cf. 0.0.3 and its references) in the diagram

$$\begin{array}{ccc} Y(i) & \longrightarrow & EG \\ \downarrow & & \downarrow \\ F(i) & \longrightarrow & BG \end{array}$$

and these terms $Y(i)$ define a $\mathfrak{C}$-diagram Y of G-spaces. Now the right-hand vertical map $EG \to BG$ above is in fact a Serre fibration with fiber G; so at each i, from the left-hand vertical map above we also get a Serre fibration

$$G \to Y(i) \to F(i),$$

and hence the usual corresponding long exact sequence in homotopy—see for example [**22**, 1.6.6]. Furthermore the fiber G is a discrete space, with no homotopy in dimensions above 0. So using the map of $Y(i)$ to contractible EG, one can show via that sequence that each connected component of $Y(i)$ is weakly contractible; and using the map of $Y(i)$ to $F(i) \simeq BH_i$, one can show that the set $\pi_0\big(Y(i)\big)$ of connected components of $Y(i)$ is isomorphic to G/H_i as a G-set. So the $\pi_0\big(Y(i)\big)$ form a $\mathfrak{C}$-diagram X of transitive G-sets, together with a map $Y \to X$ of diagrams. Furthermore X is essentially just a set-theoretic equivalent of Y: that is, when we regard the terms of X as discrete spaces in the sense of Example 3.1.2, the map $Y \to X$ gives an isomorphism of diagrams of G-spaces.

Next we may project the Borel construction $EG \underset{G}{\times} Y(i)$ on the second factor, which is just the quotient $F(i)$ in the above fibration; and this in fact gives a further Serre fibration $EG \underset{G}{\times} Y(i) \to F(i)$, this time with fiber EG—which of course is contractible and so has vanishing reduced homotopy groups. From the contractibility of the fiber, the corresponding long exact sequence in homotopy gives isomorphisms of the homotopy groups of $EG \underset{G}{\times} Y(i)$ and $F(i)$—that is, a weak equivalence. Finally from the isomorphism $Y \to X$ of diagrams in the previous paragraph, we obtain a weak equivalence $EG \underset{G}{\times} Y(i) \to EG \underset{G}{\times} X(i)$ of Borel constructions. So by Proposition 4.5.17 (ii), it follows that we have weak equivalences

$$\underrightarrow{\mathrm{Hocolim}}_{\mathfrak{C}}\big(EG \underset{G}{\times} X(-)\big) \overset{\simeq}{\longleftarrow} \underrightarrow{\mathrm{Hocolim}}_{\mathfrak{C}}\big(EG \underset{G}{\times} Y(-)\big) \overset{\simeq}{\longrightarrow} \underrightarrow{\mathrm{Hocolim}}_{\mathfrak{C}} F.$$

Thus we see that every homology decomposition F in the sense of (5.2.1) is weakly equivalent to one coming from the Borel construction of a $\mathfrak{C}$-diagram X of transitive G-sets; we recover the decomposition corresponding to F when we apply (5.2.2) and (5.2.3) to X.

So it is natural to make the following definition:

DEFINITION 5.2.4 (Homology decomposition). Let G be a finite group and R a coefficient ring. An R-homology decomposition of BG consists of a $\mathfrak{C}$-diagram X of transitive G-sets with the property that the map (5.2.3) is an R-homology equivalence.

REMARK 5.2.5. A homology decomposition in this sense *(which is the sense we shall use from now on)* gives rise via $\underrightarrow{\mathrm{Hocolim}}_{\mathfrak{C}}\big(EG \underset{G}{\times} X(-)\big)$ to a Bousfield–Kan spectral sequence as in Theorem 4.8.7, which converges to the homology of that homotopy colimit. Hence from the R-homology equivalence in Definition 5.2.4, that homology is the same as $H_*(BG; R) = H_*(G, R)$, and 4.8.7 takes the form:

$$\underset{i \in \mathfrak{C}}{\underrightarrow{\operatorname{colim}}}\, {}_s\big(H_t^G(X(i); R)\big) \Rightarrow H_{s+t}(G, R). \tag{5.2.6}$$

Also by Proposition 4.8.14, this Bousfield–Kan spectral sequence can be identified (from the E^2 page onward) with the isotropy spectral sequence for the action of G on $\underset{\mathfrak{C}}{\underrightarrow{\operatorname{Hocolim}}}\, X$, which we shall investigate in Section 5.8. In particular, we shall examine conditions under which the derived functors of colim vanish for $s > 0$, because this situation of "collapse" gives rise to particularly simple expressions for the homology of G, which are called *sharp* decompositions in Definition 5.8.1.

When the coefficient ring R is one of the usual rings related to a prime p as in Section 2.3, we can make use of earlier results such as those in that section, to indicate the following relationships among possible equivalences related to the R-homology equivalence in Definition 5.2.4:

THEOREM 5.2.7. *Let X be a $\mathfrak{C}$-diagram of transitive G-sets, where $\mathfrak{C}$ is a small indexing category. Then among the following statements, we have the implications:*
(i) $\Leftrightarrow$ (v),
(iii) $\Rightarrow$ (i) $\Rightarrow$ (ii),
(iii) $\Rightarrow$ (iv) $\Rightarrow$ (ii), *and*
(i) $\Rightarrow$ (vi).
If in addition $\mathfrak{C}$ has a finite number of arrows, then in fact (i)–(v) *are all equivalent, and imply* (vi). *(This finiteness hypothesis will hold in our later categories related to finite subgroup collections and simplicial complexes for finite groups G.)*

Here is the list of statements to be related:
(i) *The map* (5.2.3) *is an $\mathbb{F}_p$-homology equivalence.*
(ii) *The map* (5.2.3) *is an $\mathbb{F}_p$-cohomology equivalence.*
(iii) *The map* (5.2.3) *is a $\mathbb{Z}_{(p)}$-homology equivalence.*
(iv) *The map* (5.2.3) *is a $\mathbb{Z}_{(p)}$-cohomology equivalence.*
(v) *The map* (5.2.3), *followed by $\mathbb{F}_p$-completion, is a homotopy equivalence.*
(vi) *The homotopy colimit on the left of* (5.2.3) *is $\mathbb{F}_p$-good.*

PROOF. The equivalence of (i) and (v) is given by Proposition 4.2.5. The implications (i) $\Rightarrow$ (ii) and (iii) $\Rightarrow$ (iv) are given by Theorem 2.4.8. The implications (iii) $\Rightarrow$ (i) and (iv) $\Rightarrow$ (ii) are given by Proposition 2.3.5. The implication (i) $\Rightarrow$ (vi) is given by Proposition 4.2.9—since the term BG on the right side of (5.2.3) is $\mathbb{F}_p$-good by Theorem 4.3.1 (v).

By Proposition 2.1.5 (1), BG has finite $\mathbb{F}_p$- and $\mathbb{Z}_{(p)}$-homology groups. The same holds for the terms BH_i in (5.2.2). So if in addition $\mathfrak{C}$ has a finite number of arrows, then finite generation of homology groups is preserved in the Bousfield–Kan spectral sequence; and since by Proposition 4.8.14 we can identify that sequence with the isotropy spectral sequence, it follows that the homotopy colimit has finitely generated $\mathbb{F}_p$- and $\mathbb{Z}_{(p)}$-homology groups. So we can apply the reverse implications in Theorem 2.4.8, namely (ii) $\Rightarrow$ (i) and (iv) $\Rightarrow$ (iii); and in Proposition 2.3.5, namely (i) $\Rightarrow$ (iii) and (ii) $\Rightarrow$ (iv). Hence (i)–(v) are equivalent when $\mathfrak{C}$ has a finite number of arrows. □

REMARK 5.2.8. These statements cannot be improved on, even if G is the trivial group. The problem is that by Example 4.5.13, the homotopy colimit of a $\mathfrak{C}$-diagram where every space is a single point is the nerve $\mathcal{N}\mathfrak{C}$; and every space is weakly equivalent to $\mathcal{N}\mathfrak{C}$ for some suitable small category $\mathfrak{C}$ (for example the poset of its singular simplices). So take for example $\mathfrak{C} := \mathfrak{B}_{\mathbb{Q}}$, the category for the bar construction in Example 3.1.9, where $\mathbb{Q}$ denotes the infinite group given by the additive rationals. So this category has one object and an infinite number of arrows. Now $\underrightarrow{\mathrm{Hocolim}}_{\mathfrak{B}_{\mathbb{Q}}} \star = \mathcal{N}\mathfrak{B}_{\mathbb{Q}} = B\mathbb{Q}$ as in 3.1.9. By Gruenberg [**69**, Proposition 8.3.4], for any coefficients R we have $H_1(\mathbb{Q}, R) = \mathbb{Q} \otimes_{\mathbb{Z}} R$ and $H_i(\mathbb{Q}, R) = 0$ for $i > 1$. So $\tilde{H}_*(B\mathbb{Q}; \mathbb{F}_p) = \tilde{H}_*(\mathbb{Q}, \mathbb{F}_p) = 0$, whereas $H_1(\mathbb{Q}, \mathbb{Z}_{(p)}) = \mathbb{Q} \otimes_{\mathbb{Z}} \mathbb{Z}_{(p)} = \mathbb{Q}_{(p)} \neq 0$. So for G equal to the trivial group, the map

$$B\mathbb{Q} = \underrightarrow{\mathrm{Hocolim}}_{\mathfrak{B}_{\mathbb{Q}}} \star \longrightarrow BG = \star$$

is an $\mathbb{F}_p$-equivalence, and hence gives a mod p homology decomposition for the trivial group G; but the map is not a $\mathbb{Z}_{(p)}$-equivalence.

Homology decompositions based on indexing categories $\mathfrak{C}$ with a finite number of arrows are much better behaved than those for arbitrary small categories; and all the homology decompositions considered from now on will be of this form (since in practice they will always arise in connection with finite groups), so that (i)–(vi) in the above theorem will always then hold for such decompositions.

5.3. Ampleness for collections of subgroups

In this section, we shall describe the notion of ampleness for a collection of subgroups; and in the three sections after this one, we shall see that an ample collection gives rise to three natural homology decompositions of the classifying space. Our exposition is based on Dwyer—[**51**], and especially Section 7 of [**53**].

DEFINITION 5.3.1 (Collections of subgroups). Let G be a finite group. A *collection* $\mathcal{C}$ of subgroups of G is a set of subgroups which is closed under conjugation in G. In other words, if $H \in \mathcal{C}$ and $g \in G$, then also $gHg^{-1} \in \mathcal{C}$. We regard $\mathcal{C}$ as a poset, ordered by inclusion (indeed $\mathcal{C}$ is a G-poset); and hence as in Example 3.1.10, we can also regard $\mathcal{C}$ as a category in which the morphisms are the inclusions.

The *orbit category* $\mathcal{O}_{\mathcal{C}}$ of a collection $\mathcal{C}$ of subgroups is defined to be the category of transitive G-sets G/H, where H is required to be a member of the collection $\mathcal{C}$. The morphisms are the corresponding G-maps among such G-sets. These morphisms correspond roughly with composites of the $\mathcal{C}$-inclusions $H \leq H'$ appearing in G and the G-conjugations among them.

The *conjugation category* $\mathfrak{A}_{\mathcal{C}}$ of a collection $\mathcal{C}$ of subgroups is defined to be the category whose objects are the pairs (H, γ), where H is a finite group and γ is a G-conjugacy class of monomorphisms $i\colon H \to G$ such that the image $i(H)$ in G is required to be a member of the collection $\mathcal{C}$ for each $i \in \gamma$. A morphism from (H, γ) to (H', γ') consists of a group monomorphism $j\colon H \to H'$ such that $\gamma = \{i' \circ j : i' \in \gamma'\}$. Again the morphisms correspond roughly with composites of inclusions and conjugations from $\mathcal{C}$ represented in G. ◇

REMARK 5.3.2. (Set theoretic aside) As defined above, the category $\mathfrak{A}_{\mathcal{C}}$ is not a small category—since the possibilities for H do not form a set. This is a problem, because we wish to take a homotopy colimit over it. An easy solution to this technical problem is to let Ω be a fixed infinite set, and only use groups H whose elements are taken from Ω.

Recall we observed at Definition 3.3.7 that the simplicial complex $\Delta(\mathcal{C})$ (namely the complex of chains from the poset $\mathcal{C}$) has the same geometric realization as the nerve $\mathcal{NC}$ of the poset $\mathcal{C}$ regarded as a category. In [**51**, 1.2] and [**53**, 7.7], Dwyer uses the Borel construction viewpoint in the conclusion of Webb's result, Proposition 5.1.17, but replacing $\Delta(\mathcal{C})$ by $\mathcal{NC}$, to give the general definition of ampleness:

DEFINITION 5.3.3 (Ampleness). Let G be a finite group and let p be a prime dividing $|G|$. A collection $\mathcal{C}$ of subgroups of G is *ample* (or, more precisely, $\mathbb{F}_p$-*ample*) if the map from the Borel construction on the nerve of $\mathcal{C}$

$$EG \underset{G}{\times} \mathcal{NC} \to EG \underset{G}{\times} \star = BG$$

is an $\mathbb{F}_p$-homology equivalence.

Notice by Proposition 4.2.5 that if we then pass to p-completions, the homology equivalence becomes a homotopy equivalence

$$(EG \underset{G}{\times} \mathcal{NC})^\wedge_p \to BG^\wedge_p.$$

However, here we are not applying Theorem 5.2.7 (v)—since at this level of generality, we have not yet given a specific $\mathcal{C}$-diagram X of transitive G-sets. $\diamondsuit$

EXAMPLE 5.3.4. Brown's Theorem 5.1.7 on the Borel construction of $\mathcal{S}_p(G)$ shows, using this language of the nerve of $\mathcal{S}_p(G)$ as a category, that the collection $\mathcal{S}_p(G)$ is ample. We will discuss other ample collections in later sections of the chapter.

Associated to an ample collection $\mathcal{C}$ of subgroups are three different homology decompositions of BG, called the centralizer decomposition (Theorem 5.4.4), the subgroup decomposition (Theorem 5.5.4), and the normalizer decomposition (Theorem 5.6.4). In fact our results on sporadic groups in Chapter 7 will be stated using only the normalizer decomposition—but we will need to quote results referring to the other decompositions, so in the next three sections we will give a description of each of the three.

In each case, the strategy is the same: For a suitable indexing category $\mathfrak{C}$ associated to $\mathcal{C}$—not necessarily just the poset $\mathcal{C}$ itself regarded as a category—we begin with a certain $\mathfrak{C}$-diagram X of G-sets (indeed transitive G-sets); and form its Borel construction as at (5.2.2). As indicated there, we then use Proposition 4.7.6, which states that the Borel construction commutes with the homotopy colimit over $\mathfrak{C}$. Furthermore using (4.5.19), the homotopy colimit $\underset{\overrightarrow{\mathfrak{C}}}{\text{Hocolim}}\, X$ of the G-spaces can be identified as the nerve $\mathcal{N}\big(\mathsf{Gr}(X)\big)$ of the category given by the Grothendieck construction on X; and in each case, we will obtain a G-equivariant map from that nerve to the nerve $\mathcal{NC}$ of the original poset $\mathcal{C}$. This map is a homotopy equivalence which is a G-map, but not necessarily a G-homotopy equivalence (except in the case of the normalizer decomposition); however that weaker equivalence suffices by Corollary 2.5.13 to gives a homotopy equivalence of Borel constructions. So we obtain a commutative diagram

$$\begin{array}{ccccc}
\underset{\mathfrak{C}}{\underrightarrow{\mathrm{Hocolim}}}(EG \underset{G}{\times} X) & \xrightarrow{\cong} & EG \underset{G}{\times} (\underset{\mathfrak{C}}{\underrightarrow{\mathrm{Hocolim}}}\, X) & \xrightarrow{\simeq} & EG \underset{G}{\times} \mathcal{NC} \\
& \searrow & & \swarrow & \\
& & EG \underset{G}{\times} \star = BG & &
\end{array}$$

Since ampleness is defined using the right-hand diagonal map, we then deduce that $\mathcal{C}$ is ample if and only if the left-hand diagonal map is a mod p homology equivalence.

5.4. The centralizer decomposition

We summarize the viewpoint of Dwyer on the decomposition used by Jackowski and McClure in [**76**, 7.1], who worked with the collection $\mathcal{C}$ given by $\mathcal{A}_p(G)$, namely the nontrivial elementary abelian p-subgroups of G. The notation we used at Theorem 5.1.19 of $\mathfrak{A}_p(G)$ and α_{JM} will correspond to $\mathfrak{A}_{\mathcal{C}}$ and $\tilde{\alpha}_{\mathcal{C}}$ below.

Let $\mathcal{C}$ denote a collection of subgroups of a finite group G. We consider the corresponding conjugation category $\mathfrak{A}_{\mathcal{C}}$ indicated in Definition 5.3.1. We may define an $\mathfrak{A}_{\mathcal{C}}^{\mathrm{op}}$-diagram of transitive G-sets as follows: The functor

$$\tilde{\alpha}_{\mathcal{C}} \colon \mathfrak{A}_{\mathcal{C}}^{\mathrm{op}} \to G\text{-}\mathfrak{Set}$$

takes an object (H, γ) to the transitive G-set γ, namely a G-conjugacy class of monomorphisms into G; and takes an arrow $j \colon (H, \gamma) \to (H', \gamma')$ to composition with j viewed as a map from γ' to γ. This functor will play the role of "X" in Definition 5.2.4 of a homology decomposition.

Regarding sets as discrete spaces as in Example 3.1.2, we may also regard $\tilde{\alpha}_{\mathcal{C}}$ as a functor from $\mathfrak{A}_{\mathcal{C}}^{\mathrm{op}}$ to the category G-$\mathfrak{Sp}$ of G-spaces. So we may take the homotopy colimit, to get a G-space

$$\underset{\mathfrak{A}_{\mathcal{C}}^{\mathrm{op}}}{\underrightarrow{\mathrm{Hocolim}}}\, \tilde{\alpha}_{\mathcal{C}}. \tag{5.4.1}$$

We may apply the Borel construction to this space—and then use Proposition 4.7.6 on the commuting of the Borel construction with the homotopy colimit; hence from the map of the space in (5.4.1) to a point, we get a map

$$\underset{\mathfrak{A}_{\mathcal{C}}^{\mathrm{op}}}{\underrightarrow{\mathrm{Hocolim}}}\big(EG \underset{G}{\times} \tilde{\alpha}_{\mathcal{C}}(-)\big) = EG \underset{G}{\times} \big(\underset{\mathfrak{A}_{\mathcal{C}}^{\mathrm{op}}}{\underrightarrow{\mathrm{Hocolim}}}\, \tilde{\alpha}_{\mathcal{C}}\big) \to EG \underset{G}{\times} \star = BG. \tag{5.4.2}$$

Now abbreviating the Borel-construction diagram $EG \underset{G}{\times} \tilde{\alpha}_{\mathcal{C}}(-)$ by $\alpha_{\mathcal{C}}(-)$, using 2.5.19 we obtain for any $i \in \gamma$ that

$$\alpha_{\mathcal{C}}(H, \gamma) = EG \underset{G}{\times} \gamma \simeq B\big(C_G\big(i(H)\big)\big),$$

since the stabilizer of the element i in the conjugation action of G on the transitive G-set γ is $C_G\big(i(H)\big)$. So the map described in (5.4.2) takes the form

$$\text{“}\underset{(H,\gamma)\,\in\,\mathfrak{A}_{\mathcal{C}}^{\mathrm{op}}}{\underrightarrow{\mathrm{Hocolim}}}\, B\big(C_G\big(i(H)\big)\big)\text{”} = \underset{\mathfrak{A}_{\mathcal{C}}^{\mathrm{op}}}{\underrightarrow{\mathrm{Hocolim}}}\, \alpha_{\mathcal{C}} \to BG, \tag{5.4.3}$$

where i denotes a representative of the G-orbit γ. The reason for the inverted commas here is that the expression is ostensibly only a diagram in the homotopy category of spaces, whereas one needs to have a diagram of spaces and actual maps in order to take a homotopy colimit; see Warning 4.5.16 and Example 4.7.5.

THEOREM 5.4.4 (Centralizer decomposition). *A collection $\mathcal{C}$ of subgroups of a finite group G is ample if and only if the map* (5.4.3) *is a mod p homology equivalence.*

PROOF. By (4.5.19), the homotopy colimit (5.4.1) is isomorphic to the nerve of the Grothendieck construction $\mathsf{Gr}(\tilde{\alpha}_{\mathcal{C}})$. From Definition 4.5.18, an object in this latter category is determined by a pair consisting of an object (H,γ) in the opposite conjugation category $\mathfrak{A}_{\mathcal{C}}^{\mathrm{op}}$, and a member of the image of $\tilde{\alpha}_{\mathcal{C}}$ on that object: so an object in the Grothendieck construction is in fact determined by a pair (H,i), where H is a finite group, and $i\colon H \to G$ is a monomorphism (representing the G-orbit γ) with image $i(H)$ in $\mathcal{C}$. An arrow $(H,i) \to (H',i')$ is a monomorphism $j\colon H \to H'$ such that $i' \circ j = i$. The action of $a \in G$ on $\mathsf{Gr}(\tilde{\alpha}_{\mathcal{C}})$ is given by $a\big((H,i)\big) := \big(H, a(i)\big)$ where $a(i)(h) := a\big(i(h)\big)a^{-1}$.

We next indicate an equivalence of categories $\mathsf{Gr}(\tilde{\alpha}_{\mathcal{C}}) \to \mathcal{C}$ via a G-map: In the forward direction, we use the map f sending (H,i) to $i(H)$, which is equivariant with respect to the G-action given in the previous paragraph. A G-equivariant map g in the reverse direction is given by sending H to (H,j_H), where $j_H\colon H \to G$ is the inclusion map. The composition fg is the identity on $\mathcal{C}$. It remains to give a natural transformation from the identity on $\mathsf{Gr}(\tilde{\alpha}_{\mathcal{C}})$ to the other composite gf: the value of the natural transformation on the object (H,i) is the map to $\big(i(H), j_{i(H)}\big)$ given by the monomorphism i itself and its natural inclusion.

The category equivalence in the previous paragraph leads by (3.4.6) to a homotopy equivalence of their nerves via a G-map; combining this with the isomorphism in the first paragraph, we get

$$\underset{\mathfrak{A}_{\mathcal{C}}^{\mathrm{op}}}{\underrightarrow{\mathrm{Hocolim}}}\, \tilde{\alpha}_{\mathcal{C}} \;\cong\; \mathcal{N}\big(\mathsf{Gr}(\tilde{\alpha}_{\mathcal{C}})\big) \;\to\; \mathcal{N}\mathcal{C}.$$

Hence using Proposition 4.7.6 and Corollary 2.5.13, we obtain a weak equivalence of Borel constructions

$$\underset{\mathfrak{A}_{\mathcal{C}}^{\mathrm{op}}}{\underrightarrow{\mathrm{Hocolim}}}\, \alpha_{\mathcal{C}} \;\cong\; EG \underset{G}{\times} \big(\underset{\mathfrak{A}_{\mathcal{C}}^{\mathrm{op}}}{\underrightarrow{\mathrm{Hocolim}}}\, \tilde{\alpha}_{\mathcal{C}}\big) \;\to\; EG \underset{G}{\times} \mathcal{N}\mathcal{C}$$

commuting with the maps from both sides to BG. The Theorem now follows, since Definition 5.3.3 expresses ampleness via mod p equivalence in the right-hand map to BG. □

WARNING 5.4.5. Notice that the natural equivalence of gf with the identity in the above proof need not respect the G-action on $\mathsf{Gr}(\tilde{\alpha}_{\mathcal{C}})$; since the abstract group H is not necessarily preserved by i; so we do not in general get a weak G-equivalence of the corresponding nerves—cf. Warning 2.5.2.

The main result of Jackowski and McClure [**76**, 7.1] in particular establishes the second condition in Theorem 5.4.4 for the collection $\mathcal{C}$ given by $\mathcal{A}_p(G)$; so in Dwyer's language, their result shows that $\mathcal{A}_p(G)$ is ample.

5.5. The subgroup decomposition

We summarize the viewpoint of Dwyer on the decomposition used by Jackowski, McClure, and Oliver in [**77**], who worked with the collection $\mathcal{C}$ given by $\mathcal{B}_p(G)^e$;

recall this consists of the p-radical $\mathcal{B}_p(G)$ subgroups of G, as well as the trivial subgroup in the case that $O_p(G) = 1$. The notation we used at Theorem 5.1.20 of $\mathcal{R}_p(G)$ and β_{JMO} will correspond to $\mathcal{O}_C$ and $\tilde{\beta}_{\mathcal{C}}$ below.

Let $\mathcal{C}$ denote a collection of subgroups of a finite group G. We consider the corresponding orbit category $\mathcal{O}_{\mathcal{C}}$ indicated in Definition 5.3.1. We may define an $\mathcal{O}_{\mathcal{C}}$-diagram of transitive G-sets as follows: The functor

$$\tilde{\beta}_{\mathcal{C}} \colon \mathcal{O}_{\mathcal{C}} \to G\text{-}\mathfrak{Sp}$$

just takes G/H regarded as an object of $\mathcal{O}_{\mathcal{C}}$ to G/H itself—now regarded as a transitive G-set. This functor will play the role of "X" in Definition 5.2.4 of a homology decomposition.

Regarding these G-sets as discrete G-spaces in Example 3.1.2, we may take the homotopy colimit, to get a G-space

$$\underset{\mathcal{O}_{\mathcal{C}}}{\underrightarrow{\text{Hocolim}}}\, \tilde{\beta}_{\mathcal{C}}. \tag{5.5.1}$$

We may apply the Borel construction to this space—and then use Proposition 4.7.6 on the commuting of the Borel construction with the homotopy colimit; hence from the map of the space in (5.5.1) to a point, we get a map

$$\underset{\mathcal{O}_{\mathcal{C}}}{\underrightarrow{\text{Hocolim}}}(EG \underset{G}{\times} \tilde{\beta}_{\mathcal{C}}(-)) \;\cong\; EG \underset{G}{\times} (\underset{\mathcal{O}_{\mathcal{C}}}{\underrightarrow{\text{Hocolim}}}\, \tilde{\beta}_{\mathcal{C}}) \;\to\; EG \underset{G}{\times} \star = \; BG. \tag{5.5.2}$$

Now abbreviating the Borel-construction diagram $EG \underset{G}{\times} \tilde{\beta}_{\mathcal{C}}(-)$ by $\beta_{\mathcal{C}}(-)$, using Remark 2.5.19 we obtain for any $H \in \mathcal{C}$ that

$$\beta_{\mathcal{C}}(G/H) \;=\; EG \underset{G}{\times} (G/H) \;\simeq\; BH,$$

since H itself is the stabilizer of the coset H in the natural translation action of G on the transitive G-set G/H. So the map described in (5.5.2) takes the form

$$\text{“}\underset{G/H \in \mathcal{O}_{\mathcal{C}}}{\underrightarrow{\text{Hocolim}}}\, BH\text{”} \;=\; \underset{\mathcal{O}_{\mathcal{C}}}{\underrightarrow{\text{Hocolim}}}\, \beta_{\mathcal{C}} \;\to\; BG. \tag{5.5.3}$$

As before, the reason for the inverted commas is that the expression on the left ostensibly only describes a diagram in the homotopy category, as in Warning 4.5.16.

THEOREM 5.5.4 (Subgroup decomposition). *A collection $\mathcal{C}$ of subgroups of a finite group G is ample if and only if the map* (5.5.3) *is a mod p homology equivalence.*

PROOF. By (4.5.19), the homotopy colimit (5.5.1) is isomorphic to the nerve of the Grothendieck construction $\mathsf{Gr}(\tilde{\beta}_{\mathcal{C}})$. From the definition in 4.5.18, an object in this latter category is determined by a pair consisting of an object G/H in the orbit category $\mathcal{O}_{\mathcal{C}}$, and a member of the image of $\tilde{\beta}_{\mathcal{C}}$ on that object: so an object in the Grothendieck construction is in fact determined by a pair $(G/H, x)$, where $H \in \mathcal{C}$ and x is some coset yH in G/H; and an arrow $(G/H, x) \to (G/H', x')$ is a map of G-sets taking x to x'. The action of $a \in G$ is given by $a\big((G/H, x)\big) := (G/H, ax)$.

We next indicate a G-equivariant functor $f \colon \mathsf{Gr}(\tilde{\beta}_{\mathcal{C}}) \to \mathcal{C}$ which gives an equivalence of categories: namely f sends $(G/H, x)$ to the stabilizer G_x; we have equivariance with respect to the action in the previous paragraph since $(G/H, ax)$ is mapped by f to $(G/H, ax)$, and G_{ax} is the conjugate of G_x by a. A functor g in the reverse direction is defined by sending H to the pair $(G/H, H)$, where the latter H is being regarded as the identity coset in G/H; and this map is visibly G-equivariant. The composite fg is the identity on $\mathcal{C}$. It remains to give a natural transformation from the identity on $\mathsf{Gr}(\tilde{\beta}_{\mathcal{C}})$ to the other composite gf: the value of the natural transformation on an object is the natural map $(G/H, aH) \mapsto (G/aHa^{-1}, aHa^{-1})$.

The category equivalence in the previous paragraph leads by (3.4.6) to a homotopy equivalence of their nerves via a G-map; combining this with the isomorphism in the first paragraph, we get

$$\underset{\mathcal{O}_{\mathcal{C}}}{\underrightarrow{\mathrm{Hocolim}}}\, \tilde{\beta}_{\mathcal{C}} \cong \mathcal{N}\big(\mathsf{Gr}(\tilde{\beta}_{\mathcal{C}})\big) \rightarrow \mathcal{N}\mathcal{C}.$$

Hence using Proposition 4.7.6 and Corollary 2.5.13 we obtain a weak equivalence of Borel constructions

$$\underset{\mathcal{O}_{\mathcal{C}}}{\underrightarrow{\mathrm{Hocolim}}}\, \beta_{\mathcal{C}} \cong EG \underset{G}{\times} (\underset{\mathcal{O}_{\mathcal{C}}}{\underrightarrow{\mathrm{Hocolim}}}\, \tilde{\beta}_{\mathcal{C}}) \rightarrow EG \underset{G}{\times} \mathcal{N}\mathcal{C}$$

commuting with the maps from both sides to BG. As the right-hand map to BG is used in Definition 5.3.3 for ampleness, the Theorem follows. □

WARNING 5.5.5. We continue the theme of earlier Warning 5.4.5: Notice that the natural equivalence of gf with the identity in the proof need not respect the G-action on $\mathsf{Gr}(\tilde{\beta}_{\mathcal{C}})$: for the equivalence followed by G-action typically gives a nontrivial coset in the second position of the pair—whereas G-action followed by the equivalence always has a subgroup in that position. So we will not in general get a weak G-equivalence of the corresponding nerves—cf. Warning 2.5.2.

The main result of Jackowski, McClure, and Oliver in [**77**] in particular establishes the second condition in Theorem 5.5.4 for the collection $\mathcal{C}$ given by $\mathcal{B}_p(G)^e$; so in Dwyer's language, their result shows that $\mathcal{B}_p(G)^e$ is ample. (The result for $\mathcal{B}_p(G)$ itself was established by Dwyer in [**52**, 10.4]; see also Grodal [**66**, p. 448].)

5.6. The normalizer decomposition

We next summarize how Dwyer applied his viewpoint also to Webb's use of the Borel construction of a G-complex Δ in Proposition 5.1.17. Dwyer in fact considered the case where Δ is the complex $\Delta(\mathcal{C})$ of inclusion-chains from a collection $\mathcal{C}$, and we indicate that case below; we will turn to the obvious extension to a general G-complex Δ shortly thereafter, beginning at Definition 5.6.6.

The case of the complex of inclusion-chains from a collection. Let $\mathcal{C}$ denote a collection of subgroups of a finite group G. As in Example 3.1.10, we may also regard $\mathcal{C}$ as a category. In order to obtain good behavior for the quotient category modulo the G-action (cf. the discussion at Definition 2.5.5), we work in fact with the related category provided by the barycentric subdivision of the poset $\mathcal{C}$, which is defined by its inclusion-chains as we indicated at Definition 2.7.7. That is, this is just the obvious variant on the subdivision construction for a category in Example 4.5.24: namely we now let the objects of a new category $s\mathcal{C}$ be defined by the nonempty chains of just the *proper* inclusions in $\mathcal{C}$: so that we have objects of form $\sigma := (H_0 < \cdots < H_n)$ with each H_i in $\mathcal{C}$; and the arrows $\sigma \rightarrow \sigma'$ are the reverse inclusions $\sigma \supseteq \sigma'$ of chains.

Hence just as in Example 4.5.24, there is a functor $s\mathcal{C} \rightarrow \mathcal{C}$ which takes a chain of subgroups $(H_0 < \cdots < H_n)$ to its initial element H_0, and takes a containment of chains to the corresponding inclusion of initial elements. Furthermore the group G acts on both $\mathcal{C}$ and $s\mathcal{C}$ by conjugation, and the functor $s\mathcal{C} \to \mathcal{C}$ commutes with the G-action. It follows just as in Example 4.5.24 that there is an induced weak G-equivalence of nerves $\mathcal{N}(s\mathcal{C}) \to \mathcal{N}\mathcal{C}$ via a G-map. Hence it follows by Corollary 2.5.13 that these nerves have weakly equivalent Borel constructions—namely those corresponding to the complex $\Delta(\mathcal{C})$ of inclusion-chains from $\mathcal{C}$, and to the barycentric subdivision of that complex.

For Dwyer's viewpoint, we will need a related category suitably analogous to $\mathfrak{A}_C$ or $\mathcal{O}_C$—in which the objects are defined via suitable *transitive G-sets*. So we now write $s\mathcal{C}/G$ for the quotient of $s\mathcal{C}$ by the action of G; hence the objects of $s\mathcal{C}/G$ are now G-orbits of inclusion-chains from $\mathcal{C}$. For the morphisms in the orbit category $s\mathcal{C}/G$, we observe that a containment of chains of subgroups induces a restriction map on orbits; and since a finite group cannot be isomorphic to a proper subgroup of itself, at most one conjugate of a chain can be a subchain of a longer chain—so that the restriction map is well defined.

We may define an $s\mathcal{C}/G$-diagram of transitive G-sets as follows: The functor

$$\tilde{\delta}_{\mathcal{C}} : s\mathcal{C}/G \to G\text{-}\mathfrak{S}\mathfrak{p}$$

takes an object given by a G-orbit of chains of subgroups in $\mathcal{C}$ to that orbit itself—now regarded as a transitive G-set. This functor will play the role of "X" in Definition 5.2.4 of a homology decomposition.

As usual regarding sets as discrete spaces as in Example 3.1.2, we may also regard this as a diagram of G-spaces. We may take the homotopy colimit, to get a G-space

$$\underset{\overrightarrow{s\mathcal{C}/G}}{\text{Hocolim}}\, \tilde{\delta}_{\mathcal{C}}. \tag{5.6.1}$$

We may apply the Borel construction to this space—and then use Proposition 4.7.6 on the commuting of the Borel construction with the homotopy colimit; hence from the map of the space in (5.6.1) to a point, we get a map

$$\underset{\overrightarrow{s\mathcal{C}/G}}{\text{Hocolim}}\big(EG \underset{G}{\times} \tilde{\delta}_{\mathcal{C}}(-)\big) \;\cong\; EG \underset{G}{\times} \big(\underset{\overrightarrow{s\mathcal{C}/G}}{\text{Hocolim}}\, \tilde{\delta}_{\mathcal{C}}\big) \;\to EG \underset{G}{\times} \star \;=\; BG. \tag{5.6.2}$$

Now abbreviate the Borel construction diagram $EG \underset{G}{\times} \tilde{\delta}_{\mathcal{C}}(-)$ by $\delta_{\mathcal{C}}(-)$. Let $(H_0 < \cdots < H_n)$ be a representative of an object γ of $s\mathcal{C}/G$, namely a G-orbit of chains in $\mathcal{C}$. The stabilizer in G of the chain $(H_0 < \cdots < H_n)$ is the intersection $N_G(H_0) \cap \cdots \cap N_G(H_n)$ of the normalizers of the members of the chain. So using 2.5.19 we obtain

$$\delta_{\mathcal{C}}(\gamma) \;=\; EG \underset{G}{\times} \gamma \;\simeq\; B\big(N_G(H_0) \cap \cdots \cap N_G(H_n)\big).$$

So the map described in (5.6.2) takes the form

$$\text{“}\underset{\overrightarrow{\gamma \in s\mathcal{C}/G}}{\text{Hocolim}}\, B\big(N_G(H_0) \cap \cdots \cap N_G(H_n)\big)\text{”} \;=\; \underset{\overrightarrow{s\mathcal{C}/G}}{\text{Hocolim}}\, \delta_{\mathcal{C}} \;\to\; BG. \tag{5.6.3}$$

As before, the reason for the inverted commas is that the expression on the left ostensibly only describes a diagram in the homotopy category as in Warning 4.5.16.

THEOREM 5.6.4 (Normalizer decomposition). *A collection $\mathcal{C}$ of subgroups of a finite group G is ample if and only if the map (5.6.3) is a mod p homology equivalence.*

PROOF. By (4.5.19), the homotopy colimit (5.6.1) is isomorphic to the nerve of the Grothendieck construction $\mathsf{Gr}(\tilde{\delta}_{\mathcal{C}})$. From the definition in 4.5.18, an object in this latter category is determined by a pair consisting of an object in the quotient category $s\mathcal{C}/G$ (namely a G-orbit of inclusion-chains in $\mathcal{C}$), and a member of the image of $\tilde{\delta}_{\mathcal{C}}$ on that object: so an object in the Grothendieck construction is in fact determined by by a chain of subgroups $(H_0 < \cdots < H_n)$, namely just an object in $s\mathcal{C}$. The arrows correspond to the reverse containments.

So we have a natural equivalence $\mathsf{Gr}(\tilde{\delta}_{\mathcal{C}}) \cong s\mathcal{C}$ of categories; and this time (in contrast to the situation in the proofs of Theorems 5.4.4 and 5.5.4) this equivalence commutes with the G-action.

Thus this category equivalence leads by (3.4.6) to a G-homotopy equivalence of their nerves; combining this with the isomorphism in the first paragraph, we get

$$\underset{s\mathcal{C}/G}{\underrightarrow{\mathrm{Hocolim}}}\, \tilde{\delta}_{\mathcal{C}} \;\cong\; \mathcal{N}\big(\mathsf{Gr}(\tilde{\delta}_{\mathcal{C}})\big) \;\to\; \mathcal{N}(s\mathcal{C}).$$

Now recall we saw in the preliminary discussion before (5.6.1) that there is a weak G-equivalence $\mathcal{N}(s\mathcal{C}) \to \mathcal{N}\mathcal{C}$ of nerves; so using Proposition 4.7.6 and Corollary 2.5.13, we get a weak equivalence of Borel constructions

$$\underset{s\mathcal{C}/G}{\underrightarrow{\mathrm{Hocolim}}}\, \delta_{\mathcal{C}} \;\cong\; EG \underset{G}{\times} \underset{s\mathcal{C}/G}{\underrightarrow{\mathrm{Hocolim}}}\, \tilde{\delta}_{\mathcal{C}} \;\to\; EG \underset{G}{\times} \mathcal{N}\mathcal{C}$$

commuting with the maps from both sides to BG. As the right-hand map to BG is used in Definition 5.3.3 for ampleness, the Theorem follows. □

The relationship among the three decompositions. We are now in a position to indicate a connection among the centralizer, subgroup and normalizer decompositions:

THEOREM 5.6.5 (see 11.3 in Dwyer [**53**]). *There are G-equivariant functors between the three Grothendieck constructions used in the proofs of Theorems 5.4.4, 5.5.4 and 5.6.4*

$$\mathsf{Gr}(\tilde{\alpha}_{\mathcal{C}}) \xrightarrow{u} \mathsf{Gr}(\tilde{\delta}_{\mathcal{C}}) \xleftarrow{v} \mathsf{Gr}(\tilde{\beta}_{\mathcal{C}})$$

given by $u(H,i) := i(H)$ and $v(G/H,x) := G_x$. These give G-maps which afford weak equivalences among the nerves of these categories (which in turn define homotopy colimits involved in the three decompositions); but these are not necessarily weak G-homotopy equivalences.

PROOF. From the proofs of Theorems 5.4.4 and 5.5.4, we see there are G-equivariant functors u', v' in the reverse direction, which give equivalences of categories, as follows: The functor $u' \colon \mathsf{Gr}(\tilde{\delta}_{\mathcal{C}}) \to \mathsf{Gr}(\tilde{\alpha}_{\mathcal{C}})$ sends a subgroup H to (H,i), where i is the inclusion of H in G. The functor $v' \colon \mathsf{Gr}(\tilde{\delta}_{\mathcal{C}}) \to \mathsf{Gr}(\tilde{\beta}_{\mathcal{C}})$ sends H to $(G/H,H)$, where the latter H is the identity coset in G/H. However, we also recall as in those earlier proofs that the natural equivalences of the compositions $u'u$ and $v'v$ with the corresponding identity functors need not in general commute with the G-action. □

We comment that the use of the Grothendieck construction as in Example 4.5.18 has removed any explicit reference to the different indexing categories for the three functors. So the formulation in Theorem 5.6.5 emphasizes that the different functors lead to weakly equivalent Borel constructions by Corollary 2.5.13; and hence that the homotopy colimits giving the three decompositions can be regarded just as varying approaches to ampleness of a collection.

The reader is also referred to the approach in Grodal [**66**, 2.9] to this relationship among the three decompositions.

The normalizer decompositions for a general G-complex. While the treatment in Dwyer [**53**] is focused on a collection (hence poset) $\mathcal{C}$ of subgroups, we will extend that viewpoint to consider a more general admissible G-complex Δ: this is in the spirit of Webb's approach in Proposition 5.1.17, and that approach is the natural setting for the 2-local geometries we describe in Chapter 6. (For the particular geometries Δ for sporadic groups G in Chapter 7, we will always indicate a G-map giving a homotopy equivalence of Δ with the poset complex $\Delta(\mathcal{C})$ of inclusion-chains from a suitable collection $\mathcal{C}$.)

To begin, note that we can regard the inclusion-chains of subgroups from a collection $\mathcal{C}$, which give the category $s\mathcal{C}$ whose quotient defines the normalizer decomposition 5.6.4, as the simplices in the G-simplicial complex given by the poset complex $\Delta(\mathcal{C})$ in 2.7.5. So the normalizer decomposition has a natural generalization to G-actions on general finite simplicial complexes, which do not necessarily arise as complexes of inclusion-chains from collections of subgroups. (Since the centralizer and subgroup decompositions are determined by categories involving morphisms more complicated than just set-theoretic inclusions, there are no corresponding descriptions as simplicial complexes, and hence no analogous generalizations of those decompositions.)

So to parallel our earlier discussion which applied to the case of the poset complex $\Delta(\mathcal{C})$ for a collection $\mathcal{C}$, we first extend Definition 5.3.3 of ampleness to more general complexes—replacing the nerve $\mathcal{NC}$ by the simplicial set defined just by Δ itself:

DEFINITION 5.6.6 (Ampleness for G-complexes). If Δ is a finite simplicial complex with an admissible action of a finite group G, regarded as a simplicial set as in 3.1.3, we say that Δ is *ample* (or, more precisely, $\mathbb{F}_p$*-ample*) if the map of spaces

$$EG \underset{G}{\times} \Delta \to EG \underset{G}{\times} \star = BG$$

is a mod p homology equivalence.

Then we can re-state Webb's result indicated earlier, in the language of Dwyer's viewpoint:

PROPOSITION 5.6.7. *Let Δ be an admissible G-simplicial complex. If Δ^P is contractible for all p-subgroups P of G, then Δ is ample.*

PROOF. This is essentially Webb's result [**124**, 2.5.1], which we stated earlier as Proposition 5.1.17; the only difference is that there we gave Webb's statement for the Borel construction $EG \underset{G}{\times} |\Delta|$ in the context of topological spaces—whereas now we have defined ampleness for Δ in 5.6.6 above for the simplicial Borel construction $EG \underset{G}{\times} \Delta$ of Remark 3.6.9; which uses the simplicial set EG from Example 3.1.6, and the simplicial set view of Δ in Example 3.1.3. (We need the context of simplicial sets, since we will wish to apply Dwyer's homotopy-colimit approach in the further discussion below.) □

In practice we will often replace one ample complex by another which is more convenient for calculation; we will use:

LEMMA 5.6.8. *Let Δ and Δ' be finite simplicial complexes with an admissible action of a finite group G. Let $f \colon \Delta \to \Delta'$ be a G-equivariant map which induces a homotopy equivalence. Then Δ is ample if and only if Δ' is ample.*

PROOF. We emphasize that the homotopy equivalence in the hypothesis is at the level of abstract simplicial complexes and their classical geometric realizations. As we had mentioned at 3.6.9, when we regard these complexes as simplicial sets, they need not be Kan complexes; so that when we apply Corollary 2.5.13, the mapping induced by f will in general give only a weak equivalence of the simplicial Borel constructions of Δ and Δ'. But that weak equivalence still suffices to give the mod p homology equivalence of the Borel constructions: for we may apply Proposition 3.5.2 (ii) in the context of simplicial sets rather than topological spaces, since the proof involves only general homological algebra. Since ampleness is defined by mod p homology equivalence of the Borel construction with BG, the result follows. □

We will apply this result for general G-complexes; and also often just for complexes of inclusion-chains from collections. For example, Theorem 5.1.3 shows that $\mathcal{S}_p(G)$ satisfies the hypothesis of Proposition 5.6.7, so that $\mathcal{S}_p(G)$ is ample. But using the homotopy equivalences in Theorems 5.1.6 and 5.1.9, we can apply Lemma 5.6.8 to conclude that the subcollections $\mathcal{A}_p(G)$ and $\mathcal{B}_p(G)$ of $\mathcal{S}_p(G)$ are also ample—and these smaller collections are often more convenient for practical computation of the normalizer decomposition.

Next we make the obvious extension of our earlier discussion which for a collection $\mathcal{C}$ gave a diagram of transitive G-sets; this time since we are already assuming a simplicial complex Δ with an admissible G-action, we are guaranteed the needed good behavior of the quotient modulo the G-action; so it is not necessary at this initial stage to pass to a "subdivision" like $s\mathcal{C}$ in the earlier case for a collection $\mathcal{C}$. We write Δ/G for the quotient of Δ by the action of G; regarded both as an abstract simplicial complex, and as a space.

Exactly as before, we may define a Δ/G-diagram of transitive G-sets as a functor

$$\tilde{\delta}_\Delta \colon \Delta/G \to G\text{-}\mathfrak{Sp}$$

taking an orbit regarded as an object to that orbit regarded as a G-set; and this functor will play the role of "X" in Definition 5.2.4 of a homology decomposition. We regard the G-sets as discrete spaces as in Example 3.1.2. Again we may compose the Borel construction with this functor, abbreviating the result as $\delta_\Delta \colon \Delta/G \to \mathfrak{Sp}$; and we have a map

$$\text{“}\underset{\bar{\sigma}\in\Delta/G}{\underrightarrow{\operatorname{Hocolim}}}\, BG_\sigma\text{”} = \underset{\Delta/G}{\underrightarrow{\operatorname{Hocolim}}}\, \delta_\Delta \to BG \tag{5.6.9}$$

with the same caveat as before: that ostensibly the expression in inverted commas only describes a diagram in the homotopy category as in Warning 4.5.16. The indexing of the homotopy colimit is over G-orbits $\bar{\sigma}$ of simplices in Δ; and G_σ is the stabilizer of a simplex σ representing the orbit $\bar{\sigma}$.

THEOREM 5.6.10 (Normalizer decomposition for complexes). *A finite simplicial complex Δ with admissible G-action is ample if and only if the map* (5.6.9) *is a mod p homology equivalence.*

PROOF. The proof is essentially the same as the proof of Theorem 5.6.4, so we will simply emphasize the needed modifications; for example, the notion of subdivision will only arise at the final step.

This time from the definition in 4.5.18, an object of the Grothendieck construction $\mathsf{Gr}(\tilde{\delta}_\Delta)$ is determined by a pair consisting of an object $\bar{\sigma}$ in the quotient complex Δ/G and a member σ of the image of $\tilde{\delta}_C$ on that object: so an object in the Grothendieck construction corresponds just a simplex σ of Δ itself. Thus we have a natural equivalence $\mathsf{Gr}(\tilde{\delta}_\Delta) \cong \Delta$ of categories, and hence a weak equivalence of their nerves; and the natural equivalence respects the G-action, so again we have a weak G-equivalence. Applying the Borel construction, by Corollary 2.5.13 we get a weak equivalence of the homotopy colimit in (5.6.9) with the Borel construction of the nerve of Δ. But the nerve of Δ regarded as a category is weakly equivalent to Δ regarded as a simplicial set: For the former is built from inclusion-chains in the latter, so that we may again apply the basic idea of the subdivision in 4.5.24; again we get a weak G-equivalence. So again we may apply Proposition 4.7.6 and Corollary 2.5.13, to get a weak equivalence of Borel constructions; hence a weak equivalence of the homotopy colimit in (5.6.9) with the Borel construction of Δ in Definition 5.6.6 of ampleness for complexes. □

We close our discussion of the normalizer decomposition for a general complex by indicating some further concepts that will arise in practice when we apply the construction to complexes for sporadic simple groups in Chapter 7.

REMARK 5.6.11 (Flag-transitivity and indexing categories). The quotient Δ/G in (5.6.9) is not necessarily a simplicial complex, but is always a CW-complex—see Warning 2.5.6. But there is one situation which usually occurs in our applications to simple groups, where the quotient is a simplicial complex:

Assume that G acts *flag-transitively* on the complex Δ; this means (cf. later Definition 6.4.2) that the orbit complex Δ/G is just a simplex Δ^I as in Definition 2.7.1, for some finite index set I. So if the stabilizer of the vertex of Δ^I given by $i \in I$ is denoted by H_i, then the simplex whose vertices are given by the subset J of I has stabilizer given by the intersection $H_J = \bigcap_{i\in J} H_i$. Hence our homotopy colimit is indexed over the subgroup simplex $\mathcal{H}_I$ of Definition 4.6.2 determined by the vertex stabilizers H_i. And so the normalizer decomposition in (5.6.9) takes the simple form given in 4.6.4, namely

$$\text{“}\underset{\varnothing \neq J \subseteq I}{\underrightarrow{\text{Hocolim}}}\ BH_J\text{”}. \tag{5.6.12}$$

In particular we will obtain such decompositions for the examples of flag-transitive complexes Δ indicated after Definition 6.4.2, notably the model case given by the building for G of Lie type in Proposition 6.4.4. And we will also obtain this form of homotopy decomposition for each sporadic geometry Δ in our main results in Chapter 7. (The form of the corresponding cohomology decomposition will be indicated at Remark 5.8.5.)

We emphasize that a simplex of subgroups $\mathcal{H}_I$ as in Definition 4.6.2 will usually not be closed under G-conjugacy, and so will usually *not* be a collection in the sense of Definition 5.3.1. However, as we will indicate in more detail at the start of Chapter 6, our development will naturally proceed as in this section—from some initial complex Δ, often given by a collection $\mathcal{C}$ of subgroups, to an orbit simplex $\mathcal{H}_I$ of subgroups indexing the final form (5.6.12) of our decomposition. ♢

The form (5.6.9) of the normalizer decomposition emphasizes the homotopy colimit over an orbit complex (namely Δ/G). Since the normalizer decomposition is the one we use in our applications in Chapter 7, we now establish some further notation which in particular helps to describe the stabilizers G_σ which are involved.

DEFINITION 5.6.13 (The normalizer map N on a poset). Given a G-poset $\mathcal{Q}$, we let N denote the mapping taking $x \in \mathcal{Q}$ to the subgroup G_x of G which stabilizes x. The image $\mathsf{N}(\mathcal{Q})$ will then afford a subposet of the poset of subgroups of G; but note that distinct x need not in general determine distinct G_x. In particular, we emphasize that *in general* N *is only a map of sets.* However, in many favorable cases N will turn out to induce a poset map.

If Δ is a simplicial complex with an admissible G-action, by $\mathsf{N}(\Delta)$ we will mean the application of this notation of N to the poset $\mathcal{Q}(\Delta)$ of simplices of Δ and their inclusions; so that a simplex $\sigma \in \Delta$ is mapped to the simplex stabilizer G_σ. In particular, the image poset $\mathsf{N}(\Delta)$ consists of the vertex stabilizers, along with their intersections which give the stabilizers of higher-dimensional simplices.

Furthermore if G is flag-transitive on Δ, with Δ/G a simplex whose faces are stabilized by subgroups of the form H_J described in Remark 5.6.11, then it will be convenient to express this relation as:

$$\mathsf{N}(\Delta/G) = \mathcal{H}_I,$$

using the notation of the subgroup simplex $\mathcal{H}_I$ in Definition 4.6.2.

As mentioned earlier: Sometimes, but not always, the map N will in fact be a poset map, or else an order-reversing map. For example, we mentioned in Example 5.1.1 that in the case of G of Lie type, the unipotent radicals $\mathcal{U}$ determine the poset opposite to that of the parabolics $\mathcal{P}$—where the latter are the normalizers of the former; so in this case, the normalizer map N is order-reversing, and induces a poset anti-isomorphism of $\mathcal{U}$ with $\mathcal{P}$. ♢

Further remarks on the various decompositions. We close this section with some general remarks relevant to the three decompositions introduced in the last few sections.

Given an ample collection $\mathcal{C}$, it is obviously desirable (for example, for purposes of practical computation) to try to determine inside $\mathcal{C}$ a *smallest* possible subcollection $\mathcal{D}$ of $\mathcal{C}$ which is still ample.

In view of Lemma 5.6.8, one approach is be to look for a smallest possible subcollection which is either G-homotopy equivalent to $\mathcal{C}$—or at least homotopy equivalent to $\mathcal{C}$ via a G-map. For this purpose, we will often make use in Chapter 7 of the considerable literature on methods for homotopy equivalence: beginning with such results as Theorem 5.1.6 of Quillen, and proceeding through more recent methods such as those in Smith and Yoshiara [**116**] developed primarily for applications to sporadic groups.

But we will also see in the later sections of this chapter that some proper subcollections of $\mathcal{C}$ may be ample, but *not* homotopy equivalent to $\mathcal{C}$. In fact there seems to be no known characterization of ampleness expressed only in terms of homotopy types.

Finally we summarize Dwyer's viewpoint on the decompositions in this and the previous two sections, in the following proposition; again this primarily emphasizes the equivalences with ampleness that we saw in the corresponding earlier results. Essentially we have just added our notion of ampleness for a complex to the other decompositions in Dwyer's [**53**, 7.11].

PROPOSITION 5.6.14. *Let $\mathcal{C}$ be a collection of subgroups of a finite group G. Then the following are equivalent.*

(i) *$\mathcal{C}$ is ample.*

(ii) *The centralizer decomposition* (5.4.3) *is a mod p homology equivalence.*

(iii) *The subgroup decomposition* (5.5.3) *is a mod p homology equivalence.*

(iv) *The normalizer decomposition* (5.6.3) *is a mod p homology equivalence.*

If Δ is a simplicial complex with an admissible G-action, then Δ is ample if and only if the normalizer decomposition (5.6.9) *is a mod p homology equivalence.* □

We mention also that the categories involved in the decompositions in Proposition 5.6.14 have a finite number of arrows; so that we may make use of the equivalences in Theorem 5.2.7—where statement (i) in that result corresponds to ampleness.

5.7. Ampleness for the centric collection

In studying the subgroup decomposition in Theorem 5.1.20 of Jackowski, McClure, and Oliver for the collection $\mathcal{B}_p(G)^e$ from the viewpoint of sharpness (see Section 5.8), Dwyer was led to make a similar analysis for the collection of p-centric subgroups of G.

DEFINITION 5.7.1. A *p-centric* subgroup of G is a p-subgroup P of G such that the centralizer $C_G(P)$ is the product of the center $Z(P)$ of P with a group of order coprime to p; this is equivalent to the statement that $Z(P)$ is a Sylow p-subgroup of $C_G(P)$ (and indeed is the unique Sylow group of $C_G(P)$). We write $\mathcal{C}e_p(G)$ for the collection of p-centric subgroups of G. Notice when p divides the group order $|G|$ that the identity subgroup is *not* p-centric.

REMARK 5.7.2. If $P \geq Q$ are p-subgroups of G and Q is p-centric, then P is also p-centric: For note that $C_G(P)$ is a subgroup of $C_G(Q)$; so that uniqueness of the Sylow $Z(Q)$ of $C_G(Q)$ implies uniqueness of a Sylow group X of $C_G(P)$—and X lies in $Z(Q) \leq Q \leq P$, so X must in fact be $Z(P)$.

Dwyer obtained:

PROPOSITION 5.7.3 (1.19 in [**51**]). *The collection of p-centric subgroups of G is ample.*

PROOF. The proof in [**51**, 8.3] actually establishes ampleness for slightly more general collections: in fact for any collection $\mathcal{C}$ containing the p-centric subgroups, and also satisfying the closure-under-supergroups property given for the centric subgroups in Remark 5.7.2: namely that if $P \supseteq Q$ are p-subgroups of G and $Q \in \mathcal{C}$, then also $P \in \mathcal{C}$. The proof begins with the collection of all p-subgroups, and works by downward induction—removing one G-conjugacy class of non-centric subgroups at a time. □

We will indicate some further results from the literature related to centric subgroups, later on in Section 5.11. For now, we mention that the viewpoint of Proposition 5.7.3 gives an alternative proof of a classical result:

COROLLARY 5.7.4 (Swan's theorem). *Suppose that G is a finite group with an abelian (and nontrivial) Sylow p-subgroup P. Then the inclusion $N_G(P) \to G$ is a mod p homology equivalence.*

PROOF. Since P is abelian, it centralizes any subgroup of itself; so there is only one G-conjugacy class of p-centric subgroups, namely that of P itself. Since the collection of p-centric subgroups is ample by Proposition 5.7.3, the normalizer decomposition for this collection in Theorem 5.6.4 involves only the restriction of group cohomology arising from the inclusion of $N_G(P)$ into G. □

REMARK 5.7.5. For a more conventional classical proof of 5.7.4, see for example Evens [**58**, Thm. 4.2.8]. We mention further that an easy application of the Lyndon–Hochschild–Serre spectral sequence (see for example [**22**, Sec. 3.5]) for the group extension $1 \to P \to N_G(P) \to N_G(P)/P \to 1$ shows that the cohomology of $N_G(P)$ (that is, the unique term in the decomposition in 5.7.4) can in fact be "locally" computed as $H^*(P, \mathbb{F}_p)^{N_G(P)/P}$—namely the invariants of $N_G(P)/P$ acting on $H^*(P, \mathbb{F}_p)$.

The abelian Sylow hypothesis of Corollary 5.7.4 in fact puts us in the situation of Example 4.6.9, involving the "degenerate" simplex given by a single point, so that the homology decomposition involves just one subgroup. In Section 7.6, we shall apply this in the case of the Janko group J_1, which is the only sporadic group with an abelian Sylow 2-subgroup.

5.8. Sharpness for the three decompositions

For a homology decomposition given by an ample collection, as in the three decompositions in the previous sections, there is a further desirable property which appears in many of the results in the literature. Again a unified analysis of such results was given by Dwyer in [**52**] and in the exposition in Section 8 of [**53**]; Dwyer gave the property the name *sharpness*. This property allows the homology decomposition to provide information about homology of BG in terms of direct sums (in the spirit of Webb's formula in Theorem 5.1.16).

Recall from Section 4.8 (and also our discussion around (5.2.3)) that a diagram X of transitive G-sets gives rise via the homotopy colimit of its Borel construction to a Bousfield–Kan homology spectral sequence; which by Proposition 4.8.14 can be identified from the E^2 page onward with an appropriate isotropy spectral sequence from Definition 2.6.10. In particular, this identification applies to homology decompositions in the sense of Definition 5.2.4; for example, take X to be one of the functors

$$\tilde{\alpha}_{\mathcal{C}}, \tilde{\beta}_{\mathcal{C}}, \tilde{\delta}_{\mathcal{C}}, \tilde{\delta}_{\Delta}$$

in the various decompositions described in the previous sections; and recall that the homotopy colimit in (5.2.3) is for the Borel-construction functor which is denoted by the same expression with the tilde removed. Results in the literature on the collapse of this spectral sequence for various of these decompositions led to the following general definition of Dwyer [**51**, 1.20] (see also [**52**, 1.8] and [**53**, 8.1]):

DEFINITION 5.8.1. A homology decomposition X in the sense of Definition 5.2.4 is *sharp* if the corresponding Bousfield–Kan homology spectral sequence (5.2.6) satisfies $E^2_{s,t} = 0$ for $s > 0$. In other words, a sharp decomposition X is one which satisfies

$$\underset{i \in \mathfrak{C}}{\underrightarrow{\operatorname{colim}}}_s \big(H^G_t\big(X(i); \mathbb{F}_p\big)\big) = 0 \tag{5.8.2}$$

for $s > 0$; so that the spectral sequence collapses, to give isomorphisms

$$E^2_{0,*} = \underset{i \in \mathfrak{C}}{\underrightarrow{\operatorname{colim}}}\big(H^G_*\big(X(i); \mathbb{F}_p\big)\big) \cong H_*(G, \mathbb{F}_p).$$

This implies that the E^1 page gives a complex whose homology is just the group homology. Remark 4.4.12 gives an alternative description of that E^1 page in terms of the homology of isotropy subgroups for the action of G on $\mathfrak{C}$. And Proposition 5.8.4 below shows that this latter formulation gives rise to alternating sum formulas for (co)homology such as Theorem 5.1.16 (though the indexing of the sum may be more complicated, when the decomposition is over a category which is more elaborate than just a simplicial complex).

Note that the corresponding spectral sequence in cohomology with E_2 page $E_2^{s,t} = \underset{i \in \mathfrak{C}}{\underleftarrow{\lim}}{}^s\big(H^t_G\big(X(i); \mathbb{F}_p\big)\big)$ is by Corollary 2.4.6 just the graded $\mathbb{F}_p$-vector space dual of the one in homology. So in the definition of sharpness, it does not matter whether we work with homology or cohomology. ♢

EXAMPLE 5.8.3. In the situation of Swan's Theorem 5.7.4, there is only one subgroup in the decomposition; we have the situation of Example 4.6.9, in which the homotopy colimit itself degenerates to just the single value $BN_G(P)$. In particular there is only a single $\underrightarrow{\operatorname{colim}}_0$ term, and hence no higher limits for $s > 0$—so that the Bousfield–Kan spectral sequence collapses to the fiber $s = 0$. So in this case, the normalizer decomposition for the p-centric collection is sharp.

Webb's celebrated decomposition formula in Theorem 5.1.16 can be phrased in the language of sharpness:

PROPOSITION 5.8.4. *Let Δ be an admissible G-simplicial complex. Suppose that Δ^P is contractible for all p-subgroups P of G (so that by Proposition 5.6.7, Δ is ample). Then the normalizer decomposition* (5.6.9) *for Δ is sharp.*

PROOF. This is proved in Webb [**123, 124**]; see Theorem 5.1.16 and Proposition 5.1.17, and the comments following them. Webb in fact demonstrated the collapse at the E_2-page of the isotropy spectral sequence for Δ. □

REMARK 5.8.5 (Flag-transitivity and normalizer sharpness). We continue the theme of Remark 5.6.11. Namely in the examples mentioned there, and particularly in our main results in Chapter 7, we will have flag-transitive action of G on a complex Δ, Furthermore we will even have normalizer sharpness, so that the collapse of the cohomology spectral sequence for that homotopy decomposition then produces a cohomology decomposition indexed by a simplex:

$$H^*(BG; \mathbb{F}_p) = \bigoplus_{\varnothing \neq J \subseteq I} (-1)^{|J|-1}\, H^*(BH_J; \mathbb{F}_p) \tag{5.8.6}$$

Dwyer was also able to phrase results from Jackowski and McClure [**76**] and Jackowski, McClure, and Oliver [**77**] in the language of sharpness; we will summarize a number of such sharpness statements in the results in the final three sections of the chapter. For a collection $\mathcal{C}$ (in comparison with a G-complex Δ in Proposition 5.8.4 above), the definitions relevant to the corresponding decompositions are:

DEFINITION 5.8.7. A collection $\mathcal{C}$ of subgroups is
(i) *centralizer sharp* if the centralizer decomposition of Theorem 5.4.4 is sharp,
(ii) *subgroup sharp* if the subgroup decomposition of Theorem 5.5.4 is sharp,
(iii) *normalizer sharp* if the normalizer decomposition of Theorem 5.6.4 is sharp.

For example we can express the original Brown–Quillen case of Webb's formula Theorem 5.1.16 in this language:

THEOREM 5.8.8. *The collection $\mathcal{S}_p(G)$ of nontrivial p-subgroups of G is normalizer sharp.*

PROOF. This follows from Theorem 5.1.3, applying Proposition 5.8.4 to the poset complex $\Delta\big(\mathcal{S}_p(G)\big)$. □

REMARK 5.8.9. We emphasize that in contrast to ampleness for a collection $\mathcal{C}$, which is determined just via the G-simplicial complex given by the complex $\Delta(\mathcal{C})$ of inclusion-chains, sharpness is a property of the particular choice of one of the three decompositions for $\mathcal{C}$; or more precisely, of the corresponding homotopy colimit of the Borel construction for $\tilde{\alpha}_{\mathcal{C}}$, $\tilde{\beta}_{\mathcal{C}}$, or $\tilde{\delta}_{\mathcal{C}}$. This might seem unexpected: for we have been emphasizing in results like Theorem 5.6.5 and Proposition 5.6.14 that these different functors lead to weakly equivalent Borel constructions, and so have the same G-equivariant homology—so that it might appear that the validity of (5.2.3) should be independent of which of these functors we choose for X. However, notice that the expression in (5.2.3) does refer explicitly to the underlying indexing category $\mathfrak{C}$, which differs for these three functors. Correspondingly the objects $i \in \mathfrak{C}$ and also the colimits in that expression will vary for different X, so that we do indeed get different spectral sequences. That is, the weak equivalence of the Borel constructions only guarantees that the spectral sequences converge to the same limit; not that the spectral sequences themselves should agree.

Thus for example in the case of $\mathcal{C} := \mathcal{A}_p(G)$, Dwyer [**53**, 8.8] indicates that though the normalizer and centralizer decompositions for $\mathcal{A}_p(G)$ are sharp, the subgroup decomposition for $\mathcal{A}_p(G)$ need not be sharp. Similarly in case $\mathcal{C} := \mathcal{B}_p(G)$, centralizer sharpness need not hold; for example, by Grodal [**66**, 1.2] centralizer sharpness for a p-group P can only hold when the cohomology $H^*(P, \mathbb{F}_p)$ of P is equal to the P-fixed points in the cohomology of the center, $H^*\big(Z(P), \mathbb{F}_p\big)^P$, which is not true for general p-groups P. ◇

REMARK 5.8.10. As a different aspect of the theme underlying the previous remark, we can fix one of the three decompositions, but consider sharpness for different collections $\mathcal{C}$ and $\mathcal{D}$ which are related by equivalence.

So here we also emphasize that sharpness, unlike ampleness as discussed in Lemma 5.6.8, need not be preserved by a G-homotopy equivalence with another collection—essentially because the isotropy spectral sequence is not a homotopy invariant (cf. Warning 2.6.4 on Leray spectral sequences in general). ◇

In fact there is no such problem in the particular case of the normalizer decomposition, because of the following result:[7]

THEOREM 5.8.11 (7.2 in Grodal [**66**]). *If $\Delta \to \Delta'$ is a G-map inducing a homotopy equivalence of simplicial complexes with admissible G-action, then Δ is normalizer sharp if and only if Δ' is normalizer sharp.* □

WARNING 5.8.12. However to continue the theme of Remark 5.8.10, centralizer sharpness and subgroup sharpness need *not* be preserved by a G-homotopy equivalence between collections $\mathcal{C}$ and $\mathcal{D}$ of subgroups.

For example, $\mathcal{S}_p(G)$ is G-homotopy equivalent to $\mathcal{A}_p(G)$ (as we saw at Theorem 5.1.6). And we mentioned above in Remark 5.8.9 that the centralizer decomposition of $\mathcal{A}_p(G)$ need not be sharp; whereas the centralizer decomposition of $\mathcal{S}_p(G)$ is sharp (cf. Proposition 5.9.1).

Sharpness is analyzed from the viewpoint of equivalences in a methodical way for many of the standard collections from the literature in Grodal and Smith [**67**]. ◇

Finally, we mention that various methods for establishing sharpness of a decomposition are described in Dwyer [**52**], in Sections 10–14 of Dwyer [**53**], and in Grodal [**66**]. They make extensive use of Bredon cohomology and transfer arguments. But we will simply quote the relevant results, rather than trying to develop the necessary background here.

5.9. The "standard" ample G-homotopy type defined by $\mathcal{S}_p(G)$

In this section, we briefly review some of the standard collections in the G-homotopy type determined by Brown's original collection $\mathcal{S}_p(G)$. These collections will in fact suffice for our later discussion of many of the sporadic groups in Chapter 7.

It is convenient to use here the language of the G-homotopy type; but we recall that by Lemma 5.6.8 and Theorem 5.8.11, we can for purposes of ampleness and normalizer sharpness consider the larger class determined by the relation from G-maps which induce weak equivalences.

We already discussed certain important members of this G-homotopy type: We saw in Theorems 5.1.6 and 5.1.9 that $\mathcal{A}_p(G)$ and $\mathcal{B}_p(G)$ are G-homotopy equivalent to $\mathcal{S}_p(G)$; and similarly in Theorem 5.1.10 that so is the building for G, when G is of Lie type in characteristic p.

We mention one further poset from the literature which lies in this G-homotopy type; namely the subposet $\mathcal{Z}_p(G)$ of $\mathcal{A}_p(G)$ defined by Benson in [**22**, Exercise, p. 231]: this consists of those $E \in \mathcal{A}_p(G)$ satisfying[8] $\Omega_1 O_p Z\big(C_G(E)\big) = E$. One method of establishing that the inclusion of $\mathcal{Z}_p(G)$ in $\mathcal{A}_p(G)$ is a G-homotopy equivalence follows the idea in 1.6 of Quillen's paper [**104**] (and subsequent literature): namely verifying that removal from $\mathcal{A}_p(G)$ of subgroups E satisfying $\Omega_1 O_p Z\big(C_G(E)\big) > E$ is a G-homotopy equivalence. Another method is indicated in Grodal and Smith [**67**, Sec. 3]—following the idea in 1.3 of Quillen [**104**]: namely verifying that the mapping $F\colon E \mapsto \Omega_1 O_p Z\big(C_G(E)\big)$ is a poset map on $\mathcal{A}_p(G)$

[7]This result seems implicit in the treatment of Dwyer; but it is convenient to quote an explicit statement in the treatment of Grodal.

[8]Recall for a p-group X that $\Omega_1(X)$ denotes the subgroup generated by elements of order p.

which "dominates" the identity mapping via the order relation $F \geq \mathrm{Id}_{\mathcal{A}_p(G)}$—that is, $F(E) \geq E = \mathrm{Id}_{\mathcal{A}_p(G)}(E)$ for all E.

We first consider ampleness: We observed earlier at Example 5.3.4 that $\mathcal{S}_p(G)$ is ample. Furthermore in Lemma 5.6.8 we saw that a weak equivalence of $\mathcal{S}_p(G)$ with another collection via a G-map preserves ampleness. *Thus the above collections—and indeed all other collections weakly equivalent to $\mathcal{S}_p(G)$ via a G-map—are ample.*

We next consider sharpness of the three standard decompositions of the previous section, as applied to these collections. We begin by recalling some "classical" results on sharpness in the literature:

We observed earlier in Theorem 5.8.8 that $\mathcal{S}_p(G)$ is normalizer-sharp. In view of Theorem 5.8.11, *we also obtain normalizer sharpness for all other collections weakly equivalent to $\mathcal{S}_p(G)$ via a G-map,* and in particular for those indicated above.

The further methods of Dwyer and Grodal established a number of results on sharpness for some of the other decompositions for these collections. We have also indicated at Remark 5.8.9 that some cases of those decompositions need not in general be sharp.

However, it does not seem that the question of sharpness of the decompositions had been considered methodically for *all* these collections until recently: Namely during the preparation of this work, conversations of the second author with Grodal led to the methods in [**67**, Thm. 1.2]—which in particular answered the one question that had remained in the table below, as well as a similar question indicated later at Proposition 5.10.2. The table in the following Proposition is part of a larger table in [**67**].

PROPOSITION 5.9.1. *The table below summarizes the sharpness properties of the three decompositions, for the indicated ample collections in the G-homotopy type of $\mathcal{S}_p(G)$: Positive results are indicated by a reference to the original proof in the literature; "No" means "fails for at least some finite G".*

$\mathcal{C}$	*normalizer-sharp?*	*centralizer-sharp?*	*subgroup-sharp?*
$\mathcal{S}_p(G)$	[**124**, 2.5.1]	[**52**, 7.3]	[**52**, 7.4]
$\mathcal{B}_p(G)$	[**124**, 2.5.1]	*No*	[**77**, 2.14] [**52**, 10.4]
$\mathcal{A}_p(G)$	[**124**, 2.5.1]	[**76**, 2.1]	*No*
$\mathcal{C}e_p(G)$	[**66**, p. 449]	*No*	[**52**, 10.3]
$\mathcal{Z}_p(G)$	[**22**, Exer., p. 231]	[**67**, Thm. 1.2]	*No*

Further for G of Lie type in characteristic p, the identification in Remark 5.1.11 of $\mathcal{B}_p(G)$ with the unipotent radicals $\mathcal{U}$ affording the building shows (from the row for $\mathcal{B}_p(G)$ in the table) that the building is normalizer- and subgroup-sharp (but not necessarily centralizer-sharp).

In particular, each collection is normalizer-sharp; so ampleness and normalizer sharpness are inherited by any collection or complex weakly equivalent via a G-map to one of these collections.

Note that the cohomology decomposition for Lie type groups in Example 5.1.18 now appears as the building case in the Proposition.

As a brief preview of Chapter 7, where we give corresponding results for certain complexes Δ given by 2-local geometries in sporadic groups G, we mention that Δ

is weakly equivalent to $\mathcal{S}_2(G)$ via a G-map for 11 of those groups, namely:

$$M_{11}, M_{22}, M_{23}, M_{24}, J_1, J_3, J_4, McL, Co_2, Th, Ly. \tag{5.9.2}$$

In fact the first 10 groups in this list form the sublist, for sporadic groups G and the prime $p = 2$, from Table I of Smith and Yoshiara [**116**]; the 11th case given by Ly, for which the relevant geometry Δ had not been described at the time of [**116**], is treated in Section 7.26 of the present work. Thus in particular, all those complexes Δ are ample and normalizer-sharp. So we see that the G-homotopy type of $\mathcal{S}_2(G)$ already affords an ample collection for nearly half the sporadic groups. (In the case of J_1, we saw at the end of Section 5.7 that the collection of conjugates just of the Sylow normalizer is ample.)

In fact, the earlier work of Ryba, Smith, and Yoshiara [**108**] had originally verified the contractibility hypothesis of Webb's normalizer-sharpness result in Theorem 5.1.16 for the sublist of 10 groups indicated above (again excluding the then-unstudied geometry for Ly). But as had already been observed by Quillen (see for example [**104**, (a),(b) in proof of 3.1]), that contractibility hypothesis usually also leads to a homotopy equivalence with $\mathcal{S}_2(G)$. This viewpoint led to the development of the equivalence methods in [**116**], which in fact show homotopy equivalence of Δ in the above sublist with either $\mathcal{A}_2(G)$ or $\mathcal{B}_2(G)$. These equivalences proceed by removing one G-orbit at a time, and so are given by G-maps. Indeed via the observations of Thévenaz and Webb indicated in Remark 5.1.12, these even extend without difficulty to G-homotopy equivalences. We establish a corresponding notational convention:

NOTATION 5.9.3. In our later sections of Chapter 7 which treat the individual sporadic groups G, we will write $\mathcal{A}_2^-$ or $\mathcal{B}_2^-$ for a subposet of $\mathcal{A}_2(G)$ or $\mathcal{B}_2(G)$ which exhibits a G-map inducing a weak equivalence with a suitable 2-local geometry Δ for G. (See for example $G = M_{24}$ in Section 7.5, and $G = J_1$ in Section 7.6.)

We mention that in the group theory literature, for most of the larger sporadic groups, seemingly $\mathcal{A}_2(G)$ has not yet been explicitly determined.

However, recent work by various authors, especially Yoshiara in [**131**], has completed the explicit determination of $\mathcal{B}_2(G)$ for all sporadic groups, and we definitely use this information. In particular, the relevant literature citations for each group are summarized in Table 3 of [**131**]. From that source, we will indicate the relevant reference for $\mathcal{B}_2(G)$ for each group in the corresponding sections of Chapter 7.

In the remaining sections of the chapter, we will consider homotopy types other than that of $\mathcal{S}_p(G)$. Indeed it was already clear from the work of [**108**] that posets *not* homotopy equivalent to $\mathcal{S}_p(G)$ would be needed for small cohomology decompositions from local geometries for simple groups: since for those sporadic groups which were not listed in [**108**, Table 1], the then-known 2-local geometries Δ were shown (as mentioned in that paper) to be inequivalent to $\mathcal{S}_2(G)$—by the failure of Brown's congruence (5.1.4). In fact we use that elementary computation where relevant throughout Chapter 7 of the present work, in the form of the following remark:

REMARK 5.9.4 (Inequivalence via Euler characteristic). For an admissible G-complex Δ, if the reduced Euler characteristic $\tilde{\chi}(\Delta)$ is not divisible by the full p-part $|G|_p$ of the group order, then Δ lies outside the homotopy type of $\mathcal{S}_p(G)$.

Thus in order to confirm the suspicion indicated in [**116**, Conj. 1, p. 376] that the 2-local geometries Δ for the 15 sporadic groups not listed in (5.9.2) should nonetheless be ample, it is necessary to consider collections which can lie outside that standard ample G-homotopy type. We describe some such collections in the next two sections.

5.10. Ample (but potentially inequivalent) subcollections of $\mathcal{A}_p(G)$

In analyzing their proof of the properties of the centralizer decomposition for $\mathcal{A}_p(G)$ in Theorem 5.1.19, Jackowski and McClure produced in [**76**, 7.1] a list of conditions on a subcollection $\mathcal{E}$ of $\mathcal{A}_p(G)$ sufficient to guarantee the same properties for $\mathcal{E}$. In particular, the centralizer decomposition for $\mathcal{E}$ is then sharp, and hence $\mathcal{E}$ is ample; but it is not true in general that the inclusion $\mathcal{E} \to \mathcal{A}_p(G)$ is a homotopy equivalence. Furthermore, just as for $\mathcal{A}_p(G)$ in Remark 5.8.9, the subgroup decomposition for $\mathcal{E}$ need not be sharp. But the relevant part of Grodal and Smith [**67**, Thm. 1.2] shows at least that centralizer and normalizer sharpness hold for the form of $\mathcal{E}$ given by $\mathcal{E}_p(G)$ in Definition 5.10.1 below—a fact we will use in our relevant later cases of sporadic groups in Chapter 7.

In studying the mod 2 cohomology of the sporadic Conway group Co_3, Benson in [**24**, Sec. 3] axiomatized the following useful special case of the hypotheses of Jackowski and McClure:

DEFINITION 5.10.1 (The collection $\mathcal{E}_p(G)$). Assume that p divides the order of G. Let $\mathcal{E}_0$ denote the smallest subset of elements of G which

(i) contains the elements of order p in the center of a Sylow p-group P of G;

(ii) is closed under conjugation in G; and

(iii) is closed under taking products of commuting elements.

Let $\mathcal{E}_p(G)$ denote the subcollection of $\mathcal{A}_p(G)$ defined by subgroups all of whose nontrivial elements lie in $\mathcal{E}_0$.

We stress that the closure conditions in (ii) and (iii) are very strong; so it is often the case that $\mathcal{E}_0$ must contain all elements of order p, so that $\mathcal{E}_p(G)$ is then all of $\mathcal{A}_p(G)$. Of course in this section, we are primarily interested in the case when $\mathcal{E}_p(G)$ is *not* homotopy equivalent to $\mathcal{A}_p(G)$, and so in particular is a proper subcollection.

PROPOSITION 5.10.2. *$\mathcal{E}_p(G)$ is ample, and centralizer- and normalizer-sharp. Hence ampleness and normalizer sharpness hold for any Δ weakly equivalent to $\mathcal{E}_p(G)$ via a G-map.*

PROOF. Benson [**24**] observed that $\mathcal{E}_p(G)$ satisfies the requirements of Jackowski and McClure [**76**, 7.1], giving centralizer sharpness and in particular ampleness. And as mentioned in the first paragraph of the section, sharpness for the normalizer decomposition is part of [**67**, Thm. 1.2]. However, the subgroup decomposition of $\mathcal{E}_p(G)$ need not be sharp. The statements for Δ follow from Theorem 5.8.11. □

In subsequent analysis of sporadic groups beyond Co_3, as we will see in the corresponding sections of Chapter 7, this method of Benson shows that 6 of the sporadic groups, namely

$$M_{12}, J_2, HS, Suz, Co_3, Ru \tag{5.10.3}$$

have a 2-local geometry Δ which is homotopy equivalent via a G-map to $\mathcal{E}_2(G)$, which in turn is not homotopy equivalent to $\mathcal{A}_2(G)$. Correspondingly we establish the following notational convention:

NOTATION 5.10.4. We let $\mathcal{E}_2^-$ denote a proper subposet of $\mathcal{E}_2(G)$ which is homotopy equivalent via a G-map to a suitable 2-local geometry Δ for G. (See for example $G = M_{12}$ in Section 7.2.)

We also mention that in the cases in (5.10.3), and indeed fairly often for simple groups in general, there is just one conjugacy class of involutions which is represented in the center of a Sylow 2-subgroup; so this class provides a natural candidate for a minimal set $\mathcal{E}_0$ in Definition 5.10.1 above. Furthermore there is then a standard way of checking whether the involutions in $\mathcal{E}_0$ lie only in that single 2-central class: For given representatives t_1, t_2 and t_3 of conjugacy classes of involutions in G, we may use the irreducible characters $\mathrm{Irr}(G)$ (from the published character tables of known groups G) to compute the corresponding class multiplication coefficient

$$\xi(t_1,t_2,t_3) := \sum_{\chi_i \in Irr(G)} \frac{|G|}{\chi_i(1)} \frac{\chi_i(t_1)\chi_i(t_2)\chi_i(t_3)}{|C_G(t_1)||C_G(t_2)||C_G(t_3)|}, \tag{5.10.5}$$

which gives the (suitably weighted) number of conjugacy classes in G of fours-groups whose three nontrivial elements are conjugate to t_1, t_2 and t_3. (This value is of course just the structure constant for the multiplication of conjugacy-class sums in the center of the group algebra.) In the case where there is a unique 2-central class of involutions in G represented say by t, and for all involutions s not conjugate to t we have $\xi(t,t,s) = 0$, we see that any fours-group containing two conjugates of t must in fact have all three involutions conjugate to t. This is the desired property (iii) in 5.10.1 of closure under commuting conjugates; so that the involutions in $\mathcal{E}_0$ are just the G-conjugates of t, and then $\mathcal{E}_2(G)$ consists of the elementary abelian 2-subgroups all of whose involutions lie in that class. Of course the group theory languages GAP [**68**] and MAGMA contain character tables for many groups such as the sporadic groups; and the sum in (5.10.5) can be evaluated using built-in functions such as ClassMultiplicationCoefficient in GAP.

5.11. Ample (but potentially inequivalent) subcollections of $\mathcal{B}_p(G)$

In [**52**, 10.4], Dwyer established subgroup sharpness (and hence ampleness) for the collection obtained by removing from $\mathcal{S}_p(G)$ first the subgroups which are not p-centric, and then those which are not p-radical; thus he obtained the following subcollection of $\mathcal{B}_p(G)$:

DEFINITION 5.11.1. $\mathcal{B}_p^{cen}(G)$ denotes the collection of nontrivial p-subgroups which are both p-radical and p-centric.

We mention that in some groups, it happens that all members of $\mathcal{B}_p(G)$ are in fact p-centric; so that $\mathcal{B}_p^{cen}(G)$ is then all of $\mathcal{B}_p(G)$. Of course in this section, we are primarily interested in the case when $\mathcal{B}_p^{cen}(G)$ is *not* homotopy equivalent to $\mathcal{B}_p(G)$, and in particular is a proper subcollection.

Subsequently Grodal was led to consider a further subcollection $\mathcal{D}_p(G)$ of $\mathcal{B}_p^{cen}(G)$, which he called the *principal* p-radical subgroups: namely those p-centric p-subgroups H satisfying the condition $O_p\big(N_G(H)/HC_G(H)\big) = 1$, which is stronger than the condition $O_p\big(N_G(H)/H\big) = 1$ defining $\mathcal{B}_p(G)$, so that these subgroups are in particular also p-radical. Grodal's result [**66**, 9.1] shows that $\mathcal{D}_p(G)$ is both subgroup-sharp and normalizer-sharp. Furthermore his results [**66**, 1.2, 7.3] show that subgroup-sharpness and normalizer-sharpness are inherited by suitable collections lying between $\mathcal{D}_p(G)$ and $\mathcal{B}_p(G)$, including $\mathcal{B}_p(G)$ itself and also $\mathcal{B}_p^{cen}(G)$. In particular, we have the following proposition:

PROPOSITION 5.11.2 (see 1.2 and 7.3 in [**66**]). *$\mathcal{B}_p^{cen}(G)$ is ample, and subgroup- and normalizer-sharp. However, these collections need not be centralizer-sharp. Ampleness and normalizer sharpness hold for any Δ weakly equivalent to $\mathcal{B}_p^{cen}(G)$ via a G-map.*

As before, the statements for Δ follow from Theorem 5.8.11.

We focus attention in this section on $\mathcal{B}_p^{cen}(G)$. It turns out in our sporadic examples in Chapter 7 that Grodal's subcollection $\mathcal{D}_2(G)$ at $p = 2$ is in fact usually just equal to $\mathcal{B}_2^{cen}(G)$, so we do not explicitly mention $\mathcal{D}_2(G)$.

Indeed to summarize our applications to sporadic groups in Chapter 7: The homotopy equivalence methods of Smith and Yoshiara in [**116**], though originally developed to establish equivalence of a geometry Δ with $\mathcal{A}_2(G)$ or $\mathcal{B}_2(G)$, can also be applied (see the discussion in Section 4 of that paper) to obtain equivalence of 2-local geometries with $\mathcal{B}_2^{cen}(G)$ via a G-map—often affording a smaller ample collection not equivalent to $\mathcal{B}_2(G)$. The suspicion indicated in [**116**, p. 377]) that the normalizer decomposition of $\mathcal{B}_p^{cen}(G)$ should be sharp was confirmed by the subsequent work of Grodal [**66**] mentioned above. In overview, it turns out in Chapter 7 that 13 of the sporadic groups, namely

$$M_{12}, J_2, Suz, Co_1, Fi_{22}, Fi_{23}, Fi'_{24}, HN, B, M, He, Ru, O'N \tag{5.11.3}$$

exhibit a 2-local geometry Δ which is homotopy equivalent via a G-map to $\mathcal{B}_2^{cen}(G)$, which in turn is not G-homotopy equivalent to $\mathcal{B}_2(G)$. (Indeed we also mention that our treatment of J_1 in Section 7.6 via Swan's theorem, when regarded as a "decomposition" over just one subgroup, provides a further case of the use of $\mathcal{B}_2^{cen}(G)$.) We establish the following notational convention.

NOTATION 5.11.4. We let $\mathcal{B}_2^{cen-}$ denote a proper subposet of $\mathcal{B}_2^{cen}(G)$ which is homotopy equivalent via a G-map to a suitable 2-local geometry Δ for G. (See for example $G = Co_1$ in Section 7.15.)

The four groups M_{12}, J_2, Suz, and Ru in (5.11.3) also appeared in our earlier list (5.10.3), since they can be approached via subposets of $\mathcal{E}_2(G)$ as well as by subposets of $\mathcal{B}_2^{cen}(G)$. Thus combining those two lists with (5.9.2), we cover all the $11 + 6 + 13 - 4 = 26$ sporadic groups.

We mention that recent work of Maginnis and Onofrei ([**87**] and subsequent preprints) studies further interrelations among the above posets for a given G.

CHAPTER 6

2-local geometries for simple groups

In this chapter, we review some concepts from the group-theoretic literature on geometries for simple groups; and in particular, on the 2-local geometries for the sporadic groups. Much of the material could be applied to any group;[1] but so far the structures seem to be of interest mainly for simple groups.

Here is a summary of how these geometries fit into our overall treatment.

The literature on group geometries has unfortunately developed in a rather ad-hoc way. So we will begin with a rather vague definition of the geometries of interest; this will at least establish some basic notation that we can build further on.

DEFINITION 6.0.1 (An initial notion of 2-local geometry). A *2-local geometry* Δ for G over I is a simplicial G-complex Δ, where the complex is defined using suitable maximal 2-local subgroups H_i of G, and i varies over some indexing set I.

Several standard ways of constructing such a complex Δ from subgroups of G will be described in Section 6.3. (Of course in order to then apply the topological constructions described in previous chapters, we will need to pass at least implicitly from the simplicial complex to a suitable related simplicial set, such as that in Example 3.1.3.)

In the following Chapter 7 containing our main results, we will construct a particular such geometry Δ for each sporadic group G. We will then show that Δ is homotopy equivalent via a G-map to the complex of inclusion-chains in one of the usual posets of 2-subgroups of G, for example $\mathcal{S}_2(G)$, discussed in the final sections of Chapter 5—where in particular that poset affords an ample and normalizer-sharp collection in the language used there. It then follows from Lemma 5.6.8 and Theorem 5.8.11 that Δ is also ample and normalizer-sharp.

Furthermore Δ will be constructed essentially from the conjugates of the 2-local subgroups H_i given as the input in Definition 6.0.1. We will show in each case that the quotient Δ/G is a single simplex. Then it follows as in Remark 5.6.11 that the normalizer decomposition for Δ takes the simple form (5.6.12): namely we obtain an expression for BG as the homotopy colimit of a diagram of classifying spaces BH_J over nonempty subsets $J \subseteq I$, where $H_J = \bigcap_{i \in J} H_i$. Thus the homotopy colimit is over a pushout n-cube diagram for the relevant dimension n, as described in Section 4.6.

For a fuller introduction to the material on group geometries in this chapter, we refer the reader to several more extended expositions: for example, that given in Chapter 2 of the work [**115**] in preparation by the second author—especially

[1] And, we also emphasize, to p-local subgroups for any prime p: indeed there is work in the literature on sporadic geometries for odd primes. However, our interest in this work in applications to mod 2 group cohomology is behind our focus on geometries determined by 2-local subgroups.

Section 2.3, which explores a number of examples in detail; and also to Ivanov [**75**], which treats a number of the geometries of interest here. We also mention some earlier expository references on the area, such as [**113**] and [**114**].

Overview of some high points in the chapter. Here is an outline of our more limited treatment in this chapter.

The first section gives a brief sketch of some of the history of geometries for simple groups.

The second section then demonstrates, by means of two particular examples, the notion of a geometry given by suitable local subgroups: first, the building of a Lie type group as given by parabolic subgroups; and then a "similar" complex arising from maximal 2-local subgroups in the sporadic group M_{24}.

These structures lead naturally in the third section to several general ways of constructing for a group G a 2-local geometry in the sense of Definition 6.0.1—namely a simplicial complex based on suitable 2-local subgroups $H_i (i \in I)$ of G. Most of these methods can be regarded as variants of the standard method given as

- the 2-local complex $\Delta_{\mathcal{H}_I}$ of Definition 6.3.5;

though in fact we view essentially all our examples of geometries as special cases of a more general construction $\Delta_{\mathcal{A}_I}$ in Definition 6.3.2 which we call an *intersection complex*. We then discuss further examples of these various constructions for several specific groups.

The fourth section discusses the standard condition of

- flag-transitive action in Definition 6.4.2

by a group on a simplicial complex, giving a single simplex as the quotient—which we saw at Remark 5.6.11 leads to pushout hypercube diagrams for our homotopy colimits.

The fifth section describes some standard methods which we use for establishing

- homotopy equivalences for 2-local geometries,

especially with suitable subgroup collections described in Chapter 5.

The final section contains a few general remarks about certain other features of 2-local geometries.

6.1. Some history of geometries for simple groups

The reader who wishes to skip this historical section may safely go straight to the introductory Section 6.2; or even on to Section 6.3, where the general constructions of our 2-local geometries are developed.

Lie type groups and their Tits buildings. Geometry, in the sense of algebraic geometry, was the essential tool in the classification during the 1950s of the semisimple algebraic groups, defined over algebraically closed fields; from the group, one could pass to an associated root system and Weyl group, with their corresponding Lie theory and classification. A by-product of that effort was the realization that "most" of the finite simple groups could be understood in a unified way as Lie type groups over finite fields—exhibited as fixed points of suitable automorphisms of the infinite algebraic groups. In particular, the groups of Lie type over finite fields came to be viewed as the "generic" examples of finite simple groups.

But algebraic geometry is not applicable to these finite groups, since basic techniques such as the Nullstellensatz are not available for finite fields. Instead, it seemed reasonable to hope for a unified description of the groups of Lie type by means of *finite* (i.e., projective) geometry—especially since the standard matrix groups (linear, unitary, orthogonal, symmetric) were classically understood in terms of the projective geometry of the vector spaces on which their action is defined (cf. our later discussion of Example 6.4.3.) And indeed an appropriate notion of a geometry applicable to all groups of Lie type was provided during the 1960s by Tits' theory of *buildings* (see for example [**119**, Secs. 3 and 5]).

We now sketch only briefly how buildings provided the background for some later geometric developments. We won't state the detailed building axioms; instead we summarize by saying that for a fixed Weyl group W, a building of type W is a simplicial complex which is built up, in a suitably "convex" way, from copies of a subcomplex naturally associated to W called its Coxeter complex. (The *rank* of the building is that of W indicated in Example 5.1.1.) We can roughly summarize Tits' fundamental construction [**119**, Thm. 5.2] (specialized to the finite subcase) in the form:

THEOREM 6.1.1. *If G is a finite group of Lie type (with Weyl group W), then the parabolic subgroups of G determine a finite building (of type W).*

It is perhaps even more wonderful that Tits was able to establish a suitable converse; we summarize [**119**, Thm. 11.4] in the form:

THEOREM 6.1.2. *If Δ is a finite thick*[2] *building of type W, and W has rank at least 3, then Δ is the building of a group G of Lie type over a finite field (with Weyl group W).*

Indeed there are also converse statements for W of rank 1 or 2; but further hypotheses are then required.

Geometries for other simple groups. Tits' work soon inspired various efforts to obtain at least a partial understanding, from a geometric viewpoint, of simple groups which are not of Lie type—especially the sporadic groups. The reader is referred to Chapter 2 of [**115**] for a more expanded discussion than the brief historical outline given below.

How might we generalize from buildings? Seemingly not so directly: for note that we can interpret Theorem 6.1.2 as saying that we can *not* expect the building axioms, notably the detailed properties of apartments, to be helpful for the study of simple groups which are not of Lie type. (And indeed this is the reason that we did not even state those axioms.)

One might further explain this limitation by saying that the building axioms are "global"—in that a pair of apartments at any arbitrary distance in the geometry must be connected via paths related to W. In a group of Lie type, this restricts the possible relationship between any two parabolic subgroups. By contrast, one can instead consider properties of buildings which are just "local": this amounts to considering just one apartment and its immediate neighbors; or in a Lie type group, to considering just the parabolics over a single Borel subgroup.

[2]Here, "thick" is the technical condition that at least three maximal simplices lie above each submaximal simplex.

And this roughly describes the approach of Tits in his influential later "local approach" paper [**120**]: geometrically, the local information at some simplex in the building is given by the structure of its *link* (or *residue* in the language of Tits). This viewpoint strongly emphasizes an inductive feature of buildings, already visible in earlier results such as [**119**, 3.12]; namely these links are also buildings, with Weyl groups given by subgroups of W—and indeed they correspond to subdiagrams of the Dynkin diagram for W. Thus the Dynkin diagram in fact encodes certain axioms for the strictly local structure of the building. And the base case for this inductive viewpoint is given by the buildings of rank two, namely the *generalized polygons*. Furthermore the weaker hypotheses corresponding to just this local information allowed the buildings to be viewed within a wider class of geometries described only locally (within which Tits now characterized the buildings by means of suitable notions related to simple connectivity).

This wider context soon opened the door to searching for geometries for sporadic groups which would give some kind of weak analogue of 6.1.1. (It still seems visionary to hope for any converse like 6.1.2.) In pioneering work in the mid 1970s such as [**37**], Buekenhout observed that, by also allowing in the base case certain rank two geometries[3] other than generalized polygons, one could inductively obtain axioms describing simplicial complexes which he called *diagram geometries*, since they have "Dynkin-like" diagrams.[4] These diagrams turned out to axiomatize natural-seeming geometries for many other groups, notably simple groups such as sporadic groups; and often such geometries exhibit analogies with some of the other properties of buildings, such as those mentioned in Example 5.1.1. (We will further examine the general viewpoint of diagrams for geometries later in the chapter, notably at Digression 6.2.14.)

In 1979, Ronan and Smith in [**106**] focused more specifically on the subclass of diagram geometries which they called 2-*local geometries*: since these were obtained (in a fairly ad-hoc way) by studying the configuration of suitable 2-local subgroups in a group, especially in sporadic simple groups. This 2-local viewpoint, of which we have indicated a preliminary version at Definition 6.0.1, especially emphasized analogies with groups of Lie type and their buildings—since the building is afforded as in Example 5.1.1 by the poset $\mathcal{P}$ of parabolic subgroups, which are p-local subgroups for the characteristic prime p, so that the building is a p-local geometry in the obvious extension to p of the terminology in that Definition.

Further developments. Since the early 1980s, there has been much further work emphasizing this type of geometric viewpoint on simple groups, by many authors; we will not here attempt to further survey that substantial literature. However, the confluence of that literature on groups and geometries with the literature in algebraic topology on complexes of subgroups provides a major theme in [**115**]; and many further references can be found there. We mention here just a few, which are especially relevant to the present work.

Webb's expository article [**123**], also mentioned earlier in our historical summary of decompositions in Section 5.1, exerted an important influence not only

[3]See for example the rank-2 truncation of projective 3-space in the Ronan–Smith geometry given in Digression 6.2.14.

[4]These diagrams axiomatizing geometries should not be confused with the topological use of the term diagram to indicate a functor defined on an indexing category as in Definition 4.5.1.

for researchers in group cohomology, but also for researchers in groups and geometries. For example, Webb explicitly included geometries for certain sporadic groups as analogues of buildings, in the context of the work of Brown and Quillen summarized in Section 5.1. In particular, Webb indicated relations with the usual collections of p-subgroups such as $\mathcal{S}_p(G)$ described in the later sections of Chapter 5; and notably at [**123**, p. 357], he gave corresponding results on projectivity of reduced Lefschetz modules as in Theorem 5.1.14. Sporadic geometries have also been used in the explicit calculation of the ring structure of group cohomology in a number of papers in the literature, especially the extensive program beginning with the work of Adem, Maginnis and Milgram in [**3**] for M_{12}; and also in work on Co_3 by Benson in [**24**].

Webb's discussion of projective modules in [**123**] suggested the more methodical study of 2-local geometries from the viewpoint of projective modules in [**108**]; and that paper in turn led to the development of the homotopy equivalence methods in Smith and Yoshiara [**116**], and in some other papers such as Sawabe [**111**]. Such methods will be used frequently in our work on individual sporadic groups in Chapter 7.

Indeed the present work in some sense completes a treatment, for geometries for *all* the sporadic groups, of at least some of the analogies with buildings observed in the earlier papers just indicated (especially flag-transitivity and the orbit-simplex aspect of decompositions indicated in Remark 5.6.11). So to complement the above historical discussion, we now provide a preview of a some of those analogies. Within the context of the particular properties of buildings indicated at Example 5.1.1, here we are primarily motivated by those related to the topological and cohomological constructions considered in Chapter 5. So now we will briefly indicate some relevant features of our analysis of sporadic groups in Chapter 7, just in the case of M_{24}—which we will use as a running example in much of the present chapter, starting at Example 6.2.10. We will see (in Section 7.5) that the 2-local geometry $\Delta_{\mathcal{H}_I}$ of $G = M_{24}$ in 6.2.10 is homotopy equivalent to $\mathcal{S}_2(G)$ via a G-map—in analogy with Quillen's equivalence in Theorem 5.1.10. As a consequence, we obtain for example the projectivity of the reduced Lefschetz module $\tilde{L}(\Delta_{\mathcal{H}_I}; \mathbb{F}_2)$—in analogy with Theorem 5.1.14, which holds for the building in view of Quillen's equivalence. Furthermore $\Delta_{\mathcal{H}_I}$ is ample and normalizer-sharp, in Dwyer's language in 5.3.3 and 5.8.7—in analogy with these properties of the building indicated in Proposition 5.9.1. We then get for example Webb's formula over $\Delta_{\mathcal{H}_I}/G = \mathcal{H}_I$, in analogy with Example 5.1.18 for the building. We mention in particular that the 2-local complex $\Delta_{\mathcal{H}_I}$ gives the *smallest* known ample complex for M_{24}—it has dimension two, whereas the usual collections such as $\mathcal{B}_2(G)$ in the G-homotopy type of $\mathcal{S}_2(G)$ in Section 5.9 give complexes of dimension at least three.

6.2. Preview: some initial examples of local geometries

This "optional reading" section is primarily intended to indicate some natural motivation, via particular examples, for the notion of a simplicial complex defined by suitable local subgroups—the notion has so far been only briefly suggested in Definition 6.0.1, and during our historical overview in Section 6.1.

The discussion here is fairly informal; and ideally will be helpful background for the reader who has never been exposed to the overall viewpoint. But it is not required for the general constructions of such geometries in Section 6.3; the more impatient reader can skip directly to the more formal treatment there.

The foundation for the section is laid by examining certain "local" aspects of the building. After that, we indicate a simplicial complex for the Mathieu group M_{24}, which exhibits certain analogies with the building. We then conclude the section by abstracting from this case of M_{24} some features to be used in defining local geometries for the other sporadic groups.

The building as a complex on maximal parabolics. We begin with the model case afforded by the building, expanding a bit on our mention in Section 6.1 of the local approach of Tits in [**120**].

So we now assume that G is a group of Lie type in characteristic p, with Weyl group W of rank n; the reader may wish to review earlier Example 5.1.1 as background for the discussion that follows. Then it is standard (see for example the proof of Proposition 6.4.4) that for an indexing set $I := \{1, 2, \cdots, n\}$ of size n, we have distinct maximal parabolics H_i $(i \in I)$, with further parabolics $H_J = \bigcap_{i \in J} H_i$ determined by nonempty subsets $J \subseteq I$, where H_I is a Borel subgroup. (We refer to H_J and its conjugates as being of *type* J.)

DIGRESSION 6.2.1 (Desirability of a more efficient representation). Our initial view of the building for G in Example 5.1.1 was given via the complex $\Delta(\mathcal{P})$ (as in Definition 2.7.5) of inclusion-chains in the poset $\mathcal{P}$ of parabolic subgroups. We'll try to demonstrate why and how we might get a better view.

(1) One important eventual motivation is to obtain a representation of the geometry which provides a particularly natural context for flag-transitivity; but this aspect will remain at least partly in the background until our later treatment in Section 6.4. Flag-transitivity will be a feature of interest for our geometries for sporadic groups; but it cannot usually be expected for the standard ample collections such as $\mathcal{A}_2(G)$ and $\mathcal{B}_2(G)$ discussed in Chapter 5.

(2) A more immediate and elementary motivation for changing the view is that the complex in the form $\Delta(\mathcal{P})$ is computationally rather inefficient—indeed we will see that it arises as the barycentric subdivision of a more convenient smaller complex used in the local viewpoint of Tits. (The reader already content with this statement can proceed directly to that local view in Example 6.2.2 below.)

The possibility of a smaller complex may be particularly clear in the special case given by the classical matrix groups G: We work first at the level of posets: Take the projective space or polar space from the natural module V for G—cf. Example 6.4.3, and $Sp_4(2)$ in Example 6.4.7. Then the poset subdivision (Definition 2.7.7) in turn gives the larger poset of chains of relevant subspaces in V—whose stabilizers exhibit the poset $\mathcal{P}$ of parabolics. (So in this case, the normalizer set-map N of 5.6.13 induces a poset anti-isomorphism.) And then passing to the complex of chains in each poset gives an isomorphism of simplicial complexes. Indeed this geometric structure for the classical groups was an important inspiration for Tits' original geometric analysis of all Lie type groups, which led to the theory of buildings (approached via the complex $\Delta(\mathcal{P})$).

And we now turn to the possibility of a complex smaller than $\Delta(\mathcal{P})$ for a general Lie type group G in Tits' later local approach to buildings. We begin by reviewing what is involved in our standard computation: Recall that the form (5.6.9) of the normalizer decomposition for the complex $\Delta(\mathcal{P})$ involves a homotopy colimit indexed by the category for the quotient CW-complex $\Delta(\mathcal{P})/G$; here some readers may prefer the language of simplicial sets to that of complexes. So we examine that indexing category in more detail, by first describing the orbit poset $\mathcal{P}/G$. For it is standard (see Proposition 6.4.4) that representatives of $\mathcal{P}/G$ are given by the parabolics H_J over the fixed Borel subgroup H_I, so that $\mathcal{P}/G$ is indexed by the nonempty members of the power set 2^I. And then it follows that the complex $\Delta(\mathcal{P})/G$ arising from chains in $\mathcal{P}$ is indexed by subsets of 2^I, hence essentially by members of 2^{2^I}; so that our normalizer decomposition has around 2^{2^n} terms. This number is seriously unwieldy for practical computation.

Furthermore, the number should perhaps even seem unnecessary: For using the self-normalizing property of the parabolics in 5.1.1, namely $N_G(H_J) = H_J$, we see that the stabilizer of an inclusion-chain of parabolics is in fact given by the initial (i.e., smallest) member of that chain—so that only the $2^n - 1$ distinct H_J will actually occur in the normalizer decomposition, and of course each will then occur repeatedly. So it should seem desirable to be able to instead exhibit the building as a smaller simplicial complex, given essentially by the parabolics themselves—or at least, with simplices defined in terms of the maximal parabolics; see the distinction made in Remark 6.2.4. Then the orbit complex has stabilizers given by the H_J, each occurring only once. That is, we want a complex which after a barycentric subdivision gives the earlier viewpoint of the building $\Delta(\mathcal{P})$ regarded as inclusion-chains of parabolics. $\diamondsuit$

EXAMPLE 6.2.2 (The building as a p-local complex). We now essentially follow the local viewpoint of Tits in [**120**]. We define a complex Δ (i.e., an abstract simplicial complex in the sense of 2.7.3) as follows:

As vertices, we take the maximal parabolics themselves, namely the conjugates of H_i for $i \in I$. We may for clarity use the singleton-set notation $\{H_i\}$, to denote the maximal parabolic H_i viewed as a vertex. The G-action on the vertices is just given by conjugation: $g^{-1}\{H_i\}g := \{g^{-1}H_ig\}$ for $g \in G$.

Simplices of higher dimension are then defined as sets of non-conjugate(!) vertices which satisfy the following relation of mutual *incidence*:

$$\text{The maximal parabolics in a simplex contain a common Borel subgroup.} \quad (6.2.3)$$

That is, an edge is a pair of incident vertices; a 2-simplex is a triple of mutually incident vertices; and so on. Again we may use set braces to indicate these simplices.

Notice that the subset closure condition (2.7.4) for a complex is indeed satisfied, since this relation of mutual incidence, defined by containing a common Borel subgroup, is obviously inherited by any non-empty subset of a simplex.

Finally note that the subgroups H_J define a simplex $\mathcal{H}_I$ of subgroups of G as in Definition 4.6.2—namely a functor from subsets of I to subgroups of G. So since the conjugates of this subgroup simplex define our complex Δ, we see that the building Δ in this viewpoint affords an example (for general p) of a complex of locals as denoted by $\Delta_{\mathcal{H}_I}$ in our later Definition 6.3.5. (This starts to give some specificity to our initial rather vague notion in Definition 6.0.1.) $\diamondsuit$

REMARK 6.2.4 (An optimal feature: simplices realized via subgroups). We single out one special best-case aspect, which we can *not* expect to be always available in later generalizations: Note that the standard simplex $\sigma_J := \{H_i : i \in J\}$ of type J has stabilizer given by the standard parabolic $H_J = \cap_{i \in J} H_i$ of type J

over H_I. Now we had mentioned in Example 5.1.1 that in the present example given by the building, the parabolic H_J is self-normalizing;[5] and it follows that the G-conjugates of σ_J could just be identified with the conjugates of the subgroup H_J. That is, our simplices in our complex Δ could in effect be realized in terms of subgroups of G, using the poset $\mathcal{P}$ of parabolics (rather than just in terms of *sets* of subgroups of G).

In sporadic groups, by contrast, such an identification of simplices with subgroups (such as their stabilizers) would be appropriate only in some cases, where the analogues of the H_J are self-normalizing (as happens for example in M_{24}; see Remark 6.2.12). However in quite a few other sporadic cases (for example M_{11} in Section 7.1), we can get the situation where $H_J < N_G(H_J)$—so that there are $|N_G(H_J) : H_J| > 1$ different conjugates of the simplex σ_J which are stabilized by the same subgroup H_J.

We emphasize that this mild extra complication will not in fact be any obstruction in our main developments such as flag-transitivity in Section 6.4, and in such applications as our standard isomorphisms of complexes via Lemma 6.5.4.

On the other hand, there *will* be some corresponding complication when we want to compare the complexes with certain closely related posets, and here the situation will require some extra care: Namely we will review at (6.2.16) how our usual hypotheses lead to the property $N_G(\sigma_J) = H_J$; and hence when $H_J < N_G(H_J)$, to distinct conjugates of σ_J stabilized by H_J. And for that latter situation, we will develop Lemma 6.5.7 to describe the actual comparison via a suitable poset-quotient procedure, rather than via an isomorphism of complexes. Indeed we will study a number of such situations in fair detail: starting with Example 6.5.5 for Σ_5, and continuing in later applications to various sporadic geometries, such as that for M_{11} in Section 7.1, and for M_{22} in Section 7.3.

So this additional mild complication gives one important reason for the convention above of regarding simplices not as subgroups, but instead just as sets of vertices such as $\sigma_J = \{H_i : i \in J\}$. That is, we can essentially identify *vertices* with certain *maximal* local subgroups; but we will usually avoid identifying higher-dimensional simplices with proper intersections of those subgroups. $\diamondsuit$

REMARKS 6.2.5 (Some further features of the building as a local geometry). In the discussion in the next page or so, we will use the representation of the building as the complex Δ in Example 6.2.2 to focus on a number of particular properties, which will have natural analogues in our geometries for other simple groups.

First by the assumption of non-conjugacy above in a simplex (together with the fact that the H_i are indeed pairwise non-conjugate), we see that a simplex is of form $\{g_i^{-1}H_ig_i : i \in J\}$—for some nonempty subset $J \subseteq I$ (and suitable $g_i \in G$). Following Tits, we say the simplex has *type* J; that is, the type is defined by just recording which conjugacy classes occur in a simplex. Hence the types of simplices are indexed by members of $2^I \setminus \varnothing$. Furthermore, the types serve as indices for the members of the orbit complex Δ/G—in view of the standard

[5]As an alternative to the self-normalizing condition, one could instead analyze using the property that the unipotent radicals U_J are weakly closed (mentioned earlier in Example 5.1.1). In some sporadic groups, the unavailability of an identification of simplices with subgroups will correspond to failure of the analogous weak closure condition; in particular we will typically see this when the analogue of H_I fails to contain a Sylow 2-subgroup of G. This weakly-closed viewpoint can be useful further background, but is not essential, for our later development.

property of flag-transitivity of a Lie type group on its building, which we discuss at Proposition 6.4.4—and this is as desired for the simplex form (5.6.12) of the normalizer decomposition.

Next, we had mentioned that the parabolics are self-normalizing in G.[6] We especially single out the case of the maximal parabolics, where $N_G(H_i) = H_i$ implies that

$$\text{The stabilizer in } G \text{ of the vertex } \{H_i\} \text{ is given by } H_i. \qquad (6.2.6)$$

But we also had mentioned in Remark 6.2.4 that for larger index sets J, the simplex σ_J has stabilizer H_J. Indeed for a more general simplex $\{g_i^{-1}H_i g_i : i \in J\}$ of type J, any conjugate of the Borel subgroup H_I of G which is contained in each of the corresponding maximal parabolics $g_i^{-1}H_i g_i$ in fact determines a Borel subgroup in some Levi complement in each of those groups—where we regard the Levi complement as a group of Lie type in its own right. (See for example [**43**, 8.5.2] for the terminology of the Levi decomposition of a parabolic.) It then follows that the intersection of those maximal parabolics in the simplex of type J is in fact a G-conjugate of the parabolic H_J. Indeed this observation can be made inductively: for any nonempty subset $K \subseteq J$, the intersection of the $g_i^{-1}H_i g_i$ for $i \in K$ is a conjugate of H_K.

This has the further consequence that we could replace the earlier requirement for mutual incidence in a simplex of type J, given in (6.2.3) in terms of just the single subgroup H_I, by a condition based on all subsets of J:

CONDITION 6.2.7. Define incidence of vertices in a simplex σ of type J by requiring, for any nonempty $K \subseteq J$, that the conjugates from σ of the maximal parabolics H_i with $i \in K$ should intersect in a conjugate of the parabolic H_K.

Now this might seem to be an unnecessary complication—but in fact this inductive form will be more convenient in our later geometries for sporadic groups, such as Example 6.2.10 for M_{24} and the more general Definition 6.3.2. Indeed we will indicate at Remark 6.2.11 that in optimal cases, the more inductive condition will be no harder to verify than the analogue of (6.2.3); and that once verified, in some sporadic cases it is easier to then work with: for it has the technical advantage of specifying the exact structure of the intersection—as opposed to requiring that the intersection should be just *some* overgroup of H_I.

Next we reconsider the optimal feature of our complex in Remark 6.2.4, at the level of posets: Namely for the building, Condition 6.2.7 associates simplices of type J bijectively with conjugates of H_J—namely with members of the poset $\mathcal{P}$ of parabolics. If we regard a simplex of type J now as a member of the poset $\mathcal{Q}(\Delta)$ of simplices of the complex Δ and their inclusions as in Definition 2.7.1, then that association in fact affords an order-reversing isomorphism of posets[7] from $\mathcal{Q}(\Delta)$ to $\mathcal{P}$. This in turn induces an isomorphism of the corresponding poset complexes $\Delta(\mathcal{Q}(\Delta))$ and $\Delta(\mathcal{P})$ in the sense of 2.7.5. And now we recognize $\Delta(\mathcal{Q}(\Delta))$ as in 2.7.7 as the barycentric subdivision $\mathsf{Sd}(\Delta)$ of our original local complex Δ; justifying our earlier mention that Δ leads via subdivision to the usual view of the

[6]It is then standard that the permutation representation of G on the conjugates of the parabolics is equivalent to that on their cosets. So our complex of conjugates is in fact equivalent to that given by Tits in [**120**], where the complex is instead constructed from *cosets* of parabolics.

[7]This isomorphism is well-known; though the routine details might not be written down explicitly in the literature. But some observation of this nature implicitly underlies Tits' comparison of two notions of buildings in the introduction to his local approach paper [**120**].

building as inclusion-chains $\Delta(\mathcal{P})$. It follows that our construction Δ is weakly G-equivalent to $\Delta(\mathcal{P})$; and so as usual via Corollary 2.5.13, these two viewpoints afford weakly equivalent Borel constructions, and hence have weakly equivalent normalizer decompositions. ◇

An analogous complex defined on maximal 2-locals of M_{24}. We next use the sporadic Mathieu group M_{24} to demonstrate how suitable 2-local subgroups can be used in analogy with parabolic subgroups of a Lie type group—to construct a "2-local geometry" for M_{24} resembling the complex in Example 6.2.2, and exhibiting some features like those in Remarks 6.2.5.

In fact M_{24} was the first sporadic group considered by Ronan and Smith in the discussions which led to [**106**]; and such analogies inspired the construction of the particular geometry for M_{24} given at p. 284 of that work, as well as the geometries in that paper for other sporadic groups. We will also be using M_{24} to provide examples of a number of related constructions elsewhere in this chapter; and we will of course use it finally in obtaining our homotopy decomposition result for M_{24} in Section 7.5. In addition, we refer the interested reader to the more detailed treatment of this particular geometry in Smith [**115**, Ch. 2].

DIGRESSION 6.2.8 (A property of parabolic subgroups). Before starting our sporadic example, we briefly sketch one background technical feature, concerning just which local subgroups of a more general group we might regard as appropriate analogues of the parabolic subgroups in a Lie type group.

If G is a group of Lie type in characteristic p, then a parabolic subgroup P is in particular a p-local subgroup; indeed we saw at Example 5.1.1 that P is the normalizer in G of the p-subgroup given by its unipotent radical U. In fact, U affords the largest normal p-subgroup $O_p(P)$ of P; and in addition, P has no normal q-subgroup for any other prime q, nor does it have any nonabelian simple subgroup which is normal. The more precise formulation of this behavior (see for example [**64**, 2.6.5]) is customarily given in the now-standard language of Bender:

$$F^*(P) = O_p(P). \tag{6.2.9}$$

Here $F^*(P)$ denotes the *generalized Fitting subgroup* of P: which is the product of all normal q-subgroups for all primes q, and all subgroups which are quasisimple (that is, perfect, and simple modulo center) and subnormal (that is, appearing at the beginning of a chain of subgroups each normal in the next, ending with the whole group).

Local subgroups satisfying the condition (6.2.9) were exploited in many parts of the classification of the finite simple groups. Furthermore they seem to be the most natural subgroups for the purposes of geometric analysis of simple groups; and correspondingly we will focus on them in our sporadic examples to come.[8] ◇

Thus when we now consider $G := M_{24}$ for $p := 2$, we take (as our analogues of the maximal parabolics appearing in Example 6.2.2) just those maximal 2-local subgroups in G which satisfy (6.2.9). Using for example the Atlas [**47**, p. 96], we

[8]Note that some of the geometric literature, notably Ronan and Smith [**106**], instead uses the more classical terminology of 2-local subgroups which are 2-*constrained*. We will simply mention here that when $O_{p'}(P) = 1$ (which is the case for parabolic subgroups and typically in sporadic situations of interest for $p = 2$), the condition of p-constraint then does take the form (6.2.9).

see there are three G-conjugacy classes of such subgroups; we will denote representatives of these classes by H_O, H_T, and H_S. We will describe these groups in more detail shortly at Digression 6.2.13; but for now we will just mention that we choose these subscripts to indicate that these groups are the respective stabilizers in G of an *octad*, *trio*, and *sextet* in by the Steiner system $\mathcal{S}(5,8,24)$—see the Atlas [**47**, p. 94] for details of the terminology of the Steiner system, and see also our further discussion at later Example 6.3.11.

We define a corresponding index set $I := \{O, T, S\}$. We can in fact choose the H_i $(i \in I)$ so that the intersection H_I of all three is an extension of a Sylow 2-subgroup of G by a subgroup of order 3; and H_I will be our analogue of a Borel subgroup. Indeed we define the intersections for all nonempty $J \subseteq I$, namely $H_J = \bigcap_{i \in J} H_i$.

We can now proceed just as in Example 6.2.2: defining vertices of a complex via conjugates of the H_i; and we could define higher-dimensional simplices via incidence from containment of a common conjugate of H_I, just as in (6.2.3).

But it is also easy to verify the analogue of the stronger incidence condition 6.2.7. (Indeed we will detail the equivalence of the two notions of incidence later at 6.2.17 (1); roughly, the analogue works for M_{24} because the geometry is inductively built up from subgeometries obtained from buildings, as discussed for example in Digression 6.2.14.) So in fact we may as well use that more inductive relation to define incidence in our analogous complex:

EXAMPLE 6.2.10 (A complex of 2-locals for M_{24}). Assume the maximal subgroups H_i $(i \in I)$—and hence the intersections $H_J (J \subseteq I)$—described above. Then the standard 2-local geometry for M_{24} in [**106**, p. 284] can be realized as the following 2-local complex Δ:

Vertices are given by G-conjugates of the H_i; and a simplex of *type* J (having dimension $|J| - 1$) is given by a subset σ of vertices, one for each $i \in J$, which are mutually incident under the relation:

> For a simplex σ of type J, given any nonempty subset $K \subseteq J$, we require that the conjugates from σ of the H_i with $i \in K$ should intersect in a conjugate of H_K.

Just as in Example 6.2.2, we see that the incidence relation is inherited by subsets of a simplex; that is, (2.7.4) holds, so that Δ is indeed an abstract simplicial complex.

And again the intersections H_J define a subgroup simplex $\mathcal{H}_I$ as in Definition 4.6.2; where the conjugates of the H_i are used in defining simplices in our complex Δ, so that we again have an example of a complex of 2-locals $\Delta_{\mathcal{H}_I}$ as in later Definition 6.3.5. ◊

REMARK 6.2.11 (The inductive incidence, continued). We mentioned just above that we will see at 6.2.17 (1) that the present geometry for M_{24} is one of the best-behaved cases, in which we could in fact have just used the simpler incidence relation given by the analogue of (6.2.3)—although in those cases the more inductive form is really no harder to verify. However here and at the more general Definition 6.3.2, we will use the more inductive form of incidence. For a better example of how that form can be more useful, see for example the discussion of the incidence in the geometry for M_{22} in Section 7.3.

REMARK 6.2.12 (Realizing simplices via subgroups, continued). We follow up for a moment the theme of earlier Remark 6.2.4: In this particular geometry for M_{24}, just as was the case for the building in Example 6.2.2, the subgroups H_J are self-normalizing. (And again we could optionally analyze instead via the condition that the subgroups $O_2(H_J)$, denoted below by U_J, are weakly closed—as is not difficult to check.) So it would be possible to replace the standard simplex $\sigma_J := \{H_i : i \in J\}$ by the subgroup H_J. This would in effect realize the complex via actual subgroups of M_{24}, namely intersections of suitable conjugates of the H_i. However, for sporadic geometries where we can have $H_J < N_G(H_J)$ so that this realization via subgroups is not available, see Example 6.5.5, and Sections 7.1 and 7.3.

From this geometry for M_{24}, we obtain a number of properties similar to those for the building observed earlier in Remarks 6.2.5; we emphasize the following: The maximal 2-locals H_i here are in particular self-normalizing; so that our construction on conjugates of the H_i is equivalent to that on the cosets of the H_i. We also get the analogue of the condition (6.2.6) that the vertex stabilizers are the maximal 2-locals given by the conjugates of the H_i.

The next several remarks are perhaps digressive even by the informal standards of the present section. They do provide some further discussion of the particular case of M_{24}; and the details will continue our overall theme of analogies of sporadic 2-local subgroup structure with the theory of Lie type groups and buildings. But they can certainly be skipped by the reader wishing to proceed to considerations more directly related to our general constructions of sporadic geometries.

DIGRESSION 6.2.13 (Parabolic-like behavior of certain subgroups of M_{24}). We observe that for each $i \in I = \{O, T, S\}$, the maximal 2-local subgroup H_i has a structure analogous to the usual Levi decomposition of a parabolic subgroup in a group of Lie type in characteristic 2. First, we follow the Atlas notation at [**47**, p. 96] in describing the structure of these subgroups by

$$H_O \cong 2^4 : \mathrm{Alt}_8,$$
$$H_T \cong 2^6 : (L_3(2) \times \Sigma_3),$$
$$H_S \cong 2^6 : 3^{\cdot}\Sigma_6.$$

Now note that each $U_i := O_2(H_i)$ is a normal 2-subgroup of H_i, as is the case for the unipotent radical of a parabolic of a group in characteristic 2. And in analogy with the structure of a Levi complement of a parabolic, the quotient $L_i := H_i/O_2(H_i)$ is "essentially" a group of Lie type in characteristic 2—for we may regard Alt_8 as $L_4(2)$; and Σ_3 as $L_2(2)$; and $3^{\cdot}\Sigma_6$ as the triple cover of $Sp_4(2)$. Indeed for simplices of larger dimension in the complex Δ of Example 6.2.10, namely of type J with $|J| > 1$, similar structural remarks hold for U_J and L_J obtained from H_J.

This demonstrates some of the analogies of the conjugates of the H_J with the poset $\mathcal{P}$ of parabolics in a building. Correspondingly the conjugates of $U_J := O_2(H_J)$ afford a poset exhibiting certain analogies with the poset $\mathcal{U}$ of unipotent radicals in a building; for example (compare with Remark 5.1.11), the poset given by these U_J is a subposet of $\mathcal{B}_2(G)$, and this subposet inclusion turns out (see Section 7.5) to be a G-homotopy equivalence. Indeed one checks that the poset of the U_J is opposite to that of the H_J; we will consider this kind of subgroup-poset duality in more detail later at Definition 6.3.13. It also turns out (see later Example 6.4.7) that the quotient Δ/G is a single simplex afforded by the subgroup simplex $\mathcal{H}_I$; so in view of the ampleness of $\mathcal{B}_2(G)$ in Section 5.9, we obtain a simplex normalizer decomposition as in (5.6.12). This analogue of the standard decomposition 5.1.18 over a building is essentially our main result for M_{24} in Section 7.5. ♢

DIGRESSION 6.2.14 (The diagram for the 2-local geometry for M_{24}). We mention also that the embedding of the H_i in G is axiomatized by a diagram similar to the Dynkin diagram for a building, in the viewpoint of Tits and Buekenhout mentioned in Section 6.1. While we will not in the present work require such diagrams, in Chapter 7 we will follow the fairly standard practice in the literature of providing those diagrams when they are available. For a fuller discussion of the meaning of such diagrams, the reader is referred to sources such as the original [**106**] or the expository discussion in [**115**, Ch. 2]. Here we will only give, by way of example, a very brief discussion of the diagram for the 2-local geometry Δ for M_{24} in Example 6.2.10. (Also later at 6.2.17 (1), we discuss the embedding of overgroups of H_I in terms of parabolic subgroups of the H_i, which implicitly describes the links of vertices via the diagram.)

Ronan and Smith observed in [**106**] that this geometry for M_{24} can be axiomatized by the following variant of the Dynkin diagram of type C_4:

The "square node" at the right end of the diagram does *not* correspond to any subgroup of M_{24}—whereas otherwise we might have expected some subgroup such as $2^7 : Sp_6(2)$ which occurs in the case $G = Sp_8(2)$, whose building is described using the ordinary C_4 diagram. But the other three nodes do correspond, as the indexing by $i \in I$ suggests, to the subgroups H_O, H_T, and H_S in Example 6.2.10. For example, if we remove the node at O and the bonds emanating from it, we are left with the A_3 Dynkin diagram; notice that this indicates part of the structure of the group H_O—namely the quotient $L_O = H_O/O_2(H_O) \cong \mathrm{Alt}_8$ regarded as $L_4(2)$ in Digression 6.2.13, whose building has the A_3 Dynkin diagram. More precisely, at the level of axioms for geometries, the link of a vertex of type O is a *truncation* of the usual building of type A_3: the usual building would be the full projective space[9] of $L_4(2)$, but the square node indicates that the projective planes of that building are "missing" from the truncation—the truncation uses only the projective points and lines, represented here by the vertices of type T and S which are incident to O. Similar remarks apply if we remove the node of type T, or of type S. In particular, the remaining subdiagram then indicates the corresponding group structure: of L_T—viewed as $L_2(2) \times L_3(2)$, with diagram of type $A_1 \times A_2$; or of L_S (modulo $O_3(L_S)$, which is trivial on the link of H_S)—viewed as $Sp_4(2)$, with diagram of type C_2.

(We mention that there is also a "Cambridge" variant of this diagram, in which the square node convention is replaced by a "trapped node", more suitable for description of generators and relations of the H_i; and a tilde-geometry diagram, mentioned later in Section 7.5.)

We reiterate that the formalism of such diagrams is not required for this work—since in our proofs, we will just quote any background geometric details required. However, we will sometimes use the diagrams, just as a shorthand way of recording that quoted structure, for example of the quotients L_i of the H_i for $i \in I$. ♢

[9] We use the standard term *projective space* in the sense of finite geometry (cf. later Example 6.4.3); so this differs from our usual convention where "space" means simplicial set.

Choosing maximal 2-locals to define geometries for other groups. In the geometry for M_{24} in Example 6.2.10 and the discussion thereafter, we have tried to preview some features of geometries for further sporadic groups. In particular we suggested a number of geometrically significant properties of maximal 2-locals,[10] by abstracting from corresponding features of maximal parabolics in the viewpoint on the building in Example 6.2.2.

We now focus on the following specific desirable properties for subgroups H_i. These will be included "in practice"—that is, in the choice of maximal 2-locals that will constitute the first step, when we define the geometry for each sporadic group in Chapter 7. (However, they need not be included "in theory"—that is, in the more general overall construction of geometries in the next section.)

CONDITION 6.2.15 (Choice of maximal 2-locals). Let I denote a finite index set I; and let $H_i (i \in I)$ denote subgroups of G satisfying:

(a) Each H_i is a maximal 2-local subgroup of G.

(b) The H_i are pairwise non-conjugate in G.

(c) The H_i have intersection H_I with Sylow 2-subgroup of maximal order.

We indicate some of the effects of making each of these choices:

The maximality in condition (a) in particular guarantees that the H_i are self-normalizing in G; and we saw after Example 6.2.10 for M_{24} that this makes the complex on conjugates of the H_i equivalent to that on the cosets. We had also indicated before that Example that we usually restrict attention to maximal 2-locals which in fact satisfy the further restriction (6.2.9).

We saw early in the discussion of the building in Remarks 6.2.5 (and incorporated this directly into the discussion of M_{24} in Example 6.2.10) that the non-conjugacy in condition (b) results in associating a *type* to each simplex of the the complex on conjugates of the H_i—namely a subset $J \subseteq I$ which just records which conjugacy classes of maximal 2-locals are represented in the simplex. (In particular (b) will imply the type-condition 6.3.1 (a) in our more general construction of geometries in the next section.) It follows that the G-conjugacy action must preserve the types of vertices in any simplex formed from conjugates of the H_i; and in particular, now also using the self-normalizing property in the previous pargraph, we see that the standard simplex $\sigma_J := \{H_i : i \in J\}$ has stabilizer given exactly by the intersection $H_J = \cap_{i \in J} H_i$; that is:

$$N_G(\sigma_J) = H_J, \tag{6.2.16}$$

We emphasize however that it does *not* follow that the intersections H_J must also be self-normalizing. This will be important in the situation mentioned earlier in Remark 6.2.4, where we have $H_J < N_G(H_J)$—so that H_J is in fact the full stabilizer in G of each of the $N_G(H_J)/H_J$-conjugates of σ_J (which are distinct). In later Lemma 6.5.7, we will encounter some natural hypotheses for sporadic geometries under which identifying such conjugates gives a poset homotopy equivalence.

The condition 6.2.15(c) might seem particularly mysterious to the reader encountering it for the first time. And since we will make use of the consequences of that condition frequently later (usually in fairly abbreviated form), we will provide

[10]By contrast, as we mentioned in Section 6.1, in some other work in the geometric literature such as Buekenhout [**37**] and the related later developments indicated in [**115**, Ch. 2], other kinds of maximal subgroups of G can be of interest as vertex stabilizers.

in the next few pages some examples in further detail on deducing such consequences.

REMARKS 6.2.17 (More detailed comments on 6.2.15(c)). The choice of the intersection H_I with maximal 2-power order is intended to mimic the feature of the building in Example 6.2.2, where the intersection H_I is given by a Borel subgroup. In particular, the Borel subgroup contains a Sylow p-subgroup for the characteristic prime p of the Lie type group; and in our sporadic cases, usually H_I will contain either a full Sylow 2-subgroup of G (as is the case for M_{24} in Example 6.2.10), or a subgroup of small index (typically 2) in a Sylow group (as is the case for M_{22} in Section 7.3).[11]

For each sporadic group G in Chapter 7, we always proceed in the order of first emphasizing the subgroups $H_i (i \in I)$, and thereafter adjusting the choice if necessary to maximize the 2-power in the intersection H_I. But one could ask whether if we were to begin with the group H_I, should the H_i then correspondingly be uniquely determined? The answer is certainly yes in the case of the building—we had mentioned in Example 4.6.3 that a Borel subgroup H_I is contained in just one maximal parabolic H_i for each $i \in I$. But in our sporadic cases, the answer would have to be hedged to something like "usually either unique, or close to unique": For example the choice of H_i over H_I is indeed unique in the case of M_{24} (as we saw in Remark 6.2.12). However in the case of M_{11} in Section 7.1, uniqueness fails since H_I is contained in H_2 as well as in one further conjugate of H_2. This reflects[12] the situation mentioned at the end of Remark 6.2.12, namely that H_I is in this case the stabilizer of more than one maximal simplex. (See also the similar discussion for the triples geometry in Example 6.5.5, and for the various groups involving that geometry—starting with M_{22} in Section 7.3.)

With the above observations as general background, we next indicate some of the more specific ways in which Condition 6.2.15(c) can be used in our analysis.

(1) *Describing overgroups of H_I in some H_i (and hence incidence).* Above we fixed H_I, and then considered the possible uniqueness of overgroups of type H_i. But more frequently we will instead fix an H_i; and then wish to describe (and ideally even *characterize*) the overgroups of conjugates of H_I which appear in that H_i.

In particular this will lead to a description of the geometric incidence relation within the link of the vertex $\{H_i\}$; and this will justify our claim in earlier Remark 6.2.11 that the geometry for M_{24} is an example of the best-behaved cases, in which the inductive incidence relation just boils down to the analogue of (6.2.3), namely containment of a common conjugate of H_I—and that the more inductive relation is then really no harder to verify.

The rough reason that we might have any hope of describing the overgroups of H_I is that our geometries are usually inductively built up (cf. Digression 6.2.14) from links which are buildings—in which the overgroups of a Borel subgroup are just the parabolic subgroups.

---

[11]While this intersection is a standard theorem in the case of Lie type groups, our ability to choose the intersection with this large 2-power is verified separately for each sporadic group—that is, no one really knows why this should turn out to be possible.

[12]It also reflects the failure of weak closure—since $H_I \cong D_8$ in fact contains the four-group A_2 as well as a further conjugate of A_2 under a Sylow 2-subgroup $N_G(H_I)$ of G, which contains H_I with index 2.

We demonstrate this for the sporadic group $G = M_{24}$, using the subgroups H_i in Example 6.2.10 for i in the index set $I = \{O, T, S\}$.

We start with $i = S$. Here H_I contains $O_{2,3}(H_S) \cong 2^6 3$, so we set $\bar{H}_S := H_S/O_{2,3}(H_S) \cong Sp_4(2)$. Then $\bar{H}_I$ is of order 2^4, and so is a Sylow 2-subgroup of $Sp_4(2)$; in particular, it is a Borel subgroup of that Lie type group (since the group is defined over $\mathbb{F}_2$ so that a Cartan subgroup is trivial). Now the overgroups of Borel subgroups in a Lie type group are just the parabolic subgroups (as mentioned in Example 5.1.1). Hence the overgroups of H_I in H_S can be characterized as the preimages of the parabolic subgroups of $\bar{H}_S \cong Sp_4(2)$.

In particular, it follows that those G-conjugates of subgroups $H_j (j \in I \setminus \{S\})$ which are incident with H_S (in the sense defined in Example 6.2.10) must in fact intersect H_S in such a preimage—in this case, of a maximal parabolic over $\bar{H}_I$. One checks that these intersections conversely determine the corresponding conjugates of H_j. (This corresponds essentially to the weakly closed property mentioned in Example 6.2.12.) So we see that incidence with H_S in the above inductive sense in fact just reduces to the analogue of the simpler incidence (6.2.3) in the case of the building, namely containment of a common conjugate of H_I. This begins to demonstrate our claim that the stronger incidence is "no more difficult to verify" in the present geometry than the ostensibly-weaker one.

A similar analysis will apply when we turn to the case $i = T$. Here $U_T = O_2(H_T) \leq H_I$, and we set $\tilde{H}_T := H_T/U_T \cong L_2(2) \times L_3(2)$. Now $\tilde{H}_I$ is of order $2^4 3$, and hence contains a Sylow 2-subgroup of $\tilde{H}_T$; thus it contains a Borel subgroup of the (product) Lie type group $L_2(2) \times L_3(2)$. Again the overgroups of a Sylow 2-group of $\tilde{H}_I$ are characterized as the parabolic subgroups of $\tilde{H}_T$. However this time, not all of the parabolics there will arise from intersections with H_T of incident subgroups $H_j (j \in I \setminus \{T\})$; in fact those arising will determine at most a point stabilizer of $L_3(2)$, but will not involve a line stabilizer (i.e., those arising correspond to types O, S in 6.2.14). Nonetheless, these incidences in the strong sense of Example 6.2.10 again reduce just to the simpler analogue of (6.2.3), namely being determined just by containment of a common conjugate of H_I.

Finally we take $i = O$ and a similar overview emerges. Here $U_O \leq H_I$, and we set $H_O^* := H_O/U_O \cong L_4(2)$. Now H_I^* is of order $2^6 3$, and hence contains a Sylow 2-subgroup of H_O^*, which once again affords a Borel subgroup of $L_4(2)$. So again the overgroups of that Sylow group are characterized as parabolic subgroups of H_O^*. And again not all parabolics will arise from intersections with H_O of incident subgroups $H_j (j \in I \setminus \{O\})$; in fact those arising will determine at most a maximal parabolic stabilizing a point and a line in the projective space for $L_4(2)$ (corresponding to types T, S in 6.2.14). As before, the inductive incidence reduces to containing a common conjugate of H_I.

(2) *Deducing inclusion relations among subgroups A_i corresponding to the H_i.* We next demonstrate one particularly useful way in which Condition 6.2.15(c) can be used to exhibit suitable notions of "nearness"—especially when we replace a complex of 2-locals such as Example 6.2.10 by some alternative and possibly more convenient simplicial complex. (This will implicitly preview our more general construction of a geometry in later Definition 6.3.2.)

To set up this comparison, we observe that in part (1) above, we used the close relationship of H_I with a Sylow 2-subgroup, in order to more precisely analyze incidence among the subgroups H_i. And of course incidence gives the nearness relation of "distance 1", in the complex for M_{24} in Example 6.2.10—where the complex has the form we are calling $\Delta_{\mathcal{H}_I}$ in Definition 6.3.5, in which it is the subgroups H_i which are regarded as vertices in a maximal simplex. However, we will also often consider complexes based on other notions of vertices; in the complex of form $\Delta_{\mathcal{A}_I}$ in Definition 6.3.2, we consider some general G-action on a collection of groups, and as our model for alternative vertices we take subgroups A_i normalized by the chosen subgroups H_i in G. In that case, the notion of nearness will usually be different from the Borel subgroup model of containing a common H_I—for example, the incidence is often a containment relation among the A_i themselves. And indeed we can frequently establish such containment relations based on the embedding condition 6.2.15(c) for the H_i, by using Sylow subgroups of H_I in the following alternative way.

Namely we recall a standard fact from the representation theory of Lie type groups (based on the Borel fixed point theorem, cf. [**22**, Ex. (i), p. 252]): Assume that X is of Lie type in characteristic p, and let V denote an irreducible X-module over the algebraic closure $\bar{\mathbb{F}}_p$ of $\mathbb{F}_p$. (We mention that we typically work with groups defined over $\mathbb{F}_2$, and in practice it then usually suffices to consider modules over that field, rather than over the algebraic closure $\bar{\mathbb{F}}_2$.) We have:

$$\text{A Sylow } p\text{-group fixes a unique 1-subspace of the irreducible } V. \tag{6.2.18}$$

In our later treatment, when we invoke 6.2.15(c) in this context, we typically do so just in a very abbreviated way. So below, we will indicate below several representative examples in somewhat fuller detail.

For a very easy case, we take the small Lie type group $G = L_3(2)$; and use the index set $I = \{1, 2\}$ for its geometry of rank 2. But in place of the building realized directly via parabolic subgroups in the form $\Delta_{\mathcal{H}_I}$ in Example 6.2.2, we now instead use the complex of form $\Delta_{\mathcal{A}_I}$ arising from the projective space (cf. Example 6.4.3) of the natural G-module V of dimension 3 in characteristic 2. In particular, the role of A and the A_i in Definition 6.3.2 will be played by V and the V_i indicated below. Now the maximal parabolic H_2 is the stabilizer of a 2-subspace V_2 of V; and V_2 then affords the irreducible natural module for $\bar{H}_2 := H_2/U_2 \cong L_2(2)$. And the Borel subgroup $H_I = H_1 \cap H_2$ is in fact a Sylow 2-subgroup of G, so that $\bar{H}_I$ is also Sylow in $\bar{H}_2$. So it follows from (6.2.18) that $\bar{H}_I$ centralizes a unique 1-subspace V_1 of V_2. In particular V_1 is then the member of the projective space stabilized by the full maximal parabolic H_1 over H_I. That is: we have shown that for incident H_1 and H_2, we can deduce the inclusion relation $V_1 < V_2$ among the corresponding subspaces stabilized in V—as opposed to having V_1 and V_2 in some other random relationship at greater mutual distance within the poset of subspaces of V.

We turn to a larger example for the sporadic group $G = M_{24}$. This time in place of the geometry of the form $\Delta_{\mathcal{H}_I}$ in Example 6.2.10, we instead take the complex of form $\Delta_{\mathcal{A}_I}$ given in later Example 6.3.7, based on A taken as the irreducible Golay code module over $\mathbb{F}_2$. In particular we replace the earlier index set $\{O, T, S\}$ by $I = \{1, 2, 4\}$ corresponding to the $\mathbb{F}_2$-dimensions of the A_i. We can begin by directly mimicking the argument of the previous paragraph: The maximal 2-local subgroup H_2 is the stabilizer of a trio 2-subspace A_2 of A; which affords the natural irreducible module for $\bar{H}_2 := H_2/C_{H_2}(A_2) \cong L_2(2)$. The choice in the embedding condition 6.2.15(c) gives H_I of order $2^{10}3$, and in particular $\bar{H}_I$ is a Sylow 2-subgroup of $L_2(2)$. So by (6.2.18), $\bar{H}_I$ fixes a unique 1-subspace A_1 of A_2—which must be the (i.e., unique) 1-subspace of A fixed by the maximal

2-local H_1 over H_I; that is, we have deduced the inclusion relation $A_1 < A_2$. And we can make a similar argument within the sextet 4-space A_4 of A which is stabilized by the maximal 2-local H_4: Here A_4 affords the irreducible natural module for $\tilde{H}_4 := H_4/C_{H_4}(A_4) \cong Sp_4(2)$; and $\tilde{H}_I$ is now a Sylow 2-subgroup of $Sp_4(2)$. The previous argument then produces a 1-subspace fixed by $\tilde{H}_I$, which again must be the 1-space A_1 fixed by H_1. But we can make a similar argument in the 2-dimensional section $B_1^* := A_1^{\perp}/A_1$ of A_4 under irreducible action by the parabolic subgroup $\tilde{H}_{1,4} \cong 2^3 L_2(2)$ of $Sp_4(2)$: the 1-subspace fixed there by H_I has 2-dimensional preimage in A_4; and this must in fact be the unique 2-subspace A_2 of A normalized by the maximal 2-local H_2. That is, we have deduced the inclusion relations $A_1 < A_2 < A_4$.

We conclude by mentioning that in both parts (1) and (2) above for M_{24}, we have repeatedly used the fact that H_I contains a Sylow 2-subgroup of G. In cases such as M_{22} (see Section 7.3) where H_I contains a little less than a Sylow, a bit more effort can be required to mimic these arguments—but having H_I contain say a subgroup of index 2 in a Sylow already goes a long way in that direction. ◇

We have devoted the last several pages to demonstrating the details of consequences of Condition 6.2.15, for the particular choice of the H_i indicated for M_{24} in Example 6.2.10. By contrast, the process of checking that the H_i *can* be chosen as in 6.2.15 for M_{24} is quite straightforward, as it normally is for the other sporadic groups: Typically one uses for example just the list of maximal subgroups for each sporadic group in the Atlas [**47**]; possibly inductively, in the sense of also knowing the maximal subgroups of the quotients $H_i/O_2(H_i)$—which often are just groups of Lie type in characteristic 2. So we now emphasize for the future that *for the individual sporadic groups in Chapter 7, we normally just specify a choice of subgroups H_i, without detailing the routine verification of 6.2.15.*

We also mention that our choice of the H_i usually proceeds as follows: We want our geometry of 2-locals to have maximal possible *rank*—that is, we choose a set of representatives H_i of *all* the conjugacy classes of maximal 2-locals which satisfy the condition (6.2.9). In particular, this choice will satisfy (a) and (b) of Condition 6.2.15, with an index set I of maximal size. We then arrange the embedding of the H_i as in (c). And as mentioned earlier, the resulting intersection H_I can then be observed to contain either a Sylow 2-subgroup of G, or a subgroup of small index in such a Sylow group.

6.3. Some general constructions of 2-local geometries

We wish to emphasize at the outset that analogies of buildings with various sporadic geometries are verified *case by case*—we saw the case of M_{24} in our discussion of Example 6.2.10, and other sporadic groups are also treated individually. Put another way: No one has found any really general geometric theory of simple groups that implies such analogies; and it might even be futile to hope for such a theory.

On the other hand, because this work undertakes to describe geometries for *all* the sporadic groups, in this section we do attempt to provide at least some common uniform *notational* framework for discussing all those geometries. The reader is warned in advance that the theoretical content of this section is not at all deep—in particular, all proofs are essentially immediate from the definitions.

Furthermore, many of our unifying conventions are being introduced just for this work; the reader will not find them in the existing literature on sporadic geometries.

The section focuses primarily on two general constructions: One structure, which we will use for every sporadic group, is the geometry we call the 2-local complex, namely $\Delta_{\mathcal{H}_I}$ in Definition 6.3.5; we have tried to indicate its general flavor via the examples in the previous section. However, it is natural to subsume that geometry, along with some other variations which can often be more convenient, in a more general construction which we call the intersection complex; and so we present that first, as our more fundamental $\Delta_{\mathcal{A}_I}$ in Definition 6.3.2. After it, we provide a number of examples of such variations; and we also give a preliminary indication of some relations of these geometries with the various standard collections of p-subgroups described in the final sections of Chapter 5.

One might regard the variation allowed by $\Delta_{\mathcal{A}_I}$ as offering a kind of "representation theory" for a more specific simplicial complex $\Delta_{\mathcal{H}_I}$ defined by 2-local subgroups; in particular it will afford a convenient context for flag-transitive complexes. (Furthermore $\Delta_{\mathcal{A}_I}$ will often correspond to a *coefficient system* over $\Delta_{\mathcal{H}_I}$, especially in the sense used by Ronan and Smith—see for example [**22**, Sec. 7.5].)

A complex with incidence defined via intersections. The examples in Section 6.2 followed the fairly customary construction of local geometries for a group G: in which vertices of the complex are essentially *identified* with maximal 2-local subgroups H_i. But that view can sometimes be complicated in practice, in that it may require fairly detailed knowledge of the internal structure of G.

Instead, it can be convenient to obtain our simplicial complex from the action of G on some other structure, which might be more easily accessible than the action on the conjugates of the H_i. Indeed we should expect (see later Lemma 6.5.4) to obtain essentially the same complex—if we replace the vertex $\{H_i\}$ by some new vertex whose stabilizer under a more convenient action of G is still given by H_i.

As our model for an alternative structure, we will consider some group A admitting a G-action. We will then want to replace the simplex $\mathcal{H}_I$ of subgroups of G with a corresponding simplex of subgroups taken from A:

NOTATION 6.3.1. Assume that the finite group G acts on the finite group A; denote the action by conjugation. Let I denote a finite indexing set.

(a) Let $A_i (i \in I)$ denote subgroups of A, not conjugate under the G-action.

(b) For each nonempty subset $J \subseteq I$, set $A_J := \bigcap_{i \in J} A_i$.

Write $\mathcal{A}_I$ to abbreviate the collection of subgroups A_J. Also set $H_i := N_G(A_i)$.

Notice that $\mathcal{A}_I$ is indeed a simplex of subgroups of A in the sense of Definition 4.6.2. However we emphasize that (in contrast with the usual case for the H_J) we will *not* in general expect distinct index sets J to determine distinct subgroups A_J.

The final condition that $N_G(A_i) = H_i$ will be consistent with normally having chosen the desired stabilizers H_i in advance. In practice in Chapter 7 we will choose the H_i to satisfy the requirements of Condition 6.2.15—for example, 6.3.1 (a) above would be implied by 6.2.15 (b). However, that choice for the H_i is not required in the elementary formalism in this section for constructing geometries.

Now the main idea of our generalized construction is to further mimic Example 6.2.10, by using the intersections A_J to give the mutual incidence relation, which determines which sets of vertices should form a simplex:

DEFINITION 6.3.2 (Intersection complex). Assume Notation 6.3.1 (and in particular the groups A_i and A_J given there). Then we define a simplicial complex in which

(a) the vertices are given by the G-conjugates of the $A_i (i \in I)$;

(b) a set of vertices determines a simplex if and only if it is given, for some nonempty subset $J \subseteq I$, by conjugates $g_i^{-1} A_i g_i$ (for $i \in J$ and $g_i \in G$) satisfying:

For each nonempty subset $K \subseteq J$, the intersection $\bigcap_{i \in K} g_i^{-1} A_i g_i$ is some G-conjugate of A_K.

We call this construction the *intersection complex* determined by the $A_i (i \in I)$; and we denote it by $\Delta_{\mathcal{A}_I}$, in terms of the subgroup simplex $\mathcal{A}_I$ defined by the A_i in Notation 6.3.1. ♢

Just as in Example 6.2.10, we see from the inductive nature of the definition that this incidence relation is inherited by nonempty subsets of a simplex; so the subset closure condition (2.7.4) holds, and we do indeed have an abstract simplicial complex.

We briefly indicate a few features of this general construction.

We can continue the spirit of some of the notation from the Examples in the previous section: we may use set braces $\{A_i\}$, when we wish to explicitly regard A_i as a vertex rather than a subgroup; and similarly write $\{g_i^{-1} A_i g_i : i \in J\}$ for the above simplex, which we say has *type* J—indicating which G-classes of the A_i are represented in the simplex. (The set J has no repetitions in view of the non-conjugacy of the A_i assumed in Notation 6.3.1 (a).) Note that simplices of each type J do exist, since we could take all $g_i = 1$.

This notion of type essentially follows Tits in [**120**]; it has the following useful consequence for the G-action on the complex:

REMARK 6.3.3 (Preservation of type and admissible action). The incidence relation in Definition 6.3.2 is preserved by the G-action, so that an intersection complex $\Delta_{\mathcal{A}_I}$ is a G-complex. Furthermore G preserves the type of vertices, so that a simplex stabilizer in fact fixes all the vertices of the simplex; hence the G-action on the complex is admissible (Definition 2.5.5).

WARNING 6.3.4. We had mentioned above that our construction based on the subgroup simplex $\mathcal{A}_I$ guarantees for each nonempty $J \subseteq I$ that the set $\sigma_J := \{A_i : i \in J\}$ is a simplex of type J. However, we caution that the conjugates of σ_J under the various $g \in G$ need not in general give *all* the simplices of type J; for we have the freedom to conjugate each of the vertices $\{A_i\}$ of σ_J by distinct $g_i \in G$—and the result might still satisfy the incidence relation, without actually being a common conjugate under a single g. In fact the stronger property of having the conjugates of σ_J afford the unique class of simplices of each type J corresponds to the condition of flag-transitivity, which we will discuss later at Definition 6.4.2. And we emphasize that flag-transitivity will have to be *proved*, for each sporadic geometry that we introduce in Chapter 7. (On the other hand, our notation based on $\mathcal{A}_I$ is designed as a natural context for the expectation of flag-transitivity; and we will see that the property is often straightforward to verify—see Example 6.4.7.)

Various manifestations of the intersection complex. We now want to begin to substantiate our earlier mention that the general intersection complex construction of Definition 6.3.2 in fact provides a unified approach to several ostensibly different constructions of geometries which we will be encountering frequently.

In doing so, we invent some nomenclature—just in order to be able to refer to those various constructions, when they arise at many later points. So the reader should not search for any particularly deep meaning when we use below the terms 2-local complex, poset complex, and almost-poset complex; nor when we distinguish external from internal versions of these constructions, where the latter will just mean "inside G".

First, we recover complexes constructed directly from maximal local subgroups, of the general form given for the building in Example 6.2.2, and for M_{24} in Example 6.2.10:

DEFINITION 6.3.5 (The complex of 2-locals). Assume we have non-conjugate 2-local subgroups H_i of G for $i \in I$.[13] Then in Notation 6.3.1, we can take $A := G$ and $A_i := H_i$. So for nonempty $J \subseteq I$, the subgroup A_J becomes the group we have often denoted earlier by $H_J = \bigcap_{i \in J} H_i$; and the subgroup simplex $\mathcal{A}_I$ is denoted by $\mathcal{H}_I$ just as in Definition 4.6.2.

Thus the intersection complex from Definition 6.3.2 is now denoted by $\Delta_{\mathcal{H}_I}$, with vertices given by the 2-local subgroups H_i; and incidence for type J is determined by intersections giving a conjugate of H_K for $K \subseteq J$ (cf. Condition 6.2.7). We call $\Delta_{\mathcal{H}_I}$ the *2-local complex* determined by the H_i. ♢

This form $\Delta_{\mathcal{H}_I}$ will always be one of the 2-local geometries which we construct for each sporadic group in Chapter 7.

But often it will in fact be easier to instead work with some simpler intersection complex $\Delta_{\mathcal{A}_I}$. And typically we will see that $\Delta_{\mathcal{A}_I}$ is indeed just a variant of $\Delta_{\mathcal{H}_I}$, by applying Lemma 6.5.4: our notation $N_G(A_i)$ guarantees that both complexes have the same stabilizers for vertices in a maximal simplex; so the process boils downs to showing that the G-action is flag-transitive on both complexes—a procedure that we will illustrate in later Example 6.4.7.

Now to consider suitable variants $\Delta_{\mathcal{A}_I}$, we will investigate some choices for the group A other than G itself.

Perhaps the most natural alternative configuration for the group A will afford the familiar case of a poset complex—namely the complex of inclusion-chains in a suitable G-subposet of the poset of subgroups of A.

Recall first that we had mentioned, in the discussion following the definition at 4.6.2 of a simplex $\mathcal{H}_I$ of subgroups of G, that in practice our various subgroups H_J would usually be distinct, for distinct index sets $J \subseteq I$. But for our more general choices of A_i in Notation 6.3.1, the analogous distinctness of the A_J will frequently *not* hold. Indeed in our applications of the intersection complex, we will see a considerable spectrum of possibilities for distinctness of the A_J.

Furthermore the maximum non-distinctness occurs in particular in the special case where the A_i form a linear chain of inclusions: for then the intersection of any pair will just equal the smaller of the pair; so that each A_J will be equal to one of the original A_i. This extremal situation does have at least one advantage, in that

[13] Actually the H_i could even be random non-conjugate subgroups of G; but then the resulting complex would probably be of little interest. Indeed in practice, we normally take the H_i to further satisfy Condition 6.2.15.

the incidence relation defined by intersections then simplifies to being given just by inclusion; and so we recover the usual complex of chains in a poset as in 2.7.5:

PROPOSITION 6.3.6 (The poset complex subcase of the intersection complex). *Assume in Notation 6.3.1 that* $I = I_n = \{1, 2, \ldots, n\}$*; and that the subgroups* A_i *of* A *satisfy* $A_1 < A_2 < \cdots < A_n$*. Then the intersection complex* $\Delta_{\mathcal{A}_I}$ *of Definition 6.3.2 becomes just the complex* $\Delta(\mathcal{A})$ *of inclusion-chains in the poset* $\mathcal{A}$ *defined by the* G*-conjugates of the* $A_i (i \in I)$.

PROOF. Note for $i < j$ in I that we have $A_{i,j} = A_i \cap A_j = A_i$; and so since conjugates of $g^{-1}A_i g$ and $h^{-1}A_j h$ are incident in the intersection complex $\Delta_{\mathcal{A}_I}$ when they contain a common conjugate of $A_{i,j} = A_i$, we see by order considerations that $h^{-1}A_j h$ must then in fact contain $g^{-1}A_i g$. It follows that the vertices of a simplex of arbitrary type J are given just by an inclusion-chain of conjugates of the $A_k (k \in J)$. □

In our applications, it will sometimes be convenient to make the following further distinction:

When this poset complex case occurs for a group A other than a subgroup of G, we call $\Delta_{\mathcal{A}_I}$ an *external* poset complex for G. Often we will obtain such an A from an $\mathbb{F}_2 G$-module, so that $\Delta_{\mathcal{A}_I}$ is a subcomplex of the complex of chains of linear subspaces of A.

On the other hand, when the situation arises for A taken as G itself or some subgroup, so that $\mathcal{A}$ as above is a G-subposet of the poset of all subgroups of G, we call $\Delta_{\mathcal{A}_I}$ an *internal* poset complex for G. Often we will work with such an $\mathcal{A}$ given by a G-subposet of $\mathcal{A}_2(G)$—indeed frequently A_i will be defined as $\Omega_1 Z(O_2(H_i))$ for our maximal 2-locals H_i.

In the next few pages, we give some examples of such external and internal poset complexes for particular sporadic groups.

EXAMPLE 6.3.7 (External poset complexes from modules for M_{24}). We begin with some examples which provide partial analogues of the realization of buildings for classical groups via the projective or polar spaces of their natural modules, mentioned in Digression 6.2.1 (2).

We first take A to be the irreducible *Golay code module* for M_{24}; we give a summary below, referring the reader to, for example, the Atlas [**47**, p. 94] for further detail. Let Ω denote the underlying 24-set for the Steiner system $\mathcal{S}(5, 8, 24)$ in our earlier discussion leading to Example 6.2.10; and let V denote the $\mathbb{F}_2 G$-module given by the power set of Ω, namely the space spanned by the elements of Ω, in which subsets of Ω appear as sums of such elements. Then the octads (the blocks of size eight defining the Steiner system) span a 12-dimensional subspace C of the 24-space V called the (extended) *binary Golay code*; and A is the 11-dimensional quotient of C modulo the vector for the 24-set Ω. We choose the subgroups H_O, H_T, H_S as in Example 6.2.10; but (as will frequently be the case in Chapter 7) it will now be appropriate to re-label them instead as H_1, H_2, H_4—using a slight variant $I := \{1, 2, 4\}$ of the index set $I_3 = \{1, 2, 3\}$ indicated in Proposition 6.3.6: This is because when we define $A_i := C_A(O_2(H_i))$, we find that A_1, A_2, A_4 have $\mathbb{F}_2$-dimensions 1, 2, 4. Now the A_i define the subgroup simplex $\mathcal{A}_I$ of Notation 6.3.1; and in particular $N_G(A_i)$ is given by the above subgroup H_i, as in the convention for H_i in that Notation. Furthermore we had checked at 6.2.17 (2) that we in fact get an inclusion-chain $A_1 < A_2 < A_4$; and this is the hypothesis for Proposition 6.3.6.

Hence the G-poset $\mathcal{A}$ in the notation of that Proposition consists of the G-conjugates of A_1, A_2, and A_4; and the intersection complex $\Delta_{\mathcal{A}_I}$ of Definition 6.3.2 is in fact just the poset complex $\Delta(\mathcal{A})$. (In particular, this complex arises from the G-subposet $\mathcal{A}$ which exhibits the relevant *part* of the full projective space on the Golay code module A.) Thus in our further nomenclature, the intersection complex $\Delta_{\mathcal{A}_I}$ exhibits an external poset complex for M_{24}. Often this poset complex $\Delta_{\mathcal{A}_I}$ from inside the Golay code module will provide a more elementary geometric viewpoint than the 2-local complex $\Delta_{\mathcal{H}_I}$.

Next we wish to see that $\Delta_{\mathcal{A}_I}$ really is just a variant of $\Delta_{\mathcal{H}_I}$. This follows via a standard argument for isomorphism of complexes which we will use frequently in the remainder of the work. For the present case of M_{24}, we just quickly preview the details checked later: namely at Example 6.4.7 we obtain flag-transitivity of M_{24} on both complexes; and since the stabilizers of the vertices in a maximal simplex are identical, Lemma 6.5.4 then gives the isomorphism.

As an alternative to the above construction, we could equally well have chosen A to be the irreducible *Golay cocode*, or *Todd module* for M_{24}: namely the 11-dimensional space given as the quotient of the 23-dimensional subspace of V generated by even subsets of Ω, modulo the 12-dimensional code module C above. This time it is convenient to choose the index set $I := \{6,3,1\}$: for when we now define $A_i := C_A\big(O_2(H_i)\big)$ for this new module A, we find $\dim_{\mathbb{F}_2}(A_i) = i$ for each i. Furthermore using the method of 6.2.17 (2) we obtain the inclusion relations $A_6 > A_3 > A_1$. So this new $\Delta_{\mathcal{A}_I}$ also gives the poset complex subcase in Proposition 6.3.6.

Furthermore we again have $N_G(A_i)$ given by the H_i chosen earlier. Also as before, this $\Delta_{\mathcal{A}_I}$ "reproduces" the 2-local complex $\Delta_{\mathcal{H}_I}$, since again the normalizer set-map in fact induces an isomorphism of complexes: In order to apply Lemma 6.5.4, we must verify flag-transitivity of G now also on our new complex $\Delta_{\mathcal{A}_I}$; and indeed we can make an argument very similar in flavor to that demonstrated in Example 6.4.7 for the previous case of the Golay code module—as mentioned there, such arguments are essentially superficial variants of arguments based just on the geometric structure of the links of vertices described by the diagram. So again we get $\mathsf{N}(\Delta_{\mathcal{A}_I}) = \Delta_{\mathcal{H}_I}$. $\diamondsuit$

EXAMPLE 6.3.8 (An internal poset complex for M_{11}). We now consider instead the sporadic Mathieu group $G := M_{11}$; and we will anticipate here certain features of our treatment in Section 7.1.

We take as our index set $I := \{1,2\}$; and for example from the Atlas [**47**, p. 18], the G-classes of maximal 2-locals are represented by

$$H_1 \cong M_8 : \Sigma_3 \cong Q_8 : \Sigma_3,$$
$$H_2 \cong \Sigma_4 \cong 2^2 : \Sigma_3,$$

where Q_8 is the quaternion group of order 8. We mention with regard to (c) of Condition 6.2.15 that the intersection H_I is a dihedral group D_8 of order 8, which is of index two in a semidihedral Sylow 2-subgroup $N_G(H_I)$ of M_{11}.

For $i = 1,2$ we set $A_i := \Omega_1 Z\big(O_2(H_i)\big)$; then A_i is elementary abelian of rank i, and these groups define the subgroup simplex $\mathcal{A}_I$ of Notation 6.3.1. We now have a very simple case of the method of 6.2.17 (2): Since A_1 is the center of $H_I \cong D_8$, which in turn lies in $H_2 \cong \Sigma_4$, we conclude that A_1 lies in $O_2(H_2) = A_2$. Thus we have the inclusion-chain $A_1 < A_2$ for the A_i. This gives the hypothesis for

Proposition 6.3.6, with "A" taken as G itself; so the intersection complex $\Delta_{\mathcal{A}_I}$ of Definition 6.3.2 is just the complex $\Delta(\mathcal{A})$ of inclusion chains in the poset $\mathcal{A}$ of G-conjugates of the A_i. Thus in our further nomenclature, $\Delta_{\mathcal{A}_I}$ affords an internal poset complex for M_{11}. In fact, the A_i are representatives of the two G-conjugacy classes in the poset $\mathcal{A}_2(G)$ of elementary abelian 2-subgroups of G; so that $\mathcal{A} = \mathcal{A}_2(G)$.

Furthermore $N_G(A_i)$ is given by the subgroup H_i chosen at the outset. And in Section 7.1 we will follow the standard procedure of checking flag-transitivity, so that the normalizer set-map N extends in this case (much as in the previous Example 6.3.7) to an isomorphism of complexes from our internal poset complex $\Delta_{\mathcal{A}_I}$ to the 2-local complex $\Delta_{\mathcal{H}_I}$ determined by the H_i. $\diamondsuit$

We noted in the above example that the 1-simplex determined by $A_1 < A_2$ represents the unique G-orbit of maximal simplices; that is, the action of G on $\mathcal{A}_2(G)$ is flag-transitive in the sense of Definition 6.4.2. But we emphasize that this behavior of the particular poset $\mathcal{A}_2(G)$ is for the smallest sporadic group M_{11}—and in particular, that the existence of a unique conjugacy class of *maximal* members of $\mathcal{A}_2(G)$ (here represented by A_2)—is not to be expected in the larger sporadic groups (although J_1 does provide another exceptional small case).

Indeed, typically there will be more than one G-class in $\mathcal{A}_2(G)$ of subgroups which are maximal under inclusion; and then we will *not* be able to parametrize G-orbits via simple inclusion-chains as in the poset complex case of Proposition 6.3.6. Some of the A_J will then be different from any A_i; so that we will get a situation of distinctness among the A_J, which is intermediate between those of the earlier examples given by the 2-local complex and by the poset complex. However, usually the situation is not that much more complicated than a poset complex; the remaining special case of the intersection complex $\Delta_{\mathcal{A}_I}$ which we will need most frequently later is characterized by having just two maximal elements:

DEFINITION 6.3.9 (The almost-poset complex: the case of two maximal A_i). Assume in Notation 6.3.1 that I is the union of $I_{n-2} = \{1, 2, \ldots, n-2\}$ with two further indices denoted n and n'; and assume that $A_1 < A_2 < \cdots < A_{n-2}$ and $A_n > A_{n-2} < A_{n'}$—but that A_n and $A_{n'}$ are *not* related by inclusion.

Largely following the analysis in the proof of Proposition 6.3.6, we then see that the incidence relation for a pair i, j of vertex types is "usually" just inclusion—that is, unless $\{i, j\} = \{n, n'\}$, in which case incidence is determined by containing in common a conjugate of the intersection $A_{n,n'} = A_n \cap A_{n'}$, which is proper in both A_n and $A_{n'}$ (and contains A_{n-2}). Incidence for type sets of size larger than two is then inductively determined by inclusions and/or intersections.

Thus most of the incidence relations are determined by inclusions, as would be the case in a poset complex. Consequently we will refer to this subcase of the intersection complex $\Delta_{\mathcal{A}_I}$ of Definition 6.3.2, in which there are two maximal members of the A_i which are otherwise linearly ordered, as an *almost-poset complex*. We apologize for the ungraceful terminology; but hope that it is at least reasonably descriptive. $\diamondsuit$

We mention that this last variant includes a well-known classical example, namely the building for G of Lie type D_n: Here A would be an orthogonal space of dimension $2n$ and plus type, over the relevant finite field; and the A_i would be representatives of the G-classes of totally singular subspaces of A—of the following dimensions i: There are two G-classes of maximal dimension n, represented by A_n and $A_{n'}$; but their intersection $A_n \cap A_{n'} = A_{n-1}$ of dimension $n-1$ is *not* included among the remaining singular subspaces $A_1, \dots, A_{n-2}$ which also define vertices of the building.

However we will also need the almost-poset complex construction for quite a few sporadic groups, such as the following:

EXAMPLE 6.3.10 (An almost-poset complex for McL). We now consider the sporadic group $G := McL$ of McLaughlin; and we will here anticipate certain features of our treatment in Section 7.11.

For example from the Atlas [**47**, p. 100], there are three G-classes of maximal 2-locals satisfying (6.2.9); so technically we should take $n = 3$ with types 3 and $3'$ in the notation of Definition 6.3.9. However, in order to later have the property that $\dim_{\mathbb{F}_2} A_i = i$, for convenience we define our index set as $I := \{1, 4, 4'\}$. The G-classes of maximal 2-locals are represented by groups having the structure

$H_1 \cong 2^{\cdot}\mathrm{Alt}_8$,

$H_4 \cong 2^4 : \mathrm{Alt}_7$,

$H_{4'} \cong 2^4 : \mathrm{Alt}_7$.

As usual, Condition 6.2.15 holds; where in (c) the intersection H_I is a subgroup of structure $2^4 : 2^2 : \Sigma_3$—it is an extension of a Sylow 2-subgroup by a subgroup of order 3.

Setting $A_i := \Omega_1 Z\big(O_2(H_i)\big) = O_2(H_i)$, we get A_i elementary abelian of rank i; the A_i define the subgroup simplex $\mathcal{A}_I$ of Notation 6.3.1, and in particular $N_G(A_i)$ is given by the subgroup H_i chosen above. Via the usual method of 6.2.17 (2), much as in Example 6.3.8, we get the containment relation $A_1 < A_4 \cap A_{4'}$ among the A_i, where that intersection is of rank two. Thus the intersection complex $\Delta_{\mathcal{A}_I}$ of Definition 6.3.2 is an almost-poset complex as in Definition 6.3.9. And in our further nomenclature, $\Delta_{\mathcal{A}_I}$ affords an internal almost-poset complex defined by the 2-local structure of McL.

We see that the complex $\Delta_{\mathcal{A}_I}$ is *not* a subcomplex of the complex $\Delta\big(\mathcal{A}_2(G)\big)$ of inclusion-chains in the poset $\mathcal{A}_2(G)$ of elementary abelian 2-subgroups. For note that $\{A_4, A_{4'}\}$ defines a 1-simplex of $\Delta_{\mathcal{A}_I}$; whereas A_4 and $A_{4'}$ are not related by inclusion, so that they cannot both appear in any inclusion-chain from $\mathcal{A}_2(G)$, that is, in any simplex of $\Delta\big(\mathcal{A}_2(G)\big)$.

However as in earlier examples in this section, in Section 7.11 we follow the standard procedure of verifying flag-transitivity, in order to conclude via Lemma 6.5.4 that the normalizer set-map N in this case induces an isomorphism of complexes from our almost-poset complex $\Delta_{\mathcal{A}_I}$ to the 2-local complex $\Delta_{\mathcal{H}_I}$ determined by the H_i. $\diamondsuit$

As the above example may suggest: If from a 2-local complex $\Delta_{\mathcal{H}_I}$ for some G, we extract the subgroups A_i defined by $\Omega_1 Z\big(O_2(H_i)\big)$, it is then a fairly common occurrence that the internal intersection complex $\Delta_{\mathcal{A}_I}$ defined by members of $\mathcal{A}_2(G)$ has some fairly simple structure like the almost-poset complex in Definition 6.3.9. But sometimes we might need the full generality of the definition of the intersection complex. For example, to revisit once more the case of $G := M_{24}$, we use the type set $I := \{4, 6, 6'\}$ in place of $\{O, T, S\}$, and define the subgroups $A_i := \Omega_1 Z\big(O_2(H_i)\big) = O_2(H_i)$ as above; we find that A_i has rank i, as indicated by the structures of the H_i given in 6.2.13. Then we would need to define all three pairwise incidences via intersections rather than inclusions: for $A_4 \cap A_6$ has rank three, $A_4 \cap A_{6'}$ has rank two, and $A_6 \cap A_{6'}$ has rank four (but is not conjugate to A_4).[14]

Some other constructions of complexes. Naturally occurring complexes related to 2-locals need not arise only from posets of subgroups; sometimes there is a simplicial complex structure which can be viewed essentially independently of the group structure. In fact the 2-local geometry for M_{24}, which we have discussed in various viewpoints above, was originally studied in the following way:

EXAMPLE 6.3.11. (An external complex for M_{24} from the Steiner system) We recall, for example from the Atlas [**47**, p. 94] and our discussion leading up to Example 6.2.10, the Steiner system $\mathcal{S}(5, 8, 24)$ on a 24-set Ω. We let vertices of a simplicial complex Δ be given by the octads, trios, and sextets. We can replace the definition of incidence in Example 6.2.10 by a more natural set-theoretic definition, based on containment in the sense of refinement of partitions: An octad (namely a block of the Steiner system, of size 8) is incident to a trio, if that trio regarded as an 8^3 partition of Ω is a refinement of the octad regarded as a $16, 8$ partition. An octad is incident to a sextet, if that sextet as a 4^6 partition refines the $16, 8$ octad partition. Similarly a trio is incident to a sextet if the 4^6 sextet partition refines the 8^3 trio partition. But we do *not* define incidence among vertices of the same "type"—for example no incidence between two octads; thus a simplex consists of at most one representative each of our usual types $\{O, T, S\}$.

We mention in particular that the relation in (c) of Condition 6.2.15, in which H_I is here an order-3 extension of a Sylow 2-subgroup, now corresponds to mutual incidence of some initially chosen octad, trio, and sextet. In fact, up to conjugation by an element of the symmetric group Σ_{24}, we could take our initial octad O to be $\{1, 2, \ldots, 8\}$; our trio T to be O, together with the sets $\{9, 10, \ldots, 16\}$ and $\{17, 18, \ldots, 24\}$; and our sextet S to be the refinement of each octad in T into its first four and last four elements.

Much as in earlier Example 6.3.7 for M_{24}: If we apply the normalizer set-map N to Δ, we find it induces in this case an isomorphism of complexes with the usual 2-local complex $\Delta_{\mathcal{H}_I}$ defined by the H_i in Definition 6.3.5. For we can check the flag-transitivity on Δ (we mention at the start of Example 6.4.7 (1) that this involves a straightforward adjustment of the argument given there); and then we can make the obvious adaptation of Lemma 6.5.4 to Δ in place of $\Delta_{\mathcal{A}_I}$.

The particular external viewpoint in the present example is reasonably convenient for describing the geometry of M_{24} (and indeed for M_{23} and M_{22} as well). Furthermore it is not necessarily far removed from our earlier viewpoints on the geometry: For example, in the natural map of an octad to its image in the Golay code module A in Example 6.3.7, we find that the simplex given by the incident triple O, T, S maps to the inclusion-chain $A_1 < A_2 < A_4$—so the viewpoint in 6.3.7 would suffice as a poset complex equivalent to the external complex Δ defined above. $\diamondsuit$

[14] As suggested earlier: in many cases such as this one, we could regard our intersection complex $\Delta_{\mathcal{A}_I}$ as roughly corresponding to a *coefficient system* over the specific 2-local complex $\Delta_{\mathcal{H}_I}$, in the sense used by Ronan and Smith (cf. [**22**, Sec. 7.5]). And more general patterns for intersections, such as those for the three distinct pairs here, then reflect the greater freedom of general coefficient systems.

We close the section with a brief discussion of a variant of the 2-local complex $\Delta_{\mathcal{H}_I}$ of Definition 6.3.5—which is approached essentially from the viewpoint of the *opposite* poset.

First however we introduce a further piece of notation, which is also useful more generally in relating p-local geometries to the collections of p-subgroups discussed in Chapter 5. (We also often use it in parallel with the normalizer set-map N of Definition 5.6.13.)

DEFINITION 6.3.12 (The map $\mathcal{O}_p$ on a subgroup poset). Given a prime p and a G-poset $\mathcal{Q}$ consisting of finite groups, we let $\mathcal{O}_p$ denote the mapping taking $H \in \mathcal{Q}$ to the subgroup $O_p(H)$. Then the image $\mathcal{O}_p(\mathcal{Q})$ is some subposet of $\mathcal{Q}$, consisting of p-groups. But as in Definition 5.6.13, we emphasize that *in general* $\mathcal{O}_p$ *is only a map of sets.* However in many favorable circumstances, $\mathcal{O}_p$ will turn out to induce a poset map.

Typically our $\mathcal{Q}$ will consist of suitable p-local subgroups of G; so that the members $O_p(H)$ of $\mathcal{O}_p(\mathcal{Q})$ will be nontrivial p-groups, and hence we will obtain some subposet of $\mathcal{S}_p(G)$.

For example, we mentioned in Example 5.1.1 that for G of Lie type in characteristic p, the unipotent radicals $\mathcal{U}$ determine the poset opposite to that of the parabolics $\mathcal{P}$. In this case, $\mathcal{O}_p$ is order-reversing on $\mathcal{P}$; and it even induces a poset anti-isomorphism of $\mathcal{P}$ with $\mathcal{U}$. Indeed from composition with the poset anti-isomorphism afforded in this case by the normalizer set-map N as in our discussion after Definition 5.6.13, we obtain poset isomorphisms $\mathcal{P} = \mathsf{N}\big(\mathcal{O}_p(\mathcal{P})\big)$ and $\mathcal{U} = \mathcal{O}_p\big(\mathsf{N}(\mathcal{U})\big)$.

For this context, we now define a complex where the incidence relation arises not from intersection of 2-local subgroups, but instead from suitable products of 2-subgroups.

DEFINITION 6.3.13 (A "dual" variant of $\Delta_{\mathcal{H}_I}$). Assume that we are given a 2-local complex $\Delta_{\mathcal{H}_I}$ as in Definition 6.3.5; define $U_i := O_2(H_i)$ for $i \in I$. Assume that the set products $U_J := \Pi_{i \in J} U_i$ for each nonempty $J \subseteq I$ are in fact subgroups; so that they define a subgroup simplex $\mathcal{U}_I$ appropriately dual to the notion in Definition 4.6.2.

We define a complex with vertices given by the G-conjugates of the U_i; and a set of vertices determines a simplex if and only if it is given, for some nonempty subset $J \subseteq I$, by conjugates $g_i^{-1} U_i g_i$ (for $i \in J$ and $g_i \in G$) satisfying:

For each nonempty subset $K \subseteq J$, the set product $\Pi_{i \in K}\, g_i^{-1} U_i g_i$ is some G-conjugate of U_K.

We may write $\Delta_{\mathcal{U}_I}$ for this resulting complex corresponding to $\Delta_{\mathcal{H}_I}$. (As usual it does define a complex in view of (2.7.4).)

Note that the poset of simplices of $\Delta_{\mathcal{U}_I}$ is not necessarily equal to the poset obtained by applying $O_2(-)$ to the stabilizers of the simplices in $\Delta_{\mathcal{H}_I}$. But we will see that in many favorable cases, such an equality will in fact hold. ◇

We discuss briefly the form of the conditions in this definition: Usually, but not always, it will be a consequence of the embedding condition (c) in Condition 6.2.15 (using variants of the techniques demonstrated in Remark 6.2.17) that the U_i will lie in H_I. Hence for $i \neq j \in I$, it follows that U_i lies in H_I, which in turn lies in $H_j \leq N_G(U_j)$; this is the reason we can usually expect the set product of the

$U_i (i \in J)$ to in fact afford a subgroup, which lies in $O_2(H_J)$. Again usually, but not always, equality holds.

In our standard example of M_{24}, the 2-local complex $\Delta_{\mathcal{H}_I}$ of Example 6.2.10 does indeed produce a "dual" complex $\Delta_{\mathcal{U}_I}$ of 2-groups as in Definition 6.3.13. Furthermore much as in Example 6.2.12, we may identify a simplex $\{U_i : i \in J\}$ with the subgroup U_J. Thus the posets of simplices in the two complexes can be realized as posets of subgroups of G; and these posets are related by the set-maps N and $\mathcal{O}_2$, which in this case give poset anti-isomorphisms. (It is easy to check the details of flag-transitivity on $\Delta_{\mathcal{U}_I}$ as in Example 6.4.7 (2); and to correspondingly adapt Lemma 6.5.4.)

However, we mention that there are also a number of cases where we do not have the condition $U_i \leq H_I$ for all $i \in I$; but in which the product $\prod_{i \in J} U_i$ is still a subgroup. Typically in these cases we will denote the product by B_J rather than U_J; and usually we will expect B_J to define a member of the poset $\mathcal{B}_2(G)$ of 2-radical subgroups.

6.4. Flag-transitive action on geometries

In the previous section, we indicated various constructions of a simplicial complex Δ as a 2-local geometry; that is, a geometry based on maximal 2-local subgroups H_i, and the subgroup simplex $\mathcal{H}_I$ they determine in Definition 4.6.2. In particular, our constructions produce simplices (such as $\sigma_J = \{A_i : i \in J\}$ in the intersection complex $\Delta_{\mathcal{A}_I}$) with a "type" given by a nonempty subset $J \subseteq I$.

However, we also mentioned at Warning 6.3.4 that such a simplex does not necessarily represent the *only* G-orbit of simplices of type J in the quotient Δ/G. In order to have that condition of a unique orbit of type J—the desirability of which we had indicated at Remark 5.6.11—we need the further property of flag-transitive action of G on Δ. So in this section, we review that standard condition, along with some typical methods for verifying it.

It will be sufficient for the purposes of the present work to use the context given by the intersection complex $\Delta_{\mathcal{A}_I}$ in Definition 6.3.2; though of course we could equally well discuss many of the concepts in this section in considerably greater generality (for example, as in Tits [**120**]).

Complexes with type. In the previous section, we have been informally using the language of the *type* of a simplex; and we now more formally observe that our definitions do indeed produce such a structure, with some consequent properties:

DEFINITION 6.4.1 (The type mapping for $\Delta_{\mathcal{A}_I}$). Assume an intersection complex $\Delta_{\mathcal{A}_I}$ as in Definition 6.3.2. Then the complex $\Delta_{\mathcal{A}_I}$ has the *type set* I in the language of Tits in [**120**, Sec. 1.2]: The relevant *type mapping* τ is defined in the obvious way—first Notation 6.3.1 gives the vertices v of $\Delta_{\mathcal{A}_I}$ as the conjugates of the initial data $A_i (i \in I)$; so that defining $\tau(A_i) := i$ and extending by G-conjugacy,

(a) Each vertex v of $\Delta_{\mathcal{A}_I}$ has a type $\tau(v) \in I$.

Simplices σ of higher dimensions are then defined as sets of vertices which are mutually incident; note that the distinctness of the G-classes[15] given by the A_i in

[15]In fact, we usually begin our constructions with the stabilizers H_i; when we take them to satisfy the non-conjugacy in Condition 6.2.15 (b), the non-conjugacy of the A_i is a consequence.

Notation 6.3.1 (a) guarantees that vertices in σ (and hence their types) are distinct. Thus we can use (a) above to extend τ to all simplices:

$$\tau(\sigma) := \{\tau(v) : v \in \sigma\},$$

so that σ has a type which is a nonempty subset J of I. In particular,

(b) No two vertices of the same type define an edge in $\Delta_{\mathcal{A}_I}$.

We had also noted after Definition 6.3.2 that the full initial set $\sigma_I = \{A_i : i \in I\}$ is a simplex, with full type I, and so it follows that

(c) There are simplices of each nonempty type $J \subseteq I$.

However, we emphasize that although A_J is contained in each A_K with $J \subseteq K \subseteq I$, it is not automatic from the definitions that every simplex of type J is contained in a simplex of type K; we will return to this point later.

We also repeat in the present language the observation made earlier in Remark 6.3.3:

Since the G-action preserves the types of the vertices in (a), the action of G on $\Delta_{\mathcal{A}_I}$ is *type-preserving*: that is, $\tau\big((\sigma)g \big) = \tau(\sigma)$ for all simplices σ and $g \in G$. Then it follows from this fact using (b) (or just from the non-conjugacy of the A_i) that

(d) The stabilizer in G of σ fixes each vertex of σ.

That is, the G-action on $\Delta_{\mathcal{A}_I}$ is admissible (Definition 2.5.5). ♢

Flag-transitivity. Since the action of G on $\Delta_{\mathcal{A}_I}$ is type-preserving in the sense just indicated, the form of flag-transitivity which will be relevant for our purposes is:

DEFINITION 6.4.2. We say that G is *flag-transitive* on $\Delta_{\mathcal{A}_I}$ if G is transitive on the simplices of type J in $\Delta_{\mathcal{A}_I}$, for each nonempty subset $J \subseteq I$.

Notice that if flag-transitivity holds, then the orbit CW-complex $\Delta_{\mathcal{A}_I}/G$ is a simplicial complex, given just by the simplex $\sigma_I = \{A_i : i \in I\}$. In particular, any simplex is then contained in a simplex of full type I. Furthermore, we had already indicated in Remark 5.6.11 that this leads us to normalizer decompositions which are indexed by the subgroup simplex $\mathcal{H}_I$ (arising as $\mathsf{N}(\Delta_{\mathcal{A}_I}/G)$ using the language of the normalizer set-map N indicated there), so that we get the simplex form given in (5.6.12).

As a familiar example of flag-transitivity, we will review below—in our notation of $\Delta_{\mathcal{A}_I}$—the complex given by classical projective space. This complex, and suitable G-subcomplexes, will be involved "locally" (that is, as links of simplices) in many of our later geometries.

EXAMPLE 6.4.3 (Projective space and flag-transitivity). Let V denote a vector space of dimension n over a finite field F of characteristic p. Its automorphism group $G := GL_n(F)$ is a group of Lie type in characteristic p. Let $\mathbf{P}(V)$ denote the poset given by the classical *projective space*[16] of V, namely the collection of nonzero proper F-linear subspaces of V under inclusion. Then the complex $\Delta\big(\mathbf{P}(V)\big)$ of

[16]Here we are using the word "space" for this poset as is customary in the geometric literature. Of course the nerve of this poset does provide a space in the sense of simplicial set as in our usual convention.

inclusion-chains in $\mathbf{P}(V)$ consists of the classical *flags* in V—the original meaning of the term "flag" in the mathematical literature.

Of course the above flag complex for projective space provides one view of the building for $GL_n(F)$. (And indeed the prospect of finding analogies of the natural vector-space geometries for the classical matrix groups, but applicable to Lie type groups corresponding to exceptional Dynkin diagrams, provided a major motivation for Tits in developing the theory of buildings.)

If we choose a full flag, namely subspaces A_i of dimension i forming a chain $A_1 < A_2 < \cdots < A_{n-1}$, then using the type set $I := \{1, 2, \ldots, n-1\}$ we can construct the corresponding intersection complex $\Delta_{\mathcal{A}_I}$ of Definition 6.3.2. Indeed we have the poset complex subcase of Proposition 6.3.6, and we see that $\Delta_{\mathcal{A}_I}$ is just the flag complex $\Delta(\mathbf{P}(V))$ above. (In our further nomenclature, this view of the building affords an external poset complex for G.)

Now any ordered basis of V defines such a full flag of $\mathbf{P}(V)$, by taking the obvious linear spans: namely of the first vector; then of the first two vectors; and so on—through the first $n-1$ vectors. Furthermore it is a standard theorem in linear algebra that G is transitive on ordered bases; and this shows that G is transitive on full flags in V—and hence on flags of type J, for each nonempty subset $J \subseteq I$; as desired to exhibit the flag-transitivity of G. ◇

As suggested by the above example, flag-transitivity in fact holds for any G of Lie type on its building. We can state this basic observation of Tits in the following form:

PROPOSITION 6.4.4 (Flag-transitivity for buildings). *Let G be of Lie type in characteristic p; and exhibit the building Δ of G as the p-local complex based on maximal parabolics $H_i (i \in I)$ as in Example 6.2.2. Then G is flag-transitive on Δ.*

PROOF. Of course Δ has the form (for general p) of the local variant $\Delta_{\mathcal{H}_I}$ in Definition 6.3.5 of the intersection complex $\Delta_{\mathcal{A}_I}$ of Definition 6.3.2. In Remark 6.2.4, we essentially identified a simplex of type J with a G-conjugate of the parabolic H_J over the fixed Borel subgroup H_I. So since arbitrary parabolics are defined (cf. 5.1.1) as overgroups of Borel subgroups, the transitivity of G on parabolics of type J essentially follows from:

(a) the G-conjugacy of the Borel subgroups; and
(b) the characterization (mentioned in Example 4.6.3) of the parabolic subgroups above some fixed Borel subgroup H_I in the form H_J of intersections of the H_i.

See for example [**43**, Sec. 8.3] for these statements. □

It also follows that G is also flag-transitive on the barycentric subdivision as in 2.7.7 of Δ, given by the inclusion-chains from the poset $\mathcal{P}$ of parabolics—which is the more usual view of the building in Tits' original description.

In particular by Remark 5.8.5, the cohomology decomposition over the building in Example 5.1.18 is in fact indexed by the subgroup simplex $\mathcal{H}_I$ of the parabolic subgroups over H_I as in (5.8.6). It has been a main goal of the present work to obtain a similar formula for the mod 2 cohomology of each sporadic group.

Techniques for verifying flag-transitivity. In view of Remark 5.6.11, for our applications in Chapter 7 we will wish to establish flag-transitivity for each sporadic group G on some suitable 2-local geometry, in the general form $\Delta_{\mathcal{A}_I}$ of an intersection complex.

Since the property of flag-transitivity is often left implicit in the group theory literature on those geometries, in this work we will try to give at least some more explicit indication of why flag-transitivity holds in those cases. For that purpose, we now review some sufficient conditions for establishing that property.

We observe first that, since Definition 6.3.2 gives the vertices of type $i \in I$ as the G-conjugates of A_i, we automatically have transitivity of G on vertices of each type.

So it should be fairly clear from Definition 6.4.2 of flag-transitivity that we can take an inductive approach, namely: for the vertex $\{A_i\}$ of each type $i \in I$, check flag-transitivity of the stabilizer H_i of $\{A_i\}$ on the subcomplex of the simplicial complex $\Delta_{\mathcal{A}_I}$ given by the link of $\{A_i\}$.

Indeed even a little less is required:

PROPOSITION 6.4.5. *For flag-transitivity on $\Delta_{\mathcal{A}_I}$, it suffices to check flag-transitivity of H_i on the link of $\{A_i\}$—for all but one of the types $i \in I$.*

(In many cases the diagram given for G will indicate, by means of the subdiagram with the node for i removed, the particular flag-transitive geometry for H_i afforded by the link of $\{A_i\}$.)

PROOF. By our remark above about transitivity on individual types of vertices, it suffices to examine simplices defined by at least two vertices. Such a simplex must then appear in the subcomplex of $\Delta_{\mathcal{A}_I}$ given by the link of a vertex of some type i, where i runs through a subset of I containing all but one of the types. □

Notice however that it does not necessarily suffice to check just one type i: since as noted after (c) of Definition 6.4.1, it is not *a priori* clear from the definitions that any simplex is contained in a simplex of full type I. (Though we mentioned after Definition 6.4.2 that containment in a flag of full type does follow, once flag-transitivity is eventually established.)

We can ordinarily verify the above inductive requirement in a straightforward way: since the 2-local geometries $\Delta_{\mathcal{A}_I}$ of interest are built up inductively (as mentioned in our brief discussion of diagram geometries, for example in the case of M_{24} at Digression 6.2.14), using links which are geometries (such as buildings) which are known to be flag-transitive. Indeed we can typically quote one of the following standard cases:

REMARK 6.4.6 (Some conditions for flag-transitivity). For $\{A_i\}$ a vertex of an intersection complex $\Delta_{\mathcal{A}_I}$ with type set I, we get flag-transitivity of the stabilizer H_i on the link of $\{A_i\}$ if one of the following holds:

(1) $|I| = 2$, and H_i is transitive on vertices adjacent to $\{A_i\}$. Notice since $|I| = 2$ that by the Proposition, it suffices to check this condition for vertices $\{A_i\}$ of just one type i. We mention that the amalgam situation discussed after Diagram (4.5.6) is a special case: Here the simplicial complex is just the graph defined by cosets, so that the group action is defined by translation rather than by conjugation. Since in practice we choose the H_i to be maximal 2-local subgroups as in Condition 6.2.15 (a), they are self-normalizing; so as usual the representation on conjugates is equivalent to that on cosets. Since the edges on the vertex H_1 are

just the cosets in H_1 of $H_{12} = H_1 \cap H_2$, the stabilizer H_1 of that vertex is indeed transitive on those cosets.

(2) The link of $\{A_i\}$ is isomorphic to a projective space $\mathbf{P}(V)$ as in Example 6.4.3; and H_i induces $GL(V)$ (or some flag-transitive subgroup) on $\mathbf{P}(V)$. Here we observed the flag-transitivity of G in Example 6.4.3. Indeed we could replace $\mathbf{P}(V)$ by the subposet defined using only a subset of the possible dimensions of subspaces.

(3) The link of $\{A_i\}$ is isomorphic to a building, and H_i induces the group of that building on the link. Here we observed the flag-transitivity in Proposition 6.4.4. Again for the link we could use just some subcollection of the possible types in the building.

Finally we mention an obvious inductive extension of our collections of methods given so far, namely checking that:

(4) The link of $\{A_i\}$ is given by a geometry for which we have previously established flag-transitivity for H_i. $\diamondsuit$

Since in this work we are not really discussing the details of diagrams for geometries, we simply mention here (but will not make use of) the result of Aschbacher [**19**] giving flag-transitivity of a group geometry with a "string diagram" (one with no loops or branching).

Often flag-transitivity is easiest to check "directly" in the context of an external poset complex as in Proposition 6.3.6—or sometimes in an independent external complex as in Example 6.3.11. Typically we will give such arguments for individual sporadic groups in Chapter 7 in a fairly abbreviated form; so in order to illustrate the standard form of such arguments, we now indicate some examples for which we give fuller details.

EXAMPLE 6.4.7 (Flag-transitivity for some M_{24} geometries). We return to the 2-local geometry of M_{24}, in several versions.

(1) *The external poset complex of Example 6.3.7.* Recall that this geometry is based on the chain $A_1 < A_2 < A_4$ in the irreducible Golay code module A. In particular it arises there in the poset complex subcase of the intersection complex $\Delta_{\mathcal{A}_I}$. (The discussion below also uses the related language of the Steiner system, from the external complex of Example 6.3.11; that construction is based on a corresponding triple $\{O, T, S\}$ given by an incident octad, trio, and sextet, and so the combinatorial argument for flag-transitivity below can be adapted fairly straightforwardly to apply instead to 6.3.11.)

We will implement our inductive method of checking flag-transitivity in links given in Proposition 6.4.5—at each step using a relevant part of Remark 6.4.6.

We begin with the link of the vertex $\{A_4\}$, corresponding to the sextet S. Our abbreviated treatment would simply be to observe that this link affords the building for $H_4/O_{2,3}(H_4) \cong Sp_4(2)$ (as suggested by the double-bond subdiagram for node S in the diagram in Digression 6.2.14); so that flag-transitivity is covered by Remark 6.4.6 (3).[17] For a more detailed and direct treatment, one checks that this building is realized via suitable subspaces of the 4-dimensional H_4-submodule A_4 of A: Namely the sextet S is incident with 15 octads and 15 trios; and the corresponding H_4-conjugates of A_1 and A_2 are the totally isotropic linear subspaces

[17]Of course the polar space structure below is also the analogue, in symplectic space, of the full projective space appearing in 6.4.6 (2).

of dimensions 1 and 2 in A_4. These spaces (or read projectively, points and lines) define the classical *polar space* geometry (in this case, a generalized quadrangle) for the symplectic group $Sp_4(2)$; with conjugates of the chain $A_1 < A_2$ providing the maximal simplices in the link.

We proceed rather similarly for the link of $\{A_2\}$, corresponding to the trio T. In the briefer overview, that link is the join of the three points (defined by octads in T) of a projective line, with the seven points (defined by sextets refining T) of a projective plane over $\mathbb{F}_2$; and this structure is a truncation of the full product building for the Lie type group $H_2/U_2 \cong L_2(2) \times L_3(2)$ (as suggested by the subdiagram at node T in 6.2.14); so that flag-transitivity is covered using Remark 6.4.6 (3). But again, a more detailed and direct realization of the link can be obtained within the module A. Namely we let $\bar{B}_2$ denote the 6-dimensional quotient B_2/A_2—where B_2 of dimension 8 is the span of the conjugates of A_4 lying over the 2-dimensional H_2-submodule A_2 of A. Then $\bar{B}_2$ affords the tensor product of natural modules for $L_2(2)$ and $L_3(2)$—in which the three H_2-conjugates of $\bar{A}_4$ define 2-subspaces; while the seven conjugates of A_1 determine 3-subspaces such as $\bar{B}_1$—using the 5-space B_1 indicated in the argument below for the link of $\{A_1\}$. Any pair from these 2-subspaces and 3-subspaces will always intersect in some 1-subspace, which exhibits the join (complete-graph) structure of the link.

We have now checked flag-transitivity for links of two of the three types, which suffices by Proposition 6.4.5 for flag-transitivity of M_{24} on Δ.

But for completeness, we also indicate a similar argument for the link of $\{A_1\}$, determined by the octad O. In the brief form, this link consists of 15 trios and 35 sextets, exhibiting the points and lines of the projective space for $H_1/U_1 \cong L_4(2)$ (as suggested by the subdiagram at node O in 6.2.14); so that flag-transitivity is covered this time by Remark 6.4.6 (2). For a more direct and detailed realization of the link in the module A: One checks that the 15 H_1-conjugates of A_2 above A_1 span a 5-subspace B_1 of A; and in the quotient 4-space $\tilde{B}_1 := B_1/A_1$, these trio 2-spaces define projective points such as $\tilde{A}_2$. Furthermore the 4-space A_4 intersects B_1 in a 3-space B_4 over A_1, so that the 35 H_1-conjugates of A_4 determine projective lines such as $\tilde{B}_4$. Indeed the conjugates of $\tilde{A}_2$ and $\tilde{B}_4$ exhibit all the 15 points and 35 lines giving the truncation of the projective space $\mathbf{P}(V)$ indicated in Digression 6.2.14 as defining this link.

To summarize: This completes (in fact for *all* vertex types) the proof of the inductive condition 6.4.5; so we have shown flag-transitivity of $G = M_{24}$ on the 2-local geometry given in Example 6.3.7.

(2) *The 2-local complex of Example 6.2.10.* For comparison with the above "direct" proof—and also for use in the hypothesis of the standard isomorphism Lemma 6.5.4—we also indicate a proof of flag-transitivity for M_{24} on the 2-local complex $\Delta_{\mathcal{H}_I}$ of Definition 6.3.5, which we originally introduced for M_{24} at Example 6.2.10.

In principle, working with $\Delta_{\mathcal{H}_I}$ might require detailed knowledge of the subgroup structure of G. But we emphasize that our essentially-local argument for flag-transitivity below really uses just the observations we had made earlier at Remark 6.2.17 (1): that is, it uses the embedding condition 6.2.15 (c), in realizing the links recorded in the diagram in Digression 6.2.14.

In more detail: For $i = 4$, we saw at Remark 6.2.17 (1) that the H_j in the link of $\{H_4\}$ determine by intersection with H_4 the maximal parabolics (and their incidences) of $\bar{H}_4 = H_4/O_{2,3}(H_4) \cong Sp_4(2)$. This gives an isomorphism of that link with the realization of the building of $Sp_4(2)$ provided by the viewpoint of Example 6.2.2; so that flag-transitivity follows using Remark 6.4.6 (3). In the case of $i = 2$, Remark 6.2.17 (1) similarly leads to an isomorphism of the link of $\{H_2\}$ with a truncation of the building for $\tilde{H}_2 = H_2/U_2 \cong L_2(2) \times L_3(2)$ which uses only the points of the $L_3(2)$ factor; giving flag-transitivity by 6.4.6 (3). Finally for $i = 1$, 6.2.17 (1) gives an isomorphism of the link of $\{H_1\}$ with the truncation of the building for $H_1^* = H_1/U_1 \cong L_4(2)$ using only points and lines, giving flag-transitivity by 6.4.6 (2).

In particular, we see that the proof of flag-transitivity for the geometry $\Delta_{\mathcal{H}_I}$ for M_{24} is a comparatively superficial variant of the proof for the geometry $\Delta_{\mathcal{A}_I}$ in part (1) above: the only difference lies in the particular method used for identifying the links of the vertices of the types $i \in I$. $\diamondsuit$

We observe finally that Remark 6.4.6 (4) is the obvious inductive extension of our methods, and typically we do proceed inductively. For example we will see, in the case of the 2-local geometry for Co_1 in Section 7.15, that a suitable vertex has link given by the 2-local geometry of M_{24}, and its stabilizer induces M_{24}—whose flag-transitivity we have just established in Example 6.4.7; so that result is available for us to quote in treating Co_1.

6.5. Some equivalence methods for 2-local geometries

Some fundamental methods for establishing homotopy equivalences for posets and complexes are indicated in Section 1 of Quillen's paper [**104**]—indeed we had mentioned some equivalence results which were obtained using those methods during our historical discussion in Section 5.1. In this work, we will typically use some extensions of those techniques developed for the more specific context of 2-local geometries, such as those in Smith and Yoshiara [**116**]. So we review several such methods in this final section.

General methods relating posets and complexes. The following well-known elementary technique was stated for simplicial complexes in [**116**]; but it will be convenient to state it here in the context of posets:

LEMMA 6.5.1 (Cf. 1.5 in [**116**]). (1) *Suppose that x is an element of a poset $\mathcal{E}$, such that the link in $\Delta(\mathcal{E})$ of x is contractible. Then $\mathcal{E}$ is homotopy equivalent to $\mathcal{E} \setminus \{x\}$.*

(2) *If further $\mathcal{E}$ is a G-poset, and the link of x is contractible via a G_x-map, then the removal of the G-orbit of x is a homotopy equivalence via a G-map.*

We mention that assuming "G_x-contractible" in (2) then leads to the stronger conclusion of G-homotopy equivalence. And indeed the further hypothesis of G_x-contractibility will hold in most of our arguments in this chapter. However, we have chosen to use only the weaker form of equivalence in (2); since sometimes our arguments also involve relations with the centralizer or subgroup decompositions, and we saw at Warnings 5.4.5 and 5.5.5 that some of the the mappings there exhibit only the weaker condition.

REMARK 6.5.2. In the most frequent applications of Lemma 6.5.1, one of the subposets $\mathcal{E}_{<x}$ or $\mathcal{E}_{>x}$ will have a unique maximal or minimal element y, so that the link L of x is a cone on y, and hence contractible. Furthermore when $\mathcal{E}$ is a G-poset, the stabilizer G_x will preserve that cone point y, so that the poset contraction to that cone point is a G_x-equivariant map.

We will also use a natural extension of the method of Proposition 1.4 in [**116**], which is motivated by our use in this work of the intersection complex $\Delta_{\mathcal{A}_I}$ of Definition 6.3.2.

We recall first that [**116**, 1.4] is a simple application of the method of "closed sets in products" of Quillen in [**104**, Cor. 1.8]—a method which is particularly useful for homotopy equivalence of a poset with a simplicial complex.

Now we saw in Proposition 6.3.6 that in the special case where the subgroups $A_i (i \in I)$ form an inclusion-chain, the intersection complex $\Delta_{\mathcal{A}_I}$ becomes just the poset complex $\Delta(\mathcal{A})$ on the poset $\mathcal{A}$ given by the G-conjugates of the A_i. However, we will need to work with more general variants of $\Delta_{\mathcal{A}_I}$ such as the almost-poset complex of Definition 6.3.9; and we will need to establish homotopy equivalence via a G-map with the complexes $\Delta(\mathcal{E})$ of inclusion-chains in suitable corresponding posets $\mathcal{E}$ such as those in Chapter 5. In particular, it will be natural to require that $\mathcal{E}$ should contain not just the conjugates $\mathcal{A}$ of the A_i, but also the conjugates of the intersections $A_J (J \subseteq I)$ which define the incidence relation in the complex $\Delta_{\mathcal{A}_I}$. Hence the hypotheses corresponding to [**104**, Cor. 1.8] appear in the following form:

LEMMA 6.5.3. *Assume an index set I, and the intersection G-complex $\Delta_{\mathcal{A}_I}$ of Definition 6.3.2, defined as in Notation 6.3.1 by subgroups $A_i (i \in I)$ of a group A on which G acts. Assume that $\mathcal{E}$ is a G-subposet of the poset of subgroups of A, satisfying:*

(a) *$\mathcal{E}$ contains $\mathcal{A}_I$ (i.e., contains $A_J = \bigcap_{i \in J} A_i$ for each nonempty $J \subseteq I$).*

For each simplex σ of type J in $\Delta_{\mathcal{A}_I}$, say $\sigma := \{g_i^{-1} A_i g_i : i \in J, g_i \in G\}$, denote the intersection of the indicated conjugates by $E_\sigma := \bigcap_{i \in J} g_i^{-1} A_i g_i$. Then (a) implies

(a′) *$\mathcal{E}$ contains E_σ for each $\sigma \in \Delta_{\mathcal{A}_I}$.*

Now assume further for each $E \in \mathcal{E}$ that:

(b) *the subcomplex $L_E := \{\sigma \in \Delta_{\mathcal{A}_I} : E \leq E_\sigma\}$ of $\Delta_{\mathcal{A}_I}$ is contractible.*

Then $\Delta_{\mathcal{A}_I}$ is homotopy equivalent to $\Delta(\mathcal{E})$. Furthermore if each L_E is contractible via a G_E-map, then the homotopy equivalence is via a G-map.

PROOF. We recall first that the incidence relation in Definition 6.3.2 on the conjugates of the A_i defining σ requires that the intersection E_σ be a G-conjugate of A_J. Now $A_J \in \mathcal{E}$ by (a), and $\mathcal{E}$ is a G-poset by hypothesis; hence $E_\sigma \in \mathcal{E}$, so that (a′) holds.

Under the additional hypothesis (b), we simply make some straightforward adjustments to the proof of [**116**, 1.4]: We have $\mathcal{E}$ in the role of the poset "$\mathcal{A}_p(G)$" there. Then E_σ plays the role of the intersection "I_σ", so that the closed set "$\mathcal{I}$" consists of pairs (E, σ) from $\mathcal{E} \times \Delta_{\mathcal{A}_I}$ with $E \leq E_\sigma$; this set is closed in the appropriate sense, since E_σ is defined by the intersection of terms for the vertices of the simplex σ. Then the poset "$\mathcal{I}_\sigma$" for σ in the projection of $\mathcal{I}$ on $\Delta_{\mathcal{A}_I}$ is the poset $\mathcal{E}_{\leq E_\sigma}$—which is contractible, since it has unique maximal element E_σ in $\mathcal{E}$ in view of (a′). Indeed by that uniqueness, the contraction is a G_σ-map (cf. the argument in Remark 6.5.2). Hence this condition replaces the hypothesis "$O_p(I_\sigma) > 1$" which gives the corresponding contractibility in [**116**, 1.4]. Finally the role of the the complex "$\mathcal{I}_E$" of E in the projection of $\mathcal{I}$ on $\mathcal{E}$ is played by the subcomplex

$$L_E = \{\sigma \in \Delta_{\mathcal{A}_I} : E \leq E_\sigma\};$$

and this is assumed contractible by hypothesis (b). Thus [**116**, 1.4] yields the homotopy equivalence; the extension to a homotopy equivalence via a G-map follows under the indicated strengthening of (b) with a G_E-map. □

We mention that further assuming G_E-contractibility of L_E leads in fact to a G-homotopy equivalence.

Relating the 2-local complex $\Delta_{\mathcal{H}_I}$ to a variant $\Delta_{\mathcal{A}_I}$. We had mentioned, in introducing the intersection complex $\Delta_{\mathcal{A}_I}$ of Definition 6.3.2, that the generality was primarily intended as a way of encompassing many different *variants* in the literature of the 2-local complex construction $\Delta_{\mathcal{H}_I}$ of Definition 6.3.5. In this subsection, we indicate one standard sufficient condition for verifying that a given $\Delta_{\mathcal{H}_I}$ and $\Delta_{\mathcal{A}_I}$ are indeed isomorphic as simplicial complexes.

One could attempt to address this question entirely in the abstract. But in practice, it typically follows easily from checking just that our groups actions are flag-transitive. In essence, we are simply observing that the familiar fact

A transitive permutation representation is determined by a point stabilizer

extends in a straightforward way to the observation

A flag-transitive complex is determined by stabilizers of vertices in a full flag.

Notice in particular that in this statement we are indicating the *set* of subgroups (our H_i) stabilizing vertices, rather than the single subgroup giving the intersection (our H_I) of those stabilizers. Here we are again continuing the warning indicated in Remark 6.2.4; namely we will often get flag-transitivity even in the case where H_I may stabilize more than one full flag. (See for example M_{11} in Section 7.1.)

The version of the above observation that we will use in practice takes the form:

LEMMA 6.5.4. *Assume self-normalizing subgroups H_i of G, defining the 2-local complex $\Delta_{\mathcal{H}_I}$ of Definition 6.3.5. Also assume groups A_i with $N_G(A_i) = H_i$, defining the intersection complex $\Delta_{\mathcal{A}_I}$ of Definition 6.3.2. Finally assume that the action of G on both complexes is flag-transitive.*

Then the normalizer set-map N of Definition 5.6.13 in fact induces an isomorphism of simplicial complexes from $\Delta_{\mathcal{A}_I}$ to $\Delta_{\mathcal{H}_I}$. (In this situation, we just write $\mathsf{N}(\Delta_{\mathcal{A}_I}) = \Delta_{\mathcal{H}_I}$.)

PROOF. We mention first that our hypothesis that the H_i should be self-normalizing corresponds to our preference for working with complexes consisting of G-conjugates; it could be released if the complexes are instead defined via G-cosets. But we will in practice always have the self-normalizing condition, because of choosing the H_i to be maximal 2-local subgroups as in Condition 6.2.15 (a).

We first view N just as a set-map, from the Cartesian product of the G-conjugates of the $A_i (i \in I)$, to the product of the conjugates of the H_i. The map is G-equivariant just from its definition; and since we have $N_G(A_i) = H_i$, the above familiar fact about transitive permutation representations guarantees that N induces a set bijection on these products. In particular, the map remains injective when we restrict to the subset of the first product given by simplices.

Next by flag-transitivity of G on $\Delta_{\mathcal{A}_I}$, any simplex in the first product, say of type $J \subseteq I$, is G-conjugate to the standard simplex $\{A_j : j \in J\}$—whose image under N is in turn the standard simplex $\{H_j : j \in J\}$ of $\Delta_{\mathcal{H}_I}$, and G-conjugates of this simplex are by definition simplices. So by G-equivariance, the restriction of N to the subset $\Delta_{\mathcal{A}_I}$ of the first product is a simplicial map.

Finally by flag-transitivity of G on $\Delta_{\mathcal{H}_I}$, any simplex in the second product is G-conjugate to a standard simplex $\{H_j : j \in J\}$—which we just saw is the image under N of the standard simplex $\{A_j : j \in J\}$ of $\Delta_{\mathcal{A}_I}$. So by G-equivariance, the restriction of N surjects $\Delta_{\mathcal{A}_I}$ on $\Delta_{\mathcal{H}_I}$.

Similar arguments in the other direction apply to the set-map inverse of the restriction of N; completing the proof. □

A quotient relationship arising in some sporadic geometries. For our main theorems for each sporadic group in Chapter 7, we need to produce a flag-transitive 2-local geometry, say of form $\Delta_{\mathcal{H}_I}$, along with a G-map inducing a homotopy equivalence to one of the usual ample and normalizer-sharp collections indicated at the end of Chapter 5. So in general, it will suffice to work for example with homotopy equivalence methods such as those indicated at the start of this section.

But we have also mentioned that in many cases, we can in fact obtain statements stronger than just homotopy equivalence, such as isomorphisms at the level of posets or complexes; the previous Lemma 6.5.4 affords a standard class of such isomorphisms.

However, it is important to caution[18] that we will also encounter a number of situations which appear to be very similar to the hypothesis of Lemma 6.5.4—but where the relationship turns out to be "not quite" an isomorphism of complexes, but instead a homotopy equivalence via a quotient map of posets, whose kernel induces only a comparatively small number of identifications.

As a brief initial summary: Sometimes we will compare a flag-transitive complex, say of form $\Delta_{\mathcal{H}_I}$, with a suitable poset $\mathcal{Q}$ of subgroups of G; and for all $i \in I$ we might get $N_G(Q_i) = H_i$—where the Q_i are representatives of the G-conjugacy classes of minimal members in the poset $\mathcal{Q}$. This resembles the condition $N_G(A_i) = H_i$, which gives the bijection at the level of vertices of the complexes in the setup for Lemma 6.5.4. However we may not get any analogous isomorphism for the poset $\mathcal{Q}$—since we may not have the analogue of the second flag-transitivity hypothesis in that result. This can arise from the fact that members of $\mathcal{Q}$ are actual subgroups of G; whereas simplices in $\Delta_{\mathcal{H}_I}$ are just *sets* of conjugates of the subgroups H_i. That is, here we are again continuing the theme of the warning indicated in Remark 6.2.4; and we will develop Lemma 6.5.7 for the situation where we can identify certain simplices, so that in the quotient we do get a natural correspondence with actual subgroups.

Hence in this final part of the section, we will be demonstrating some examples of this situation, and techniques for dealing with it.

We will begin with a geometry related to the (non-sporadic) group Σ_5: partly because this small example is very easy to analyze; but more importantly because the geometry is involved as a link in the geometries for a fair number of the sporadic groups (as was observed in Ronan and Smith [**106**]), starting with M_{22} in Section 7.3.

---

[18]We are grateful to Bob Oliver and Jesper Grodal for emphasizing this distinction—which seems to be glossed over in the group-geometries literature to date (including the original version of this work).

EXAMPLE 6.5.5 (The triples geometry for Σ_5). Set $G := \Sigma_5$. We emphasize first of all that the geometry we want is *not* the usual building geometry for the commutator subgroup $G' \cong \mathrm{Alt}_5$ regarded as $L_2(4)$ or $\Omega_4^-(2)$: That building has rank 1, and as a complex consists just of 5 isolated vertices corresponding to the Sylow 2-subgroups of G' (that is, to the points $\{1, 2, 3, 4, 5\}$ permuted by Σ_5). Instead we will need a related geometry of rank 2 (that is, a bipartite graph).

We will at least initially follow the original language of Ronan and Smith [**106**, p. 286] (see also [**115**, Ch. 2]): The triples geometry Δ has one type of vertices given by the 5 points above, and the other type given by the $\binom{5}{3} = 10$ triples of the points. Incidence is just inclusion of a point in a triple; there are $\binom{5}{1}\binom{4}{2} = 30$ edges in the graph, i.e., simplices of dimension 1.

Note that Δ is an external complex, using the language of 6.3.11 of the present work. But it might in fact be slightly more natural to realize Δ as an external complex based on the complements of the above sets: This corresponds to working in the 4-dimensional natural orthogonal module V for G regarded as $O_4^-(2)$; which we can realize as the even subsets from the 5 points. Thus the 5 individual points now become the complementary 4-subsets, giving the singular vectors of the orthogonal space V; and the 10 triples become the 2-subsets, giving the nonsingular vectors of V. The set-theoretic containment relation is now orthogonality in V.

Flag-transitivity of G on Δ (in either version above) is easily verified via Remark 6.4.6 (1); for example, the stabilizer Σ_4 of a point is transitive on the $\binom{4}{2} = 6$ triples containing that point.

We can realize Δ also in the form of a 2-local complex $\Delta_{\mathcal{H}_I}$, using index set $I := \{1, 2\}$ as follows: The stabilizer H_1 of the point 1 is the subgroup Σ_4 on the remaining four points $\{2, 3, 4, 5\}$; and the stabilizer H_2 of the triple 123 is the subgroup $\Sigma_3 \times \Sigma_2$ on the corresponding partition into the parts $\{1, 2, 3\}$ and $\{4, 5\}$. Thus the intersection $H_I = H_1 \cap H_2$ takes the form $\Sigma_1 \times \Sigma_2 \times \Sigma_2$ of a four-subgroup of G. Notice in particular that H_I is in fact the stabilizer of *two* maximal simplices, namely $1 \in 123$ and $1 \in 145$. Thus we are in the more complicated situation mentioned in Remark 6.2.4, in which simplices must be regarded only as sets of vertices, and cannot be identified with subgroups H_J. Finally, note that flag-transitivity of G on $\Delta_{\mathcal{H}_I}$ is again easy via 6.4.6 (1): for $H_1 \cong \Sigma_4$ transitively permutes the 6 pairs from $\{2, 3, 4, 5\}$; and if ab is such a pair with complement cd, then that pair determines a conjugate of H_2 on $\{1, c, d\}$ and $\{a, b\}$—which is incident with H_1 via their intersection in the conjugate of H_I inducing the Σ_2 on both $\{c, d\}$ and $\{a, b\}$. So $\Delta_{\mathcal{H}_I}$ is isomorphic to Δ by the standard method[19] of Lemma 6.5.4.

Observe that the involution generating $O_2(H_2)$ is a transposition of Σ_5; and in particular is an outer automorphism of the commutator subgroup $\mathrm{Alt}_5 \cong L_2(4)$— where all local subgroups in the subgroup $L_2(4)$ are contained in parabolic subgroups, and hence in subgroups like H_1. That is, our ability to describe a geometry of rank 2, rather than just 1, which is determined by maximal *local* subgroups— actually requires the use of the larger group Σ_5 rather than the smaller group $L_2(4)$ of Lie type. (We will encounter this situation in various sporadic groups later.)

Next for a further poset "$\mathcal{Q}$" of subgroups of G to compare with the complex Δ, we take the 2-radical subgroups $\mathcal{B}_2(G)$: The minimal members are the 5 regular four-subgroups (such as $B_1 := \langle (2, 3)(4, 5), (2, 4)(3, 5) \rangle$) along with the 10 subgroups

[19] i.e., the analogue with the external complex Δ replacing an intersection complex $\Delta_{\mathcal{A}_I}$.

generated by a transposition (such as $B_2 := \langle (4,5) \rangle$). The maximal members are the 15 Sylow 2-subgroups (there are three in each conjugate of $H_1 \cong \Sigma_4$).

We now observe that the normalizers $N_G(B_1)$ and $N_G(B_2)$ are precisely the subgroups H_1 and H_2 introduced above with respect to the complexes Δ and $\Delta_{\mathcal{H}_I}$. In particular this gives a set bijection of the vertices of Δ with the minimal members of $\mathcal{B}_2(G)$. So we might expect to have some analogue of Lemma 6.5.4; say an isomorphism at the level of posets, from the poset of simplices of Δ to $\mathcal{B}_2(G)$. But this cannot be the case: for Δ has 30 distinct 1-simplices, whereas $\mathcal{B}_2(G)$ has only 15 maximal members.

On the other hand, these numbers do suggest that there might be a simple quotient relationship obtained by identifying certain pairs of 1-simplices; and indeed we see there is an obvious candidate for such pairs—namely the pair $1 \in 123$ and $1 \in 145$ stabilized by H_I above (and of course its G-conjugates). However, the quotient after this identification will no longer have the structure of a simplicial complex. We could in fact pass to the language of CW-complexes; but it seems more elementary to just work at the level of posets. We let $\mathcal{R}$ denote the poset of simplices $\mathcal{Q}(\Delta)$ from Δ, and $\bar{\mathcal{R}}$ the result of the above identifications. Then the image $\bar{\mathcal{R}}$ is still a poset, and the natural projection map p is a poset map—since the identifications are made within the single "level" of $\mathcal{R}$ given by the maximal elements, among which there can be no poset relations.

And we now check that p in fact induces a homotopy equivalence of posets—using[20] Quillen's standard homotopy fiber theorem [**104**, 1.6]: for each $\bar{r} \in \bar{\mathcal{R}}$, we will need to show that the preimage $p^{-1}(\bar{\mathcal{R}}_{\leq \bar{r}})$ is contractible. First for $\bar{r}$ a minimal element, $\bar{\mathcal{R}}_{\leq \bar{r}}$ is just the point $\bar{r}$, and we saw above that p induces a set bijection at the minimal level, so that the preimage is the point given by the unique corresponding minimal member $r \in \mathcal{R}$. Next for $\bar{r}$ maximal, using the G-equivariance of p we may assume that $\bar{r}$ is of form $(1 \in 123) \equiv (1 \in 145)$; and then $\bar{\mathcal{R}}_{\leq \bar{r}}$ also contains the minimal elements corresponding to 1, 123, and 145. The preimage then just un-identifies this pair of 1-simplices, resulting in a subposet of $\mathcal{R}$ having the form:

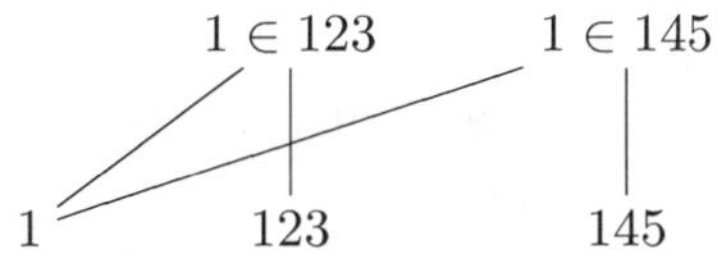

which is contractible—for example by first proceeding up from the triples, and then down to 1. This completes the proof of the homotopy equivalence of $\mathcal{R}$ with $\bar{\mathcal{R}}$

We single out one technical feature: namely that the variation in the simplices being identified is confined to *one* type of vertex, i.e., the triples such as 123 and 145. As a result those simplices are in the *star* of the non-varying vertex of the remaining type, the point 1. The needed contractibility is then an easy consequence of the subconfiguration given by the preimage within this star.

[20] It would be possible here to instead use the language of categories related by a cofinal functor (as in the Cofinality Theorem 4.5.22); but again it seems more elementary to remain at the level of posets. Furthermore this distinction will reflect a difference in our applications: We tend to use the current poset method when two simplices of the *same* type have the same stabilizer (here H_I); while we use the Cofinality Theorem in the situation where we have two different types $J \subset J' \subseteq I$ for which we have a coincidence $H_J = H_{J'}$ of stabilizing subgroups.

Finally, it is easy to verify that the quotient poset $\bar{\mathcal{R}}$ is isomorphic with the poset $\mathcal{B}_2(G)$: At the minimal level, we already indicated a natural set bijection with images such as B_1 and B_2 above; and we now map the identified pair $(1 \in 123) \equiv (1 \in 145)$ to the Sylow 2-subgroup $B_1 B_2 \cong D_8$ normalizing that pair, observing that this is indeed a poset map.

So by composition of the maps in the two previous paragraphs, we obtain our promised poset homotopy equivalence of $\mathcal{R} = \mathcal{Q}(\Delta)$ with $\mathcal{B}_2(G)$. $\diamondsuit$

Next we will generalize from certain aspects of the above discussion, to establish a more general result covering most of the cases where a similar situation arises among geometries for some sporadic group.

REMARK 6.5.6. It seems likely that such relationships among geometries could be accounted for, just at the level of homotopy equivalence, by the classical nerve of the covering which arises from minimal elements of the relevant posets. However because of the prominence of the poset of centric radical subgroups in the recent literature, we will try to indicate the actual relationship among the posets in somewhat more detail.

We will state our result in the language of the 2-local complex $\Delta_{\mathcal{H}_I}$, which is typically most convenient; but analogues will hold for suitable other forms of intersection complex $\Delta_{\mathcal{A}_I}$. For a type given by a subset $J \subseteq I$, recall the notation in the complex $\Delta_{\mathcal{H}_I}$ for the standard simplex of type J given by $\sigma_J = \{H_i : i \in J\}$.

By way of a preliminary overview: we may wish (especially as our eventual goal is a Borel construction affording the normalizer decomposition) to essentially identify certain distinct simplices having the same normalizer. Indeed in the notation used below, we will specify certain types denoted by subsets $K^+ \subseteq J$, and identify the simplices having the same stabilizer H_{K^+} as σ_{K^+}—using the analysis previewed earlier around (6.2.16). The identified set will then have stabilizer $N_G(H_{K^+})$ rather than H_{K^+}. So in the language of our analysis preceding Example 6.5.5, we will relate the poset of simplices of $\Delta_{\mathcal{H}_I}$ to a poset whose members are actual subgroups of G.

The main hypothesis for the result concerns the subgroup $N_G(H_I)$. And the result will typically be used in the case where H_I is Sylow in certain simplex stabilizers, but is proper in $N_G(H_I)$—while $N_G(H_I)$ in turn is Sylow in G and in other stabilizers. The set J below will also usually be a singleton, just as the distinguished vertex type $\{k\}$ is.

LEMMA 6.5.7. *Assume a 2-local complex $\Delta_{\mathcal{H}_I}$ as in Definition 6.3.5. Assume that the H_i satisfy Condition 6.2.15, and that G is flag-transitive on $\Delta_{\mathcal{H}_I}$. Assume further that there is a distinguished index $k \in I$, and a distinguished subset $J \subseteq I \setminus \{k\}$, such that for each K with $J \subseteq K \subseteq I \setminus \{k\}$, setting $K^+ := K \cup \{k\}$, we have*

$$N_G(I) \text{ lies in } H_K\text{, but is a supplement to } H_{K^+} \text{ in } N_G(H_{K^+}). \tag{6.5.8}$$

Then

(1) For each such K, the set of simplices with type K^+ and stabilizer H_{K^+} consists of the joins $\sigma_K * v$, where v runs over the $N_G(H_I)$-orbit containing $\{H_k\}$. (Indeed this set is an $N_G(H_{K^+})$-orbit; and the transitivity of the subgroup $N_G(H_I)$ is then equivalent to the supplement condition in (6.5.8).*)*

(2) Let $\mathcal{R}$ denote the poset $\mathcal{Q}(\Delta_{\mathcal{H}_I})$ of simplices of $\Delta_{\mathcal{H}_I}$, and let $\bar{\mathcal{R}}$ denote the quotient obtained by identifying the simplices in (1) (for each K^+, and extending by G-conjugacy); let p denote the quotient map. Then p induces a homotopy equivalence of posets from $\mathcal{R}$ onto $\bar{\mathcal{R}}$.

PROOF. The main purpose of the flag-transitive hypothesis is to use G-conjugacy to reduce questions about arbitrary simplices of some type $L \subseteq I$ to consideration of the standard simplex σ_L. The purpose of assuming 6.2.15 is to guarantee the property indicated in (6.2.16) that $N_G(\sigma_L) = H_L$. Indeed we saw it then follows that the set of simplices with stabilizer H_L consists of the $N_G(H_L)/H_L$-conjugates of σ_L (which are distinct). We will use these facts, along with (6.5.8), to obtain the characterization of various simplices given in conclusion (1).

Consider any K with $J \subseteq K \subseteq I \setminus \{k\}$. Now H_{K^+} in particular stabilizes $\sigma_{K^+} = \sigma_K * \{H_k\}$. Furthermore the set S of simplices with type K^+ and stabilizer H_{K^+} includes the $N_G(H_{K^+})/H_{K^+}$-orbit X containing σ_{K^+}. But the members of S are G-conjugate by flag-transitivity, and so are conjugate in the normalizer $N_G(H_{K^+})$ of their common stabilizer; hence we conclude that the $N_G(H_{K^+})$-orbit X covers all of S, as claimed. Next $N_G(H_{K^+}) = N_G(H_I)H_{K^+}$ by (6.5.8), so that the subgroup $N_G(H_I)$ must in fact be transitive on X. Furthermore (6.5.8) also gives $N_G(H_I) \leq H_K = N_G(\sigma_K)$; so by transitivity we conclude that every member of X has σ_K its face of type K. Thus X consists of the joins $\sigma_K * v$, where v runs over the $N_G(I)$-orbit Y containing $\{H_k\}$. This establishes (1)—in particular we have seen that (6.5.8) implies the transitivity of $N_G(H_I)$; while transitivity conversely implies the supplement condition just by a standard Frattini argument (that is, a transitive subgroup is supplemented by a point stabilizer).

We turn to the proof of (2). Observe first that the poset $\mathcal{R}$ is naturally ranked, with all simplices of type L with a fixed value of $|L|$ appearing on the same level. Now any identifications made in our process are within a fixed type; and the inclusion in the quotient $\bar{\mathcal{R}}$ is induced directly from that in $\mathcal{R}$, so that the relation in the quotient remains antisymmetric and transitive. Thus $\bar{\mathcal{R}}$ is also a poset, and p is a morphism of posets (surjective by definition).

We show that p is a homotopy equivalence by verifying Quillen's condition [**104**, 1.6]—in an obvious extension of the procedure indicated at the end of Example 6.5.5: We want to show for each $\bar{r} \in \bar{\mathcal{R}}$ that the preimage $p^{-1}(\bar{\mathcal{R}}_{\leq \bar{r}})$ is contractible. Let L denote the type of any preimage r of $\bar{r}$. If L does not contain J^+, then $\bar{r}$ contains no identified simplices, so that r is unique; and indeed this injectivity also applies to all faces of r. So the preimage is just $\mathcal{R}_{\leq r}$, trivially contractible in view of its unique maximal element r. Thus it remains to treat the case where $L = K^+$ for some $J \subseteq K \subseteq I \setminus \{k\}$. Using flag-transitivity we may assume that the preimage of $\bar{r}$ contains the simplex $r = \sigma_{K^+} = \sigma_K * \{H_k\}$. So by (1), the full preimage of $\bar{r}$ consists of the joins $\sigma_K * v$ as v varies over the $N_G(I)$-orbit Y containing $\{H_k\}$. Furthermore we may apply to all the faces of these simplices a similar analysis with regard to containing J^+; to conclude that the preimage $p^{-1}(\bar{\mathcal{R}}_{\leq \bar{r}})$ is just the poset of faces of the subcomplex $Z := \sigma_K * Y$ (where there are no edges among the vertices of Y). Now the link in Z of each v is just the simplex σ_K, which is certainly contractible. So as in [**116**, 1.5], the removal from Z of the star of v is a homotopy equivalence (indeed a strong deformation retraction); and the result of the removal is contractible σ_K, so we conclude that the complex Z is contractible. Contractibility then extends to the barycentric subdivision of Z, which is just the complex of chains taken from the poset of faces of Z—which in turn is the definition of contractibility of our preimage poset.[21] This case completes the proof of contractibility, and hence of (2). □

We can now replace the relevant parts of the discussion of the triples geometry for $G = \Sigma_5$ in Example 6.5.5 with a briefer verification of the hypotheses for Lemma 6.5.7: Recall we had realized the geometry also in the form $\Delta_{\mathcal{H}_I}$—with index set $I := \{1,2\}$; and we had $H_1 \cong \Sigma_4$, $H_2 \cong \Sigma_3 \times \Sigma_2$, and $H_I \cong \Sigma_2 \times \Sigma_2$. We had checked flag-transitivity; and the requirements of 6.2.15 are immediate. Now as our distinguished index, we take $k := 2$; and as our distinguished subset, we take $J := I \setminus \{2\} = \{1\}$. Thus in particular, J^+ is the full index set I; so in verifying (6.5.8), the only candidate for "K" is J itself, so that we are concerned just with $H_J = H_1$ and $H_{J^+} = H_I$. In the present example for Σ_5, H_I is a four-group; and its normalizer $N_G(H_I)$ is dihedral of order 8, and Sylow in G. And in particular the Sylow $N_G(H_I)$ does lie in H_1, while it is vacuously a supplement to H_I in itself. Thus we may apply 6.5.7 to obtain the poset homotopy equivalence of $\mathcal{R}$ with $\bar{\mathcal{R}}$. And of course the ultimate purpose is to compose this equivalence with our observation in Example 6.5.5 that $\bar{\mathcal{R}}$ is in turn poset-isomorphic to $\mathcal{B}_2(G)$.

The lemma is also relevant for various sporadic geometries in which the triples geometry arises as the link of some nonempty simplex σ; these groups are

$$M_{22}, J_4, HS, Co_2, B$$

with certain mild variants arising in M_{23}, Fi_{22}, Fi_{23}, and Ru. In these cases, the relevant subset J^+ will be a proper subset of the full index I; see for example M_{22} in Section 7.3. The situation where $J^+ = I$ in Lemma 6.5.7 also occurs arises for M_{11} in Section 7.1. However this geometry does not seem to be involved as a link in any geometry for a larger sporadic group. Still other variants of the method will arise for Fi'_{24}, and for $U_4(2){:}2$ which is involved in several sporadic geometries.

We remark that the viewpoint of Lemma 6.5.7 seems to help explain some possible interrelations among a number of phenomena which had only been tied together observationally in the previous literature on group geometries: such as configurations involving the triples geometry; stabilizers normalizing a 2-subgroup which is not weakly closed; and the notion of a 2-split cover in Ronan and Smith [**106**, p. 286].

6.6. Final remarks on 2-local geometries

In this final section, we briefly mention various further points arising from the literature on 2-local geometries for sporadic groups.

We first re-emphasize that 2-local geometries are not really a well-defined class of geometries—this reflects the fact that there is no uniform theory of sporadic groups, to provide a foundation for an axiomatic development. Instead, these geometries are defined by using various fairly uniform, but still somewhat ad-hoc,

[21] For a more direct but gradual argument for contractibility, just in terms of posets: Elements w in the preimage poset $\mathcal{T}$ which satisfy $w \geq v$ for some $v \in Y$, but are not themselves maximal, have $\mathcal{T}_{>w}$ contractible by virtue of a unique maximal member $\sigma_K * v$; and so they can be removed (proceeding upward one level at a time) in view of 6.5.1 (1). After that, the maximal elements $x = \sigma_K * v$ themselves can be similarly removed: since in the remaining poset $\mathcal{T}'$, we see that $\mathcal{T}'_{<x}$ has unique maximal element σ_K. So we have contracted the preimage poset to the poset of faces of σ_K, which again has unique maximal element σ_K, and so is contractible.

principles—exhibited for example in our earlier discussion of various approaches to a geometry for M_{24} defined using 2-local subgroups. Historically geometries were developed for each group, in order to better understand the structure of that group; what is perhaps amazing is that there could be any even moderately uniform analogies among all the sporadic groups—in particular, with the model case of buildings.

Of course these analogies are rough and limited; and in particular, several different but closely related geometries may be candidates for being considered "the" 2-local geometry for a particular sporadic group G. Indeed we will see (for example for M_{23} in Section 7.4) that our particular focus on group cohomology and decompositions can suggest replacing the geometry which is more usual in group theory with a slight variant which is more suitable for cohomology.

We mention one further aspect, which will arise in our discussion of the literature (or lack thereof) on geometries for certain groups: Some of the sporadic groups determine a geometry of rank just two; and usually that rank two geometry is not involved in the geometry of any larger sporadic group, *so usually there is no diagram defined for those groups with a geometry of rank 2.*

Part 2

Main results on sporadic groups

In Chapter 7 we state our main results on homotopy decompositions for the 26 sporadic groups G. In each case, we first provide background information on suitable aspects of the 2-local subgroup structure of G, including the context for our particular diagram of subgroups—and then lead up in a fairly standard way to the relevant decomposition result. Finally when available we also provide further information, from the literature as known to us at this time, on the ring structure of the mod 2 group cohomology of G.

Chapter 8 contains various technical details: these are primarily group-theoretic arguments establishing certain of the results in Chapter 7, in cases where we postponed those proofs in order to preserve a comparatively elementary level of exposition of the main results.

CHAPTER 7

Decompositions for the individual sporadic groups

Each section of this chapter will deal with one of the 26 sporadic groups; the group is indicated in the title of the section, and it will be abbreviated by G within that section. The treatment in each case will culminate in a theorem giving a homotopy decomposition of the 2-completed classifying space of G.

Common aspects of the treatment within each section. Though of course the individual details will differ, we proceed for each group G in a fairly uniform manner—ordinarily emphasizing the following features along the way:

As group-theoretic background for a discussion of the 2-local geometry of G, we begin by distinguishing

- certain maximal 2-local subgroups $H_i (i \in I)$ of G,

which then determine a subgroup simplex $\mathcal{H}_I$ as in Definition 4.6.2. The groups H_i are described in the Atlas [**47**]; and we usually assume implicitly the straightforward verification that they satisfy the requirements in Condition 6.2.15. The indexing set I is typically *not* chosen in the general form $I_n = \{1, 2, \ldots, n\}$ of Definition 4.6.1; instead usually the indices $i \in I$ are chosen to provide a reminder of the structure of H_i—most often as in Example 6.3.10, so that i is the rank of the elementary abelian 2-group $\Omega_1 Z\big(O_2(H_i)\big)$ typically denoted by A_i.

Next as further geometric background, we indicate something of the structure, in our particular group G, of two standard collections of 2-subgroups indicated in Section 5.9: Namely we give a description of

- the poset $\mathcal{A}_2(G)$ of nontrivial elementary abelian 2-subgroups,

when this is relevant, and is available from the literature. And similarly we give a description from the literature of

- the poset $\mathcal{B}_2(G)$ of nontrivial of 2-radical subgroups.

Since the published lists of members of $\mathcal{B}_2(G)$ use different notation for the various cases of G, we will also when relevant provide remarks for translating those results into the more uniform notation of this work. (The groups H_i will ordinarily arise as stabilizers of suitable members of these posets.)

We then specify our geometry of choice; namely we use the H_i to give a description of

- a particular 2-local geometry Δ for G,

with a reference when available to the appropriate earlier literature. In principle, Δ could in each case be chosen as the 2-local complex $\Delta_{\mathcal{H}_I}$ determined by the conjugates of the subgroup simplex $\mathcal{H}_I$ as in Definition 6.3.5. But often we can in fact realize Δ in a more elementary way, using one of the other variants of the intersection complex $\Delta_{\mathcal{A}_I}$ of Definition 6.3.2, as in the examples in the latter part of Section 6.3. Then in order to obtain our homotopy decomposition over Δ, we indicate

- a homotopy equivalence via a G-map of Δ with some collection

from Sections 5.9–5.11, where that earlier collection is known to be ample and normalizer sharp. Hence it follows by results in those sections (quoting 5.6.8 and 5.8.11) that the complex Δ is also ample and normalizer sharp. In addition, we sketch a proof that

- G acts flag-transitively on Δ.

So it follows by Remark 5.6.11 that the normalizer decomposition for Δ takes the special form in (5.6.12), of a homotopy colimit over the simplex category $\mathfrak{D}_I$ underlying the subgroup simplex $\mathsf{N}(\Delta/G) = \mathcal{H}_I$.

These properties of the normalizer decomposition for Δ then deliver the cohomology decomposition which is our primary goal; and we collect these various conclusions in a

- main result for the particular sporadic group G:

By the equivalence of (i) and (v) in Theorem 5.2.7, the ampleness of the normalizer decomposition gives an expression for the 2-completed classifying space $BG_2^{\wedge}$ of G—as the homotopy colimit, indexed by a pushout diagram of the appropriate dimension in the form (5.6.12), of the classifying spaces for the members of the subgroup simplex $\mathcal{H}_I$. And the sharpness of the decomposition then via Remark 5.8.5 delivers a simplex cohomology decomposition of form (5.8.6): namely a formula for the group cohomology $H^*(BG, \mathbb{F}_2)$ as the alternating sum of the cohomology of the members of the subgroup simplex $\mathcal{H}_I$. (In a few cases, there are further cancellations, resulting in a still smaller indexing diagram than the simplex; we can make that reduction at the level of our homotopy colimit using the Cofinality Theorem 4.5.22.)

Finally we conclude each section with a discussion of any other information known to us from the literature about

- the ring structure of the group cohomology $H^*(BG, \mathbb{F}_2)$

for that particular group G.

Other general remarks. For completeness we state a decomposition for each sporadic group; but we emphasize that a number of those decompositions were already known in the earlier literature. In particular, we will refer at various points to the book of Adem and Milgram [**6**, Ch. 8], and to various papers in their extensive program determining the cohomology ring for various sporadic groups. Similarly a number of decompositions follow from results on homotopy equivalences in the group theory literature on 2-local geometries, such as Smith and Yoshiara [**116**].

However, many of the results are established here for the first time. In some cases, the proofs are comparatively brief. But when lengthier proofs are required for any of the steps indicate above, we will generally postpone the details to a section corresponding to that particular group in the subsequent Chapter 8.

In the course of our proofs, we will quote various standard facts about the structure of the sporadic groups. When no explicit reference is given, the reference is implicitly to the Atlas [**47**]—for example, for features such as conjugacy classes and the structure of maximal subgroups, *we primarily follow the conventions of the Atlas* [**47**]; but see the variations mentioned after (0.0.1).

For a fixed group G, we may also compare our chosen geometry Δ with various other ample complexes and collections, from the viewpoint of indicating which gives the "smallest known" decomposition—where the measure of size is typically the dimension of the complex, or of the complex of inclusion-chains from the collection. In particular, since for a decomposition we assume the use of more than one subgroup, it is implicit that any geometry Δ of dimension one (corresponding to a simplex $\mathcal{H}_I$ of subgroups with index set I of rank two) is smallest possible for such purposes. (Only in the case of J_1 in Section 7.6 will we actually encounter a "decomposition" using only one subgroup.)

Finally, we indicated above that we will relate Δ to some standard collection by means of a G-map which determines a homotopy equivalence. We mention that in the proofs in this chapter, it is possible in *most* cases to show that the indicated map in fact induces the stronger condition of a G-homotopy equivalence, using standard techniques such as those given in Thévenaz and Webb [**118**]. However, we chose to phrase our proofs in terms of the weaker condition using only a G-map: since at least some of the time, we also need to consider relations with the centralizer and subgroup decompositions—and we saw at Warnings 5.4.5 and 5.5.5 that some of the mappings involved exhibit only that weaker condition. Furthermore the proofs just of the weaker condition are slightly shorter and may have a somewhat more intuitive feel.

7.1. The Mathieu group M_{11}

For $G := M_{11}$, we take as our index set $I := \{1, 2\}$—corresponding to the G-classes of maximal 2-local subgroups in the Atlas [**47**, p. 18] represented by

$H_1 \cong M_8 : \Sigma_3 \cong Q_8 : \Sigma_3 \cong 2^{\cdot}\Sigma_4$, and
$H_2 \cong \Sigma_4 \cong 2^2 : \Sigma_3$,

where Q_8 denotes a quaternion group of order 8; the remaining member of the subgroup simplex $\mathcal{H}_I$ as in Definition 4.6.2 determined by the H_i is

$$H_I = H_{12} = H_1 \cap H_2 \cong D_8,$$

a dihedral subgroup of order 8, which is of index two in a semidihedral Sylow 2-subgroup $N_G(H_I)$ of G. Note also that although H_2 is a maximal 2-local subgroup of G, it is not a maximal subgroup of G: it is proper in a non-local subgroup M_{10}. (We mention for later use that the semidihedral subgroup $N_G(H_I)$ fuses the two classes of 4-groups in its dihedral group H_I.)

It would now be possible to just summarize our treatment of $G = M_{11}$ by saying that the 2-local geometry is obtained from the collection $\mathcal{A}_2(G)$; and then to proceed essentially immediately to our main results for this case such as Theorem 7.1.1.

However since this is the first sporadic group that we treat, it seems reasonable to instead provide a more fully detailed development of our overall notation and methods. Those details will still remain fairly uncomplicated in this small case; but they will give an indication of the similar structure of our development even in the larger cases (where we may be omitting the verification of the more routine details).

Obtaining the 2-local geometry from $\mathcal{A}_2(G)$. So we begin by exhibiting a 2-local geometry corresponding to the above H_i. As just indicated, that geometry turns out in the present case to be just the complex of inclusion-chains in the poset

$\mathcal{A}_2(G)$ of elementary abelian 2-subgroups. And that poset is very easy for us to describe below; again using for example the Atlas [**47**].

Namely we now recall various details for M_{11} which we had described at earlier Example 6.3.8: Our 2-local geometry will be an intersection complex $\Delta_{\mathcal{A}_I}$, as in Definition 6.3.2; as the vertex groups required from Notation 6.3.1, for $i \in I$ we choose

$$A_i := \Omega_1 Z\big(O_2(H_i)\big).$$

Then in fact A_1 and A_2 are of ranks 1 and 2; and the A_i afford representatives of the two G-classes in $\mathcal{A}_2(G)$, with $N_G(A_i) = H_i$ as needed. Furthermore we had verified the inclusion relation $A_1 < A_2$, giving the hypothesis for Proposition 6.3.6, so that the intersection complex $\Delta_{\mathcal{A}_I}$ is just the poset complex $\Delta\big(\mathcal{A}_2(G)\big)$—and in particular, $\Delta_{\mathcal{A}_I}$ is an internal poset complex in our nomenclature.

We now observe further that $H_2 = N_G(A_2) \cong \Sigma_4$ induces Σ_3 on the three involutions in A_2; hence (cf. Remark 6.4.6 (1)) G is flag-transitive[1] on the poset complex $\Delta\big(\mathcal{A}_2(G)\big)$—which here is just a graph, bipartite on vertices given by conjugates of A_1 and of A_2. In particular since the chain $A_1 < A_2$ has stabilizer $H_1 \cap H_2 = H_{12}$, we have $\mathsf{N}\big(\ \Delta\big(\mathcal{A}_2(G)\big)\ /\ G\ \big) = \mathcal{H}_I$, using the language of the normalizer set-map N in Definition 5.6.13.

We can also introduce the variant of $\Delta_{\mathcal{A}_I}$ given by the 2-local complex $\Delta_{\mathcal{H}_I}$ determined by the H_i in Definition 6.3.5. Much as in the previous paragraph, it is easy to check flag-transitivity of G on $\Delta_{\mathcal{H}_I}$ via Remark 6.4.6 (1): For example, there are three conjugates of H_1 incident to H_2; indeed these are just the centralizers of the three involutions in A_2, so the transitivity of H_2 on the latter gives transitivity on the former.[2] Thus we have the hypotheses of Lemma 6.5.4—so that the set-map N in this case extends to an isomorphism of complexes from $\Delta_{\mathcal{A}_I}$ to $\Delta_{\mathcal{H}_I}$. We write $\mathsf{N}(\Delta_{\mathcal{A}_I}) = \Delta_{\mathcal{H}_I}$—that is, we have this relationship for the original complexes, and not just for the simplex given by their G-orbits.

Digression: the relationship with $\mathcal{B}_2(G)$. As indicated above, we will be using $\mathcal{A}_2(G)$ for our main results; so it would not really be necessary for us to now describe $\mathcal{B}_2(G)$. However we will do so, following our convention of providing for each sporadic group some comparison and contrast of these two posets.

The determination of the poset $\mathcal{B}_2(G)$ of 2-radical subgroups for $G = M_{11}$ is given in An and Conder [**13**, 2B]: The G-classes of 2-radical subgroups are represented by $B_1 := O_2(H_1) \cong Q_8$ (with A_1 affording the center of the Q_8); $B_2 := O_2(H_2) = A_2$; and a Sylow 2-subgroup of G, which we can take to be $B_I := N_G(H_I)$. We had mentioned earlier that B_I is semidihedral of order 16; it follows that H_I is the unique subgroup $\Omega_1(B_I)$ of B_I which is isomorphic to D_8.

In particular, we see that the set-map $\mathcal{O}_2$ of Definition 6.3.12 establishes a bijection between the vertices of $\Delta_{\mathcal{H}_I}$ and the minimal members of the poset $\mathcal{B}_2(G)$; although G is not flag-transitive on the complex $\Delta\big(\mathcal{B}_2(G)\big)$ of chains from $\mathcal{B}_2(G)$. So

[1] But we caution that flag-transitivity of G on the full elementary poset $\mathcal{A}_2(G)$ (so that the orbit complex $\Delta\big(\mathcal{A}_2(G)\big)/G$ is just a single simplex) can only be expected in a comparatively small group such as M_{11}. In larger groups, the orbit complex from $\mathcal{A}_2(G)$ is typically much more complicated than a simplex—and in some cases is not even yet known; and then we will instead work with a 2-local geometry determined by some structure considerably smaller than $\mathcal{A}_2(G)$.

[2] More generally, we could methodically analyze overgroups of H_I and the incidence among the H_i by a suitable adaptation of the techniques in 6.2.17 (1); but that level of detail does not seem needed here.

from the partial resemblance to the hypotheses of 6.5.4, we might be led to expect that the map $\mathcal{O}_2$, slightly modified to change the image of H_I from $O_2(H_I) = H_I$ to $N_G(H_I) = B_I$, might induce a poset isomorphism from the poset of simplices of $\Delta_{\mathcal{H}_I}$ to $\mathcal{B}_2(G)$—in which case $\mathcal{B}_2(G)$ would be just the poset subdivision of $\mathcal{A}_2(G)$, in view of the isomorphism of $\Delta_{\mathcal{H}_I}$ with $\Delta(\mathcal{A}_2(G))$ discussed earlier. But this is not the case: for example, $\Delta_{\mathcal{H}_I}$ has 990 1-simplices, while $\mathcal{B}_2(G)$ has only 495 maximal members. The actual relationship of the two posets is just a little subtler: in fact it is another example of the quotient situation in Lemma 6.5.7; so we will indicate some further details—especially since there will be a fairly similar situation in a number of other sporadic groups, typically when one of the chosen H_i does not contain a Sylow 2-subgroup of G.

First we will look a little more concretely at what will be involved in that quotient process. Recall that a simplex of type $I = \{1, 2\}$ corresponds to an inclusion $A_1 < A_2$ in the $\Delta_{\mathcal{A}_I}$ version of the geometry. But in fact the dihedral group $H_I \cong D_8$ with center A_1 contains a further four-subgroup A_2' with $A_1 < A_2'$. Further H_I is the stabilizer of both these simplices of type I; and as we had mentioned at the outset, they are fused in the semidihedral overgroup $N_G(H_I) = B_I$. (This statement about transitivity on the pair of conjugates gives an example of the equivalence indicated in 6.5.7 (1) with the supplement condition in (6.5.8).) There are 990 simplices of type I; compared to 495 conjugates of the Sylow group $N_G(H_I)$ which is a maximal member of $\mathcal{B}_2(G)$. (Thus the numbers of simplices of types 1, 2, 12 are 165, 330, and 990—compare with 5, 10, and 30 in the case of the triple geometry for Σ_5 in earlier Example 6.5.5.) Such a pair of simplices will be identified when we pass from $\mathcal{R}$ with $\bar{\mathcal{R}}$ in the notation of Lemma 6.5.7.

Before applying that lemma, we observe that for this particular case of M_{11}, the two posets being compared are the standard collections $\mathcal{A}_2(G)$ and $\mathcal{B}_2(G)$; so in fact we know independently that they are homotopy equivalent, since they both lie in the standard G-homotopy type in Section 5.9 determined by $\mathcal{S}_2(G)$. Of course this general equivalence is indirect, in the sense that it proceeds via the theorems 5.1.6 of Quillen and 5.1.9 of Bouc, showing they are each homotopy equivalent to $\mathcal{S}_2(G)$. So one feature of applying Lemma 6.5.7 here will be to establish a poset homotopy equivalence from $\mathcal{A}_2(G)$ directly to $\mathcal{B}_2(G)$. (Again such a direct mapping should probably not be expected away from small cases such as M_{11}.)

As suggested above, the present case for M_{11} will exhibit some formal analogy with the case of the triples geometry for Σ_5—especially with the re-considered view of that geometry given immediately after 6.5.7. Namely we will exhibit a poset homotopy equivalence of $\Delta(\mathcal{A}_2(G))$ with $\mathcal{B}_2(G)$, which is in some sense very close to a poset isomorphism: we first pass to a quotient poset by identifying certain pairs of simplices, and then the quotient will give an isomorphism with $\mathcal{B}_2(G)$.

So we proceed to verifying the hypotheses of 6.5.7. We choose the form $\Delta_{\mathcal{H}_I}$ of our geometry. We had checked flag-transitivity; and we assumed at the outset (as indicated in the introduction to the chapter) the choice of H_i satisfying 6.2.15. Now as our distinguished index, we take $k := 2$; and as our distinguished subset, we take $J := I \setminus \{2\} = \{1\}$. Thus in particular, J^+ is the full index set I; so in verifying (6.5.8), the only candidate for "K" is J itself, so that we are concerned just with $H_J = H_1$ and $H_{J^+} = H_I$. In the present example, H_I is dihedral of order 8; and its normalizer $N_G(H_I) = B_I$ is semidihedral of order 16, and Sylow in G. Now the Sylow $N_G(H_I)$ does lie in H_1, while it is vacuously a supplement to H_I in itself. Thus we may apply 6.5.7 to obtain the poset homotopy equivalence of the posets denoted by $\mathcal{R}$ with $\bar{\mathcal{R}}$ in that result.

Of course our ultimate purpose is to compose this equivalence with the easily-verified observation that the composition of set-maps N followed by $\mathcal{O}_2$ extends to a poset isomorphism of $\bar{\mathcal{R}}$ with $\mathcal{B}_2(G)$—where the above identified pair in $\bar{\mathcal{R}}$ is normalized by $N_G(H_I)$ (as opposed to H_I). In particular, the number of maximal elements in $\bar{\mathcal{R}}$ is now 495 rather than 990, as desired to match the number in $\mathcal{B}_2(G)$.

Main results for M_{11}. We now return to our primary theme, with our 2-local geometry obtained from the collection $\mathcal{A}_2(G)$.

The posets and complexes discussed so far are of rank two; and so as mentioned at the end of Chapter 6, they do not seem to be much treated in the earlier literature (such as [**106**]) on group geometries—the most relevant reference for a 2-local geometry for M_{11} is probably Buekenhout [**40**, end of Sec. 3], describing essentially what we are here calling the 2-local complex $\Delta_{\mathcal{H}_I}$. Of course here we are primarily using instead the version given by the poset complex $\Delta\big(\mathcal{A}_2(G)\big)$.

The collection $\mathcal{A}_2(G)$ is ample and normalizer-sharp as we saw in Proposition 5.9.1; so the same conditions hold for the complex $\Delta\big(\mathcal{A}_2(G)\big)$ giving our 2-local geometry. We mention that in the present case of M_{11}, our 2-local geometry turns out to actually be *equal* to the poset complex $\Delta\big(\mathcal{A}_2(G)\big)$ of the standard collection $\mathcal{A}_2(G)$—and not just homotopy equivalent to that standard collection via some G-map, as will typically be the case for most of the larger sporadic groups we will treat.

We now summarize the effects of the above discussion, in our main result for the case of M_{11}; the decomposition for this group is not new, in view of earlier results which we will indicate after the statement.

THEOREM 7.1.1. *For M_{11}, the 2-local complex $\Delta_{\mathcal{H}_I}$ and the collection $\mathcal{A}_2(M_{11})$ are ample and normalizer-sharp. In particular, for the groups in $\mathsf{N}(\Delta_{\mathcal{H}_I}/G) = \mathcal{H}_I$, the natural group surjection from the amalgamated free product $H_1 *_{H_{12}} H_2$ onto M_{11} induces a mod 2 cohomology isomorphism, with a simplex cohomology decomposition as in* (5.8.6). *The corresponding map from the homotopy colimit of the pushout square diagram* (4.5.6) *to BM_{11}, followed by 2-completion, induces a homotopy equivalence.*

PROOF. These assertions follow from various parts of the discussion above. In particular, since we saw that G is flag-transitive on $\mathcal{A}_2(G)$, Remark 5.6.11 shows that the diagram of spaces for the homotopy colimit of the normalizer decomposition for $\Delta_{\mathcal{A}_I}$ takes the simplex form (5.6.12): namely the pushout square $\mathcal{H}_2$ corresponding to $I = \{1, 2\}$ as in Diagram (4.5.6). We got ampleness and normalizer sharpness from Proposition 5.9.1. Definition 5.3.3 of ampleness then gives the mod 2 cohomology isomorphism; while the sharpness gives the cohomology decomposition as in (5.8.6). Finally the equivalence of (i) and (v) in Theorem 5.2.7 then gives the homotopy equivalence with the 2-completed classifying space $BG_2^\wedge$. □

However, we recall that H_1 contains a Sylow 2-subgroup S of G, which is semidihedral of order 16; while a Sylow 2-subgroup of H_2 is dihedral of order only 8—so that in particular, the Sylow condition in the Benson–Wilkerson result Lemma 4.5.10 fails for H_2. Thus we cannot, in this case, alternatively deduce the final homotopy equivalence in the result directly from Proposition 5.6.14.

The mod 2 cohomology ring of M_{11} was computed by Benson and Carlson in [**27**], using methods different from the above decomposition. They prove that

$$H^*(M_{11}, \mathbb{F}_2) = \mathbb{F}_2[x, y, z]/(x^2y + z^2)$$

where the degrees of the indicated cohomology classes are given by $|x| = 3$, $|y| = 4$, and $|z| = 5$. The cohomology is detected on elementary abelian 2-subgroups. (For the terminology of *detection*, see for example Adem and Milgram [**6**, IV.4.2].)

In particular the ring structure above determines the dimensions of the cohomology groups in the various degrees. These dimensions are customarily recorded in the *Poincaré series* for the cohomology—namely the polynomial

$$\sum_{i=0}^{\infty} \dim_{\mathbb{F}_2} H^i(M_{11}, \mathbb{F}_2)\, t^i$$

in a variable t, whose i-th coefficient records the dimension of the ith cohomology group.

Working independently at around the same time, Webb in [**122**, 7.8, 7.9] used the cohomology decomposition which we have stated in Theorem 7.1.1, to obtain the Poincaré series for the cohomology

$$\sum_{i=0}^{\infty} \dim_{\mathbb{F}_2} H^i(M_{11}, \mathbb{F}_2)\, t^i \;=\; \frac{1+t^5}{(1-t^3)(1-t^4)},$$

though he did not determine the ring structure.

7.2. The Mathieu group M_{12}

For $G := M_{12}$, we take as our index set $I := \{1, 2\}$, corresponding to G-classes of maximal 2-local subgroups in the Atlas [**47**, p. 33] with structure:

$H_1 \cong M_8 : \Sigma_4 \cong Q_8 : \mathrm{Aut}(Q_8) \cong 2^{1+4}_+ : \Sigma_3$, and

$H_2 \cong 4^2 : D_{12} \cong 4^2 : (2 \times \Sigma_3)$.

The remaining member of the subgroup simplex $\mathcal{H}_I$ as in Definition 4.6.2 determined by the H_i is $H_I = H_{12}$, which is a Sylow 2-subgroup of M_{12}, of order 2^6.

The structure of $\mathcal{A}_2(G)$ is substantially more complicated for M_{12} than for the smaller group M_{11} in the previous section; in particular, the orbit complex $\Delta(\mathcal{A}_2(G))/G$ is not a simplex. The detailed structure of that orbit complex is determined by Adem, Maginnis, and Milgram in [**4**, p. 106], where the authors use that orbit complex as a basis for their computation of the mod 2 cohomology ring of M_{12} (see also [**6**, VIII.3.9]). They essentially use the general form (5.6.9) of the normalizer decomposition over that fairly intricate orbit space; but eventually they reduce to an easier decomposition which is over a 1-simplex as in (5.6.12), namely the pushout square determined by $\mathcal{H}_I$ above.

A smaller poset $\mathcal{E}_2^-$ equivalent to $\mathcal{A}_2(G)$. We will now indicate a more direct alternative approach to that 1-simplex decomposition over $\mathcal{H}_I$. This will demonstrate our methods involving the subposet $\mathcal{E}_2(G)$ of $\mathcal{A}_2(G)$ described in Definition 5.10.1; in fact we will produce a further subposet $\mathcal{E}_2^-$, which will provide an internal poset complex, in the sense of Proposition 6.3.6, determined by the 2-local subgroups H_i.

As usual we set $A_i := \Omega_1 Z(O_2(H_i))$; and we have A_i elementary abelian of rank i, with $N_G(A_i) = H_i$. There are two conjugacy classes of involutions in G, denoted by $2A$ and $2B$ in the Atlas [**47**], with only those of class $2B$ lying in the center of a Sylow 2-subgroup. Thus A_1 is generated by an element of class $2B$; and we have $H_1 = C_G(A_1) = N_G(A_1)$ of order 192. And since H_I is Sylow in G, an easy application of the method of Remark 6.2.17 (2) gives the inclusion relation $A_1 < A_2$.

Next using the transitivity of the Σ_3 in H_2 on the three involutions of A_2, we see that all those involutions also lie in class $2B$. It is easy to calculate from the character table of M_{12} (for example, using the ClassMultiplicationCoefficient function in GAP) that

$$\xi(2B, 2B, 2A) = 0$$

for the structure constant defined in (5.10.5). So as discussed in the latter part of Section 5.10, the collection $\mathcal{E}_2(G)$ of Definition 5.10.1 is given by the elementary abelian 2-subgroups whose nontrivial elements are in class $2B$; and the collection is ample and normalizer-sharp using Proposition 5.10.2. Now we checked above that $\mathcal{E}_2(G)$ contains the G-classes of A_1 and A_2. In addition it contains two further G-classes; these are denoted in [**4**, 1.4 and 1.3], by R_1 and K, of ranks two and three. The class of A_2 can be distinguished from that of R_1 by the condition that A_2 is the set product of A_1 and a conjugate $g^{-1}A_1g$, where

$$g^{-1}A_1g \leq O_2\big(C_G(A_1)\big) = O_2(H_1).$$

We have that $N_G(A_2) = H_2$, which is also of order 192 but not isomorphic to H_1. We now let $\mathcal{E}_2^-$ denote the further subposet of $\mathcal{E}_2(G)$ obtained by taking the G-conjugates just of A_1 and A_2. Since $A_1 < A_2$, we see as in Proposition 6.3.6 that the poset complex $\Delta(\mathcal{E}_2^-)$ affords the intersection complex $\Delta_{\mathcal{A}_I}$ defined by the A_i, and so is an internal poset complex determined by the H_i.

We claim further that $\mathcal{E}_2^-$ is homotopy equivalent via a G-map to $\mathcal{E}_2(G)$; but we postpone to Section 8.1 the lengthier argument for this fact. It now follows by Proposition 5.10.2 that $\mathcal{E}_2^-$ is also ample and normalizer-sharp.

For flag-transitivity of G on $\Delta(\mathcal{E}_2^-)$, we have a very easy case of the method of Example 6.4.7 (1): we had seen above that the Σ_3 in H_2 transitively permutes the three G-conjugates of A_1 in A_2, as required for Remark 6.4.6 (1). In particular we have

$$\mathsf{N}(\ \Delta(\mathcal{E}_2^-)/G\ ) = \mathcal{H}_I.$$

We mention also that essentially this same observation about the Σ_3 in H_2 gives the flag-transitivity of G on the 2-local complex $\Delta_{\mathcal{H}_I}$, again an easy case—now using the viewpoint of 6.4.7 (2). Thus by applying Lemma 6.5.4 to $\Delta(\mathcal{E}_2^-)$ regarded in the above viewpoint as an intersection complex $\Delta_{\mathcal{A}_I}$, we obtain an isomorphism with the complex $\Delta_{\mathcal{H}_I}$.

Digression: The relationship with $\mathcal{B}_2^{cen}(G)$. In fact, it is even easier to obtain an ample collection determined by the H_i, if we work instead with the subcollection $\mathcal{B}_2^{cen}(G)$ of $\mathcal{B}_2(G)$ given by the members which are 2-centric. Of course $\mathcal{B}_2^{cen}(G)$ is also ample and normalizer-sharp, as we saw in Proposition 5.11.2.

The poset $\mathcal{B}_2(G)$ of 2-radical subgroups of M_{12} is given in An and Conder [**13**, 2D]. Among the five 2-radical conjugacy classes given there, just three are 2-centric: we get two classes of minimal members, represented by $B_1 := O_2(H_1)$ (with center A_1) and $B_2 := O_2(H_2)$ (with center A_2); and a single class of maximal members, represented by $B_I := H_I$.

Now (cf. the case for M_{11} in the previous section) the set-map $\mathcal{O}_2$ induces a bijection of the vertices of $\Delta_{\mathcal{H}_I}$ with the minimal members of $\mathcal{B}_2^{cen}(G)$; but G is not flag-transitive on $\Delta(\mathcal{B}_2^{cen}(G))$. And this time, we do not get even the mild complication which we encountered with M_{11}; for here we also get a bijection of the 1-simplices of $\Delta_{\mathcal{H}_I}$, such as $\sigma_I = \{H_1, H_2\}$ which has stabilizer H_I, with the maximal members such as $B_I = H_I = O_2(H_I)$ of $\mathcal{B}_2^{cen}(G)$. This reflects the fact that for M_{12}, we have the self-normalizing condition $N_G(H_J) = H_J$ for all types $J \subseteq I$. We consider some consequences.

For example, we are in the best-behaved situation that we had for M_{24} in Example 6.2.12, so that we can identify the 1-simplex σ_I with the subgroup H_I. This then realizes the poset $\mathcal{Q}(\Delta_{\mathcal{H}_I})$ of simplices of $\Delta_{\mathcal{H}_I}$ as a poset of subgroups of G. And now the set-map $\mathcal{O}_2$ of the previous paragraph is order-reversing on this poset, and so in this case extends to a poset anti-isomorphism of $\mathcal{Q}(\Delta_{\mathcal{H}_I})$ with $\mathcal{B}_2^{cen}(G)$. Indeed the normalizer set-map N in this case affords the inverse map. (Cf. the duality viewpoint of 6.3.13.)

Alternatively we could regard the condition $N_G(H_J) = H_J$ for all types J as giving the vacuous case of satisfying the hypotheses for a version of Lemma 6.5.7—in which no actual further identifications are being made, and hence the quotient process just reduces to the identity, giving the poset (anti-)isomorphism with $\mathcal{B}_2^{cen}(G)$.

Also since we had also observed that $\Delta_{\mathcal{H}_I}$ arises as $\Delta(\mathcal{E}_2^-)$, we see by examining the above anti-isomorphism one level deeper that $\mathcal{B}_2^{cen}(G)$ is isomorphic to the poset barycentric subdivision of $\mathcal{E}_2^-$.

Main results for M_{12}. We now return to our original focus on the poset $\mathcal{E}_2^-$. The poset has reduced Euler characteristic $-2^4 \cdot 31$ (just as for M_{11} in Section 7.1); this is not divisible by $|G|_2 = 2^6$, and so in view of Remark 5.9.4, $\mathcal{E}_2^-$ is not in the standard ample homotopy type represented by $\mathcal{A}_2(G)$.

In particular, this is why the particular 2-local geometry $\mathcal{E}_2^-$ was not discussed in Smith and Yoshiara [**116**], which focused instead on the standard ample G-homotopy type. But a 2-local geometry for M_{12} was considered in Buekenhout [**38**, Sec. 7] and in Ronan and Stroth [**107**, p. 62]; it was defined essentially via the above 2-local complex $\Delta_{\mathcal{H}_I}$. For our present purposes, it will be slightly more elementary to choose as our 2-local geometry here the variant of that 2-local complex (isomorphic via 6.5.4) given by the poset complex $\Delta(\mathcal{E}_2^-)$.

The discussion above leads via our standard viewpoint, much as in the case of M_{11} in the previous section, to our main result in the case of M_{12}:

THEOREM 7.2.1. *For $G = M_{12}$, the 2-local complex $\Delta_{\mathcal{H}_I}$ and the collection $\mathcal{E}_2^-$ are ample and normalizer-sharp. In particular, for the groups in $\mathsf{N}(\Delta_{\mathcal{H}_I}/G) = \mathcal{H}_I$, the natural group surjection from the amalgamated free product $H_1 *_{H_{12}} H_2$ onto M_{12} induces a mod-2 cohomology isomorphism, with a simplex cohomology decomposition as in (5.8.6). The corresponding map from the homotopy colimit of the pushout square diagram (4.5.6) to BM_{12}, followed by 2-completion, induces a homotopy equivalence.*

PROOF. The argument is similar to that for M_{11} in Theorem 7.1.1: However this time we obtained ampleness and normalizer sharpness from Proposition 5.10.2 (or alternatively from Proposition 5.11.2). Also since now both maximal 2-locals

H_i contain the Sylow 2-subgroup $H_I = H_{12}$ of M_{12}, as required in the Benson–Wilkerson result Lemma 4.5.10, we could in fact apply Proposition 5.6.14 instead of Theorem 5.2.7 for the final homotopy equivalence. □

As mentioned earlier, the statements in Theorem 7.2.1 were originally established earlier as a consequence of the extensive calculations with $\Delta(\mathcal{A}_2(G))/G$ of Adem, Maginnis and Milgram **[4]**, in their computation of the mod 2 cohomology ring of M_{12}. Another proof of the cohomology decomposition was given in Benson and Wilkerson **[29]**, by showing that the alternating sum of the characters of the permutation modules corresponding to the terms of the cohomology decomposition, namely

$$\chi[G/H_{12}] - \chi[G/H_1] - \chi[G/H_2] + \chi[G/G],$$

is a linear combination of characters in a non-principal 2-block of M_{12}. For this condition can be used in place of the condition of projectivity of the reduced Lefschetz module at the step of Webb's argument given by our (5.1.15)—the functors $\mathrm{Ext}^*_{\mathbb{F}_pG}(-, \mathbb{F}_p)$ still vanish: since there can be no non-zero homomorphisms or non-split extensions between a member of a non-principal block and the trivial module $\mathbb{F}_p$ in the principal block.[3]

There seems to be a minor error in the presentation of the mod 2 cohomology ring of M_{12} given in Adem, Maginnis and Milgram **[4]**, and copied in Adem and Milgram **[6]**, Section VIII.3. Below, we choose a new set of generators according to the following table, as in Benson and Carlson **[28]**.

	degree	name in notation of [4]	Sq^1	Sq^2	Sq^4
α	2	α	x	α^2	
x	3	x	0	αx	
y	3	y	α^2	$\alpha x + \alpha y$	
z	3	$x + y + z$	0	γ	
β	4	$\beta + \alpha^2$	0	$\delta + \alpha\beta + z^2$	β^2
γ	5	$\gamma + \alpha y$	z^2	0	$\beta\gamma + z\delta + z^3$
δ	6	$\delta + x^2 + \alpha^3 + \alpha\beta + z^2$	ε	$\alpha\delta$	$\beta\delta + z\varepsilon$
ε	7	$\varepsilon + x\beta + \alpha^2 x$	0	0	$\beta\varepsilon + \alpha^2\varepsilon$

In terms of these new generators, the cohomology is given by the following theorem. The relation (xiv) below contradicts the corresponding formula in **[4]** for ε^2, but appears nonetheless to be correct.

---

[3]A similar argument on a non-principal 2-block for J_2 will arise in Section 7.7. Indeed for all 15 sporadic groups G in (5.10.3) and (5.11.3) for which the 2-local geometry Δ is not homotopy equivalent to $\mathcal{S}_2(G)$, it seems that the reduced Lefschetz module involves an indecomposable in a suitable non-principal 2-block of G of small positive defect. No general explanation for this phenomenon seems known.

THEOREM 7.2.2. *The ring $H^*(M_{12}, \mathbb{F}_2)$ is generated by elements α, x, y, z, β, γ, δ and ε (with degrees as above), and the following is a complete set of relations:*

(i)	$\alpha z = 0$	(ii)	$y^2 = \alpha^3 + x^2 + xy$
(iii)	$xz = 0$	(iv)	$yz = 0$
(v)	$\alpha\gamma = 0$	(vi)	$x\gamma = 0$
(vii)	$y\gamma = 0$	(viii)	$x\delta = 0$
(ix)	$y\delta = \alpha\varepsilon$	(x)	$x\varepsilon = 0$
(xi)	$y\varepsilon = \alpha^2\delta$	(xii)	$\gamma^2 = z\varepsilon + z^2\beta$
(xiii)	$\gamma\varepsilon = z^2\delta$	(xiv)	$\varepsilon^2 = z\gamma\delta + z\beta\varepsilon + \alpha\delta^2$.

The Poincaré series for the cohomology is

$$\sum_{i=0}^{\infty} \dim_{\mathbb{F}_2} H^i(M_{12}, \mathbb{F}_2)\, t^i = \frac{(1+t^3)^2}{(1-t^2)(1-t^3)(1-t^4)}.$$

The cohomology of the double cover $2M_{12}$ was computed by Benson and Carlson [**28**]. It has 15 generators and 67 relations.

7.3. The Mathieu group M_{22}

For $G := M_{22}$, we take as our index set $I := \{H, O, Q\}$, corresponding to G-classes of maximal 2-local subgroups in the Atlas [**47**, p. 39] with structure:

$H_H \cong 2^4 : \mathrm{Alt}_6$,

$H_O \cong 2^3 : L_3(2)$, and

$H_Q \cong 2^4 : \Sigma_5$.

As usual the H_i determine the 2-local complex $\Delta_{\mathcal{H}_I}$ of Definition 6.3.5. We mention that H_O does not contain a full Sylow 2-subgroup of G. This time we have a subgroup simplex $\mathcal{H}_I$ as in Definition 4.6.2 of dimension 2; the common intersection $H_I = H_{H,O,Q}$ of order 2^6 is of index two in a Sylow 2-subgroup $N_G(H_I)$ of G.

The usual diagram for $\Delta_{\mathcal{H}_I}$ (which can be used to describe the intersections in $\mathcal{H}_I$, cf. Digression 6.2.14) is given in [**106**, p. 286] as:

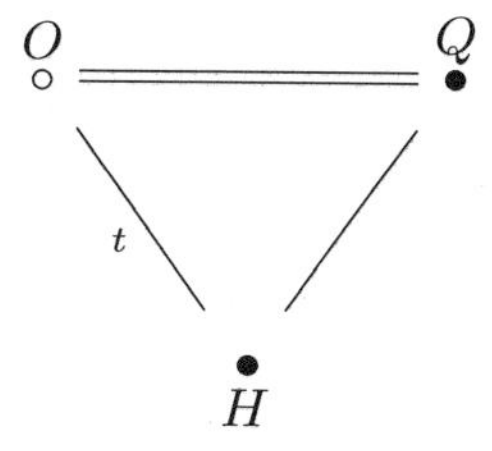

Here the "hollow node" at O is a convention indicating that H_O does not contain a full Sylow 2-subgroup of G. Also the label "t" indicates that the link of $\{H_Q\}$ is the triples geometry for Σ_5, which we have described in some detail in Example 6.5.5. (This subdiagram for Σ_5 will appear in some further diagrams later; indeed the whole diagram for M_{22} also appears in geometries for some larger groups—see for example Co_2 in Section 7.14.)

As usual we set $U_J := O_2(H_J)$ for each nonempty $J \subseteq I$.

Some details of the 2-local geometry $\Delta_{\mathcal{H}_I}$. Our entire treatment of M_{22} can essentially be deduced directly from our treatment of M_{24} in Section 7.5, since the present complex $\Delta_{\mathcal{H}_I}$ can be embedded in the complex similarly defined for M_{24} in Example 6.2.10. However, there are sufficient variations that it seems reasonable to provide a discussion of this relationship in greater detail.

The external complex Δ. It may be clearest to first discuss this natural embedding using instead the viewpoint of the external complex for M_{24} given at Example 6.3.11 by the Steiner system $\mathcal{S}(5,8,24)$ on a 24-set Ω—which we will now adapt for the subgroup M_{22}. In particular, we mention that in the index set I for M_{22} here, the letters H, O, Q stand for "hexad, octad, quintet", which we will define below in terms of corresponding structures for M_{24}.

We consider a Steiner system $\mathcal{S}(3,6,22)$ for M_{22} as follows: We fix two points of Ω, say a, b; so that $\Omega^- := \Omega \setminus \{a,b\}$ provides a 22-set for the action of the subgroup M_{22}. From the blocks of size 8 in $\mathcal{S}(5,8,24)$ (i.e., the octads for M_{24}), we select those containing a, b—which define 6-sets in Ω^-, giving the blocks in $\mathcal{S}(3,6,22)$; these are the *hexads* for M_{22}. The hexads will afford one type (H) of vertex in our external complex Δ. But we will also make use of those octads of M_{24} which do *not* meet $\{a,b\}$; these determine 8-sets from Ω^-, and they are the *octads* for M_{22}. The octads define a further type (O) of vertices in Δ; the incidence with hexads is via being disjoint in Ω^-. Finally from the sextets of M_{24}, we take those determined by a 4-subset of Ω containing a, b, so that they determine a 2-subset of Ω^-; the original sextet partitions the 22 points into the 2-set and 5 tetrads, and this configuration is then called a *quintet*. The quintets give the final type (Q) of vertices in Δ; their incidence with hexads and octads is induced directly from that for M_{24}—namely when a hexad or octad is assembled from two parts of the quintet partition.

One obvious difference from the external complex for M_{24} in Example 6.3.11 is that the trios of M_{24} no longer play any explicit role in the present complex Δ for M_{22}. However they are still at least in the background, in that the incidence of a hexad H and an octad O is via containment of the corresponding octads of M_{24} in a common trio there. Indeed we see in this case that the complement of $H \cup O$ in Ω^- is the third octad O' of that trio; and we have $H_{H,O} = H_{H,O'}$ given by the subgroup of the trio stabilizer which normalizes all three of its octads. This natural pairing of H, O and H, O' will be relevant for certain identifications when we later compare Δ with $\mathcal{B}_2(G)$ via the method of Lemma 6.5.7.

As an illustration of a more elementary computation with Δ, we check flag-transitivity of the action of M_{22}. The arguments will essentially be deduced from those for M_{24} in Example 6.4.7 (1); but now with a slight change in viewpoint: There we worked in the 11-dimensional Golay code module A, where we had realized octads, trios and sextets as certain subspaces. Here in effect we work in the 10-dimensional M_{22}-submodule B of A which is perpendicular to the 2-set $\{a,b\}$; though we will able to phrase much of our discussion directly in terms of the combinatorics of the hexads, octads, and quintets, rather than in terms of spaces in the submodule. Again, the argument will be conducted within links of vertices, which are described by the above diagram for the geometry. We work first in the link of a quintet Q: Since Q has a 2-set and five tetrads, we get $\binom{5}{1} = 5$ hexads and $\binom{5}{2} = 10$ octads. Furthermore their incidence is just that for the points and triples of the triples geometry in Example 6.5.5 (as suggested by the subdiagram for Q in the diagram). And we had observed there the flag-transitivity of $H_Q/U_Q \cong \Sigma_5$. Next we consider the link of a hexad H: Now in the calculation for the corresponding octad of M_{24} in Example 6.4.7 (1), we had obtained 15 trio 1-spaces and 35 sextet 2-spaces of the 4-space $\tilde{B}_1$. On restricting to the submodule B for M_{22}, we obtain: 30 octads, in complementary pairs O, O' for the 15 trios as above; but just 15 quintets, since we now only use sextets determined by tetrads on a, b. Indeed these latter 15 correspond to totally isotropic 2-spaces in a symplectic form on the space $\tilde{B}_4$. Now the subgroup $H_H/U_H \cong \mathrm{Alt}_6$ of Σ_6 remains flag-transitive on the building for $\Sigma_6 \cong Sp_4(2)$ given by the polar space, namely the 15 1-spaces and 15 isotropic 2-spaces (as suggested by the double-bond subdiagram at H in the diagram). So it only remains to observe that the stabilizer in Alt_6 of the 2-space for Q is a subgroup Σ_4 acting naturally on the 4 tetrads of Q not in H—and in particular there are elements interchanging pairs of complementary octads such as O and O' on those 4 tetrads. This establishes flag-transitivity in the link of H. And so since we now have transitivity for two of the three types, this suffices by Proposition 6.4.5 for flag-transitivity of M_{22} on Δ. (We could also work in the link of an octad O; but the argument in that case is perhaps less transparent in the external complex Δ—instead we will explore that link in the context of the 2-local complex $\Delta_{\mathcal{H}_I}$ below.)

Identifying Δ with $\Delta_{\mathcal{H}_I}$. We now turn to showing that our external complex Δ really does exhibit our original 2-local complex $\Delta_{\mathcal{H}_I}$. It is essentially implicit in our introduction of Δ above that the subgroups H_H, H_O, H_Q—which we chose at the outset in order to define $\Delta_{\mathcal{H}_I}$—are in fact the stabilizers of and incident hexad H, octad O, and quintet Q. But we can also now analyze the incidence in $\Delta_{\mathcal{H}_I}$—via a suitable modification of the arguments in the description of the incidence for M_{24} in Remark 6.2.17 (1).

It is perhaps easiest to start in the link of H_O: Here $U_O \leq H_I$ and we set $\bar{H}_O := H_O/U_O \cong L_3(2)$. Now although H_I is proper in a Sylow 2-subgroup of G, it is in fact Sylow in H_O; and so $\bar{H}_I$ is Sylow in $\bar{H}_O$. Hence just as in the discussion in 6.2.17 (1): $\bar{H}_I$ in particular affords a Borel subgroup of the Lie type group $L_3(2)$; and the overgroups of H_I in H_O are determined by parabolic subgroups of $L_3(2)$—including the intersections $H_{O,j}$ for $j \in I \setminus \{O\}$. It remains to verify that each conjugate of the intersection $H_{O,j}$ under H_O in turn determines a unique H_j incident to H_O; and now we encounter a slight variation from the argument in 6.2.17 (1): This time U_O is *not* in fact weakly closed—but the situation remains under control, as follows. If both H_j and its conjugate by some g contain $H_{O,j}$, then H_j contains both $H_{O,j}$ and its g^{-1}-conjugate, and in particular contains the corresponding conjugates of U_O. But in that case those conjugates of U_O are in fact conjugate under the Sylow group $N_G(H_I)$ of G (see for example the discussion below of the structure of $H_{H,O}$); and since $j \neq O$ we see that the Sylow $N_G(H_I)$ lies in the stabilizer H_j, so that the conjugate H_j is indeed uniquely determined. That is, we have verified that the link of H_O just affords the building for $L_3(2)$ (as suggested by the single-bond subdiagram at O in the diagram).

For the remaining links of types $j \in \{Q, O\}$, H_I is not Sylow in H_j. Correspondingly we will not attempt to completely characterize the overgroups of H_I as we did in 6.2.17 (1); instead it will be advantageous to take advantage of the more inductive form of the incidence in Definition 6.3.5—that is, using the form appearing in Example 6.2.10, rather than mimicking the form of (6.2.3).

We next consider the link of H_Q. Here we have $U_Q \leq H_I$, and we set $\tilde{H}_Q := H_Q/U_Q \cong \Sigma_5$. This time we are not working in terms of a Lie type group and building; so we will analyze the incidence more directly than we did in the case of H_O above. First we consider a hexad H incident to Q. We have $\tilde{H}_{Q,H} \cong \Sigma_4$; and such subgroups correspond with the five points permuted by Σ_5. Furthermore since U_Q is weakly closed (as was the case in 6.2.17 (1)), H_H is the only conjugate of itself above $H_{Q,H}$, so that the above correspondence gives five conjugate of H_H in the link of H_Q. We turn to the case of an octad O incident to Q. This time we have $\tilde{H}_{Q,O} \cong \Sigma_2 \times \Sigma_3$, and such subgroups correspond to the 10 triples from the five points; and again as U_Q is weakly closed, the correspondence gives 10 conjugates of H_O in the link of H_Q. Finally $\tilde{H}_I$ is a four-subgroup of both $\tilde{H}_{Q,H}$ and $\tilde{H}_{Q,H}$, containing a unique involution which is a product of a pair of transpositions. This gives the incidence in the link of H_Q—and so we see that link is just the triples geometry of Example 6.5.5.

Finally we consider the link of H_H. We set $\hat{H} := H_H/U_H \cong \mathrm{Alt}_6$. Then $\hat{H}_I \cong D_8$ is Sylow in $\mathrm{Alt}_6 \cong Sp_4(2)'$, and arguing in this index-2 subgroup of a Lie type group, we can again characterize the overgroups of $\hat{H}_I$ as the intersection with Alt_6 of parabolic subgroups. However this time we have $U_H \not\leq H_I$, so that it can be touchier to use preimages to deduce consequences about overgroups of H_I. There is no real complication when we consider a quintet Q incident with H. For $\hat{H}_{H,Q} \cong \Sigma_4$, and each of the 15 H_H-conjugates of this subgroup arises as the stabilizer in $Sp_4(2)$ of a totally isotropic 2-space in the natural symplectic module U_H. And U_H is weakly closed and lies in $H_{H,Q}$—so that the usual argument as in 6.2.17 (1) gives us 15 corresponding conjugates of H_Q incident with H_H. But more care is needed as we next discuss the case of an octad O incident with H. Indeed it will be useful (now and for later purposes) to give some extra detail about the structure of the stabilizer $H_{H,O}$ of the pair. Here $U_H \cap U_O$ is of order 2—let z denote a generator. Then $H_{H,O}$ lies in the centralizer of z. In fact $H_{H,O}$ is of index 2 in $C_G(z)$, and $C_G(z) = N_G(H_{H,O})$. Furthermore $O_2(H_{H,O}) = U_{H,O}$ is extraspecial, with two conjugate maximal elementary abelian subgroups of rank 3 given by $U_H \cap H_O$ and $U_H \cap H_{O'}$—where O' is the third octad in the trio determined by H and O as described earlier. Finally $\hat{H}_{H,O} \cong \Sigma_4$; while $N_G(H_{H,O})$ is an extension of $H_{H,O}$ by an element interchanging O and O' (and hence exchanging the above maximal abelian subgroups of $U_{H,O}$). This element can in fact be chosen from $N_G(H_I)$, which is Sylow in G; so that $N_G(H_{H,O}) = H_{H,O}N_G(H_I)$. To complete our discussion of the link of H_H, we observe that each of the 15 H_H-conjugates of $\hat{H}_{H,O}$ arises as the stabilizer of a vector like z in U_H of rank 4. But this time $\hat{H}_{H,O}$ does not determine a unique overgroup conjugate to H_H—in particular $H_{H,O}$ does not contain the weakly closed subgroup U_H. However just two subgroups H_O and $H_{O'}$ lie over $H_{H,O}$ (and we have observed that these are conjugate under the Sylow group $N_G(H_I)$). Thus we have 15 conjugates of H_H, and 30 conjugates of H_O, in the link of H_H, with $\hat{H}_I$ determining the incidence as in the usual polar space for $Sp_4(2)$—this is the structure called a "2-split cover" of that polar space in Ronan and Smith [**106**, p. 286].

We now obtain the flag-transitivity of M_{22} on $\Delta_{\mathcal{H}_I}$ as a corollary of the discussion just completed: For flag-transitivity on the link of H_O follows from our identification of that link with the building for the $L_3(2)$ in H_O; while flag-transitivity on the link of H_Q follows using our identification with the triples geometry discussed in Example 6.5.5. (Indeed our argument in the previous paragraph even gives flag-transitivity in the remaining link of type H). Thus we may apply Proposition 6.4.5 for flag-transitivity of M_{22}. In particular we have $\mathsf{N}(\Delta_{\mathcal{H}_I}/G) = \mathcal{H}_I$.

Finally from the flag-transitivity of M_{22} on both Δ and $\Delta_{\mathcal{H}_I}$, we obtain an isomorphism of those simplicial complexes by the obvious analogue of Lemma 6.5.4.

The relationship with $\mathcal{A}_2(G)$. We now turn to connections with the poset $\mathcal{A}_2(G)$ of elementary abelian 2-subgroups of M_{22}. We mention first that the orbit complex $\Delta(\mathcal{A}_2(G))/G$ has been described by Adem and Milgram in [**7**]. But we will focus here on the viewpoint of complexes defined by 2-local structure.

For the purposes of our main result for M_{22} in Theorem 7.3.1 below, it would now be sufficient just to quote Smith and Yoshiara [**116**, 2.8], which shows that Δ is homotopy equivalent to $\mathcal{A}_2(G)$—where as usual that equivalence is in fact via a G-map. It follows using Proposition 5.9.1 that $\Delta_{\mathcal{H}_I}$ is also ample and normalizer-sharp.

However, in order to further compare and contrast with the case of M_{24}, we now include a briefer mention of some other possible approaches to the 2-local geometry of M_{22} which are related to $\mathcal{A}_2(G)$; but for which we have not tried to verify all details.

If we set $A_i := \Omega_1 Z(O_2(H_i))$ for $i \in I$, then we obtain representatives of the G-classes which are maximal under inclusion in $\mathcal{A}_2(G)$—of ranks 4,3,4 for $i = H, O, Q$ as we indicated at the outset.

Furthermore we have $N_G(A_i) = H_i$—so that it would in fact be possible to approach the poset $\mathcal{A}_2(G)$ via the viewpoint of the intersection complex $\Delta_{\mathcal{A}_I}$ defined from these A_i. However since there are three rather than two maximal members, that approach is more complicated than the almost-poset complex with two maximals in Definition 6.3.9; and correspondingly it is not our preferred approach to a geometry for M_{22}. But one might complete such an approach by verifying flag-transitivity of M_{22} on $\Delta_{\mathcal{A}_I}$, in order to obtain an isomorphism with $\Delta_{\mathcal{H}_I}$ via Lemma 6.5.4.

Similarly we also mention that the A_i also afford the 2-subgroups $U_i = O_2(H_i)$, which could be used in the construction of the complex $\Delta_{\mathcal{U}_I}$ of Definition 6.3.13 "dual" to $\Delta_{\mathcal{H}_I}$. Again verifying flag-transitivity would show via 6.5.4 that the normalizer set-map N extends to an anti-isomorphism of the complexes $\Delta_{\mathcal{U}_I}$ and $\Delta_{\mathcal{H}_I}$.

Furthermore we could essentially restrict to M_{22} the various other alternative geometric constructions for M_{24} indicated in Section 6.3. For example, the external poset complexes in Example 6.3.7 for M_{24} lead to somewhat similar complexes for M_{22}: As suggested earlier during our construction of the external complex Δ, we could produce a corresponding external poset complex on the 10-dimensional Golay code module for M_{22}; and using suitable subspaces of dimensions 1, 1, 4 for H, O, Q, we get an intersection complex—though not a poset complex, since the incidence relation among the two types of 1-spaces cannot be inclusion (instead it is orthogonality of isotropic and non-isotropic vectors). Or, in the corresponding 10-dimensional cocode module for M_{22}, using subspaces of dimensions 4, 3, 1 we get an almost-poset complex as in Definition 6.3.9—since the two maximal members of dimensions four and three intersect in a space of dimension two. This latter complex could also be constructed using a suitable space of dimension five in place of the space of dimension four.

Finally with regard to historical sequence, we mention that the usual 2-local geometry for M_{22} as given in Ronan and Smith [**106**, p. 286] is essentially the external complex Δ, constructed as above from the Steiner system $\mathcal{S}(3,6,22)$. A different viewpoint on geometries for M_{22} is given by the approach of Ivanov and Shpectorov, for example in [**75**]: they emphasize minimal 2-locals rather than maximal 2-locals—and essentially replace the triples geometry for Σ_5 by the related Petersen graph, in their theory of *Petersen geometries*. (The Petersen graph has 10 vertices and 15 faces, and so is instead the geometry of transpositions and the 4-subgroups containing them in Σ_5.)

Digression: the relationship with $\mathcal{B}_2(G)$. Although we state Theorem 7.3.1 just in terms of $\Delta_{\mathcal{H}_I}$, using the above homotopy equivalence with $\mathcal{A}_2(G)$, it may be illuminating in this case to also describe the close connection with the poset $\mathcal{B}_2(G)$ of 2-radical subgroups. Of course from Proposition 5.9.1, $\mathcal{B}_2(G)$ is also ample and normalizer-sharp.

Now in general we know via the results of Quillen and Bouc (cf. Section 5.9) that $\mathcal{A}_2(G)$ and $\mathcal{B}_2(G)$ are always G-homotopy equivalent for any G. But we will see in the present case of $G = M_{22}$ that there is a quotient relationship of $\Delta_{\mathcal{H}_I}$ with $\mathcal{B}_2(G)$ which is "just short of isomorphism"; much as we saw in the case of the triples geometry for Σ_5 in Example 6.5.5—and indeed arising here essentially because that geometry appears as a link in the geometry for M_{22}.

The determination of the poset $\mathcal{B}_2(G)$ of 2-radical subgroups of M_{22} is given in An and Conder [**13**, 2F]. There are seven G-conjugacy classes of members; and all are centric.

These seven classes are in fact in natural bijection with the members H_J of the subgroup simplex $\mathcal{H}_I$. Indeed more precisely: for each of the seven nonempty type subsets $J \subseteq I$, the subgroup $B_J := O_2\big(N_G(H_J)\big)$ is a member of $B_2(G)$; and this subgroup can also be exhibited as the product $\prod_{j\in J} U_i$. For convenience of comparison, we mention that the seven rows in [**13**, 2F] correspond respectively to our type subsets:

$$O, H, Q, OQ, OH, HQ, OHQ.$$

However, for some types J, B_J does not lie in our original subgroup H_J; and in those cases, we will have $U_J := O_2(H_J) < B_J$. The pattern of such indices J will appear in our application of Lemma 6.5.7 below But first we will approach this relationship from a more general viewpoint.

The above bijection is just between the orbit poset $\mathcal{B}_2(G)/G$ and the G-orbits on the poset of simplices of $\Delta_{\mathcal{H}_I}$; and in fact it fails to pull back to a bijection of the original posets.

In some sense, this failure arises because H_O does not contain a Sylow 2-subgroup of G; but that does not precisely explain the distinction, since $U_O = B_O$ and $U_{O,Q} = B_{O,Q}$ are in fact members of $\mathcal{B}_2(G)$; and we do get a bijection if we restrict to their conjugates. Instead, the distinction is actually visible for the types given by $\{O, H\}$ and its over-set $\{O, H, Q\} = I$; and it arises essentially via our earlier observation that incident O and H determine a further O' incident to H. Here $H_{O,H} = H_{O',H}$, so we get a 2:1 correspondence rather than a bijection; and $U_{O,H}$ is of index 2 in the member $B_{O,H}$ of $\mathcal{B}_2(G)$. We have $B_{O,H} = U_{O,H}U_H = U_O U_H = U_{O'}U_H$; and $B_{O,H}$ is the largest normal 2-subgroup of $N_G(H_{O,H})$, which in turn is just $C_G(z)$. Similarly $H_I = H_{O,H,Q} = H_{O',H,Q}$

gives a 2:1 correspondence—but now the corresponding member B_I of $\mathcal{B}_2(G)$ is the Sylow 2-group $N_G(H_I)$, in which $H_I = U_I$ is contained with index 2. For these cases, in order to obtain an actual bijection at the level of posets, we will need to be able to identify the indicated pairs. corresponding to O and O'.

So we set up to apply Lemma 6.5.7; this application will involve more simplex types than our first application to the triples geometry for Σ_5 immediately following the lemma. Here we take $k := O$ as our distinguished type, and $J := \{H\}$ as our distinguished subset of $I \setminus \{O\}$. Hence we must verify (6.5.8) for the types K given by $\{H\}$ and its unique over-set $\{H, Q\}$ in $I \setminus \{O\}$. And this has in fact already been done in the course of our earlier discussion: First the Sylow group $N_G(H_I)$ does indeed lie in H_H and $H_{H,Q}$. So we turn to considering the types K^+ given by $\{H, O\}$ and $\{H, Q, O\} = I$. For the former, we had seen that $N_G(H_I)$ is a supplement to $H_{H,O}$ in $N_G(H_{H,O}) = C_G(z)$; and for the latter, $N_G(H_I)$ is vacuously a supplement to H_I in itself. (We remark that the former supplement property amounts essentially to the transitivity of the subgroup $N_G(H_I)$ on the pair O, O' of octads determined by incident H and O—as indicated in 6.5.7 (1).) This completes the verification of the hypotheses; so 6.5.7 shows that the indicated identifications define a poset homotopy equivalence from the poset of faces $\mathcal{R}$ of $\Delta_{\mathcal{H}_I}$ to the relevant quotient $\bar{\mathcal{R}}$.

Furthermore our earlier remarks have indicated a bijection of $\bar{\mathcal{R}}$ with $\mathcal{B}_2(G)$—which indeed is inclusion-preserving, and hence a poset isomorphism. Thus our quotient procedure gives a poset homotopy equivalence from the poset of faces of $\Delta_{\mathcal{H}_I}$ to $\mathcal{B}_2(G)$.

Main results for M_{22}. We now return from the above digression to our main theme. From that earlier discussion, via our usual viewpoint we obtain the following theorem.

THEOREM 7.3.1. *For $G = M_{22}$, the 2-local complex $\Delta_{\mathcal{H}_I}$ and the collection $\mathcal{A}_2(G)$ are ample and normalizer-sharp; so we get a simplex cohomology decomposition over $\mathsf{N}(\Delta_{\mathcal{H}_I}/G) = \mathcal{H}_I$ as in* (5.8.6). *In particular, for the groups in $\mathcal{H}_I$, the map from the homotopy colimit of the pushout cube diagram* (4.6.5) *to BM_{22}, followed by 2-completion, is a homotopy equivalence.*

PROOF. The argument is much as for M_{11} in Theorem 7.1.1, except that our geometry is now of rank three rather than rank two. □

The obvious analogue in fact holds for the automorphism group $\mathrm{Aut}(M_{22}) = M_{22}.2$; in which case the incident hexad, octad, and quintet now have stabilizers $2^4{:}\Sigma_6$, $2^4{:}L_3(2)$, and $2^5{:}\Sigma_5$.

The mod 2 cohomology ring of M_{22} was described by Adem and Milgram [**7**]; see also Adem and Milgram [**6**], Section VIII.5. (We mention that they use the above decomposition, which was originally established as a consequence of the earlier work of Ryba, Smith and Yoshiara [**108**, 3.3].) In [**7**] the authors first determine a large subring, and then write the cohomology as a module over this subring. This gives incomplete information about the ring structure, but at least determines the Poincaré series of the cohomology.

There is an interesting feature of the cohomology of a Sylow 2-subgroup of M_{22}, which also happens to be a Sylow 2-subgroup for M_{23} and McL: this Sylow 2-subgroup $H_I =: P$ is a semidirect product of a Sylow 2-subgroup Q of $L_3(4)$ by an automorphism t of order 2, which acts via a field automorphism. It turns out that the minimal projective resolution of $\mathbb{F}_2$ as an $\mathbb{F}_2Q$-module can be made into an exact sequence of $\mathbb{F}_2P$-modules. This has the consequence in the context of module cohomology that we have an isomorphism of $\mathbb{F}_2$-complexes

$$H^*(P, \mathbb{F}_2) \cong H^*\big(\langle t \rangle, H^*(Q, \mathbb{F}_2)\big).$$

In other words, in the spectral sequence for the extension of $\langle t \rangle$ by Q, there are no nonzero differentials, and no ungrading problems. This phenomenon also occurs for the cohomology of a general wreath product, by a theorem of Nakaoka [**99**]. Here is a conjecture which generalizes these two cases of the wreath product and of P above for M_{22}:

CONJECTURE 7.3.2. Suppose that H is a group of automorphisms of a p-group Q, and that k is a field of characteristic p, and set $G := Q \rtimes H$. Suppose further that there is a central series

$$1 =: Q_0 \le Q_1 \le \cdots \le Q_s := Q$$

such that for each i, Q_i/Q_{i-1} is an elementary abelian p-group, which when regarded as an $\mathbb{F}_pH$-module, is projective. Then the minimal resolution of k as a kQ-module extends to an exact sequence of kG-modules, so that $H^*(G,k) \cong H^*\big(H, H^*(Q,k)\big)$ as k-complexes.

7.4. The Mathieu group M_{23}

Our treatment of $G := M_{23}$ is fairly similar to that for M_{22} in Section 7.3—in particular, we will use the viewpoint of the relationship with M_{24}. Furthermore we recall that M_{23} contains M_{22} with odd index, so that the two groups have the same Sylow 2-subgroup. So our discussion of M_{23} can be somewhat briefer in places where we have analogies with structures for M_{22}; and we can then emphasize those details which do differ for M_{23}.

We again take as our index set $I := \{H, O, Q\}$; but now H stands for "heptad" rather than "hexad".

The corresponding G-classes of maximal 2-locals (cf. the Atlas [**47**, p. 71]) are represented by:

$H_H \cong 2^4 : \mathrm{Alt}_7$,
$H_O \cong 2^3 : L_3(2)$ (just as for M_{22}), and
$H_Q \cong 2^4 : (3 \times \mathrm{Alt}_5) : 2$.

As usual the H_i determine a subgroup simplex $\mathcal{H}_I$ as in Definition 4.6.2, and a 2-local complex $\Delta_{\mathcal{H}_I}$ as in Definition 6.3.5. As in the case of M_{22}, we have H_I of order 2^6, and of index two in $N_G(H_I)$ which is a Sylow 2-subgroup of M_{23}.

However this time, although H_O is maximal among 2-local subgroups, it is not maximal just among subgroups of G—since it is proper in a maximal subgroup $G_O \cong \mathrm{Alt}_8$ (and indeed in a further maximal subgroup M_{22}). This corresponds to the fact that while G_O is the full stabilizer in G of an octad O, our subgroup H_O is actually the stabilizer of a further structure we will call an augmented octad, and will denote by O^+ in our development below. Thus we apologize for the potential ambiguity of the expression H_O; we have used it here in order to emphasize the agreement with the corresponding subgroup of M_{22} in the previous section.

Our complex above for M_{23} has apparently not arisen in the literature on group geometries; and in particular it seems to require some extensions beyond the conventions in Ronan and Smith [**106**], in order to provide a diagram summarizing the structure of links of vertices. To that end, we propose the following diagram—which can actually be used (modulo our further discussion later in the section) for two different complexes based on either H_O or G_O.

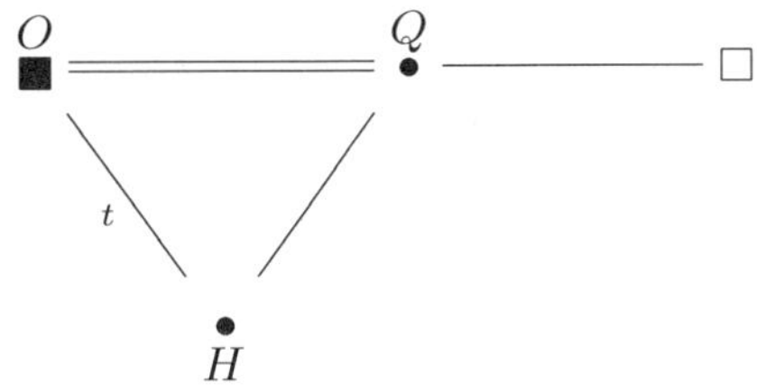

The C_3-subdiagram at H corresponds to the Neumaier geometry for the group Alt_7, determined by subgroups $L_3(2)$ and $2^2 : \Sigma_3 \times 2$ and Alt_6; see for example [**115**, Ch. 2]. (This geometry will appear in a number of other diagrams later.) The unfilled square node on the right essentially extends the convention used for M_{24} in Section 7.5: namely we do not define objects in the geometry corresponding to this node—but the node is useful for indicating truncated geometries involved in links of the objects in the geometry. The filled square node at O is a new convention we are introducing: it indicates that although the subdiagram with O removed corresponds to the group $G_O \cong \mathrm{Alt}_8$, for the complex $\Delta_{\mathcal{H}_I}$ we are in fact only using the *2-local* subgroup H_O determined by the *unfilled* square node in the diagram for G_O.

We could contrast these conventions with several other possible choices leading to geometries for M_{23}:

First, unlike the situation for M_{24}, here M_{23} actually does contain a subgroup, namely M_{22}, corresponding to the unfilled square node. So we could instead use an ordinary node $\bullet$ there—and then we would obtain a geometry of rank 4. Of course it would not be a fully 2-local geometry, in view of the use of the non-local subgroup M_{22}. This rank-4 geometry does not seem to be particularly useful.

A second alternative is to use the unfilled square node there, but at O to instead use a hollow ordinary node $\circ$, as in the usual diagram for M_{22} in Section 7.3; so we would still be ignoring the subgroup M_{22}, but we would be using G_O in place of H_O. This gives a certain rank-3 external complex Δ, which we will further develop in a moment as part of our historical discussion of geometries for M_{23}; again Δ is not a fully 2-local geometry, since it uses the non-local subgroup $G_O \cong \mathrm{Alt}_8$.

However for our main work with the 2-local complex $\Delta_{\mathcal{H}_I}$, here we will adopt the indicated convention of the filled and unfilled square nodes; since it seems to give a good description of our particular chosen 2-local geometry, which in turn seems most appropriate for purposes of cohomology.

Non-local vs local complexes for M_{23}. Our earlier remark that we use the subgroup H_O rather than G_O indicates an important difference between two otherwise similar-seeming geometries for M_{23}; these two geometries are described only rather briefly in Smith and Yoshiara [**116**, 2.7]. It seems useful to provide here a more extensive comparison and contrast; and in particular, it may be easiest to lead up to our desired complex $\Delta_{\mathcal{H}_I}$ by first examining the somewhat more elementary combinatorial structure of the other complex, which we will denote here by Δ.

(Digression:) The external complex Δ from the Steiner system. This complex Δ for M_{23} is the one which is actually considered second in the treatment in [**116**]. One approach to Δ would be to replace the above choice of H_O with the larger group $G_O \cong \mathrm{Alt}_8$; and then proceed with the usual construction as in Definition 6.3.5 of a complex $\Delta_{\mathcal{G}_I}$ based on conjugates of a subgroup simplex $\mathcal{G}_I$ determined by the new choice of maximal subgroups H_H, G_O, and H_Q. (We wouldn't call this a "2-local complex", since G_O is not a 2-local.)

But a more direct approach to Δ is to obtain an external complex by restricting from the one given for M_{24} in Example 6.3.11; this requires only a few slight and fairly obvious modifications from our discussion for M_{22} in Section 7.3. This time we fix just a single point a from the 24-set Ω for M_{24}; so that now $\Omega^- := \Omega \setminus \{a\}$ affords as 23-set for the action of M_{23}. Again the *octads* O are just those octads for M_{24} which do not contain a; and similarly the *heptads* H are the remaining octads for M_{24}—namely those which do contain a, and so determine 7-subsets of Ω^- (giving the blocks of a Steiner system $\mathcal{S}(4,7,23)$). Finally we take *all* of the sextets for M_{24}—this is because a must lie in some tetrad of each such sextet; and again we use the terminology of a *quintet* Q, since that tetrad determines a 3-subset of Ω^-, leaving five tetrads. The incidence among these three types is just induced from that for M_{24}, and indeed has essentially the same description as in the case of M_{22}: incidence between a heptad and and octad is by being disjoint; but between a heptad (or octad) and a quintet, by being built up from two parts of the quintet partition.

Also just as in the case of M_{22}: The trios of M_{24} do not appear explicitly in Δ; but they are still in the background, in that an incident heptad-octad pair H, O determines a further octad O' which with them completes a trio for M_{24}. We have $H_{H,O} = H_{H,O'}$ given by the stabilizer of that trio and its octads; and this will be relevant for identifications when we later make comparisons with $\mathcal{B}_2(G)$.

We next show flag-transitivity of M_{23} on Δ. First, just as for M_{22} in Section 7.3, the link of a quintet Q is the flag-transitive triples geometry for a Σ_5 from $H_Q/U_Q \cong (3 \times \mathrm{Alt}_5)2$. (This is as suggested by the subdiagram at Q of our diagram—where we recall that for Δ, we use a version of the diagram having instead a hollow ordinary node $\circ$ at O.) Next, the link of an octad O just reduces to that for the case of M_{24}, described in Example 6.4.7 (2): For given one of the 15 trios T on O for M_{24}, we just mentioned that the other octads in T are for M_{23} a further octad O' and a heptad H incident with O. Hence the link of O contains those 15 heptads, together with the 35 sextets on O—giving the truncation of projective 4-space over $\mathbb{F}_2$ as in that Example, with flag-transitive action by $G_O \cong \mathrm{Alt}_8 \cong L_4(2)$. (This is as suggested by the subdiagram at O.) Now our verification for these two of the three types of links suffices by Proposition 6.4.5 for flag-transitivity. But it will also be of interest to examine the link of the remaining vertex type, namely a heptad H: Just as above we consider the 15 trios and 35 sextets (in contrast to the case of M_{22} where we only used 15 of those sextets); and now it is standard that $H_H/U_H \cong \mathrm{Alt}_7$ is a subgroup of $\mathrm{Alt}_8 \cong L_4(2)$ which remains flag-transitive on that truncation of projective space. (This truncation of projective space happens to agree with the truncation to types O and Q of the Neumaier geometry, as suggested by the subdiagram at H of our diagram.) So it only remains to observe that, much as for M_{22}, the stabilizer in Alt_7 of a quintet induces a Σ_4 which in particular interchanges O and O'; completing the proof of flag-transitivity on the 30 octads and 35 quintets of this link.

We also get flag-transitivity of M_{23} on our complex in the form $\Delta_{\mathcal{G}_I}$—it is straightforward to adapt the above external-complex arguments for M_{23} into group-theoretic arguments on the 2-local complex, much as for M_{22} in Section 7.3. Consequently we get the usual isomorphism via the method of Lemma 6.5.4 of Δ with $\Delta_{\mathcal{G}_I}$.

Now it is observed in Smith and Yoshiara [**116**, p. 350] that $\Delta \cong \Delta_{\mathcal{G}_I}$ satisfies the fixed-point contractibility hypothesis for Webb's formula 5.1.16—so that we get a cohomology decomposition over the orbit simplex $\Delta_{\mathcal{G}_I}/G \cong \mathcal{G}_I$. Thus $\Delta_{\mathcal{G}_I}$ would be at least adequate or our purposes of group cohomology—and may well be the most natural complex from the viewpoint of group theory alone.

Indeed from experience with many other group geometries, we might be led to expect that the contractibility condition would lead to a proof that Δ is in the standard homotopy type represented by $\mathcal{A}_2(G)$ and $\mathcal{B}_2(G)$. But this is not the case (as is observed in [**116**] by comparison of Euler characteristics). The essential difference is that the stabilizer $G_O \cong \mathrm{Alt}_8$ is not a 2-local subgroup, and hence does not satisfy the hypothesis for the equivalence method in Proposition 1.4 of that paper.

So we will now turn to the first geometry for M_{23} considered in [**116**, 2.7], which seems perhaps more natural from the viewpoint of group cohomology (and in due course, from the viewpoint of the poset $\mathcal{B}_2(G)$).

The 2-local complex $\Delta_{\mathcal{H}_I}$. The brief discussion in [**116**, 2.7] essentially assumes the viewpoint of the maximal 2-locals H_i, with incidence determined by containment of conjugates of appropriate H_J; and this could be developed by making appropriate adjustments to the discussion of the corresponding subgroups for M_{22} in Section 7.3. However it also seems illuminating to approach $\Delta_{\mathcal{H}_I}$ by means of an external complex which requires only a fairly slight further construction, building on the complex Δ discussed just above.[4]

Recall that we wish now to use the subgroup $H_O \cong 2^3L_3(2)$ rather than the full octad stabilizer $G_O \cong \mathrm{Alt}_8$; and that there are 15 such subgroups for each octad O. Now H_O is in fact the stabilizer in $G_O \cong L_4(2)$ of one of the 15 nonzero vectors in the natural $\mathbb{F}_2$-module—and these are in correspondence with the 15 points of $\Omega^- \setminus O$. So we now introduce the additional terminology[5] of an *augmented octad* O^+ for the pair O, x—consisting of O and a further point x not in O. And we will be using the augmented octads in place of the octads themselves, as a new type of vertex in an external complex Δ^+. In particular our subgroup H_O is now more appropriately denoted as the stabilizer H_{O^+} of O^+. (And we could also adjust our diagram by using O^+ in place of O.) We define incidence of O^+ with a quintet Q whenever O is incident with Q (by being composed of two of the tetrads from Q), while x is in the 3-subset from Q. And O^+ is incident with a heptad H whenever O is incident with H (by being disjoint), while $x \in H$.

To check flag-transitivity of G on Δ^+, we require only minor adaptations of the arguments above on Δ. In the link of a quintet Q, there are now three augmented

[4]We are grateful to Bob Oliver for suggesting this development.

[5]We could alternatively refer to this 9-set say as a *nonad*; but we prefer to primarily emphasize the underlying octad O.

octads O^+ for each incident pair O, H—one for each of the points in the 3-subset from Q; and in the stabilizer $3 \times 2 \times 2$ in H_Q/U_Q of $\{O, H\}$, the factor of order three is transitive on that 3-subset. In the link of a heptad H, there are similarly three augmented octads O^+ for each incident pair O, Q—again one for each point in the 3-subset from Q; and as before the stabilizer Σ_4 in H_H/U_H of O, Q contains an element of order three transitive on the 3-subset. As usual the verification on these two of the three types suffices for flag-transitivity of M_{23} on Δ^+ by Proposition 6.4.5. (The argument in the link of an augmented octad $O^+ = O, x$ is perhaps less transparent: A standard calculation shows that of the 15 heptads and 35 quintets on O, exactly 7 each have their 7-set or 3-set containing x. These have the incidence pattern of the projective plane over $\mathbb{F}_2$, on which $H_{O^+}/U_{O^+} \cong L_3(2)$ is flag-transitive.)

As usual, these arguments can be adapted into more group-theoretic language to demonstrate flag-transitivity of M_{23} on the 2-local complex $\Delta_{\mathcal{H}_I}$; and we get the isomorphism of Δ^+ with $\Delta_{\mathcal{H}_I}$ by the method of Lemma 6.5.4. In particular we have $\mathsf{N}(\Delta_{\mathcal{H}_I}/G) = \mathcal{H}_I$.

The relationship with $\mathcal{A}_2(G)$. We are not aware of any published determination of $\mathcal{A}_2(G)$, or of the orbit complex $\Delta(\mathcal{A}_2(G))/G$.

It is shown in Smith and Yoshiara [**116**, p. 349] that $\Delta_{\mathcal{H}_I}$ is homotopy equivalent to $\mathcal{A}_2(G)$; and as usual one sees that this is via a G-map. This statement will in fact be sufficient for the purposes of our main result for M_{23} in Theorem 7.4.1 below.

However, since the terse treatment in [**116**, p. 349] only implicitly indicated the use of H_O to extend from Δ to the form of the complex we are here calling Δ^+, it seems natural to now give a summary of that argument explicitly referring to Δ^+. We fix as usual our maximal flag H, Q, O^+ of Δ^+—where $O^+ = O, x$ is an augmented octad, arising from the octad O of Δ and the point x in the heptad H; and as before we let O' denote the remaining octad of the trio of M_{24} determined by H and O. The steps in the proof in [**116**] consider elementary abelian 2-subgroups of $G = M_{23}$, in order of increasing size—beginning with $P = \langle z \rangle$ of order 2. By G-conjugacy we can take $P = U_H \cap U_Q \cap U_{O^+}$. The proof then describes the subcomplex there denoted by "$\mathcal{I}_P$", of vertices v with $P \leq U_v$; and we will now expand a little. First, the centralizer $C_G(P)$ is the extension of U_H by a particular subgroup $L \cong L_3(2)$ of $H_H/U_H \cong A_7$—where we emphasize that this L is *not* the subgroup $L_3(2)$ of A_7 which is visible in our diagram, namely that lying in the stabilizer G_O. Instead the orbit of O under our L is nontrivial—and indeed the orbit under L of the edge $O - Q$ of Δ is the subconfiguration indicated in [**116**] of seven octad pairs like O, O' and seven quintets, with the incidence pattern of of a projective plane over $\mathbb{F}_2$. But we now wish to translate this description of $\mathcal{I}_P$ into the language of Δ^+—since [**116**] only mentions the conjugates of some $H_O \cong 2^3 \colon L_3(2)$, and we wish to emphasize that this choice of H_O within G_O corresponds to the stabilizer of the augmented octad $O^+ = O, x$—which we will again denote more appropriately by H_{O^+}. Now the stabilizer in L of O^+ is in particular the stabilizer in L of x. It follows that the orbit under L of O^+ contains 7 pairs such as O^+ and $(O')^+ = O', x$. Thus the subcomplex $\mathcal{I}_P$ can be described in Δ^+ as the join of H (the unique heptad such that $P \leq U_H$) with the above configuration, which we will now denote by K, of 7 quintets and 7 pairs of augmented octads. (In particular $H * K$ is a cone on H, and so contractible.) In the remainder of the argument in [**116**], we consider a proper overgroup of P; we will denote this overgroup here by R (rather than

"Q" as in [**116**], to avoid conflict with our notation here of Q for a quintet). We want to demonstrate the contractibility of $\mathcal{I}_R$. As there, we are done if $R \leq U_H$, using the uniqueness of U_H over P; se we may assume that $R \not\leq U_H$, so that $\mathcal{I}_R$ has no heptads, and thus consists only of augmented octads and quintets. We next consider the case that $R \cap U_H = P$. Since $R > P$, we may assume that $R \geq P\langle y \rangle$, where the image of y is an involution of L. By conjugacy in L, we may assume that the image of R lies in one of two unipotent radicals—which are transvection subgroups on the natural module, corresponding either to the octad pair O, O' or to the quintet Q—and whose intersection is generated by the image of y. Indeed we may assume that the former transvection subgroup is the stabilizer of O^+, since we saw already for M_{22} in Section 7.3 that the intersection $U_{O^+} \cap U_{(O')^+}$ is just P, which does not contain y. So it follows in the various cases for R that $\mathcal{I}_R$ is either the edge $O^+ - R$ or one of its vertices; and in any case is contractible. So we are now reduced to the case where $R \cap U_H$ has order at least 2^2. Since $R \not\leq U_H$, we have $R \cap U_H$ of order exactly 2^2, with R of order at least 2^3. In particular $R \not\leq U_{O^+}$ so that $\mathcal{I}_R$ consists only of quintets; and indeed by conjugacy in L we may assume that the image of R lies in the unipotent radical for Q, with $\mathcal{I}_R$ given just by the single point Q, certainly contractible. This completes our summary of the proof in [**116**], adapted here to the language of the complex Δ^+.

Relations of $\Delta_{\mathcal{H}_I}$ with some other complexes. Just as for M_{22}, the $A_i := \Omega_1 Z\big(O_2(H_i)\big)$ for $i \in I$ are maximal in $\mathcal{A}_2(G)$, with $N_G(A_i) = H_i$; and the intersection complex $\Delta_{\mathcal{A}_I}$ as in Definition 6.3.2 is fairly complicated. Again the A_i afford the $U_i = O_2(H_i)$ in the construction of the complex $\Delta_{\mathcal{U}_I}$ of Definition 6.3.13. One could check flag-transitivity, in order to conclude by the method of Lemma 6.5.4 that this complex is dual to $\Delta_{\mathcal{H}_I}$ by extending the normalizer set-map N to these complexes. Indeed as for M_{22}, presumably one can extend some of the above arguments to show that the A_i give all the classes of maximal members of $\mathcal{A}_2(G)$, so that $\Delta_{\mathcal{A}_I}$ would give one approach to $\mathcal{A}_2(G)$.

This time from the use of the code module of dimension 11, we would not get an external complex view of $\Delta_{\mathcal{H}_I}$—that is, no analogue of that for the code module in Example 6.3.7 for M_{24}, or of that for M_{22} in Section 7.3—since here H_O and G_O determine the same 1-subspace in the module, whereas we need to distinguish the conjugates of H_O inside a fixed G_O. However we do get such an external complex from the cocode module of dimension 11, using subspaces of dimensions 6, 3, 1; these spaces form an inclusion-chain, so we even get an external *poset* complex as in Example 6.3.7.

Digression: the relationship with $\mathcal{B}_2(G)$. As was the case for M_{22} in the previous section: We actually proceed using the somewhat indirect homotopy equivalence above of $\Delta_{\mathcal{H}_I}$ with $\mathcal{A}_2(G)$; but there is actually a very close relationship of $\Delta_{\mathcal{H}_I}$ with $\mathcal{B}_2(G)$ at the level of posets.

The determination of the poset $\mathcal{B}_2(G)$ of 2-radical subgroups of M_{23} is given in An and Conder [**13**, 2H]. The situation is exactly parallel to that for the previous case of M_{22}: The seven G-classes (all centric) represented in $\mathcal{B}_2(G)$ are in bijection with the subgroups H_J of the subgroup simplex $\mathcal{H}_I$; indeed for each nonempty $J \subseteq I$, $B_J := N_G\big(O_2(H_J)\big)$ is a member of $\mathcal{B}_2(G)$. But since H_O (we return to this notation for the subgroup which we have been more appropriately denoting by $H_{O^+} \cong 2^3 : L_3(2)$) does not contain a Sylow 2-group, neither does H_I; indeed

H_I is of index 2 in a Sylow group $N_G(H_I)$, which is the 2-radical subgroup B_I corresponding to type I. Furthermore the notation in [**13**, 2H] is essentially the same as for M_{22}. (In particular, the seven rows there correspond respectively to exactly the same types O, H, Q, OQ, OH, HQ, OHQ as in the previous section.) There are just a few differences, namely that some normalizers are slightly larger in the case of M_{23}. Again this bijection of types does not pull back to an isomorphism of complexes, but instead to a quotient relationship at the level of posets.

Namely in order to apply Lemma 6.5.7, we can quote essentially verbatim the proof for M_{22} in the previous section. Again we set $k := O$ and $J := \{H\}$. Again H_H and $H_{H,Q}$ contain the Sylow $N_G(H_I)$, while it is a supplement to H_{K^+} in $N_G(H_{K^+})$ for $K^+ = \{H, O\}$ and $\{H, Q, O\} = I$. So we conclude that the mapping of the poset of simplices of $\Delta_{\mathcal{H}_I}$ to its quotient by the relevant identifications is a homotopy equivalence of posets. And we recognize the quotient as $\mathcal{B}_2(G)$.

Main results for M_{23}. We now return from the above digression to our main theme. From that earlier discussion, via our usual viewpoint we obtain the following result.

THEOREM 7.4.1. *For $G = M_{23}$, the 2-local complex $\Delta_{\mathcal{H}_I}$ and the collection $\mathcal{A}_2(G)$ are ample and normalizer-sharp. In particular, we obtain from the groups in $\mathsf{N}(\Delta_{\mathcal{H}_I}/G) = \mathcal{H}_I$ a simplex cohomology decomposition as in* (5.8.6)*; and the map from the homotopy colimit of the pushout cube diagram* (4.6.5) *to BM_{23}, followed by 2-completion, is a homotopy equivalence.*

PROOF. The argument follows that for M_{22} in Theorem 7.3.1. □

The mod 2 cohomology ring of M_{23} was computed by Milgram [**94**] as a module over a large subring (see also Adem and Milgram [**6**], Section VIII.5). An interesting feature of this calculation is that $H_i(M_{23}, \mathbb{Z}) = 0$ for $1 \leq i \leq 4$. It had not been known previously that there exists a nontrivial finite group G for which the first three integral homology groups are zero; cf. [**6**, p. 272].

7.5. The Mathieu group M_{24}

In Sections 6.2 and 6.3, we had already used the group M_{24} extensively, to provide examples of 2-local geometries; so our treatment here will refer back to those examples as needed.

For $G := M_{24}$, we take as our index set $I := \{O, T, S\}$, corresponding to the G-classes of maximal 2-locals in the Atlas [**47**, p. 96] which satisfy (6.2.9):

$H_O \cong 2^4{:}\mathrm{Alt}_8$,
$H_T \cong 2^6{:}(L_3(2) \times \Sigma_3)$, and
$H_S \cong 2^6{:}3{\cdot}\Sigma_6$.

The H_i determine the subgroup simplex $\mathcal{H}_I$ as in Definition 4.6.2, and the 2-local complex $\Delta_{\mathcal{H}_I}$ as in Definition 6.3.5; the details for this case of M_{24} were described in Example 6.2.10. We recall that the common intersection $H_I = H_{O,T,S}$ of order $2^{10}3$ contains a Sylow 2-subgroup of G with index three.

We recall from 6.2.14 that the usual diagram for this geometry is

We are not aware of any published determination of $\mathcal{A}_2(G)$, or of the orbit complex $\Delta(\mathcal{A}_2(G))/G$. However, we can recall our remarks for M_{24} after Example 6.3.10 (which are much as for M_{22} and M_{23} in the previous two sections): Setting $A_i := \Omega_1 Z(O_2(H_i))$ for $i \in I$, we obtain representatives of the maximal members of $\mathcal{A}_2(G)$, with $N_G(A_i) = H_i$; but the intersection complex $\Delta_{\mathcal{A}_I}$ is fairly complicated. Again the A_i afford the $U_i = O_2(H_i)$ in the construction of $\Delta_{\mathcal{U}_I}$ in Definition 6.3.13. By checking flag-transitivity, one could check via the method of Lemma 6.5.4 that this complex is "dual" to $\Delta_{\mathcal{H}_I}$ by extending the normalizer set-map N in the forward direction, and the set-map $\mathcal{O}_2$ in the reverse direction.

Recall we have already described at Example 6.3.11 an external complex Δ. We checked various flag-transitivity conditions at Example 6.4.7, to conclude via the method of Lemma 6.5.4 that Δ is isomorphic to $\Delta_{\mathcal{H}_I}$. And we gave versions of $\Delta_{\mathcal{H}_I}$ as an external poset complex from the code or cocode module in Example 6.3.7.

The determination of the poset $\mathcal{B}_2(G)$ of 2-radical subgroups for M_{24} is given in An and Conder [**13**, 2J]. Up to conjugacy, there are 13 subgroups, all 2-centric; These include the seven subgroups $U_J = O_2(H_J)$ of the members H_J of $\mathcal{H}_I$. Thus $\mathcal{O}_2(\Delta_{\mathcal{H}_I})$ is a (proper) subposet of $\mathcal{B}_2(G)$.

The usual 2-local geometry for M_{24} described in Ronan and Smith [**106**, p. 284] is the external complex Δ indicated above, though the viewpoint of $\Delta_{\mathcal{H}_I}$ was also implicit there. See also the viewpoint on M_{24} via *minimal parabolic geometries* in Ronan and Stroth [**107**]; and especially via *tilde geometries* in Ivanov and Shpectorov, for example [**75**]. (We also mention that we indicate a different possible approach to a geometry for M_{24}, having rank four, in our background discussion of the geometry for Fi'_{24} in Section 8.5.)

It is shown in Smith and Yoshiara [**116**, 2.8] (seemingly this fact was first established by Ronan, private communication, unpublished) that Δ is homotopy equivalent to $\mathcal{A}_2(G)$; as usual this is via a map which can be seen to be G-equivariant. Alternatively, Sawabe in [**111**, 6.2] applies his general equivalence method to the case of $G = M_{24}$, to show that the inclusion of $\mathcal{O}_2(\Delta_{\mathcal{H}_I})$ in $\mathcal{B}_2(G)$ is a homotopy equivalence. So we follow the notational convention mentioned in Section 5.9, and write $\mathcal{B}_2^-$ for this equivalent subposet $\mathcal{O}_2(\Delta_{\mathcal{H}_I})$. In fact we recover the poset of simplices of the 2-local complex $\Delta_{\mathcal{H}_I}$ as $\mathsf{N}(\mathcal{B}_2^-)$. (As was the case for M_{12} in Section 7.2, here we have $N_G(H_J) = H_J$ for all types J, in effect giving the vacuous case of the hypotheses Lemma 6.5.7, from which no actual identifications are made.) Of course from Proposition 5.9.1, $\mathcal{B}_2^-$ and $\Delta_{\mathcal{H}_I}$ are ample and normalizer-sharp.

We verified the flag-transitivity of G on $\Delta_{\mathcal{H}_I}$ in Example 6.4.7; and we have $\mathsf{N}(\Delta_{\mathcal{H}_I}/G) = \mathcal{H}_I$.

As usual this discussion leads in our viewpoint to the following theorem.

THEOREM 7.5.1. *For $G = M_{24}$, the collection $\mathcal{B}_2^-$ and the 2-local complex $\Delta_{\mathcal{H}_I}$ are ample and normalizer-sharp. In particular, we obtain from the groups in $\mathsf{N}(\Delta_{\mathcal{H}_I}/G) = \mathcal{H}_I$ a simplex cohomology decomposition as in* (5.8.6)*; and the map from the homotopy colimit of the pushout cube diagram* (4.6.5) *to BM_{24}, followed by 2-completion, is a homotopy equivalence.*

PROOF. The argument follows that for M_{22} in Section 7.3. □

We mention that the poset $\mathcal{B}_2^-$ of rank three (giving rise to the complex $\Delta_{\mathcal{H}_I}$ of dimension two) affords the smallest known ample collection for M_{24}; for example, $\mathcal{B}_2(G)$ itself gives a complex $\Delta(\mathcal{B}_2(G))$ of dimension three.

Dikran Karagueuzian has computed (as yet unpublished) that the mod 2 cohomology ring of a Sylow 2-subgroup of M_{24} is detected on the elementary abelian 2-subgroups; this is part of a joint work with Adem and Milgram and others on the cohomology of M_{24} itself, now in preparation.

7.6. The Janko group J_1

We had mentioned at (5.9.2) that the work of Ryba, Smith, and Yoshiara in [**108**] shows that J_1 can be treated via a suitable subposet of $\mathcal{A}_2(G)$. However, we will only summarize that approach, a bit later in the section; since we wish to focus primarily on the approach via Swan's Theorem 5.7.4, as mentioned earlier at the end of Section 5.7—this in fact affords the smallest possible decomposition for J_1.

Let $G := J_1$. We emphasize that J_1 is the only sporadic group with a Sylow 2-subgroup which is *abelian*. Indeed that Sylow group A_3 is elementary abelian of order 2^3, with normalizer $H_3 := N_G(A_3)$ of structure

$$H_3 \cong 2^3{:}7{:}3$$

(cf. the Atlas [**47**, p. 30]). So in view of Swan's Theorem 5.7.4, the restriction maps

$$H^*(G, \mathbb{F}_2) \to H^*(H_3, \mathbb{F}_2) \to H^*(A_3, \mathbb{F}_2)^{H_3/A_3},$$

where the latter superscript denotes invariants under H_3/A_3, are isomorphisms.

Thus, as in Example 4.6.9, we may regard the cohomology of G as being "decomposed" over the single subgroup H_3. To express this in our usual viewpoint, we take index set $I := \{3\}$ of size 1, corresponding just to the G-class of H_3; the single subgroup H_3 also coincides with the subgroup simplex $\mathcal{H}_I$ of Definition 4.6.2. The corresponding 2-local complex $\Delta_{\mathcal{H}_I}$ determined by H_3 as in Definition 6.3.5 is then the degenerate case of a simplicial complex given by a discrete set of vertices, with no simplices of higher dimensions; and the orbit complex Δ/G is given by a single vertex with stabilizer H_3. In particular, the cohomology isomorphism in the previous paragraph shows that the complex $\Delta_{\mathcal{H}_I}$ is trivially ample and normalizer-sharp: in this case, the homotopy colimit in (5.6.9) is over the single vertex Δ/G, and consists of BH_3 with no further construction needed.

We also recall our observation in the proof of 5.7.4, that the abelian hypothesis shows that the Sylow group (here A_3) represents the unique G-class of 2-centric subgroups. Since we see in the case of J_1 that A_3 is also 2-radical, the normalizer set-map N is an isomorphism of the sets $\mathcal{B}_2^{cen}(G)$ and $\Delta_{\mathcal{H}_I}$ (which in this case have only trivial poset or complex structure); so that we can also regard the approach via Swan's Theorem as a case of Proposition 5.11.2.

This discussion leads via our standard viewpoint to the following theorem.

THEOREM 7.6.1. *For $G = J_1$, the inclusion map $H_3 \to J_1$ is a mod 2 cohomology equivalence; and hence the corresponding map $BH_3 \to BJ_1$, followed by 2-completion, becomes a homotopy equivalence in view of Proposition 4.2.5. In particular, the complex $\Delta_{\mathcal{H}_I}$ for $I = \{3\}$ given by the G-conjugates of H_3 is ample and normalizer-sharp.*

PROOF. As mentioned above, we can proceed via either 5.7.4 or 5.11.2. □

For completeness, and for comparison with our discussion in the sections on the other sporadics, we now very briefly summarize the approach for J_1 via $\mathcal{A}_2(G)$

mentioned at (5.9.2). Here we would start with the larger index set $I^+ := \{1,3\}$, meaning that in addition to H_3 we also use the subgroup

$$H_1 \cong 2 \times \mathrm{Alt}_5,$$

where H_1 is the centralizer of an involution, which generates A_1 of order 2. We get $A_1 < A_3$ by an easy application of the method of 6.2.17 (2), or even just by Sylow's theorem. It is not hard to check by the method of (1) and (2) of 6.4.6 that G is flag-transitive on all of $\mathcal{A}_2(G)$; and hence on the subposet $\mathcal{A}_2^-$ determined by the conjugates just of of A_1 and A_3; so in particular, $\mathsf{N}(\Delta(\mathcal{A}_2^-)/G) = \Delta_{\mathcal{H}_I}$. Furthermore Smith and Yoshiara in [**116**, 2.10] show that $\mathcal{A}_2^-$ is homotopy equivalent to $\mathcal{A}_2(G)$, where as usual the map can be seen to be G-equivariant; the argument there is essentially via Remark 6.5.2, since one finds that A_2 lies in a unique conjugate of A_3. This reduction to the subposet $\mathcal{A}_2^-$ is parallel to "cancellation" arguments in the treatment of J_1 using cohomology decompositions beginning with $\Delta(\mathcal{A}_2(G))/G$, in Webb [**122**, 7.10] and in Adem and Milgram [**6**, V.2]. The final decomposition over $\mathcal{A}_2^-$ using Proposition 5.9.1 is then over a pushout square (4.5.6).

The classical 2-local geometry for J_1 discussed in [**116**, 2.10] is essentially the 2-local complex $\Delta_{\mathcal{H}_{I^+}}$ for $I^+ = \{1,3\}$, which we recover in the above language via the normalizer map as $\mathsf{N}(\Delta(\mathcal{A}_2^-))$.

The determination of the poset $\mathcal{B}_2(G)$ of 2-radical subgroups for J_1 is given in Dade [**49**, 10.5]: up to conjugacy these are the subgroups A_1 and A_3, so that $\mathcal{B}_2(G) = \mathcal{A}_2^-$. So in particular $\mathcal{A}_2^-$ is ample and normalizer-sharp by Proposition 5.9.1.

The mod 2 cohomology ring of J_1 up to degree 21 was computed by Chapman [**44**]; and in fact his answer gives a complete set of generators and relations (see for example Appendix B of [**23**], or Section VIII.2 of Adem and Milgram [**6**]). We have

$$\begin{aligned} H^*(J_1, \mathbb{F}_2) &= H^*(2^3, \mathbb{F}_2)^{7:3} \\ &= \mathbb{F}_2[c_2, c_1, c_0, y_1, y_2]/(y_1^4 + c_2^3 + c_1^2 + y_1^2 c_1 + y_2 c_0, y_2^2 + c_1 c_2 + y_1^2 c_2 + y_1 c_0) \end{aligned}$$

where degrees of generating classes are given by $|c_2| = 4$, $|c_1| = 6$, $|c_0| = 7$, $|y_1| = 3$, and $|y_2| = 5$. The polynomial subring generated by c_2, c_1, and c_0 is a copy of the Dickson invariants[6] of rank three. The cohomology ring is a free module over this subalgebra with eight generators $y_1^i y_2^j$ where $0 \le i \le 3$ and $0 \le j \le 1$. So it is a Cohen-Macaulay ring (cf. [**22**, 5.4.9]); and it has Poincaré series

$$\sum_{i=0}^{\infty} \dim_{\mathbb{F}_2} H^i(J_1, \mathbb{F}_2)\, t^i = \frac{(1+t^3)(1+t^5)(1+t^6)}{(1-t^4)(1-t^6)(1-t^7)}$$

7.7. The Janko group J_2

For $G := J_2$, we take as our index set $I := \{1,2\}$, corresponding to the G-classes of maximal 2-locals in the Atlas [**47**, p. 42] which satisfy (6.2.9):

$H_1 \cong 2_-^{1+4} : \mathrm{Alt}_5$, and

$H_2 \cong 2^{2+4} : (3 \times \Sigma_3)$.

---

[6]The *Dickson algebra* of rank n is given by the polynomials in n variables over $\mathbb{F}_2$ which are invariant under the natural action of $GL_n(\mathbb{F}_2)$; see for example Adem and Milgram [**6**, Sec. III.2].

The H_i as usual determine a subgroup simplex $\mathcal{H}_I$ as in Definition 4.6.2, and a 2-local complex $\Delta_{\mathcal{H}_I}$ as in Definition 6.3.5. The intersection $H_I = H_{12}$ of order $2^7 3$ contains a Sylow 2-subgroup of G with index three.

The rather complicated structure of the orbit complex $\Delta(\mathcal{A}_2(G))/G$ is determined in Carlson, Maginnis and Milgram [**41**]; this structure can be used in the form (5.6.9) of the normalizer decomposition for $\mathcal{A}_2(G)$, as the basis for making further reductions via cancellations. Here we will instead present a smaller decomposition directly, using the subposet $\mathcal{E}_2(G)$ of $\mathcal{A}_2(G)$ indicated in Definition 5.10.1.

We observe first that setting $A_i := \Omega_1 Z\big(O_2(H_i)\big)$ for $i \in I$, we obtain members A_i of rank i in $\mathcal{A}_2(G)$, with $N_G(A_i) = H_i$. Furthermore we get $A_1 < A_2$ by an easy application of the method of 6.2.17 (2). Now G has two conjugacy classes of involutions, denoted by $2A$ and $2B$ in the Atlas [**47**]; those in the center of a Sylow subgroup are in class $2A$, and the others are in class $2B$. In particular, the involution generating A_1 is in class $2A$; and from the transitivity of a Σ_3 in H_2 on the three involutions of A_2, those involutions also lie in class $2A$. We calculate from the character table of J_2 (for example using the ClassMultiplicationCoefficient function in GAP) the value

$$\xi(2A, 2A, 2B) = 0$$

of the structure constant in (5.10.5). So as discussed in the latter part of Section 5.10, the collection $\mathcal{E}_2(G)$ in 5.10.1 is given by elementary abelian 2-subgroups whose nontrivial elements are in class $2A$; and the collection is ample and normalizer-sharp using Proposition 5.10.2.

We saw above that the conjugates of A_1 and A_2 lie in $\mathcal{E}_2(G)$; and in fact $\mathcal{E}_2(G)$ consists just of those two classes. To see this, one computes $\xi(2A, 2A, 2A)$ to obtain the class-algebra multiplication coefficient of 10; so the corresponding 10 conjugates of A_1 commuting with A_1 determine five conjugates of A_2 over A_1: these conjugates of A_2 all lie in $O_2(H_1) \cong 2_-^{1+4}$, they are pairwise non-commuting, and are conjugate under the Alt_5 in H_1. In particular, there is no elementary subgroup of rank three consisting entirely of involutions of class $2A$; and A_2 represents the only such class of subgroups of rank two. So $\mathcal{E}_2(G)$ is determined by the conjugates of A_1 and A_2.

Furthermore the five conjugates above of A_2 over A_1 are all conjugate in $H_1 = N_G(A_1)$; so we conclude also that G is flag-transitive on $\Delta\big(\mathcal{E}_2(G)\big)$ (cf. Remark 6.4.6 (1)). In particular $\mathsf{N}\big(\ \Delta\big(\mathcal{E}_2(G)\big)/G\ \big) = \mathcal{H}_I$. On adjusting this proof slightly to apply also to $\Delta_{\mathcal{H}_I}$, we see as usual via Lemma 6.5.4 that the normalizer set-map N extends to an isomorphism of complexes from $\Delta\big(\mathcal{E}_2(G)\big)$ to $\Delta_{\mathcal{H}_I}$.

We observe since $A_1 < A_2$ that the intersection complex $\Delta_{\mathcal{A}_I}$ determined by the A_i is just the poset complex $\Delta\big(\mathcal{E}_2(G)\big)$; that is, $\mathcal{E}_2(G)$ affords an internal poset complex for J_2 as in Proposition 6.3.6.

The poset $\mathcal{E}_2(G)$ has reduced Euler characteristic $-2^5 \cdot 23$ which is not divisible by $|G|_2 = 2^7$, and so by Remark 5.9.4, $\mathcal{E}_2(G)$ is not in the standard ample G-homotopy type represented by $\mathcal{A}_2(G)$.

Since it was not clear if $\mathcal{B}_2(G)$ had been determined in the earlier literature (it may have been computed in unpublished work of Dade), Yoshiara provides in [**131**, Table 4] the list of four[7] G-conjugacy classes in this poset; of these, three are 2-centric, namely $O_2(H_1)$, $O_2(H_2)$, and $O_2(H_{12})$. Again by adjusting the earlier

[7]Here we are in fact using a later correction to Yoshiara's original list of five groups, observed by Jianbei An: namely that the subgroup denoted $F_2^{(2)}$ there is not radical.

argument for flag-transitivity to apply to $\mathcal{B}_2^{cen}(G)$, we see via the usual method of Lemma 6.5.4 that the set-map $\mathcal{O}_2$ extends to an isomorphism of complexes from $\Delta_{\mathcal{H}_I}$ to $\mathcal{B}_2^{cen}(G)$, with the extension of the normalizer set-map N providing the isomorphism in the reverse direction.

A natural 2-local geometry Δ for J_2 is considered in [**107**, p. 63] and [**40**, Sec. 6], essentially as the 2-local complex $\Delta_{\mathcal{H}_I}$ in our present language; we observed earlier that the normalizer set-map N extends to an isomorphism of $\Delta(\mathcal{E}_2(G))$ with $\Delta_{\mathcal{H}_I}$ (and this isomorphism is in fact via a G-map).

This discussion establishes via our usual approach the following decomposition, which was originally established by Maginnis [**86**] using a somewhat different approach.

THEOREM 7.7.1. *For $G = J_2$, the 2-local complex $\Delta_{\mathcal{H}_I}$ and the collection $\mathcal{E}_2(J_2)$ are ample and normalizer-sharp. In particular, for the groups in $\mathsf{N}(\Delta_{\mathcal{H}_I}/G) = \mathcal{H}_I$, the natural group surjection from the amalgamated free product $H_1 *_{H_{12}} H_2$ onto J_2 induces a mod-2 cohomology isomorphism, and we obtain a simplex cohomology decomposition as in* (5.8.6). *The corresponding map from the homotopy colimit of the pushout square diagram* (4.5.6) *to BJ_2, followed by 2-completion, is a homotopy equivalence.*

PROOF. The proof follows that for M_{12} in Theorem 7.2.1. □

As in the case of M_{12} in Theorem 7.2.1, the cohomology decomposition could also be proved by showing that the alternating sum of the characters of the permutation modules

$$\chi[G/H_{12}] - \chi[G/H_1] - \chi[G/H_2] + \chi[G/G]$$

is a linear combination of characters in non-principal blocks. In fact, it is the sum of χ_{16}, χ_{17} and χ_{19} in the notation of the Atlas [**47**, p. 43]. These characters are all in the 2-block of G whose Brauer correspondent is the non-principal block of the centralizer of an element of class $2B$: This centralizer is isomorphic to $2^2 \times \mathrm{Alt}_5$, and has a unique block of defect two, given by the Steinberg module of Alt_5 regarded as $L_2(4)$; this affords the standard sufficient conditions in Brauer's Extended First Main Theorem ([**48**, 64.10]) to guarantee the existence of the indicated block of G of defect two.

The mod 2 cohomology ring of J_2 was computed by Carlson, Maginnis and Milgram [**41**]. Their answer is given as a module over a large subring, so they do not determine the entire ring structure.

7.8. The Janko group J_3

For $G := J_3$, we take as our index set $I := \{1, 2, 4\}$, corresponding to G-classes of maximal 2-locals in the Atlas [**47**, p. 82]:

$H_1 \cong 2_-^{1+4} : \mathrm{Alt}_5$,

$H_2 \cong 2^{2+4} : (3 \times \Sigma_3)$, and

$H_4 \cong 2^4 : (3 \times \mathrm{Alt}_5)$.

As usual the H_i determine a subgroup simplex $\mathcal{H}_I$ as in Definition 4.6.2, and a 2-local complex $\Delta_{\mathcal{H}_I}$ as in Definition 6.3.5. We observe that H_4 does not contain a Sylow 2-subgroup of G; the common intersection H_I is of order $2^6 3$. (Note that groups with structure H_1 and H_2 also occur in the case of J_2 in Section 7.7.)

Although the geometry is of rank three, it is not usually assigned a diagram; probably this is because the link of H_4 is neither the building for $L_2(4)$ nor the triples geometry of rank two for Alt_5, which for example we saw earlier for M_{22} in Section 7.3 (though the link of H_1 does exhibit that triples geometry structure).

We are not aware of whether $\mathcal{A}_2(G)$ or the orbit complex $\mathcal{A}_2(G)/G$ have been determined. Here it will be natural for us to work instead with the subposet $\mathcal{A}_2^-$ of $\mathcal{A}_2(G)$ determined by the G-conjugates of the subgroups $A_i := \Omega_1 Z\big(O_2(H_i)\big)$ for $i \in I$, where A_i has rank i, with $N_G(A_i) = H_i$. We get $A_1 < A_2 < A_4$ via the usual method of 6.2.17 (2). So we see that the intersection complex $\Delta_{\mathcal{A}_I}$ determined by the A_i becomes just the internal poset complex $\Delta\big(\mathcal{A}_2^-\big)$ as in Proposition 6.3.6.

We postpone to Section 8.2 a sketch of the proof of two further facts: namely that A_4 represents the unique G-conjugacy class of maximal members of $\mathcal{A}_2(G)$; and that $\mathcal{A}_2^-$ is homotopy equivalent to $\mathcal{A}_2(G)$ via a G-map.

We also now describe a coincidence among the subgroups H_J for certain distinct index sets J—this situation in not usual in our construction of $\mathcal{H}_I$ for most of the other sporadic groups, and will have further consequences later. Namely we have $H_{14} = H_{124}$: This follows since the Alt_5 in H_4 acts on A_4 as $SL_2(4)$; and the conjugates in A_4 of A_2 appear as the five 1-subspaces over $\mathbb{F}_4$, each of which is determined by any one of the three conjugates of A_1 of rank one which it contains.

The poset $\mathcal{B}_2(G)$ of 2-radical subgroups of J_3 is determined in Kotlica [**82**, 2.3.1]. If for $i \in I$ we set $U_i := O_2(H_i)$, then representatives of the four nontrivial G-classes in $\mathcal{B}_2(G)$ (all centric) are given by U_1, U_2, U_4, and the product $U_1U_2U_4 =: U_{124}$ which is Sylow in G. We mention also that we have the inclusion[8] $U_4 \leq U_2$. For this reason, the complex $\Delta\big(\mathcal{B}_2(G)\big)$ of inclusion-chains is of dimension two rather than three; but it is rather different from $\Delta\big(\mathcal{A}_2^-\big)$, which also has dimension 2—and we choose $\mathcal{A}_2^-$ as our preferred 2-local geometry for J_3, since it seems simpler to work with.

In fact this 2-local geometry for J_3 was described in the group theory literature by Yoshiara in [**127**, Sec. 2], as a rank three extension of a rank two geometry given in Ronan and Stroth [**107**, p. 64]; Yoshiara's description was essentially as the 2-local complex $\Delta_{\mathcal{H}_I}$ in our present language. Using our earlier remark that $N_G(A_i) = H_i$, and the flag-transitivity on $\mathcal{A}_2^-$ below along with the usual adaptation to $\Delta_{\mathcal{H}_I}$ as well, we obtain our standard observation using Lemma 6.5.4 that the normalizer set-map N extends to an isomorphism of complexes from $\Delta(\mathcal{A}_2^-)$ to $\Delta_{\mathcal{H}_I}$.

The analogue for $\Delta_{\mathcal{H}_I}$ of the projectivity result of Theorem 5.1.14 was obtained in Ryba, Smith and Yoshiara [**108**]; and that gives as a consequence the cohomology decomposition in Theorem 7.8.1 below. Further in Smith and Yoshiara [**116**, 2.11], it was shown that $\Delta_{\mathcal{H}_I}$ is homotopy equivalent to $\mathcal{A}_2(G)$—and as usual the equivalence can be seen to be via a G-map. Indeed the proof that we are postponing to Section 8.2 of the homotopy equivalence of $\mathcal{A}_2^-$ with $\mathcal{A}_2(G)$ can be regarded as a mild variant of the argument in [**116**, 2.11], adapted for our present language and context. Also as mentioned in [**116**, 2.11], one can check that $\mathcal{A}_2^-$ coincides with the poset $\mathcal{Z}_2(G)$ of Benson [**22**, Exer., p. 231]. (Recall the latter was described in Section 5.9 as yet another example of a poset in the standard ample G-homotopy

[8]Again this situation is not usual in the corresponding constructions for most other sporadic groups, where distinct $i \in I$ usually determine U_i not related by inclusion.

type defined by $\mathcal{S}_p(G)$.) Thus we can conclude, applying one of these equivalences in the context of Proposition 5.9.1, that $\Delta_{\mathcal{H}_I}$ is ample and normalizer-sharp.

Further G is flag-transitive on $\Delta(\mathcal{A}_2^-)$: we sketch an argument here, along the lines of [**127**, 2.2]. First, we describe the link in $\mathcal{A}_2^-$ of A_4. This contains five conjugates of A_2, which afford the 1-subspaces of A_4 regarded as a 2-space over $\mathbb{F}_4$; and also contains the 15 rank one subgroups of A_4, three lying in each conjugate of A_2. The action of the Alt_5 in H_4 on this link is flag-transitive, since an Alt_4 stabilizing A_2 is transitive on the three conjugates of A_1 that it contains. Next, the link of A_2 is the join of the three conjugates of A_1 in A_2, with the $|H_2 : H_{24}| = 2$ conjugates of A_4 over A_2: the two latter conjugates are determined by the structure of

$$U_2 = O_2(H_2) = O_2\big(C_G(A_2)\big) = A_4 C_{U_1}(A_2) \cong 2^{2+4},$$

which is that of a Sylow 2-subgroup of $L_3(4)$. In particular, an involution in $H_2 \backslash H_{24}$ acts as a graph automorphism on such a Sylow of $L_3(4)$, and so interchanges those two conjugates of A_4—indeed such an involution can be chosen in $A_4U_1 \setminus U_2$. Hence a subgroup Σ_3 of H_2 with Alt_3 nontrivial on A_2 is flag-transitive on that link. We have now checked the condition in the links of two of the three types, which suffices for flag-transitivity by Proposition 6.4.5. In particular $\mathsf{N}\big(\Delta(\mathcal{A}_2^-)/G\big) = \mathcal{H}_I$.

To summarize the above discussion, our usual approach now leads to the decomposition described in the following theorem (cf. also the diagram in the introduction of [**41**]).

THEOREM 7.8.1. *For $G = J_3$, the 2-local complex $\Delta_{\mathcal{H}_I}$ and the collection $\mathcal{A}_2^-$ are ample and normalizer-sharp. In particular, we obtain from the groups in $\mathsf{N}(\Delta_{\mathcal{H}_I}/G) = \mathcal{H}_I$ a simplex cohomology decomposition as in* (5.8.6)*; and the map from the homotopy colimit of the pushout cube diagram* (4.6.5) *to BJ_3, followed by 2-completion, is a homotopy equivalence.*

PROOF. This follows the alternative argument for J_1, made after Theorem 7.6.1 for the corresponding poset "$\mathcal{A}_2^-$" there; but of course now with a geometry of rank three rather than two. □

However, it also follows from the coincidence $H_{14} = H_{124}$ mentioned earlier that we can obtain our cohomology decomposition over a smaller diagram than the pushout cube (4.6.5) used above.

We first discuss this phenomenon at the level of the cohomology decomposition: On applying cohomology functors to the form (5.6.12) of the normalizer decomposition over the pushout cube (4.6.5), we see that since $H_{14} = H_{124}$, we get identical terms with opposite sign in the alternating sum expression of Webb's formula in Theorem 5.1.16, so that those terms cancel. This cancellation is mentioned in the treatment in Carlson, Maginnis and Milgram [**41**]. But this cancellation at the level of cohomology can also be viewed at the more topological level of the simplicial complex $\Delta(\mathcal{A}_2^-)$ which is our focus in this work: For it follows from our discussion above that a chain of type 14 has its link given by a unique vertex of type 2. So using the standard technique of Lemma 6.5.1 (indeed a special case of Remark 6.5.2), $\Delta(\mathcal{A}_2^-)$ is homotopy equivalent via a G-map to the subcomplex with the star of all such chains removed (that is, with all chains of types 14 and 124 removed). Hence we can consider the homotopy colimit over the corresponding proper subdiagram of the pushout cube.

Next we discuss this reduction at the level of the homotopy colimit and its indexing category. The fact that $H_{14} = H_{124}$ implies that the corresponding classifying spaces, which are related in our homotopy colimit (4.6.5) using mappings from the indexing category that might in general be nontrivial, are in fact actually identical; and we can take account of this by corresponding simplifications to the indexing category: That is, we can replace the arrow $\{1,2,4\} \to \{1,4\}$ in the simplex category $\mathfrak{D}_{\{1,2,4\}}$ with an isomorphism; and then we can identify the two objects giving the endpoints of that arrow. This also then identifies the arrows $\{1,4\} \to \{1\}$ and $\{1,4\} \to \{4\}$ with the two composites giving the arrows from $\{1,2,4\}$ via $\{1,2\}$ and $\{2,4\}$ to those endpoints, Thus our homotopy colimit can be written over the resulting category, which we will call $\mathfrak{D}_a$, and which has the following diagram:

$$\begin{array}{ccccccccc}
\{1\} & & & & \{2\} & & & & \{4\} \\
& \nwarrow & & \nearrow & & \nwarrow & & \nearrow & \\
& & \{1,2\} & & & & \{2,4\} & & \\
& & & \nwarrow & & \nearrow & & & \\
& & & & \{1,2,4\} & & & &
\end{array}$$

Now let $\mathfrak{D}_b$ denote the subcategory of $\mathfrak{D}_a$ obtained by further removing the vertex $\{1,2,4\}$ and the two arrows originating there; namely the category given by the following subdiagram:

$$\begin{array}{ccccccc}
\{1\} & & & \{2\} & & & \{4\} \\
& \nwarrow & \nearrow & & \nwarrow & \nearrow & \\
& \{1,2\} & & & & \{2,4\} &
\end{array} \tag{7.8.2}$$

We will use the Cofinality Theorem 4.5.22 to reduce our homotopy colimit over $\mathfrak{D}_a$ to one over $\mathfrak{D}_b$. Let i denote the inclusion of $\mathfrak{D}_b$ into $\mathfrak{D}_a$. For each object Y of $\mathfrak{D}_a$, we consider the category $Y{\downarrow}i$ of Definition 4.5.21—which is determined using maps in $\mathfrak{D}_a$ from Y to images under i: if Y is other than $\{1,2,4\}$, then $Y = i(Y)$ itself is such an image, and Id_Y is such a map; and we see then that (Y, Id_Y) is an initial object in $Y{\downarrow}i$, so that the nerve $\mathcal{N}(Y{\downarrow}i)$ is contractible—this nerve essentially arises from the "poset" given by those objects in the diagram (7.8.2) for $\mathfrak{D}_b$ which are at or above Y, and this poset has unique minimal member Y. In case $Y = \{1,2,4\}$, we see that Y is initial in $\mathfrak{D}_a$, but *not* an image under i; so the category $Y{\downarrow}i$ just corresponds to the full diagram (7.8.2) for $\mathfrak{D}_b$—and this time in the poset view, there are two minimal members rather than one. But the resulting nerve $\mathcal{N}(Y{\downarrow}i)$ is still visibly contractible. Thus we have shown that the inclusion map i is right cofinal. So using the Cofinality Theorem 4.5.22, we may take our homotopy colimit instead over $\mathfrak{D}_b$, which is isomorphic to the category $\mathfrak{D}_{2+2}$ described at (4.5.11).

The homotopy colimit over this category is the classifying space of a double amalgamated product. So we have the following theorem:

THEOREM 7.8.3. *The natural surjection from the amalgamated free product*

$$H_1 *_{H_{12}} H_2 *_{H_{24}} H_4 \to J_3$$

induces a mod-2 cohomology isomorphism. The corresponding map from the homotopy colimit of the diagram of type $\mathfrak{D}_{2+2}$ *of classifying spaces to* BJ_3*, followed by 2-completion, is a homotopy equivalence.*

The cohomology ring of J_3 was described by Carlson, Maginnis and Milgram in [**41**]. Their answer is again given as a module over a large subring, so they do not determine the entire ring structure. The first displayed equation on page 148 of [**41**] is an expression at the level of cohomology of Theorem 7.8.3.

7.9. The Janko group J_4

For $G := J_4$, we take as our index set $I := \{1, 3, 10, 11\}$, corresponding to G-classes of maximals 2-locals in the Atlas [**47**, p. 190]:

$H_1 \cong 2^{1+12}{:}3{\cdot}M_{22}{:}2$,
$H_3 \cong 2^{3+12}{\cdot}\big(\Sigma_5 \times L_3(2)\big)$,
$H_{10} \cong 2^{10}{:}L_5(2)$, and
$H_{11} \cong 2^{11}{:}M_{24}$.

As usual the H_i determine a subgroup simplex $\mathcal{H}_I$ as in Definition 4.6.2, and a 2-local complex $\Delta_{\mathcal{H}_I}$ as in Definition 6.3.5. The common intersection H_I is of order $2^{20}3$, and in particular a Sylow 2-subgroup of H_I is of index 2 in a Sylow 2-subgroup of $N_G(H_I)$ (and of G).

The usual diagram for the geometry $\Delta_{\mathcal{H}_I}$ is given in [**106**, p. 287] as

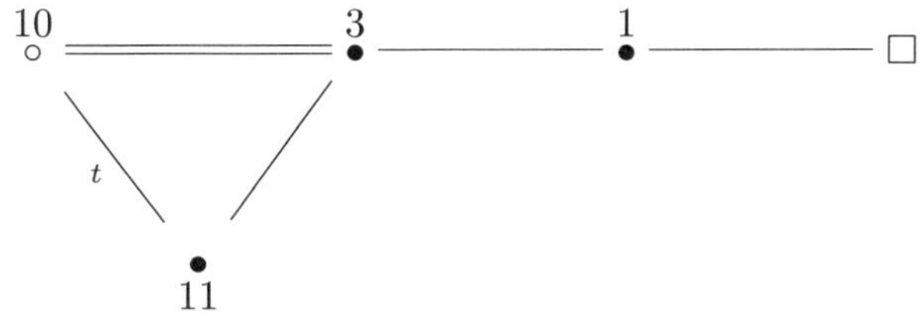

Notice that this diagram involves, as links of the vertices 1 and 11, the diagrams for M_{22} and M_{24} given in Sections 7.3 and 7.5. The link of H_{11} is in fact a 2-split cover of the usual geometry for M_{24}—as we will discuss later, in our argument for flag-transitivity of G on $\Delta_{\mathcal{H}_I}$. We set $U_i := O_2(H_i)$ for $i \in I$.

The usual 2-local geometry for J_4 given in [**106**, p. 287] is essentially the 2-local complex $\Delta_{\mathcal{H}_I}$ in our present language.

Some associated members of $\mathcal{A}_2(G)$. We are not aware of any determination of $\mathcal{A}_2(G)$. However, we now indicate certain particular members of $\mathcal{A}_2(G)$, which will be relevant for our further discussion of $\Delta_{\mathcal{H}_I}$.

In fact, we will go on to sketch a possible variant approach to a geometry for J_4, in terms of the corresponding subposet of $\mathcal{A}_2(G)$; for which we emphasize that we have not attempted to verify all details.

Namely we define an intersection complex $\Delta_{\mathcal{A}_I}$, which in fact exhibits an almost-poset complex as in Definition 6.3.9. For $i \in I$, we set $A_i := \Omega_1 Z\big(O_2(H_i)\big)$; then A_i is a member of $\mathcal{A}_2(G)$ of rank i, with $N_G(A_i) = H_i$. Using the method of 6.2.17 (2), we obtain

$$A_1 < A_3 < A_{10} \cap A_{11} =: A_6$$

where A_6 is of rank six. In particular, there are two maximal members A_{10} and A_{11} of the subposet $\mathcal{A}$ of $\mathcal{A}_2(G)$ defined by these classes, giving the structure of an almost-poset complex as in Definition 6.3.9.

To connect this geometry with the more usual 2-local geometry for J_4, it should be routine to adjust the proof (given later in the section) of flag-transitivity on $\Delta_{\mathcal{H}_I}$

to also apply to Δ_{A_I}; and then to conclude as usual by Lemma 6.5.4 that the normalizer set-map N extends to an isomorphism of complexes from the almost-poset complex Δ_{A_I} to the 2-local complex $\Delta_{\mathcal{H}_I}$.

Also later in the section, we will indicate a homotopy equivalence via a G-map of $\Delta_{\mathcal{H}_I}$ with $\mathcal{B}_2(G)$; so that Δ_{A_I} should also lie in the standard G-homotopy type, along with $\mathcal{A}_2(G)$ itself. But it also seems likely that there should be a more direct proof of this: namely using a sequence of reductions of the type in Lemma 6.5.1, and along the lines of the argument in [**116**, 2.11], to show that the inclusion of $\mathcal{A}$ in $\mathcal{A}_2(G)$ should induce a poset homotopy equivalence via a G-map, and hence also of the complexes Δ_{A_I} and $\Delta(\mathcal{A}_2(G))$. But we have not attempted to check this, as the argument would probably be quite lengthy. Similarly it seems likely (as in the case of J_3 in Section 7.8) that $\mathcal{A}$ coincides with the poset $\mathcal{Z}_2(G)$ of Benson [**22**, Exer., p. 231], which lies in the standard G-homotopy type as we saw in Section 5.9.

Flag-transitivity of J_4 on $\Delta_{\mathcal{H}_I}$. The incidence in the 2-local complex $\Delta_{\mathcal{H}_I}$ can be analyzed via the method demonstrated for M_{24} in Remark 6.2.17 (1), especially as modified for M_{22} in Section 7.3. In summary, one just verifies that the links of vertices are the geometries indicated in the diagram.

We can then establish flag-transitivity of G on $\Delta_{\mathcal{H}_I}$ in the usual inductive way—for example as we did for M_{24} in Example 6.4.7 (2).

First, the link of the vertex H_{10}, indicated by the subdiagram with 10 removed, is a truncated Dynkin diagram of type A_4—corresponding to the projective space of the natural module V of dimension five for $H_{10}/U_{10} \cong L_5(2)$. Indeed the vertices of types $i = 11, 3, 1$ correspond (via intersection of A_i with A_{10} in the notation of the members of $\mathcal{A}_2(G)$ defined above) to spaces of dimensions $6, 3, 1$—in the exterior square of V, of dimension 10. So in fact they just correspond to the projective points, lines, and planes of the projective space $\mathbf{P}(V)$. Thus we can apply Remark 6.4.6 (2) to obtain flag-transitivity of H_{10} on this link.

Next, the link of H_3 as indicated by the diagram is a join: of the triples geometry for Σ_5 of Example 6.5.5, with a truncation to points of the projective plane over $\mathbb{F}_2$. This product admits natural action by the quotient $H_3/U_3 \cong \Sigma_5 \times L_3(2)$. We get flag-transitivity (i.e., just transitivity) of the $L_3(2)$ factor by 6.4.6 (2); and can apply the observation in 6.4.6 (4) to the other factor, since we showed flag-transitivity of Σ_5 on the triples geometry at Example 6.5.5.

Finally, the link of the vertex H_1 indicated by the diagram is the 2-local complex for $H_1/U_1 \cong M_{22}$ from Section 7.3: Indeed the vertices of types $i = 10, 11, 3$ correspond (via the image in U_1/A_1 of the intersection of A_i with U_1) to $\mathbb{F}_4$-subspaces of dimensions $3, 3, 1$ in a 6-dimensional faithful $\mathbb{F}_4$-irreducible for $H_1/U_1 \cong 3^{\cdot}M_{22}$; and these spaces provide an intersection complex for the 2-local geometry of M_{22}—which is in fact an external almost-poset complex as in Definition 6.3.9, since the two maximal spaces of dimension 3 intersect in a subspace of dimension 2. Furthermore the arguments in Section 7.3 for flag-transitivity of M_{22} on its 2-local complex depend essentially only on the links indicated in its diagram, so that they can be easily adjusted to demonstrate flag-transitivity of M_{22} on this almost-poset complex. (That is, the almost-poset complex is then isomorphic to the 2-local complex by the usual Lemma 6.5.4.) Thus we are in effect applying Remark 6.4.6 (4), since we had made the essential flag-transitivity argument for M_{22} in Section 7.3.

We have now checked flag-transitivity on links of three of the four types, which suffices by Proposition 6.4.5 for flag-transitivity for J_4. In particular we have $\mathsf{N}(\Delta_{\mathcal{H}_I}/G) = \mathcal{H}_I$.

Thus treating the link of U_{11} is not strictly necessary to handle the present case of J_4. However we will sketch the details, which require only a little new work; since similar arguments are used in treating some of the other sporadic groups. That link, as indicated by the diagram, is "essentially" the 2-local geometry from Section 7.5 for $H_{11}/U_{11} = M_{24}$: Indeed the vertices of types $i = 10, 3, 1$ correspond (via intersection of A_i with A_{11}) to the spaces of dimensions $6, 3, 1$ in the cocode module, exhibiting the external poset complex version of the geometry given in Example 6.3.7. That is, these correspond to octads, trios, and sextets of M_{24}. However, by "essentially" we mean that the hollow node at type 10 refers to a 2-split cover—namely the usual doubling we have seen for complexes involving the triples geometry (such as that for M_{22}). Here it means that since only a subgroup of order 2^{10} from U_{11} lies in H_{10}, we actually get two vertices of type 10 corresponding to a single octad for M_{24}. Group-theoretically we have two conjugates A_{10} and A'_{10}, which intersect in a rank-6 subgroup of A_{11}, and whose product is a subgroup 2^{6+8} of $N_G(H_{10,11})$ which we will see again later. So it is necessary to check that the stabilizer in H_{11} of a maximal simplex, namely of type $10, 3, 1$, in fact interchanges A_{10} and A'_{10}. And this holds because $N_G(H_I)$ contains all of U_{11}, rather than just that part lying in H_{10}; where an element of U_{11} does indeed interchange this pair. So we can in effect apply Remark 6.4.6 (4) to this link—using the flag-transitivity established for M_{24} in Example 6.4.7, along with this further argument.

The relationship of $\Delta_{\mathcal{H}_I}$ with $\mathcal{B}_2(G)$. Smith and Yoshiara in [**116**, 2.12] showed that $\Delta_{\mathcal{H}_I}$ is homotopy equivalent to $\mathcal{B}_2(G)$, and as usual this equivalence can be seen to be via a G-map. In particular, it follows from Proposition 5.9.1 that $\Delta_{\mathcal{H}_I}$ is ample and normalizer-sharp. And this statement will in fact suffice for our main results on J_4 in Theorem 7.9.1 below.

However, it may be of interest to digress briefly, in order to examine the relationship with $\mathcal{B}_2(G)$ in a little more detail. In fact the equivalence proof in [**116**, 2.12] dealt only with $\mathcal{B}_2(G)$ in the abstract; the explicit determination of the poset $\mathcal{B}_2(G)$ of 2-radical subgroups for J_4 was only subsequently obtained by Yoshiara in [**128**, Thm. 17]. There are 27 conjugacy classes in $\mathcal{B}_2(G)$, all 2-centric.

By contrast, the poset of simplices of $\Delta_{\mathcal{H}_I}$ has only the $2^{|I|} - 1 = 15$ conjugacy classes determined by the subgroup simplex $\mathcal{H}_I$. Indeed for each nonempty type set $J \subseteq I$, we get a member $O_2\big(N_G(H_J)\big)$ of $\mathcal{B}_2(G)$; and we let $\mathcal{B}_2^-$ denote this subset of $\mathcal{B}_2(G)$ determined by these particular 15 classes out of the total of 27 classes. In the notation of [**128**], one obtains this subset as follows: From the subgroups there with first index E, remove the final six (those containing an index $\tilde{U}$). The remaining eight then correspond to type sets J here that begin with 1—and continue with a subset from our indices $\{3, 10, 11\}$ which correspond respectively to indices $\{Q, O', H\}$ there. From those there with first index L, remove the final four (those containing an index Z). The remaining four then correspond to J beginning with 3, and continuing with one of 10, 11, or $\{10, 11\}$, corresponding to t, v, or d there. From those there with first index A or V, remove the final one (namely those with further index p or $\square$). The remaining three then correspond to J given by 10, 11, and $\{10, 11\}$.

It seems likely, though we have not attempted to verify it in detail, that the equivalence method of Sawabe in [**111**] could be extended to establish that the inclusion of $\mathcal{B}_2^-$ in $\mathcal{B}_2(G)$ induces a homotopy equivalence.

Furthermore it is in fact the poset $\mathcal{B}_2^-$ which has a close relationship with our 2-local complex $\Delta_{\mathcal{H}_I}$. Just as was the case for M_{22} in Section 7.3, the bijection of conjugacy classes does not pull back to a bijection at the level of posets, for the poset $\mathcal{R}$ of simplices of $\Delta_{\mathcal{H}_I}$ with $\mathcal{B}_2^-$; instead there is a quotient process from identifying certain pairs, which gives a homotopy equivalence of posets. This comes from an application of Lemma 6.5.7, in higher rank than the earlier application for M_{22}; we summarize the details more briefly: we take $k := 10$ as our distinguished type, and $J := \{11\}$ as our distinguished subset of $I \setminus \{10\}$. Hence we must verify (6.5.8) for the types K given by $\{11\}$ and its over-sets $\{11, 1\}$, $\{11, 3\}$, and $\{11, 3, 1\}$ in $I \setminus \{10\}$. So we check that $N_G(H_I)$ does indeed lie in each such H_K. Then we turn to considering the types $K^+ = K \cup \{10\}$; where we check that $N_G(H_I)$ does indeed supplement H_{K^+} in $N_G(H_{K^+})$. As a sample of these latter verifications, note that $H_{10,11}$ has the structure $2^{10}\colon \big(2^4\colon L_4(2)\big)$—which is of index two in $N_G(H_{10,11}) \cong 2^{6+8}\colon \big(2 \times L_4(2)\big)$. The extra involution commuting with $L_4(2)$ can be chosen from U_{11}, and switches two conjugates of A_{10} in 2^{6+8}. The cases for larger types K^+ involve subconfigurations within this $L_4(2)$. So Lemma 6.5.7 gives the poset homotopy equivalence of $\mathcal{R}$ with the quotient $\bar{\mathcal{R}}$ determined by the identification of orbits of length 2 for types K^+.

Finally we combine the resulting poset homotopy equivalence from $\mathcal{R}$ to the quotient $\bar{\mathcal{R}}$ with the observation that $\bar{\mathcal{R}}$ just reproduces the poset $\mathcal{B}_2^-$.

Main results for J_4. The above discussion leads via our usual approach to the following theorem.

THEOREM 7.9.1. *For $G = J_4$, the collection $\mathcal{B}_2^-$ and the 2-local complex $\Delta_{\mathcal{H}_I}$ are ample and normalizer-sharp. In particular, we obtain from the groups in $\mathsf{N}(\Delta_{\mathcal{H}_I}/G) = \mathcal{H}_I$ a simplex cohomology decomposition as in (5.8.6); and the map from the homotopy colimit of the pushout 4-cube diagram (4.6.7) to BJ_4, followed by 2-completion, is a homotopy equivalence.*

PROOF. This follows the argument for M_{24} in Theorem 7.5.1; but of course now for a geometry of rank four rather than three. □

We mention that $\Delta_{\mathcal{H}_I}$ of dimension three gives a smaller decomposition than $\mathcal{A}_2(G)$ or $\mathcal{B}_2(G)$, whose complexes of inclusion-chains have dimension at least four.

The size of J_4 seems to be beyond the reach of current techniques for explicit calculation of the mod 2 cohomology ring.

7.10. The Higman–Sims group HS

For $G := HS$, we take as our index set $I := \{1, 3, 4\}$, corresponding to the G-classes of maximal 2-locals in the Atlas [**47**, p. 80] which satisfy (6.2.9):

$H_1 \cong (4{\cdot}2^4){:}\Sigma_5$,

$H_3 \cong 4^3{:}L_3(2)$, and

$H_4 \cong 2^4{:}\Sigma_6$.

As usual the H_i determine a subgroup simplex $\mathcal{H}_I$ as in Definition 4.6.2, and a 2-local complex $\Delta_{\mathcal{H}_I}$ as in Definition 6.3.5. Here H_4 does not contain a Sylow

2-subgroup of G; and the common intersection H_I of order 2^8 is of index two in a Sylow group $N_G(H_I)$.

The diagram for the geometry $\Delta_{\mathcal{H}_I}$ given in [**106**, p. 288] is

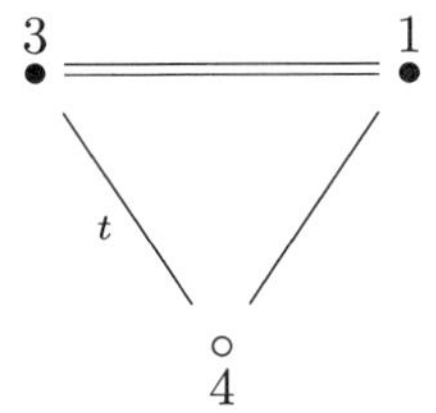

Notice that this diagram is similar to that for M_{22} in Section 7.3—except that the order of the filled and hollow nodes in the link of H_1, namely the triples geometry for Σ_5 (see Example 6.5.5), has now been reversed.

The approach via the intersection complex $\Delta_{\mathcal{A}_I}$. We are not aware of any determination of $\mathcal{A}_2(G)$. However, we will now approach the geometry of HS using certain members of $\mathcal{A}_2(G)$.

For $i \in I$, we set $A_i := \Omega_1 Z\big(O_2(H_i)\big)$; the A_i are members of $\mathcal{A}_2(G)$ of rank i, with $N_G(A_i) = H_i$. Via the usual method of 6.2.17 (2) we get $A_1 < A_3 \cap A_4 =: A_2$, where A_2 is of rank two. We let $\mathcal{A}$ denote the subposet of $\mathcal{A}_2(G)$ defined by the G-classes of the A_i for $i \in I$. The intersection complex $\Delta_{\mathcal{A}_I}$ defined by the A_i is in fact an almost-poset complex as in Definition 6.3.9, since both A_3 and A_4 are maximal under inclusion. In due course, we will conclude via the usual method of Lemma 6.5.4 that $\Delta_{\mathcal{A}_I}$ is isomorphic to the 2-local complex $\Delta_{\mathcal{H}_I}$.

As always, we will wish to relate $\Delta_{\mathcal{A}_I}$ to one of the standard collections from the final sections of Chapter 5. In summary, our procedure for HS will be somewhat similar to that for M_{12} in Section 7.2, in that we will use the poset $\mathcal{E}_2(G)$ defined in Definition 5.10.1; in fact we will reduce to a proper subposet $\mathcal{E}_2^-$, defined by the classes of the above A_i for $i \in I$, along with the class of A_2 of rank two. Then we will see that $\mathcal{E}_2^-$ is homotopy equivalent to the almost-poset complex $\Delta_{\mathcal{A}_I}$ via a G-map.

The relationship with $\mathcal{E}_2(G)$. There are two conjugacy classes of involutions in G, denoted by $2A$ and $2B$ in the Atlas [**47**]; the involutions in the center of a Sylow 2-subgroup lie in class $2A$. Then the involution generating A_1 is in class $2A$; and for $i = 2, 4$, the transitivity of a subgroup Σ_3 or $Sp_4(2)$ in H_i on the involutions in A_i shows that those involutions also lie in class $2A$. From the character table of HS (for example using the ClassMultiplicationCoefficient function in GAP), we compute the value

$$\xi(2A, 2A, 2B) = 0$$

of the structure constant defined in (5.10.5). Hence the collection $\mathcal{E}_2(G)$ of Definition 5.10.1 is given by the elementary abelian 2-subgroups whose nontrivial elements are in class $2A$; and the collection is ample and normalizer-sharp by Proposition 5.10.2. We saw above that A_1, A_2, and A_4 lie in $\mathcal{E}_2(G)$.

We next mention some further structural details related to the A_i: First $U_1 := O_2(H_1)$ is the central product of $Z(U_1) \cong \mathbb{Z}/4$ with an extraspecial 2-subgroup 2^{1+4}; the product is over the intersection A_1 of order 2. The action of the Σ_5 in H_1 on the 4-space $\tilde{U}_1 := U_1/Z(U_1)$ is that of $\Sigma L_2(4)$. (Compare with the case of the group "A_4" for $G = J_3$ in Section 7.8.) Among the details developed later in Section 8.3, we mention that A_4 is generated by an involution b lying outside the commutator subgroup H_1' of H_1, along with the involutions in U_1 that b centralizes; and such subgroups are conjugate in H_1. Since $H_1 = C_G(A_1)$, A_4 therefore represents the unique G-class in $\mathcal{E}_2(G)$ of rank four. Next letting P_3 denote the preimage in U_1 of a 1-subspace over $\mathbb{F}_4$ in $\tilde{U}_1$ under the action of H_1, we recover A_3 as a suitable H_1-conjugate of $\Omega_1(P_3)$; and again from the structure of H_1 we see that A_3 is also maximal in $\mathcal{E}_2(G)$ under inclusion. We defined A_2 as $A_3 \cap A_4$, so in particular, $A_1 < A_2 < A_3$. In fact all 4-subgroups of A_3 are conjugate to A_2, from the action of the $L_3(2)$ in H_3. From our characterization above in terms of 1-spaces over $\mathbb{F}_4$, we see that A_3 is the unique G-conjugate of itself above A_2, and hence $N_G(A_2) \leq N_G(A_3) = H_3$, so that $N_G(A_2) = N_{H_3}(A_2)$. Indeed we have that

$$A_3 = U_a \cap U_b \text{ for conjugates of } U_1 \text{ corresponding to } a, b \text{ generating } A_2. \quad (7.10.1)$$

We can also check that there are exactly two conjugates of A_4 above A_2, essentially following the argument for the analogous fact in the case of J_3 in Section 7.8.

We now define $\mathcal{E}_2^-$ to be the subposet of $\mathcal{E}_2(G)$ given by the G-conjugates of A_i for $i \in \{1, 2, 3, 4\}$. We claim that $\mathcal{E}_2^-$ is homotopy equivalent to $\mathcal{E}_2(G)$ via a G-map; but we postpone to Section 8.3 the lengthier details of a proof of this fact. Using the claim, we obtain via Proposition 5.10.2 that $\mathcal{E}_2^-$ is also ample and normalizer-sharp.

The poset $\mathcal{E}_2^-$ has reduced Euler characteristic $2^7 \cdot 3^2 \cdot 37$, which is not divisible by $|G|_2 = 2^9$, and so by Remark 5.9.4, $\mathcal{E}_2^-$ is not in the standard ample G-homotopy type represented by $\mathcal{A}_2(G)$.

We now turn our attention to the subposet $\mathcal{A}$ of $\mathcal{E}_2^-$ defined by the G-classes just of A_i for $i \in I = \{1, 3, 4\}$; and we consider the almost-poset complex $\Delta_{\mathcal{A}_I}$ which they define as indicated earlier. We claim now that $\Delta_{\mathcal{A}_I}$ is homotopy equivalent to $\Delta(\mathcal{E}_2^-)$ via a G-map; but again we postpone to Section 8.3 the lengthier details of a proof of this fact. Thus $\Delta_{\mathcal{A}_I}$ is also ample and normalizer-sharp by Proposition 5.10.2.

Finally we show that G is flag-transitive on $\Delta_{\mathcal{A}_I}$. For $i \in I$ as usual we define $U_i := O_2(H_i)$. First, the vertices of types $i = 1, 3$ in the link of A_4 correspond (by intersection with A_4) to the points and isotropic lines of the polar space A_4, which is the building for the symplectic group $Sp_4(2) \cong \Sigma_6 \cong H_4/U_4$; so we can apply Remark 6.4.6 (3) for flag-transitivity on this link. Next, the vertices of types $i = 3, 4$ in the link of A_1 in fact afford a copy of the triples geometry for $H_1/U_1 \cong \Sigma_5$ from Example 6.5.5—where we had established flag-transitivity, so that we are really applying Remark 6.4.6 (4). We have now checked the links of two of the three types, which suffices by Proposition 6.4.5 for the flag-transitivity of G.[9] In particular we have $\mathsf{N}(\Delta_{\mathcal{A}_I}/G) = \mathcal{H}_I$.

As usual this flag-transitivity argument really involves only the geometry of the links indicated in the diagram; so we can adjust the argument to also obtain flag-transitivity of G on $\Delta_{\mathcal{H}_I}$. So by the standard Lemma 6.5.4 we see that the normalizer set-map N extends to an isomorphism of complexes from $\Delta_{\mathcal{A}_I}$ to the 2-local complex $\Delta_{\mathcal{H}_I}$.

---

[9]We could alternatively argue using the vertices of types $i = 1, 4$ in the link of A_3, which correspond (by intersection with A_3) to the points and lines of the projective space $\mathbf{P}(A_3)$; though in fact the lines appear in a "2-split cover" as described in [**106**, p. 288]. So we can essentially apply Remark 6.4.6 (2) for flag-transitivity on this link—subject to adjusting for the 2-split cover for $L_3(2)$, much as we did for the similar cover for $Sp_4(2)'$ in the case of M_{22} in Section 7.3.

Digression: the relationship with $\mathcal{B}_2^{cen}(G)$. The above homotopy equivalences of $\Delta_{\mathcal{A}_I}$ and $\mathcal{E}_2^-$ are in fact sufficient for the purposes of our main results for HS in Theorem 7.10.2 below. However, as in other cases involving the triples geometry for Σ_5, it seems interesting to examine the close relationship at the level of posets with $\mathcal{B}_2^{cen}(G)$.

The poset $\mathcal{B}_2(G)$ of 2-radical subgroups of HS is determined in Yoshiara [**130**, Prop. 12], and also in Hassan and Horváth [**72**]. There are nine conjugacy classes, of which seven are 2-centric. The latter seven classes are in a natural bijection with the subgroups B_J given by the products of the U_i for $i \in J$, for subsets $J \subseteq I$; and these subgroups give $O_2\big(N_G(H_J)\big)$. For convenience of comparison, we mention that the final seven rows in [**130**, Table IV] correspond respectively to our type sets $4, 3, 34, 1, 13, 14, 134$. Again we have some similarity with the situation for M_{22} in Section 7.3; this time it is U_4 which is not weakly closed in a Sylow group. And we find, for the particular type sets $J = 34, 134$, that $O_2(H_J)$ is in fact of index two in the member B_J of $\mathcal{B}_2^{cen}(G)$.

As for M_{22}, this bijection of G-orbits does not pull back to an isomorphism of posets, from the poset of simplices of $\Delta_{\mathcal{H}_I}$ to $\mathcal{B}_2^{cen}(G)$; again there is instead a quotient relationship via the identification of certain pairs, which will give a poset homotopy equivalence via Lemma 6.5.7. The overall setup is similar to that for M_{22}: Here we take $k := 4$ as our distinguished type, and $J := \{3\}$ as our distinguished subset of $I \setminus \{4\}$. Hence we must verify (6.5.8) for the types K given by $\{3\}$ and its unique over-set $\{3,1\}$ in $I \setminus \{4\}$. And the Sylow group $N_G(H_I)$ does indeed lie in H_3 and $H_{3,1}$. The remaining types to consider are $K^+ = \{3,4\}$ and $\{3,4,1\} = I$. Certainly $N_G(H_I)$ is a supplement to H_I in itself. So it only remains to observe that the Sylow 2-subgroup $N_G(H_I)$ is also a supplement to $H_{3,4} \cong 2^{2+4}\colon \Sigma_3$ in $N_G(H_{3,4}) \cong 2^{2+4}\colon (2 \times \Sigma_3)$—where the further involution interchanges the two conjugates of H_4 in 2^{2+4}. This completes the verification of the hypothesis of Lemma 6.5.7, giving the poset homotopy equivalence of $\mathcal{R}$ with $\bar{\mathcal{R}}$; we then check that $\bar{\mathcal{R}}$ simply reproduces the poset $\mathcal{B}_2^{cen}(G)$.

Main results for HS. We return from the above digression to our main theme.

The usual 2-local geometry for HS in [**106**, p. 288] is essentially the 2-local complex $\Delta_{\mathcal{H}_I}$ in our present language. We have already mentioned that here we prefer to use the form $\Delta_{\mathcal{A}_I}$ of the geometry, which we saw is isomorphic with $\Delta_{\mathcal{H}_I}$.

So our earlier discussion leads via our usual approach to the following theorem.

THEOREM 7.10.2. *For $G = HS$, the collection $\mathcal{E}_2^-$ and the almost-poset complex $\Delta_{\mathcal{A}_I}$ are ample and normalizer-sharp. In particular, we obtain from the groups in $\mathsf{N}(\Delta_{\mathcal{A}_I}/G) = \mathcal{H}_I$ a simplex cohomology decomposition as in* (5.8.6)*; and the map from the homotopy colimit of the pushout cube diagram* (4.6.5) *to $B(HS)$ followed by 2-completion is a homotopy equivalence.*

PROOF. This follows the argument for M_{12} in Theorem 7.2.1, of course now for a geometry of rank three rather than two. □

The cohomology ring of the Sylow 2-subgroup $N_G(H_I)$ of HS has been computed by Adem, Carlson, Karagueuzian and Milgram in [**1**]. The Poincaré series is

$$\sum_{i=0}^{\infty} \dim_{\mathbb{F}_2} H^i\big(\mathrm{Syl}_2(HS), \mathbb{F}_2\big)\, t^i = \frac{(1+t)^2(1-t+t^2)(1+2t-t^5)}{(1-t)^2(1-t^4)(1-t^8)}.$$

The ring has a minimal set of 17 generators, in degrees 1, 1, 1, 2, 2, 2, 2, 3, 3, 3, 3, 4, 4, 5, 6, 7, 8; and there are 79 relations. The relations, as well as the action of the Steenrod algebra (see for example [**22**, Sec. 4.4]) on the cohomology of the Sylow, are given explicitly in [**1**]. But as far as we are aware, the mod 2 cohomology of HS itself has not been computed.

7.11. The McLaughlin group McL

We had already used McL in Example 6.3.10, to give an example of an almost-poset complex; we will recall this below as needed.

In particular, for $G := McL$, we took as our index set $I := \{1, 4, 4'\}$, corresponding to the G-classes of maximal 2-locals in the Atlas [**47**, p. 100] satisfying (6.2.9):

$H_1 \cong 2^{\cdot}\mathrm{Alt}_8$,

$H_4 \cong 2^4 : \mathrm{Alt}_7$, and similarly

$H_{4'} \cong 2^4 : \mathrm{Alt}_7$.

As usual the H_i determine a subgroup simplex $\mathcal{H}_I$ as in Definition 4.6.2, and a 2-local complex $\Delta_{\mathcal{H}_I}$ as in Definition 6.3.5. Here the intersection H_I is a subgroup $2^{2+4} : \Sigma_3$, which is an extension of a Sylow 2-subgroup by a subgroup of order 3.

The natural diagram for $\Delta_{\mathcal{H}_I}$ is the obvious rank-3 truncation of the rank-4 diagram given in [**107**, p. 62]:

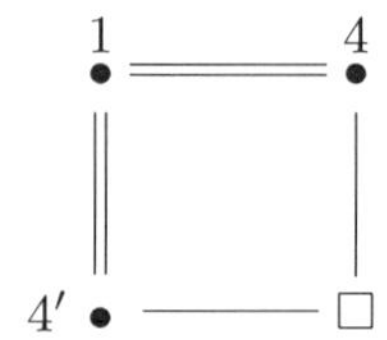

Note that in contrast to the case for M_{24} in Section 7.5, this time McL actually does contain a subgroup (namely $U_4(3)$) corresponding to the square node; but here for the purposes of mod 2 cohomology, we have just chosen to ignore that non-local subgroup. The C_3 diagrams in the links of 4 and $4'$ correspond to the Neumaier geometry for Alt_7, as in the case of M_{23} in Section 7.4.

Properties of certain elementary 2-subgroups. Again it will be convenient to approach the geometry $\Delta_{\mathcal{H}_I}$ for McL by means of a variant related to $\mathcal{A}_2(G)$.

For $i \in I$ we set

$$A_i := \Omega_1 Z(O_2(H_i)) = O_2(H_i);$$

then A_i is a member of $\mathcal{A}_2(G)$ of rank i, with $N_G(A_i) = H_i$. The intersection complex $\Delta_{\mathcal{A}_I}$ is then the internal almost-poset complex of Example 6.3.10. In due course, we will conclude via the usual method of Lemma 6.5.4 that $\Delta_{\mathcal{A}_I}$ is isomorphic to the 2-local complex $\Delta_{\mathcal{H}_I}$.

The orbit complex $\Delta(\mathcal{A}_2(G))/G$ of elementary abelian 2-subgroups for McL is described in Entz and Pahlings in [**57**]. However this orbit complex has certain complications: There are unique conjugacy classes of ranks one and two, represented respectively by A_1 and $A_2 := A_4 \cap A_{4'}$ (where the latter intersection, and further inclusion relations below, are obtained via the method of 6.2.17 (2)). But there are two conjugacy classes each of members of ranks three and four: those of rank four are represented by A_4 and $A_{4'}$; we may take analogously-denoted representatives for those of rank three so that $A_2 < A_3 < A_4$ and $A_2 < A_{3'} < A_{4'}$. Now one can show that G acts transitively on inclusion-chains among these classes; but since there are several classes of ranks three and four, the resulting G-orbit complex $\Delta(\mathcal{A}_2(G))/G$ is not a simplex.

However, to summarize our treatment below of geometries related to $\mathcal{A}_2(G)$: we will be able to obtain a simplex for the quotient, when we pass instead to the almost-poset complex $\Delta_{\mathcal{A}_I}$, defined as in Definition 6.3.9 by the A_i for $i \in I = \{1, 4, 4'\}$—with two maximal members A_4 and $A_{4'}$. The treatment has a number of similarities with that for HS in Sections 7.10 and 8.3, but is substantially easier.

We now begin the details of that treatment; in particular, we start by deriving some of the structural facts indicated above. There is just one G-conjugacy class of involutions, represented by a generator of A_1. In $H_1/A_1 \cong \mathrm{Alt}_8$, only 2-central involutions ("long" involutions, i.e., with permutation type 2^4) lift to involutions of the double cover $2\mathrm{Alt}_8 = H_1$. It follows that there is a unique class of 4-subgroups of G; and of course we have chosen our representative $A_2 > A_1$. Again in Alt_8, there are two conjugacy classes of maximal elementary subgroups containing only 2-central involutions of that quotient: namely the unipotent radicals 2^3 of the two maximal parabolics $2^3 : L_3(2)$ for the end nodes of the Dynkin diagram for $L_4(2) \cong \mathrm{Alt}_8$. It now follows that there are two conjugacy classes of maximal members of $\mathcal{A}_2(G)$ of rank four, which are in fact conjugate in $\mathrm{Aut}(McL)$; and we have chosen our representatives A_4, $A_{4'}$ to contain A_2. Indeed we see that each of these contains a representative of rank three (which we have chosen as A_3, $A_{3'}$ containing A_2). These give the remaining two classes in $\mathcal{A}_2(G)$ indicated in **[57]**—there is just one class in each of A_4 and $A_{4'}$, since the flag-transitivity of $L_4(2)$ on its projective space is in fact inherited by its subgroup Alt_7 appearing in H_4 and $H_{4'}$. Also since the unipotent radicals 2^3 of $2^3 : L_3(2)$ indicated above have trivial intersection with their distinct conjugates (they are full transvection subgroups on the natural module for $L_4(2)$), we see that A_4 is the unique conjugate of itself over A_3 or A_2; and similar uniqueness holds for $A_{4'}$ over $A_{3'}$ and A_2. In particular, $N_G(A_3) \leq N_G(A_4) = H_4$, and

$$N_G(A_2) \leq N_G(A_4) \cap N_G(A_{4'}) = H_{4,4'}.$$

For nonempty $J \subseteq I = \{1, 4, 4'\}$ we set $U_J := \prod_{i \in J} U_i$ where $U_i := O_2(H_i)$.

Initial reduction to $\mathcal{A}_2^-$. We now have an easier analogue of the first equivalence proof in Section 8.3 in the case of HS; recall that argument removed the subgroups "B_3" using the uniqueness condition denoted by (c) there. Similarly here, our observation above that A_3 is contained in a unique conjugate of A_4 (and similarly with $3, 4$ replaced by $3', 4'$) shows that we are in the usual situation of Remark 6.5.2: so we may apply Lemma 6.5.1 to remove the G-conjugates of A_3 and $A_{3'}$; hence $\mathcal{A}_2(G)$ is homotopy equivalent via a G-map to the subposet $\mathcal{A}_2^-$ defined by the G-conjugates of the A_i for $i \in \{1, 2, 4, 4'\}$. In particular, $\mathcal{A}_2^-$ is ample and normalizer-sharp by Proposition 5.9.1.

Since A_4 and $A_{4'}$ are maximal under inclusion in $\mathcal{A}_2(G)$, we see that the complex $\Delta(\mathcal{A}_2^-)$ of inclusion-chains is of dimension 2, with maximal simplices having types 124 and $124'$; but there are no simplices of full type $1244'$ in that poset complex, so that the orbit space $\Delta(\mathcal{A}_2^-)/G$ is still not a simplex.

Final reduction to $\Delta_{\mathcal{A}_I}$. As in the case of HS in Section 7.10, we will remedy this by using the almost-poset complex $\Delta_{\mathcal{A}_I}$ based on the A_i with types in the subset $I := \{1, 4, 4'\}$.

We obtain flag-transitivity of G on $\Delta_{\mathcal{A}_I}$ by the usual inductive method: For the vertex A_4, a conjugate of $A_{4'}$ incident with A_4 in fact intersects A_4 in a conjugate of A_2 of rank two. We see that the link of A_4 is given by 15 projective points conjugate to A_1, and 35 projective lines conjugate to A_2; namely the points and lines of the projective space $\mathbf{P}(A_4)$. Further H_4 contains Alt_7, which as mentioned earlier is a flag-transitive subgroup of $L_4(2)$. Thus we may apply Remark 6.4.6 (3); and the same argument applies to the link of $A_{4'}$. So we have checked flag-transitivity in the links of two of the three types, which suffices by Proposition 6.4.5 for flag-transitivity of G. In particular $\mathsf{N}(\Delta_{\mathcal{A}_I}/G) = \mathcal{H}_I$,

As usual the above argument for flag-transitivity on $\Delta_{\mathcal{A}_I}$ involves essentially just the geometry of the links indicated by the diagram; and so we can adjust to obtain flag-transitivity also on $\Delta_{\mathcal{H}_I}$. So by the standard Lemma 6.5.4 we see that the normalizer set-map N extends to an isomorphism of complexes from $\Delta_{\mathcal{A}_I}$ to the 2-local complex $\Delta_{\mathcal{H}_I}$.

We also obtain an easier analogue of the final equivalence proof in Section 8.3 for HS; namely here we will see that $\Delta_{\mathcal{A}_I}$ is homotopy equivalent via a G-map to $\Delta(\mathcal{A}_2^-)$: For $\mathcal{A}_2^-$ now plays the role of "$\mathcal{E}$" in Lemma 6.5.3; in particular, $\mathcal{A}_2^-$ contains the intersection $A_2 = A_4 \cap A_{4'}$. Indeed in the notation of that result, the intersections "E_σ" for σ of types 14 or 14′ or 144′ are conjugates of A_1, while those of type 44′ are conjugates of A_2. Now we must show for each $E \in \mathcal{A}_2^-$ that the subcomplex

$$L_E := \{\sigma \in \Delta_{\mathcal{A}_I} : E \leq E_\sigma\}$$

is contractible via a G_E-map. For $E = A_4$ or $A_{4'}$, we see by maximality of those groups that L_E is the single vertex A_4 or $A_{4'}$; so we are in the usual situation of Remark 6.5.2 for contractibility via a G_E-map. For $E = A_1$, we see that the subcomplex L_{A_1} is the join of the single vertex A_1 with the subcomplex determined by all G-conjugates of A_4 and $A_{4'}$ containing A_1; again this is contractible via a G_E-map as in Remark 6.5.2. Finally for $E = A_2$, we find that the subcomplex L_{A_2} is the join of the unique vertices A_4 and $A_{4'}$ of those types over A_2, and again we have the situation of Remark 6.5.2 giving contractibility via a G_E-map. This completes the verification of the hypotheses of Lemma 6.5.3, and hence the proof of the desired homotopy equivalence via a G-map. In particular, $\Delta_{\mathcal{A}_I}$ is ample and normalizer sharp by Proposition 5.9.1.

Some other geometric structures. The poset $\mathcal{B}_2(G)$ of 2-radical subgroups for McL is determined in Murray's thesis [**98**]; since that thesis is likely to remain unpublished, Yoshiara has provided in [**131**, Table 5] the list of its eight conjugacy classes. Furthermore Maginnis and Onofrei in [**88**] indicate homotopy equivalences (which in fact proceed via G-maps) of $\mathcal{B}_2(G)$ with a certain natural subposet (other than $\mathcal{B}_2^{cen}(G)$) and then of that subposet with 2-local geometry $\mathcal{A}_2^-$. We also mention that of the eight indicated conjugacy classes in $\mathcal{B}_2(G)$, the class given by A_1 is not 2-centric. Among the seven centric classes are $U_4 = A_4$, $U_{4'} = A_{4'}$, and $U_{44'} = U_{144'} = A_4A_{4'} \cong 2^{2+4}$. The remaining classes are preimages in H_1 of parabolics of $L_4(2)$ which involve the middle node of the Dynkin diagram. Thus the poset $\mathcal{B}_2^{cen}(G)$ is somewhat different from $\mathcal{A}_2^-$.

A group geometry of rank four for $G = McL$ was given in Ronan and Stroth [**107**, p. 62] (see also Buekenhout [**39**, Sec. 5, no. 52]); but this geometry is not fully 2-local, since one vertex has stabilizer given by the simple group $U_4(3)$ (compare with the situation for M_{23} in Section 7.4). The usual 2-local geometry Δ for $G = McL$ is correspondingly the rank three subgeometry discussed in [**108**, 3.6]; namely the "truncation" obtained by ignoring the subgroup $U_4(3)$. This 2-local geometry is essentially the 2-local complex $\Delta_{\mathcal{H}_I}$ in our present language.

Main results for McL. The analogue for $\Delta_{\mathcal{H}_I}$ of the projectivity result in Theorem 5.1.14 was obtained in Ryba, Smith and Yoshiara [**108**]; hence the cohomology decomposition below then followed as in Theorem 5.1.16. Further in Smith and Yoshiara [**116**, 2.9], it was shown that $\Delta_{\mathcal{H}_I}$ is homotopy equivalent to $\mathcal{A}_2(G)$—we have essentially reproduced that argument above, but in the context of $\mathcal{A}_2^-$ and $\Delta_{\mathcal{A}_I}$.

The discussion above leads via our usual approach to the following theorem.

THEOREM 7.11.1. *For $G = McL$, the collection $\mathcal{A}_2^-$ and the almost-poset complex $\Delta_{\mathcal{A}_I}$ are ample and normalizer-sharp. In particular, we obtain from the groups in $\mathsf{N}(\Delta_{\mathcal{A}_I}/G) = \mathcal{H}_I$ a simplex cohomology decomposition as in* (5.8.6)*; and the map from the homotopy colimit of the pushout cube diagram* (4.6.5) *to $BMcL$, followed by 2-completion, is a homotopy equivalence.*

PROOF. The argument for $\Delta_{\mathcal{A}_I}$ follows the case of J_3 in Theorem 7.8.1; the argument for $\mathcal{A}_2^-$ is similar to that for J_1 sketched after Theorem 7.6.1, though in this case we have two maximal simplices in the orbit complex $\Delta(\mathcal{A}_2^-)/G$. □

The mod 2 cohomology ring of McL was computed by Adem and Milgram in [**9**], as a module over a large subring. The result is expressed in terms of the cohomology of H_4 and $H_{4'}$. The non-local subgroup $U_4(3)$, which we had mentioned as appearing in the rank four geometry of Ronan and Stroth, is also used in that calculation. The Poincaré series given in [**9**] is

$$\sum_{i=0}^{\infty} \dim_{\mathbb{F}_2} H^i(McL, \mathbb{F}_2)\, t^i =$$

$$\frac{2(1+t^{18}+t^{20}+t^{21}+t^{24}+t^{25}+t^{27}+t^{45})}{(1-t^8)(1-t^{12})(1-t^{14})(1-t^{15})} - \frac{(1-t^7)(1-t^{11})}{(1-t^8)(1-t^{12})}.$$

7.12. The Suzuki group Suz

For $G := Suz$, we take as our index set $I := \{1, 2, 4\}$, corresponding to G-classes of maximal 2-local subgroups in the Atlas [**47**, p. 131] satisfying (6.2.9):

$H_1 \cong 2_-^{1+6} \cdot U_4(2)$,

$H_2 \cong 2^{2+8} : (\mathrm{Alt}_5 \times \Sigma_3)$, and

$H_4 \cong 2^{4+6} : 3 \cdot \mathrm{Alt}_6$.

As usual the H_i determine a subgroup simplex $\mathcal{H}_I$ as in Definition 4.6.2, and a 2-local complex $\Delta_{\mathcal{H}_I}$ as in Definition 6.3.5. The common intersection H_I of order $2^{13}3$ contains a Sylow 2-subgroup with index three.

The usual diagram for $\Delta_{\mathcal{H}_I}$ is given in [**106**, p. 288] as:

The double bond diagrams at 1 and 4 both represent generalized quadrangles; but that in the link of 4 is of type $Sp_4(2)$, while that in the link of 1 is of type $U_4(2)$—in particular, the geometry at node 4 in the link of 1 corresponds to $L_2(4)$ rather than $L_2(2)$.

We now begin an approach to the geometry of Suz via a subposet of $\mathcal{E}_2(G)$.

We are not aware of any treatment of $\mathcal{A}_2(G)$. In fact we will proceed (much as in the case of M_{12} in Section 7.2) via a further subposet $\mathcal{E}_2^-$ of the subposet $\mathcal{E}_2(G)$ of $\mathcal{A}_2(G)$. However, we also mention that we will later indicate an easier approach via $\mathcal{B}_2^{cen}(G)$.

Reducing to $\mathcal{E}_2(G)$. For $i \in I$ we set $A_i := \Omega_1 Z\big(O_2(H_i)\big)$; then A_i is a member of $\mathcal{A}_2(G)$ of rank i, with $N_G(A_i) = H_i$. We get $A_1 < A_2 < A_4$ via the usual method of 6.2.17 (2).

There are two conjugacy classes of involutions, denoted by $2A$ and $2B$ in the Atlas [**47**]; of these, the involutions in the center of a Sylow 2-subgroup lie in class $2A$. In particular, a generator of A_1 lies in class $2A$; and for $i = 2, 4$, it follows from the transitivity of a subgroup Σ_3 or Alt_6 in H_i on the involutions of A_i that those involutions also lie in class $2A$. Using the ClassMultiplicationCoefficient function in GAP, from the character table of Suz we can compute the value

$$\xi(2A, 2A, 2B) = 0$$

for the structure constant defined in (5.10.5). Hence the collection $\mathcal{E}_2(G)$ in Definition 5.10.1 is given by the elementary abelian 2-subgroups whose nontrivial elements are in class $2A$; and the collection is ample and normalizer-sharp by Proposition 5.10.2.

Some technical details of $\mathcal{E}_2(G)$. The subgroups in $\mathcal{E}_2(G)$ are determined in Wilson [**126**, Sec. 2.4]; up to conjugacy they are subgroups of A_4, and we now indicate certain details from that treatment: We may choose A_2 of rank two in $\mathcal{E}_2(G)$ such that $A_1 < A_2 < U_1 := O_2(H_1)$. We may also choose B_2 of rank two in $\mathcal{E}_2(G)$ such that $A_1 < B_2 \not\leq U_1$. Then $A_4 = B_2 Z\big(C_{U_1}(B_2)\big)$ is of rank four, and lies in $\mathcal{E}_2(G)$. Further A_4 represents the unique conjugacy class of maximal members of $\mathcal{E}_2(G)$; and any member of $\mathcal{E}_2(G)$ of rank three is G-conjugate to $A_3 := A_4 \cap U_1$. Indeed A_4 is determined by A_3 and a conjugate of A_1 which determines a 2-central involution of H_1/U_1.

In fact we will now sketch a proof of the uniqueness just indicated for the conjugacy class of A_4 of rank four in $\mathcal{E}_2(G)$—since we will be needing various related uniqueness statements. The arguments involve the group $U_4(2)$ in H_1, and are much like those using the group Alt_5 in the analogous involution centralizer "H_1" for J_3 in the first part of Section 8.2; so here we will outline those similar steps somewhat more briefly. The involutions in $U_1 \setminus A_1$ are H_1-conjugate, and each with A_1 determines a conjugate of A_2; indeed since the three involutions of A_2 are H_2-conjugate, we see conversely that any G-conjugate of A_2 over A_1 is an H_1-conjugate, and hence is characterized by the property of lying in U_1. Now let B denote some maximal member of $\mathcal{E}_2(G)$. By conjugacy we may assume that $A_1 \leq B$, so that $B \leq C_G(A_1) = H_1$. Certainly A_1 is not maximal since we saw above that $A_4 > A_1$, so we have $B > A_1$. Next U_1 has 2-rank three, and its subgroups of rank three are H_1-conjugate to A_3 described above, which is not maximal since we saw above that $A_3 < A_4$; so we conclude that $B \not\leq U_1$, and hence at least one involution $x \in B$ determines an involution $\bar{x}$ lying in $\bar{H}_1 := H_1/U_1 \cong U_4(2)$. We now claim that this involution $\bar{x}$ must be 2-central in $\bar{H}_1$, and indeed that x must lie in some conjugate of A_4: First $\tilde{U}_1 := U_1/A_1$ affords the natural orthogonal module for the $\Omega_6^-(2) \cong U_4(2)$ in H_1; and the 27 singular vectors of $\tilde{U}_1$ determine conjugates of $\tilde{A}_2$, hence 54 conjugates of a generator a_1 of A_1 lying in $U_1 \setminus A_1$. Next we saw above that an involution y lying in A_4 but not in U_1 determines a 2-central involution $\bar{y}$ of $\bar{H}_1$; then each of the 45 $\bar{H}_1$-conjugates of $\bar{y}$ determines 8 conjugates of y lying in $\bar{y}$ regarded as the coset yU_1, namely the members of $y\Omega_1(C_{U_1}(y))$. These are in fact the 8 points of the $Sp_4(2)$ generalized quadrangle A_4 which are not collinear with A_1 via lines of type A_2. Thus we have $45 \cdot 8 = 360$ conjugates of a_1 in $H_1 \setminus U_1$. On the other hand, using the ClassMultiplicationCoefficient function in GAP, from the character table of Suz we can compute the value $\xi(2A, 2A, 2A) = 414$ for the structure constant defined in (5.10.5), which says that a_1 commutes with 414 distinct G-conjugates of itself; and since we have already indicated $54 + 360 = 414$ such conjugates in $C_G(A_1) = H_1$, we have thus already described all the conjugates of a_1 in H_1. This establishes the claim: namely the involution x of B must determine a 2-central involution $\bar{x}$, and further x lies in one of the 45 indicated conjugates of A_4, which we may as well take to be A_4. And now we are done: for by maximality, B must also contain $C_{A_4}(x) = A_3$; and then $B = \langle x \rangle A_3 = A_4$. That is, we have established that A_4 of rank 4 represents the unique conjugacy class of members of $\mathcal{E}_2(G)$.

We can now obtain some further uniqueness statements, which we will wish to use in homotopy equivalence arguments. By the previous paragraph, up to conjugacy we can embed any member of $\mathcal{E}_2(G)$ in A_4. Now A_4 affords the natural symplectic module for the $Sp_4(2)' \cong \mathrm{Alt}_6$ in H_4. So since H_4 has just one orbit on 3-subspaces of A_4 (the orthogonal spaces to the 1-spaces represented by A_1, which form a single orbit), there is just one class of members of $\mathcal{E}_2(G)$ of rank three, represented by A_3. Similarly there are just two H_4-conjugacy classes on 2-subspaces of A_4—namely the totally isotropic 2-subspaces represented by A_2, and the non-isotropic 2-subspaces represented by B_2 described earlier; so these represent the G-classes of members of $\mathcal{E}_2(G)$ of rank two. We can write B_2 as $A_1\langle x \rangle$ in our earlier language, and A_3 as B_2A_2. In particular, the final part of our argument in the previous paragraph shows that A_4 is the unique maximal member of $\mathcal{E}_2(G)$ above B_2, and hence also the unique maximal member above A_3.

Reduction to $\mathcal{E}_2^-$. The uniqueness statements in the previous sentence now lead to a homotopy equivalence via a G-map of $\mathcal{E}_2(G)$ with the subposet $\mathcal{E}_2^-$ determined just by the G-conjugates of A_1, A_2, and A_4; the argument parallels that for J_3 in the second half of Section 8.2, so we just summarize it as follows: The upper subposet $\mathcal{E}_2(G)_{>A_3}$ consists just of A_4 as in Remark 6.5.2; so we can remove the G-class of A_3 by Lemma 6.5.1. Then in the poset $\mathcal{E}'$ with no members of rank 3, the upper subposet $\mathcal{E}'_{>B_2}$ now consists just of A_4, so that we can similarly remove the conjugates of B_2. In view of this homotopy equivalence via a G-map, we see by Proposition 5.10.2 that $\mathcal{E}_2^-$ is ample and normalizer-sharp.

The poset $\mathcal{E}_2^-$ has reduced Euler characteristic $2^{10} \cdot 4091$ which is not divisible by $|G|_2 = 2^{13}$, and so by Remark 5.9.4 it is not in the standard ample G-homotopy type represented by $\mathcal{A}_2(G)$.

Since $A_1 < A_2 < A_4$, we see that the intersection complex $\Delta_{\mathcal{A}_I}$ defined by the A_i is just the internal poset complex $\Delta(\mathcal{E}_2^-)$ as in Proposition 6.3.6.

We now obtain flag-transitivity of G on $\Delta(\mathcal{E}_2^-)$ by the usual inductive method: First, the link of A_4 consists of the 15 points and 15 isotropic lines in the polar space for A_4, which is the building for the $Sp_4(2)' \cong \mathrm{Alt}_6$ in H_4. Next for the link of A_1, we see in the orthogonal 6-space $\tilde{U}_1 := U_1/A_1$ that incident conjugates of A_2 determine projective points, and incident conjugates of A_4 determine singular lines (via intersection with U_1, giving conjugates of A_3), in the polar space affording the building for the $\Omega_6^-(2) \cong U_4(2)$ in H_1. So in both links, we may apply Remark 6.4.6 (3) to obtain flag-transitivity. We have now checked the condition in links of two of the three types, which suffices for flag-transitivity of G by Proposition 6.4.5. In particular, $\mathsf{N}\big(\Delta(\mathcal{E}_2^-)/G\big) = \mathcal{H}_I$.

As usual the above flag-transitivity argument really involves only the geometry of the links indicated in the diagram; so we can adjust the argument to also obtain flag-transitivity of G on $\Delta_{\mathcal{H}_I}$. So by the standard Lemma 6.5.4 we see that the normalizer set-map N extends to an isomorphism of complexes from $\Delta(\mathcal{E}_2^-)$ to the 2-local complex $\Delta_{\mathcal{H}_I}$.

An alternative approach via $\mathcal{B}_2^{cen}(G)$. The poset $\mathcal{B}_2(G)$ of nontrivial 2-radical subgroups is given in Yoshiara [**129**, Prop. 2]. There are 10 conjugacy classes, of which seven are 2-centric; and indeed the seven classes in $\mathcal{B}_2^{cen}(G)$ are represented by the $U_J := \prod_{i \in J} U_i$ for nonempty $J \subseteq I$, where $U_i := O_2(H_i)$.

Indeed we see that the complex $\Delta_{\mathcal{U}_I}$ determined by the U_i as in Definition 6.3.13 has $\mathcal{B}_2^{cen}(G)$ as its poset of simplices. It is straightforward to adjust our earlier proof of flag-transitivity on $\Delta_{\mathcal{H}_I}$ to apply to $\Delta_{\mathcal{U}_I}$; so that by the standard Lemma 6.5.4, the set-map N extends to an isomorphism of $\Delta_{\mathcal{U}_I}$ with the 2-local complex $\Delta_{\mathcal{H}_I}$. Hence the poset of simplices of $\Delta_{\mathcal{H}_I}$ is isomorphic to $\mathcal{B}_2^{cen}(G)$.

We also mention that Sawabe in [**111**, 6.5] uses his equivalence method to provide an abstract approach for this particular equivalence.

Main results for Suz. The usual 2-local geometry was given in [**106**, p. 288], essentially as the 2-local complex $\Delta_{\mathcal{H}_I}$ (but also as an external poset complex from certain subspaces of the complex Leech lattice).

This discussion via our usual approach leads to the following theorem.

THEOREM 7.12.1. *For $G = Suz$, the 2-local complex $\Delta_{\mathcal{H}_I}$ and the collections $\mathcal{E}_2^-$ and $\mathcal{B}_2^{cen}(G)$ are ample and normalizer-sharp. In particular, we obtain from the groups in $\mathsf{N}\big(\Delta_{\mathcal{H}_I}/G\big) = \mathcal{H}_I$ a simplex cohomology decomposition as in* (5.8.6)*; and the map from the homotopy colimit of the pushout cube diagram* (4.6.5) *to $BSuz$, followed by 2-completion, is a homotopy equivalence.*

PROOF. The argument for $\mathcal{E}_2^-$ follows that for HS in Theorem 7.10.2; the argument for $\mathcal{B}_2^{cen}(G)$ is similar to that for M_{22} in Theorem 7.3.1, using Proposition 5.11.2 for ampleness and normalizer-sharpness. □

We are not aware of any computation of the mod 2 cohomology ring of Suz.

7.13. The Conway group Co_3

This group was considered in detail in Benson [**24**]; in particular, that paper established the method we described in Section 5.10. We will now review the usual features of that method for the present case.

For $G := Co_3$, we take as our index set $I := \{1, 2, 4\}$, corresponding to maximal 2-local subgroups in the Atlas [**47**, p. 134]:

$H_1 \cong 2^{\cdot}Sp_6(2)$,
$H_2 \cong 2^{2+6}3^{1+2}2^2$, and
$H_4 \cong 2^4{\cdot}L_4(2)$.

As usual the H_i determine a subgroup simplex $\mathcal{H}_I$ as in Definition 4.6.2, and a 2-local complex $\Delta_{\mathcal{H}_I}$ as in Definition 6.3.5. The common intersection H_I of order $2^{10}3$ contains a Sylow 2-subgroup with index three.

The diagram for $\Delta_{\mathcal{H}_I}$ given in Ronan and Stroth [**107**, p. 68] is

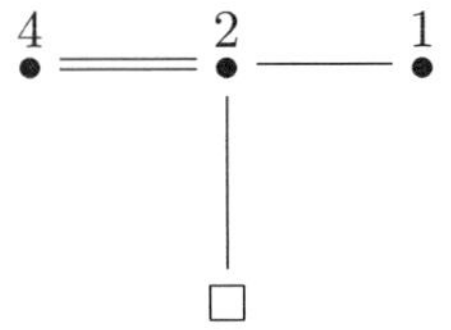

As usual the convention is that no objects in the geometry correspond to the square node.

We are not aware of any treatment of $\mathcal{A}_2(G)$. But as in the case of Suz in Section 7.12, our treatment begins with a further subposet $\mathcal{E}_2^-$ of the subposet $\mathcal{E}_2(G)$ of $\mathcal{A}_2(G)$.

For $i \in I$ we set $A_i := \Omega_1 Z\big(O_2(H_i)\big)$; then A_i is a member of $\mathcal{A}_2(G)$ of rank i, with $N_G(A_i) = H_i$. Via the usual method of 6.2.17 (2) we get $A_1 < A_2 < A_4$.

There are two conjugacy classes of involutions in G, denoted $2A$ and $2B$ in the Atlas [**47**]; of these, the involutions in the center of a Sylow 2-subgroup lie in class $2A$. Thus A_1 is generated by a member of class $2A$; and for $i = 2, 4$, from the transitivity of a suitable subgroup Σ_3 or $L_4(2)$ of H_i on the involutions of A_i, we see that those involutions also lie in class $2A$.

From the character table of $Co3$ (for example using the ClassMultiplicationCoefficient function in GAP), we can compute the value

$$\xi(2A, 2A, 2B) = 0$$

for the structure constant defined in (5.10.5). Hence the collection $\mathcal{E}_2(G)$ in Definition 5.10.1 is given by the elementary abelian 2-subgroups whose nontrivial elements are in class $2A$; and the collection is ample and normalizer sharp by Proposition 5.10.2. These subgroups are determined in Benson [**24**, Sec. 3]; up to conjugacy they are subgroups of A_4, and we indicate various details: We have chosen A_2 of rank two in $\mathcal{E}_2(G)$ such that $A_1 < A_2$; the image of A_2 in $H_1/A_1 \cong Sp_6(2)$ lies in the class denoted $2B$ for that group in the Atlas [**47**, p. 46], and A_2 represents the unique conjugacy class of members of $\mathcal{E}_2(G)$ of rank two. We may embed A_2 in A_3 of rank three, and A_3 represents the unique G-class in $\mathcal{E}_2(G)$ of rank three. We emphasize that A_4 is then determined as the unique member of $\mathcal{E}_2(G)$ properly containing A_3: namely we take the preimage of the subgroup 2^3 of $H_1/A_1 \cong Sp_6(2)$ given by the 2-central involutions only, in the unipotent radical 2^6 of a parabolic subgroup $2^6L_3(2)$ for the end node on a double bond in the Dynkin diagram of type C_3 for $Sp_6(2)$. In particular, A_4 of rank four represents the unique G-class of maximal members of $\mathcal{E}_2(G)$.

Now (much as for Suz in Section 7.12) the above remark on uniqueness of A_4 of rank four above A_3 gives the situation of Remark 6.5.2, so that we may apply Lemma 6.5.1 to remove the class of A_3: that is, $\mathcal{E}_2(G)$ is homotopy equivalent via a

G-map to the subposet $\mathcal{E}_2^-$ determined by the G-conjugates just of A_1, A_2, and A_4. In particular, $\mathcal{E}_2^-$ is ample and normalizer-sharp by Proposition 5.10.2.

The poset $\mathcal{E}_2^-$ has reduced Euler characteristic $2^7 \cdot 393583$ which is not divisible by $|G|_2 = 2^{10}$, and so by Remark 5.9.4 it is not in the standard ample G-homotopy type represented by $\mathcal{A}_2(G)$.

Since $A_1 < A_2 < A_4$, we see that the intersection complex $\Delta_{\mathcal{A}_I}$ determined by the A_i for $i \in I$ is just the external poset complex $\Delta(\mathcal{E}_2^-)$ as in Proposition 6.3.6.

We obtain flag-transitivity of G on $\Delta(\mathcal{E}_2^-)$ by the usual inductive method: First, the link of A_4 is the set of points and lines in the projective space $\mathbf{P}(A_4)$ for the $L_4(2)$ in H_4, so we may apply Remark 6.4.6 (2) to get flag-transitivity on this link. Next, The link of A_1 is given from the structure of $H_1/A_1 \cong Sp_6(2)$: the incident conjugates of A_2 determine involutions in class $2B$ for $Sp_6(2)$; and the incident conjugates of A_4 determine subgroups 2^3 of such involutions in the unipotent radicals of parabolics $2^6L_3(2)$ in $Sp_6(2)$. So this link is isomorphic to the collection of projective points and projective planes, in the polar space which is the building for the $Sp_6(2)$ for H_1; hence we may apply Remark 6.4.6 (3) for flag-transitivity. We have now checked the condition in links of two of the three types, which suffices for flag-transitivity of G by Proposition 6.4.5. In particular, $\mathsf{N}(\Delta(\mathcal{E}_2^-)/G) = \mathcal{H}_I$,

As usual the above flag-transitivity argument really involves only the geometry of the links indicated in the diagram; so we can adjust the argument to also obtain flag-transitivity of G on $\Delta_{\mathcal{H}_I}$. So by the standard Lemma 6.5.4 we see that the normalizer set-map N extends to an isomorphism of complexes from $\Delta(\mathcal{E}_2^-)$ to the 2-local complex $\Delta_{\mathcal{H}_I}$.

The poset $\mathcal{B}_2(G)$ of nontrivial 2-radical subgroups is given in An [**12**, 2C]. There are 14 conjugacy classes, of which 10 are 2-centric. Of the latter, six appear as U_J for nonempty $J \subseteq I$, where we define $U_J := \prod_{i\in J} U_i$ using $U_i := O_2(H_i)$. However, notice that $A_1 = U_1$ is not 2-centric. Thus the poset of simplices of the complex $\Delta(\mathcal{E}_2^-)$ is somewhat different from the poset $\mathcal{B}_2^{cen}(G)$. However, we mention that Maginnis and Onofrei in [**87**] show that our 2-local geometry $\mathcal{E}_2^-$ is homotopy equivalent (indeed via a G-map) to a certain subposet of $\mathcal{B}_2(G)$ suitably determined using the involutions of type $2A$.

A 2-local geometry for Co_3 is given in Ronan and Stroth [**107**, p. 68] from the viewpoint of minimal parabolic geometries; of course the maximal subgroups generated by these 2-locals are the above H_i for $i \in I$, and that description via maximal subgroups is essentially the 2-local complex $\Delta_{\mathcal{H}_I}$ in our present language.

This discussion leads via our usual approach to the following theorem.

THEOREM 7.13.1. *For $G = Co_3$, the 2-local complex $\Delta_{\mathcal{H}_I}$ and the collection $\mathcal{E}_2^-$ are ample and normalizer-sharp. In particular, we obtain from the groups in $\mathsf{N}(\Delta_{\mathcal{H}_I}/G) = \mathcal{H}_I$ a simplex cohomology decomposition as in* (5.8.6)*; and the map from the homotopy colimit of the pushout cube diagram* (4.6.5) *to BCo_3, followed by 2-completion, is a homotopy equivalence.*

PROOF. The argument follows that for HS in Theorem 7.10.2. □

Although the cohomology ring of Co_3 has not been calculated, it is proved in [**24**] that there is a map of classifying spaces $BCo_3 \to BDI(4)$, where $BDI(4)$ is the space constructed by Dwyer and Wilkerson [**56**], with the property that the homotopy fiber has finite mod 2 cohomology. It follows that $H^*(Co_3, \mathbb{F}_2)$ has the rank-4 Dickson invariants $\mathbb{F}_2[x_8, x_{12}, x_{14}, x_{15}]$ (the subscripts denote the degrees of the generators) as a subalgebra over which it is finitely generated as a module.

7.14. The Conway group Co_2

For $G := Co_2$, we take as our index set $I := \{1, 4, 5, 10\}$, corresponding to maximal 2-locals in the Atlas [**47**, p. 154]:

$H_1 \cong 2^{1+8}_+ : \mathrm{Sp}_6(2)$,
$H_4 \cong 2^{4+10}(\Sigma_3 \times \Sigma_5)$,
$H_5 \cong (2^{1+6}_+ \times 2^4)L_4(2)$, and
$H_{10} \cong 2^{10} : M_{22} : 2$.

As usual the H_i determine a subgroup simplex $\mathcal{H}_I$ as in Definition 4.6.2, and a 2-local complex $\Delta_{\mathcal{H}_I}$ as in Definition 6.3.5. The subgroup H_5 does not contain a full Sylow 2-subgroup of G. The common intersection H_I of order 2^{17} is of index two in a Sylow group $N_G(H_I)$ of G. As usual we set $U_i := O_2(H_i)$ for $i \in I$.

The diagram for $\Delta_{\mathcal{H}_I}$ given in Ronan and Smith [**106**, p. 286] is

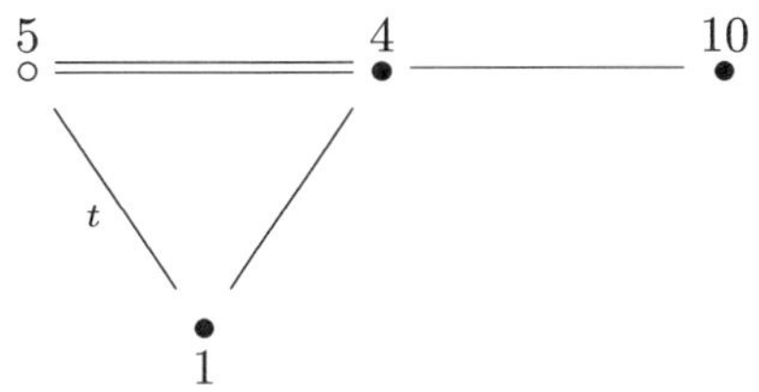

The subdiagram in the link of 10 is that for M_{22} in Section 7.3.

Our overall treatment will be parallel to that given for the rank-four geometry for J_4 in Section 7.9.

Some associated members of $\mathcal{A}_2(G)$. We are not aware of any determination of $\mathcal{A}_2(G)$. However, we now indicate certain particular members of $\mathcal{A}_2(G)$, which will be relevant for our further discussion of $\Delta_{\mathcal{H}_I}$.

In fact, we will even sketch a possible variant approach to a geometry for Co_2, in terms of the corresponding subposet of $\mathcal{A}_2(G)$, for which we have not attempted to verify all details.

For $i \in I$, we set $A_i := \Omega_1 Z\big(O_2(H_i)\big)$. Then A_i is a member of $\mathcal{A}_2(G)$ of rank i, with $N_G(A_i) = H_i$. We may define the intersection complex $\Delta_{\mathcal{A}_I}$ using these A_i as in Definition 6.3.2. We mention that the intersections of A_5 with A_4 and A_{10} have rank two and three, respectively; so the structure is slightly more complicated than an almost-poset complex as in Definition 6.3.9.

To connect this geometry with the more usual 2-local geometry for Co_2, it should be routine to adjust the proof (given later in the section) of flag-transitivity on $\Delta_{\mathcal{H}_I}$ to also apply to $\Delta_{\mathcal{A}_I}$; and then to conclude as usual by Lemma 6.5.4 that the normalizer set-map N extends to an isomorphism of complexes from the intersection complex $\Delta_{\mathcal{A}_I}$ to the 2-local complex $\Delta_{\mathcal{H}_I}$.

Also later in the section, we will indicate a homotopy equivalence via a G-map of $\Delta_{\mathcal{H}_I}$ with $\mathcal{B}_2(G)$; so that $\Delta_{\mathcal{A}_I}$ should also lie in the standard G-homotopy type, along with $\mathcal{A}_2(G)$ itself. But it also seems likely that there should be a more direct proof of this: namely using a sequence of reductions of the type in Lemma 6.5.1, to show that the inclusion of $\mathcal{A}$ in $\mathcal{A}_2(G)$ should induce a poset homotopy equivalence via a G-map, and hence also of the complexes Δ_{A_I} and $\Delta(\mathcal{A}_2(G))$. But we have not attempted to check this, as the argument would probably be quite lengthy.

A variant of the above approach would be to replace A_5 with $C_1 := \Phi(O_2(H_5))$, which is of rank one, but not conjugate to A_1. The resulting intersection complex then has two classes which are minimal under inclusion, represented by A_1 and C_1. In particular, the link of A_{10} in this variant complex is the external complex for M_{22} from the code module indicated in Section 7.3.

Flag-transitivity of $Co2$ on $\Delta_{\mathcal{H}_I}$. The incidence in the 2-local complex $\Delta_{\mathcal{H}_I}$ can be analyzed via the method demonstrated for M_{24} in Remark 6.2.17 (1), especially as modified for M_{22} in Section 7.3. In summary, one just verifies that the links of vertices are the geometries indicated in the diagram.

We can then establish flag-transitivity of G on $\Delta_{\mathcal{H}_I}$ in the usual inductive way:

First, the link of H_1 is "essentially" isomorphic to the polar space giving the building for the quotient $H_1/U_1 \cong Sp_6(2)$: For the vertices of types $i \in \{5, 10, 4\}$ determine spaces of dimensions $4, 2, 1$—via the images in 8-dimensional U_1/A_1 of C_1, $A_{10} \cap U_1$, $A_4 \cap U_1$, in the notation of the members of $\mathcal{A}_2(G)$ defined above. These give an external poset complex as in Proposition 6.3.6 from the 8-dimensional spin module for $Sp_6(2)$; by comparison, in the natural 6-dimensional symplectic module for this group, the corresponding spaces are projective points, lines, and planes. So we may apply Remark 6.4.6 (3) to that building. However, by "essentially" above, we mean that the hollow node in the diagram at type 5 refers to a 2-split cover: thus for the vertex H_4 in our link of H_1, there are actually two incident conjugates H_5 and $H_{5'}$. Correspondingly, to complete our proof of flag-transitivity on this link, we need to observe that if, in the line stabilizer $(2^2 \times 2^{1+4}) \colon (\Sigma_3 \times \Sigma_3)$ in $Sp_6(2)$, we take the further stabilizer Σ_2 of a plane (corresponding to the pair H_5, $H_{5'}$), then that Σ_2 in fact interchanges H_5 and $H_{5'}$.

Next, the link of H_5 is the projective space $\mathbf{P}(V_4)$ for $H_5/U_5 \cong L_4(2)$ where V_4 is the subgroup 2^4 of involutions in A_5 lying in the class denoted $2A$ in the Atlas [**47**]; we check for $i \in \{1, 4, 10\}$ that $A_i \cap V_4$ is of rank $1, 2, 3$. So we may apply Remark 6.4.6 (2) for flag-transitivity on this link.

Finally the link of H_{10} may be described using earlier remarks: viewed as subgroups of A_{10}, and using the subgroup C_1 corresponding to H_5 mentioned earlier, we obtain the 2-local geometry for $H_{10}/U_{10} \cong M_{22}$, in the viewpoint of the external complex from the code module of dimension 10 given in Section 7.3 for M_{22}; so by adjusting to this external complex the argument for flag-transitivity of M_{22} in that section, we may apply Remark 6.4.6 (4) for flag-transitivity on this link.

We have now verified flag-transitivity in links of three of the four types, which suffices by Proposition 6.4.5 for flag-transitivity of G. In particular, we have $\mathsf{N}(\Delta_{\mathcal{H}_I}/G) = \mathcal{H}_I$,

The usual 2-local geometry Δ for G is given in Ronan and Smith [**106**, p. 286], essentially as an external poset complex as in Proposition 6.3.6 arising from the Leech lattice (cf. the Atlas [**47**, p. 180]) modulo 2. But see also Ivanov and Shpectorov, for example [**75**], for the viewpoint of Petersen geometries.

By adjusting the above proof of flag-transitivity on $\Delta_{\mathcal{H}_I}$ to apply also to Δ, using the standard Lemma 6.5.4 we see that the normalizer set-map N extends to an isomorphism of complexes from this external poset complex Δ to the 2-local complex $\Delta_{\mathcal{H}_I}$.

The relationship with $\mathcal{B}_2(G)$. Smith and Yoshiara [**116**, 2.13] gave a proof that Δ is homotopy equivalent to the poset $\mathcal{B}_2(G)$ (considered in the abstract); and as usual that homotopy equivalence can be seen to be G-equivariant. Thus Δ and hence $\Delta_{\mathcal{H}_I}$ are ample and normalizer-sharp by Proposition 5.9.1.

The above homotopy equivalence is sufficient for the purposes of our main results for Co_2 in Theorem 7.14.1 below. However, we will digress briefly to examine the more precise relationship at the level of posets. In fact the equivalence proof in [**116**, 2.13] dealt only with $\mathcal{B}_2(G)$ in the abstract; the explicit determination of the poset $\mathcal{B}_2(G)$ of 2-radical subgroups for Co_2 was only subsequently obtained by Yoshiara in [**128**, Thm. 19], and independently in An and O'Brien [**14**, (4D)]. There are 15 conjugacy classes, and they are all 2-centric.

Now the poset of simplices of $\Delta_{\mathcal{H}_I}$ has the $2^{|I|}-1 = 15$ conjugacy classes of subgroups H_J determined by the subgroup simplex $\mathcal{H}_I$. The 15 classes from $\mathcal{B}_2(G)$ are in natural bijection with these: indeed $B_J := O_2\big(N_G(H_J)\big)$ is a member of $\mathcal{B}_2(G)$; and we may also exhibit B_J as $\prod_{i \in J} U_i$.

For convenience of comparison, we indicate the translation from indices used in [**128**, Table XI] to the type sets J used here. The eight subgroups[10] with subscript H there correspond to type sets J beginning with 1, and continuing with a subset of $\{10, 4, 5\}$ corresponding respectively to superscripts $\{p, l, \pi\}$ there. The four subgroups with subscript O' there correspond to type sets J beginning with 5, and continuing with a subset of $\{4, 1\}$ corresponding respectively to superscripts $\{l, \pi\}$ there. The first three rows there correspond to our type sets $10, 4, \{10, 4\}$.

In particular, H_I is of index 2 in a Sylow group $N_G(H_I)$, which is the 2-radical subgroup B_I corresponding to type I. As in the earlier cases for M_{22} in Section 7.3 and for J_4 in Section 7.9, which involve the triples geometry for Σ_5 (Example 6.5.5), this bijection of G-orbits does not quite pull back to an isomorphism of the poset of simplices of $\Delta_{\mathcal{H}_I}$ with $\mathcal{B}_2(G)$; instead there is a quotient relationship from identifying certain pairs, which gives a homotopy equivalence of posets. This comes from an application of Lemma 6.5.7, and the overall setup is much like that for J_4; again we summarize the details fairly briefly: We take $k := 5$ as our distinguished type, and $J := \{1\}$ as our distinguished subset of $I \backslash \{5\}$. Hence we must verify (6.5.8) for the types K given by $\{1\}$ and its over-sets $\{1,4\}$, $\{1,10\}$, and $\{1,4,10\}$ in $I \backslash \{5\}$. So we check that $N_G(H_I)$ does indeed lie in each such H_K. Then we turn to considering the types $K^+ = K \cup \{5\}$; where we check that the Sylow 2-subgroup $N_G(H_I)$ does indeed supplement H_{K^+} in $N_G(H_{K^+})$. As a sample of these latter verifications, note that $H_{5,1}$ has the structure $\big(2^{1+6} \times (2 \times 2^3)\big) \colon \big(2^3 \colon L_3(2)\big)$—which is of index two in

$$N_G(H_{5,1}) \cong 2^{2+6}\Big(2^6 \colon \big(2 \times L_3(2)\big)\Big),$$

The extra involution centralizing $L_3(2)$ can be chosen from U_1, and switches two conjugates of C_1 in $2^2 = A_1C_1$. The cases for larger types K^+ are subconfigurations of this one. So Lemma 6.5.7 gives the poset homotopy equivalence of $\mathcal{R}$ with the quotient $\bar{\mathcal{R}}$ determined by the identification of orbits of length 2 for types K^+.

Finally we combine the poset homotopy equivalence from $\mathcal{R}$ to $\bar{\mathcal{R}}$ with the observation that $\bar{\mathcal{R}}$ just reproduces the poset $\mathcal{B}_2(G)$.

---

[10]We take this opportunity to correct a slight misprint in [**128**, Table XI]: In the two rows with subscript H and superscript pl or $p\pi$, the structure of "R" ends with a 2^2—which should actually be a 2^3.

Main results for Co_2. The above discussion via our usual approach leads to the following theorem.

THEOREM 7.14.1. *For $G = Co_2$, the 2-local complex $\Delta_{\mathcal{H}_I}$ and the collection $\mathcal{B}_2(G)$ are ample and normalizer sharp. In particular, we obtain from the groups in $\mathsf{N}(\Delta_{\mathcal{H}_I}/G) = \mathcal{H}_I$ a simplex cohomology decomposition as in* (5.8.6)*; and the map from the homotopy colimit of the pushout* 4*-cube diagram* (4.6.7) *to BCo_2, followed by* 2*-completion, is a homotopy equivalence.*

PROOF. The argument follows that for M_{22} in Theorem 7.3.1, except in this case we are in rank four rather than three. □

7.15. The Conway group Co_1

For $G := Co_1$, we take as our index set $I := \{1, 2, 4, 11\}$, corresponding to maximal 2-locals in the Atlas [**47**, p. 183] satisfying (6.2.9):

$H_1 \cong 2_+^{1+8} \cdot \Omega_8^+(2)$,
$H_2 \cong 2^{2+12} : (\Sigma_3 \times L_4(2))$,
$H_4 \cong 2^{4+12} \cdot (3 \cdot \Sigma_6 \times \Sigma_3)$, and
$H_{11} \cong 2^{11} : M_{24}$.

The common intersection H_I of order $2^{21}3$ contains a Sylow 2-subgroup of G with index three. As usual the H_i determine a subgroup simplex $\mathcal{H}_I$ as in Definition 4.6.2, and a 2-local complex $\Delta_{\mathcal{H}_I}$ as in Definition 6.3.5.

The usual diagram for $\Delta_{\mathcal{H}_I}$ is given in [**106**, p. 284] as

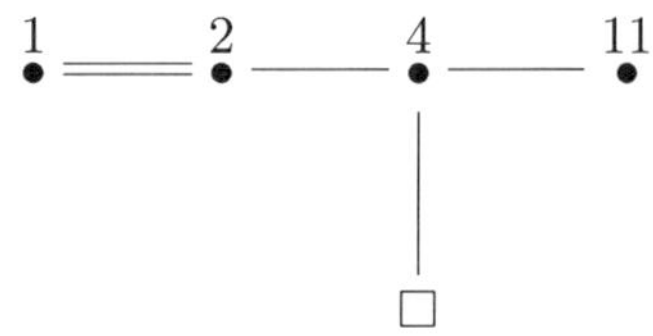

The subdiagram in the link of 11 is that for M_{24} in Section 7.5. As usual we set $U_i := O_2(H_i)$ for $i \in I$.

We are not aware of any determination of $\mathcal{A}_2(G)$. However (cf. the case of Co_2 in Section 7.14), one possible approach is by means of suitable members of $\mathcal{A}_2(G)$: For $i \in I$, we set $A_i := \Omega_1 Z\big(O_2(H_i)\big)$, and obtain a member A_i of $\mathcal{A}_2(G)$ of rank i, with $N_G(A_i) = H_i$. Using the usual method of 6.2.17 (2) we get

$$A_1 < A_2 < A_4 < A_{11}.$$

We see then that the intersection complex $\Delta_{\mathcal{A}_I}$ determined by the A_I is just an internal poset complex as in Proposition 6.3.6. In particular, the link in $\Delta_{\mathcal{A}_I}$ of A_{11} is just the 2-local geometry for $H_{11}/A_{11} \cong M_{24}$, viewed as the external poset complex in Example 6.3.7 arising from the code module.

Flag-transitivity of G on $\Delta_{\mathcal{H}_I}$ follows closely the case of Co_2 in Section 7.14: The incidence in the 2-local complex $\Delta_{\mathcal{H}_I}$ can be analyzed via the method demonstrated for M_{24} in Remark 6.2.17 (1). In summary, one just verifies that the links of vertices are the geometries indicated in the diagram. Then we can consider the various individual links. First, the link of H_1 is this time a part of the polar space which is the building for $\Omega_8^+(2)$ in H_1; namely for types $i \in 2, 4, 11$, the images in U_1/A_1 of the intersections of A_i with U_1 give totally singular spaces of dimensions $1, 2, 4$ in the natural orthogonal module of dimension 8. Next, the link of H_4 is the join of a polar space for the quotient $Sp_4(2) \cong \Sigma_6$ in H_4 (given by projective points and lines from conjugates of A_1 and A_2 in A_4) with a projective line for $L_2(2) \cong \Sigma_3$ in H_4, corresponding to three conjugates of A_{11} above A_4. In these two cases, we may apply Remark 6.4.6 (3) for flag-transitivity on the respective link. Finally the link of H_{11} viewed in A_{11} is, as mentioned earlier, the 2-local geometry for the M_{24} in H_{11}, viewed as the external poset complex in Example 6.3.7 arising from the code module. Since we checked flag-transitivity for M_{24} on that complex in Example 6.4.7 (1), we may apply Remark 6.4.6 (4). We have now checked the condition in links of three of the four types, which suffices by Proposition 6.4.5 for flag-transitivity of G. In particular, we have $\mathsf{N}(\Delta_{\mathcal{H}_I}/G) = \mathcal{H}_I$,

As usual the above argument for flag-transitivity depends essentially only on the links in the diagram, and so is easily adapted to apply to $\Delta_{\mathcal{A}_I}$ also. So by the standard Lemma 6.5.4, we see that the normalizer set-map N extends to an isomorphism of complexes from $\Delta_{\mathcal{A}_I}$ to the 2-local complex $\Delta_{\mathcal{H}_I}$.

The usual 2-local geometry Δ for G is given in [**106**, p. 284], by means of an external poset complex as in Proposition 6.3.6, arising from the Leech lattice modulo two. But see also Ivanov and Shpectorov, for example [**75**], for the viewpoint of tilde geometries. Again we can adapt the flag-transitivity argument to apply to Δ, so that Lemma 6.5.4 also extends the set-map N to an isomorphism of complexes from Δ to $\Delta_{\mathcal{H}_I}$.

The poset $\mathcal{B}_2(G)$ of 2-radical subgroups of Co_1 is determined in Sawabe [**109**, Table I, p. 132]. There are 30 conjugacy classes, of which three are not 2-centric: namely the ones denoted there by V, $V : \langle\sigma\rangle$ and F. The remaining 27 conjugacy classes of 2-centric 2-radical subgroups all have the property that their normalizers contain a Sylow 2-subgroup. In fact these normalizers all lie in one of H_i for $i \in I$ above. Furthermore in [**111**, 6.3], Sawabe applies his equivalence method to Co_1 to show that $\mathcal{B}_2^{cen}(G)$ is homotopy equivalent to the subposet given by the members of $\mathcal{O}_2(\Delta_{\mathcal{H}_I})$—which has 15 conjugacy classes, represented by the $B_J := \prod_{i \in J} U_i$ for nonempty $J \subseteq I$. As usual the equivalence can be seen to be via a G-map. So following Notation 5.11.4 we will write $\mathcal{B}_2^{cen-}$ for the poset $\mathcal{O}_2(\Delta_{\mathcal{H}_I})$. Furthermore we recover the poset of simplices of $\Delta_{\mathcal{H}_I}$ as $\mathsf{N}(\mathcal{B}_2^{cen-})$. (As was the case for M_{24} in Section 7.5, here we have $N_G(H_J) = H_J$ for all types J, in effect giving the vacuous case of the hypotheses Lemma 6.5.7, from which no actual identifications are made.) So both $\mathcal{B}_2^{cen-}$ and $\Delta_{\mathcal{H}_I}$ are ample and normalizer-sharp by Proposition 5.11.2.

For completeness, we now provide a more detailed guide for translating between our notation H_J and the member of $\mathcal{B}_2^{cen-}$ of which H_J is the normalizer, in [**109**, Table I, p. 132]: The subgroups R, Q_1, Q, E there correspond to our U_j for $j = 1, 2, 4, 11$. The 15 subgroups $R.P_i$ contain U_1, and correspond to unipotent radicals of $\Omega_8^+(2)$; we take only those seven corresponding to subsets of our indices $2, 4, 11$. Similarly the seven subgroups $Q_1.N_i$ contain U_2, and correspond to unipotent radicals of $\Sigma_3 \times L_4(2)$; we take only those three corresponding to subsets of our indices $4, 11$. Finally we take $Q : S$ corresponding only to the set $\{4, 11\}$ of two indices. This accounts for the $4 + 7 + 3 + 1 = 15$ members of $\mathcal{B}_2^{cen-}$.

The poset $\mathcal{B}_2^{cen-}$ has reduced Euler characteristic $-2^{18} \cdot 23^2 \cdot 751$ which is not divisible by $|G|_2 = 2^{21}$, and so by Remark 5.9.4 it is not in the standard ample G-homotopy type of $\mathcal{B}_2(G)$.

The discussion above leads as usual to the following theorem.

THEOREM 7.15.1. *For $G = Co_1$, the collection $\mathcal{B}_2^{cen-}$ and the 2-local complex $\Delta_{\mathcal{H}_I}$ are ample and normalizer-sharp. In particular, we obtain from the groups in $\mathsf{N}(\Delta_{\mathcal{H}_I}/G) = \mathcal{H}_I$ a simplex cohomology decomposition as in* (5.8.6)*; and the map from the homotopy colimit of the pushout 4-cube diagram* (4.6.7) *to BCo_1, followed by 2-completion, is a homotopy equivalence.*

PROOF. The argument follows that for Suz in Theorem 7.12.1; except that we are in rank four rather than three, and we use a homotopy equivalence via a G-map in Proposition 5.11.2 for ampleness and sharpness. □

We mention that the poset $\mathcal{B}_2^{cen-}$ with complex $\Delta(\mathcal{B}_2^{cen-})$ of inclusion-chains of dimension three affords the smallest known ample collection for Co_1; the corresponding complexes for $\mathcal{A}_2(G)$ and $\mathcal{B}_2^{cen}(G)$ have dimension at least four.

7.16. The Fischer group Fi_{22}

For $G := Fi_{22}$, we take as our index set $I := \{1, 5, 6, 10\}$, corresponding to maximal 2-local subgroups in the Atlas [**47**, p. 163]:

$H_1 \cong (2 \times 2_+^{1+8} : U_4(2)) : 2$,
$H_5 \cong 2^{5+8} : (\Sigma_3 \times \mathrm{Alt}_6)$,
$H_6 \cong 2^6 : Sp_6(2)$, and
$H_{10} \cong 2^{10} : M_{22}$.

As usual the H_i determine a subgroup simplex $\mathcal{H}_I$ as in Definition 4.6.2, and a 2-local complex $\Delta_{\mathcal{H}_I}$ as in Definition 6.3.5. The group H_1 is the centralizer of a 2-central involution, in the class denoted $2B$ in [**47**, p. 163]. Here H_6 does not contain a Sylow 2-subgroup of G; and the common intersection H_I of order 2^{15} is of index four in a Sylow group. As usual we set $U_i := O_2(H_i)$ for $i \in I$.

We propose the following diagram for $\Delta_{\mathcal{H}_I}$:

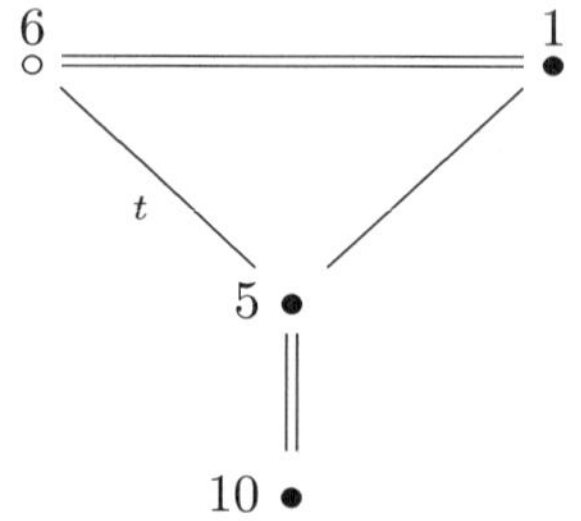

The subdiagram in the link of 1 indicates a certain geometry of rank 3 for $H_1/U_1 \cong U_4(2) : 2$; this may not have been described in the literature, so we present some details in Section 8.4.

The subdiagram in the link of 10 is a 2-split cover of the usual geometry for $H_{10}/U_{10} \cong M_{22}$ in Section 7.3. The situation is rather similar to that for J_4 in Section 7.9, where the link of H_{11} is a 2-split cover of the M_{24} geometry—and that cover is described in the flag-transitivity argument in that section. In the present case, the splitness arises because a Sylow 2-group of H_6 has index 4 in a Sylow group of G, whereas a Sylow of an octad stabilizer has index 2 in a Sylow of M_{22}. We will revisit this link during our flag-transitivity argument later in the section.

A 2-local geometry of rank three for G is described in [**107**, p. 69], with minimal parabolics leading to the maximal 2-local subgroups H_1, H_5, and H_{10} indicated above. However for our cohomological purposes here, it seems important to work instead with the rank four extension defined by these subgroups and the further maximal 2-local H_6; that is, with the 2-local complex $\Delta_{\mathcal{H}_I}$ as above.

Some associated members of $\mathcal{A}_2(G)$. We are not aware of any determination of $\mathcal{A}_2(G)$. However, we now indicate certain particular members of $\mathcal{A}_2(G)$, which will be relevant for our further discussion of $\Delta_{\mathcal{H}_I}$.

In fact, we will even sketch a possible variant approach to a geometry for Fi_{22}, in terms of the corresponding subposet of $\mathcal{A}_2(G)$, for which we have not attempted to verify all details.

For $i \in I$, as usual we set $A_i := \Omega_1 Z\big(O_2(H_i)\big)$; except for $i = 1$, where we take $A_1 := \Phi\big(O_2(H_1)\big)$. We obtain subgroups A_i of rank i in $\mathcal{A}_2(G)$, with $N_G(A_i) = H_i$. (For some purposes, it may also be convenient to replace A_5 of rank five with a subgroup B_4 of rank four given by its involutions in class $2B$.) Using the method of 6.2.17 (2), we get: $A_1 < A_5 \cap A_6$, with $A_5 \cap A_6$ of rank two; while $A_5 < A_{10}$; and finally $A_6 \cap A_{10}$ is of rank three. Thus the intersection complex $\Delta_{\mathcal{A}_I}$ is more complicated than an almost-poset complex as in Definition 6.3.9.

To connect this geometry with the more usual 2-local geometry for Fi_{22}, it should be possible to adjust the proof (given later in the section) of flag-transitivity on $\Delta_{\mathcal{H}_I}$ to also apply to Δ_{A_I}; and then to conclude as usual by Lemma 6.5.4 that the normalizer set-map N extends to an isomorphism of complexes from the intersection complex Δ_{A_I} to the 2-local complex $\Delta_{\mathcal{H}_I}$.

We also mention that the method of $\mathcal{E}_2(G)$ as in Definition 5.10.1 just fails to be available: for there are three conjugacy classes of involutions, the 2-central ones in Atlas class $2B$, and two noncentral classes $2A$ and $2C$—and from the character table of Fi_{22} (for example using the ClassMultiplicationCoefficient function in GAP), we can compute the value

$$\xi(2B, 2B, 2A) = 0$$

for the structure constant of (5.10.5). However, we have non-zero values

$$\xi(2B, 2B, 2C) = \frac{1}{2^{12}.3}, \qquad \xi(2C, 2C, 2A) = \frac{1}{2^{12}.3^2}.$$

We know of no method allowing the class $2C$ to be ignored; if that class could be removed, it would be possible to work just with elementary abelian subgroups whose elements lie in class $2B$.

Flag-transitivity of Fi_{22} on $\Delta_{\mathcal{H}_I}$. We begin with a preliminary remark about one distinctive feature of this geometry. As usual we have set $U_i = O_2(H_i)$ for $i \in I$; but as in some other cases where H_I does not contain a Sylow 2-subgroup of G (such as M_{22} in Section 7.3), we have the situation where one such "vertex kernel" U_{10} does not lie in the the stabilizer of an incident vertex H_6. And so for certain type sets J which strictly contain $\{6\}$, we will get $O_2(H_J) < \prod_{i \in J} U_i =: B_J$ (where we expect the latter to be a member of $\mathcal{B}_2(G)$). The distinctive aspect is that (for example) $B_{1,6} = U_1 U_6$ contains an involution from U_6 inducing an *outer* automorphism of the subgroup $U_4(2)$ of H_1/U_1, and centralizing a subgroup Σ_6; and so this element lies in $H_1 \setminus U_1$.

The incidence in the 2-local complex $\Delta_{\mathcal{H}_I}$ can be analyzed via the method demonstrated for M_{24} in Remark 6.2.17 (1), especially as modified for M_{22} in Section 7.3. In summary, one just verifies that the links of vertices are the geometries indicated in the diagram—including the rank 3 geometry for $H_1/U_1 \cong U_4(2):2$ which is described in Section 8.4.

We can then establish flag-transitivity of G on $\Delta_{\mathcal{H}_I}$ in essentially the usual inductive way:

The link of H_6 is the polar space for $Sp_6(2)$; in the language of the members of $\mathcal{A}_2(G)$ introduced above, we may exhibit the link by the intersections with A_6, which have ranks $1, 2, 3$. So we may apply Remark 6.4.6 (3) for flag-transitivity on this link.

The link of H_1 is the rank 3 geometry for $U_4(2){:}2$ described in Section 8.4. (It is exhibited here as an external complex on the natural module for $U_4(2)$ afforded by U_1/A_1; we use the images in that module of the intersections with U_1 of A_i for $i \in \{5, 6, 10\}$—which have ranks $2, 4, 4$ (that is, $\mathbb{F}_4$-dimensions $1, 2, 2$), corresponding to types $\{P, T, \Sigma\}$ in the notation of Section 8.4.) We establish flag-transitivity of $U_4(2){:}2$ on this geometry in Section 8.4; so we obtain flag-transitivity of H_1 on its link in $\Delta_{\mathcal{H}_I}$.

The link of H_{10} is exhibited by the intersections with A_{10} of the A_i for $i \in \{1, 6, 5\}$—which have ranks $1, 3, 5$. We have an analogue of the situation for the link of H_{11} in J_4, in the flag-transitivity argument in Section 7.9: The link is "essentially" the 2-local geometry for M_{22}, realized as the external poset complex for the cocode module indicated in Section 7.3; and we can adjust to this external complex the argument for flag-transitivity of M_{22} in that section. However by "essentially" we mean there is a doubling, since only a subgroup 2^9 of U_{10} lies in H_6; this means that U_{10} interchanges two conjugates A_6 and A_6', which intersect in a rank-3 subgroup of A_{10}, and whose product is a subgroup 2^{3+6} of $N_G(H_{10,6})$ which we will see again later. This further conjugacy completes the proof of flag-transitivity of H_{10} on its link. (An alternative argument proceeds via the subgroup $2^4 : \Sigma_5$ of $H_{10,1}$—where we establish flag-transitivity of this group on the link of "H_Σ" in the geometry for $U_4(2){:}2$ in Section 8.4.)

This in turn completes the verification of the condition for three of the four types of links in $\Delta_{\mathcal{H}_I}$, which suffices for the flag-transitivity of G by Proposition 6.4.5. In particular, we have $\mathsf{N}(\Delta_{\mathcal{H}_I}/G) = \mathcal{H}_I$.

The relationship with $\mathcal{B}_2^{cen}(G)$. The poset $\mathcal{B}_2(G)$ of 2-radical subgroups of Fi_{22} is determined in Kitazume and Yoshiara [**80**, Thm. 15]; it had also been determined by An and O'Brien in [**18**]. There are 16 conjugacy classes, of which all except the first given there, generated by the $2A$-involution and with centralizer $2U_6(2)$, are 2-centric. The 15 classes in $\mathcal{B}_2^{cen}(G)$ represented by the subgroups $B_J = \prod_{j \in J} U_j$ are in natural bijection with the nonempty type sets $J \subseteq I$. We give the translation from the notation in [**80**, Table 4] into our type sets J: The eight subgroups with subscript 2 there correspond to type sets J beginning with 1, and continuing with a subset of $\{5, 10, 6\}$ corresponding respectively to superscripts $\{p, l, \emptyset'\}$ there. The four subgroups with subscript 3 there correspond to type sets J beginning with 5; for the three of them following V_3 itself, our type J continues with one of $10, 6, \{10, 6\}$, corresponding respectively to superscripts $(1, 0), (0, 2), (1, 2)$ there. Finally the three members following the non-centric case V_1 there correspond to our type sets $6, 10, \{6, 10\}$.

But this bijection of orbit posets does not quite pull back to an isomorphism of the poset $\mathcal{R}$ of simplices of $\Delta_{\mathcal{H}_I}$ with $\mathcal{B}_2^{cen}(G)$; instead there is a quotient relationship determined by certain identifications, which gives a homotopy equivalence of posets via a G-map.

Since H_6 does not contain a Sylow 2-group, neither does H_I. Indeed H_I is of index 2 in $N_G(H_I)$, which in turn is of index 2 in a Sylow 2-subgroup S of G—where the latter gives the class in $\mathcal{B}_2^{cen}(G)$ corresponding to H_I. Orbits under $N_G(H_I)$ and S will determine the identifications, in a mild extension of the method of Lemma 6.5.7. The argument is very similar to that for $U_4(2):2$ at the end of Section 8.4, so we will indicate the proof by comparison, while contrasting the minor differences.

This time we parallel the setup for 6.5.7 using distinguished type $k = 6$ and distinguished subset $J = \{5\}$ of $I \setminus \{6\}$. Now there are 4 choices for K with $J \subseteq K \subseteq I \setminus \{6\}$; in each we have $S \leq H_K$, and for $K^+ = K \cup \{6\}$ we make identifications corresponding to the S-orbit of length 4 on H_6. And this time there are two further types L not of the form K^+ for which we make identifications, namely $\{10, 6\}$ and $\{1, 10, 6\}$—with the identifications corresponding instead to the orbit of $N_G(H_I)$ of length 2. (We implicitly saw this orbit earlier in our remark that U_{10} interchanges two conjugates A_6 and A_6' whose product is a subgroup 2^{3+6} in $N_G(H_{10,6})$.) These are the identifications that give the quotient $\bar{\mathcal{R}}$ of $\mathcal{R}$ the poset structure of $\mathcal{B}_2^{cen}(G)$.

Now we obtain contractibility of the preimages $p^{-1}(\bar{\mathcal{R}}_{\leq \bar{r}})$ just following the argument in Section 8.4; in particular for the further types L indicated above, the $N_G(H_I)$-orbits are again contained in the S-orbits, and so are included in the arguments made for types strictly larger than L.

And then the contractibility delivers the poset homotopy equivalence of $\mathcal{R}$ with the quotient $\bar{\mathcal{R}}$, which we recognize as $\mathcal{B}_2^{cen}(G)$, in this adjusted version of the method of Lemma 6.5.7.

The poset $\mathcal{B}_2^{cen}(G)$ has reduced Euler characteristic $-2^{16} \cdot 19 \cdot 229$ which is not divisible by $|G|_2 = 2^{17}$, and so by Remark 5.9.4 it is not in the standard ample G-homotopy type represented by $\mathcal{B}_2(G)$.

Main results for Fi_{22}. The above discussion leads as usual to the following theorem.

THEOREM 7.16.1. *For $G = Fi_{22}$, the 2-local complex $\Delta_{\mathcal{H}_I}$ and the collection $\mathcal{B}_2^{cen}(G)$ are ample and normalizer-sharp. In particular, we obtain from the groups in $\mathsf{N}(\Delta_{\mathcal{H}_I}/G) = \mathcal{H}_I$ a simplex cohomology decomposition as in (5.8.6); and the map from the homotopy colimit of the pushout 4-cube diagram (4.6.7) to BFi_{22}, followed by 2-completion, is a homotopy equivalence.*

PROOF. The argument follows that for Suz in Theorem 7.12.1; except that we are in rank four rather than three. □

7.17. The Fischer group Fi_{23}

The treatment of $G := Fi_{23}$ is fairly similar in outline to that for Fi_{22} in Section 7.16; the relationship arises from that of M_{23} with M_{22}.

This time we take as our index set $I := \{1, 6, 7, 11\}$, corresponding to maximal 2-local subgroups in the Atlas [**47**, p. 177] (see also Flaass [**61**]):

$H_1 \cong (2^2 \times 2_+^{1+8})(3 \times U_4(2))2$,

$H_6 \cong 2^{6+8}:(\Sigma_3 \times \mathrm{Alt}_7)$,
$H_7 \cong 2^7 : Sp_6(2)$, and
$H_{11} \cong 2^{11} \cdot M_{23}$.

These subgroups are closely related to those defining the geometry for Fi_{22} in the previous section.

As usual the H_i determine a subgroup simplex $\mathcal{H}_I$ as in Definition 4.6.2, and a 2-local complex $\Delta_{\mathcal{H}_I}$ as in Definition 6.3.5. Here the maximal 2-local H_7 is not a maximal subgroup of G, since it is proper in a subgroup $Sp_8(2)$; in fact it is a parabolic subgroup there, which shows that the extension of 2^7 by $Sp_6(2)$ splits. In particular, H_7 does not contain a Sylow 2-subgroup of G; and the common intersection H_I of order 2^{16} is of index four in a Sylow group. As usual we set $U_i := O_2(H_i)$ for $i \in I$.

We propose as a suitable diagram for $\Delta_{\mathcal{H}_I}$ the following:

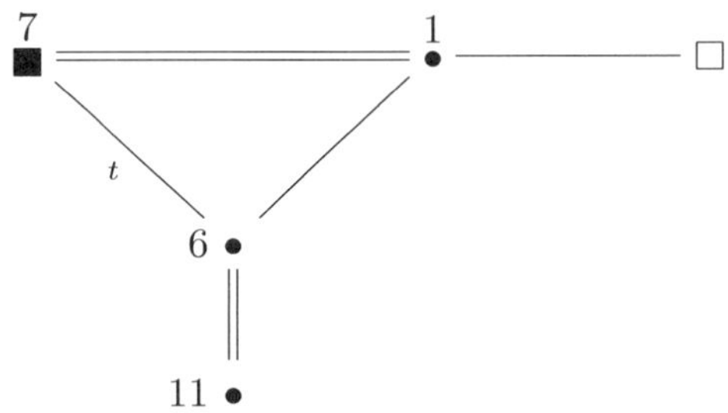

In parallel with the case of the link of 10 and M_{22} for Fi_{22} in the previous section, the subdiagram in the link of 11 here represents a 2-split cover of the usual geometry of M_{23} in Section 7.4; and we extend the conventions adopted there. In particular, the filled square node at 7 means that we do not use the full maximal subgroup $Sp_8(2)$ of Fi_{23} corresponding to the C_4-subdiagram in the link of 7; we only use its maximal 2-local subgroup $H_7 = 2^7 : Sp_6(2)$—which arises from the maximal parabolic of $Sp_8(2)$ determined by the right-hand end node in that C_4 Dynkin diagram, which we have indicated with an unfilled square node. (It would be possible to make an analysis in parallel to that for M_{23} in Section 7.4, where proceeding from Δ to Δ^+ associated to each octad O a set of 15 augmented octads O^+; in the present case, the factor in the final complex $\Delta_{\mathcal{H}_I}$ would be 63 rather than 15).

A 2-local geometry of rank four for Fi_{23} is described in Ronan and Stroth [**107**, p. 69]. As in the case of Fi_{22} in Section 7.16, for Fi_{23} it again seems important for our cohomological purposes to add consideration of the further maximal 2-local subgroup H_7; but it is also natural then to *omit* the maximal 2-local $2.Fi_{22}$ (the centralizer of an involution in class $2A$) indicated in [**107**, p. 72], which does not satisfy (6.2.9). We thus obtain the rank four geometry 2-local complex $\Delta_{\mathcal{H}_I}$.

Some associated members of $\mathcal{A}_2(G)$. We are not aware of any determination of $\mathcal{A}_2(G)$. However, we now indicate certain particular members of $\mathcal{A}_2(G)$, which will be relevant for our further discussion of $\Delta_{\mathcal{H}_I}$.

Indeed as in the previous section for Fi_{22}, we will even sketch a possible variant approach to a geometry for Fi_{23}, in terms of the corresponding subposet of $\mathcal{A}_2(G)$, for which we have not attempted to verify all details.

For $i \in I$, we set $A_i := \Omega_1 Z\big(O_2(H_i)\big)$; again except for $i = 1$, where we set $A_1 := \Phi\big(O_2(H_1)\big)$. We obtain subgroups A_i of rank i in $\mathcal{A}_2(G)$, with $N_G(A_i) = H_i$.

Using the method of 6.2.17 (2), we get $A_1 < A_6 \cap A_7$ of rank two; with $A_6 < A_{11}$, and $A_7 \cap A_{11}$ of rank three. Thus the intersection complex $\Delta_{\mathcal{A}_I}$ is more complicated than an almost-poset complex as in Definition 6.3.9.

To connect this geometry with the more usual 2-local geometry for Fi_{23}, it should be possible to adjust the proof (given later in the section) of flag-transitivity on $\Delta_{\mathcal{H}_I}$ to also apply to Δ_{A_I}; and then to conclude as usual by Lemma 6.5.4 that the normalizer set-map N extends to an isomorphism of complexes from the intersection complex Δ_{A_I} to the 2-local complex $\Delta_{\mathcal{H}_I}$.

Flag-transitivity of Fi_{23} on $\Delta_{\mathcal{H}_I}$. We make the appropriate adjustments to the preliminary remark we made in the discussion of flag-transitivity for Fi_{22} in the previous section. Again we have the situation where a "vertex kernel" U_{11} does not lie in the the stabilizer of an incident vertex H_7. And so for certain type sets J which strictly contain $\{7\}$, we will get $O_2(H_J) < \prod_{i \in J} U_i =: B_J$ (where we expect the latter to be a member of $\mathcal{B}_2(G)$). The distinctive aspect is that (for example) $B_{1,7} = U_1U_7$ contains an involution of U_7 inducing an *outer* automorphism of the subgroup $U_4(2)$ of H_1/U_1, and centralizing a subgroup Σ_6; and so this element lies in $H_1 \setminus U_1$.

The incidence in the 2-local complex $\Delta_{\mathcal{H}_I}$ can be analyzed via the method demonstrated for M_{24} in Remark 6.2.17 (1), especially as modified for M_{22} in Section 7.3. In summary, one just verifies that the links of vertices are the geometries indicated in the diagram. In particular, the link of H_1 is determined by the rank 3 geometry for the subgroup $U_4(2){:}2$ of H_1/U_1 which is described in Section 8.4—in a way which we will indicate in more detail below.

We can then establish flag-transitivity of G on $\Delta_{\mathcal{H}_I}$ in essentially the usual inductive way; we essentially adapt the route used for Fi_{22} in Section 7.16.

This time, the previous "A_6" for Fi_{22} is now replaced by A_7 when we examine the link of H_7; here intersections of the other A_i with A_7 again exhibit the polar space of $Sp_6(2)$.

Furthermore the previous "A_{10}" is now replaced by A_{11} when we examine the link of H_{11}; and A_{11} affords the appropriate cocode module for M_{23}, in which the intersections of the other A_i with A_{11} exhibit an external poset complex construction of the usual 2-local geometry for M_{23}—though again, just as for Fi_{22} in the previous section, we have a 2-split cover. So as before we can first adjust for this external poset complex our proof of flag-transitivity for M_{23} in Section 7.4. We then extend this to our 2-split cover by adjusting the argument for Fi_{22}: this time U_{11} interchanges two conjugates A_7 and A_7', which intersect in a rank-4 subgroup of A_{11}, and whose product is a subgroup 2^{4+6} of $N_G(H_{11,7})$ which we will see again later.

The proof for the link of H_1 also requires only a small adjustment: This link is "essentially" the rank 3 geometry for $U_4(2){:}2$ of Section 8.4—but where there are now 3 conjugates of H_7 incident with H_1 in Fi_{23} corresponding to each vertex of type T in that geometry; this is because of the "augmented octad" considerations we had mentioned early in the section. So to extend our proof of flag-transitivity in Section 8.4 to this actual structure of the link, we just need to observe that the normal subgroup of order 3 in $H_1/U_1 \cong (3 \times U_4(2))$ is transitive on those triples. (This is of course similar to the argument for flag-transitivity made in the link of a quintet for M_{23} in Section 7.4.)

Thus we have checked the condition for links of three of the four types, which suffices for flag-transitivity of G by Proposition 6.4.5. In particular, we have $\mathsf{N}(\Delta_{\mathcal{H}_I}/G) = \mathcal{H}_I$.

The relationship with $\mathcal{B}_2^{cen}(G)$. The poset $\mathcal{B}_2(G)$ of 2-radical subgroups of Fi_{23} is determined in An and O'Brien [**15**, Table 3]. There are 19 conjugacy classes, of which the first four listed there are not 2-centric.

The remainder of our discussion now exactly follows the overall outline of that for Fi_{22} in Section 7.16; we indicate the minor modifications, which mainly correspond to replacing the previous types $5, 6$ by the present types $6, 7$

The remaining 15 members in $\mathcal{B}_2^{cen}(G)$ are in natural bijection with the groups B_J for nonempty $J \subseteq I$. The rows in [**15**, Table 3] correspond respectively to our type sets J given by:

$$7, 11, 1, \{7,1\}, \{7,11\}, 6, \{6,11\}, \{11,1\}, \{7,6\}, \{1,6\},$$
$$\{11,7,1\}, \{11,6,1\}, \{6,7.1\}, \{6,7,11\}, \{1,6,7,11\} = I.$$

But this bijection of orbit posets does not quite pull back to an isomorphism of the poset $\mathcal{R}$ of simplices of $\Delta_{\mathcal{H}_I}$ with $\mathcal{B}_2^{cen}(G)$; instead there is a quotient relationship determined by certain identifications, which gives a homotopy equivalence of posets via a G-map.

Since H_7 does not contain a Sylow 2-group, neither does H_I. Indeed H_I is of index 2 in $N_G(H_I)$, which in turn is of index 2 in a Sylow 2-subgroup S of G—where the latter gives the class in $\mathcal{B}_2^{cen}(G)$ corresponding to H_I. Orbits under $N_G(H_I)$ and S will determine the identifications, in a mild extension of the method of Lemma 6.5.7.

The argument is very similar to that for Fi_{22} in Section 7.16. Again we parallel the setup for 6.5.7; this time using distinguished type $k = 7$ and distinguished subset $J = \{6\}$ of $I \setminus \{7\}$. There are 4 choices for K with $J \subseteq K \subseteq I \setminus \{7\}$; in each we have $S \leq H_K$, and for $K^+ = K \cup \{7\}$ we make identifications corresponding to the S-orbit of length 4 on H_7. Again there are two further types L not of form K^+ for which we make identifications, this time given by $\{11,7\}$ and $\{1,11,7\}$—with the identifications corresponding instead to the orbit of $N_G(H_I)$ of length 2. (We implicitly saw this orbit earlier in our remark that U_{11} interchanges two conjugates A_7 and A_7' whose product is a subgroup 2^{4+6} in $N_G(H_{11,7})$.) These are the identifications that give the quotient $\bar{\mathcal{R}}$ of $\mathcal{R}$ the poset structure of $\mathcal{B}_2^{cen}(G)$.

Now we obtain contractibility of the preimages $p^{-1}(\bar{\mathcal{R}}_{\leq \bar{r}})$ by just following the argument as in Section 8.4; in particular for the further types L indicated above, the $N_G(H_I)$-orbits are again contained in the S-orbits, and so are included in the arguments made for types strictly larger than L.

And then the contractibility delivers the poset homotopy equivalence of $\mathcal{R}$ with the quotient $\bar{\mathcal{R}}$, which we recognize as $\mathcal{B}_2^{cen}(G)$, in this adjusted version of the method of Lemma 6.5.7.

The poset $\mathcal{B}_2^{cen}(G)$ has reduced Euler characteristic $-2^{15} \cdot 23 \cdot 13906159$ which is not divisible by $|G|_2 = 2^{18}$, and so by Remark 5.9.4 it is not in the standard ample G-homotopy type represented by $\mathcal{B}_2(G)$.

Main results for Fi_{23}. The above discussion leads as usual to the following theorem.

THEOREM 7.17.1. *For $G = Fi_{23}$, the collection $\mathcal{B}_2^{cen}(G)$ and the 2-local complex $\Delta_{\mathcal{H}_I}$ are ample and normalizer-sharp. In particular, we obtain from the groups in $\mathsf{N}(\Delta_{\mathcal{H}_I}/G) = \mathcal{H}_I$ a simplex cohomology decomposition as in* (5.8.6)*; and the map from the homotopy colimit of the pushout 4-cube diagram* (4.6.7) *to BFi_{23}, followed by 2-completion, is a homotopy equivalence.*

PROOF. The argument follows that for Fi_{22} in Theorem 7.16.1. □

7.18. The Fischer group Fi_{24}'

Let $G := Fi_{24}'$. The treatment of this case is somewhat similar to that for Fi_{22} and Fi_{23} in the previous two sections 7.16 and 7.17. However, it also involves some further complications—especially since its geometry does not seem to be describable via the usual conventions of the group theory literature for diagram geometries.

We take as our index set $I := \{1, 3, 6, 8, 11\}$, corresponding to maximal 2-local subgroups in the Atlas [**47**, p. 207]:

$H_1 \cong 2_+^{1+12} \cdot 3U_4(3)2$,
$H_3 \cong 2^{3+12}(\mathrm{Alt}_6 \times L_3(2))$,
$H_6 \cong 2^{6+8}(\Sigma_3 \times L_4(2))$,
$H_8 \cong 2^8 : \Omega_8^-(2)$, and
$H_{11} \cong 2^{11} \cdot M_{24}$.

As usual the H_i determine a subgroup simplex $\mathcal{H}_I$ as in Definition 4.6.2, and a 2-local complex $\Delta_{\mathcal{H}_I}$ as in Definition 6.3.5. As in the case of "H_7" for Fi_{23} in Section 7.17, the maximal 2-local H_8 is not a maximal subgroup of G; it is proper in a subgroup $\Omega_{10}^-(2)$. In particular, H_8 does not contain a Sylow 2-subgroup of G; and the common intersection H_I of order $2^{20}3$ has Sylow group of index two in a Sylow group of G. As usual we set $U_i := O_2(H_i)$ for $i \in I$.

As in the earlier cases of Fi_{22} and Fi_{23}, it is convenient to set $B_J := \prod_{j \in U} U_i$ for nonempty $J \subseteq I$; and again similar remarks hold here, such as $U_1 \not\leq H_8$, and $B_{1,8} = U_1 U_8$ contains an involution of U_8 inducing an outer automorphism—this time of the subgroup $3U_4(3)$ of H_1/U_1, which centralizes a subgroup $U_4(2)$, so that this element lies in $H_1 \setminus U_1$. However, a distinctive feature of Fi_{24}' (compared to the other Fischer groups) is that not all of the products B_J are members of $\mathcal{B}_2(G)$; this feature will add a complication to our later discussion of homotopy equivalences.

Ronan and Smith in [**106**, p. 285] give a diagram for a rank-4 geometry for Fi_{24}' which excludes H_8; however, for the purposes of mod 2 cohomology, it seems clear that we need to use the rank-5 geometry $\Delta_{\mathcal{H}_I}$ which includes H_8—and this geometry does not seem to be readily describable via the usual conventions of diagram geometries. So below we present a modification of the original Ronan–Smith diagram, which adopts several fairly desperate new conventions, in order to try to codify at least some description of the rank-5 geometry:

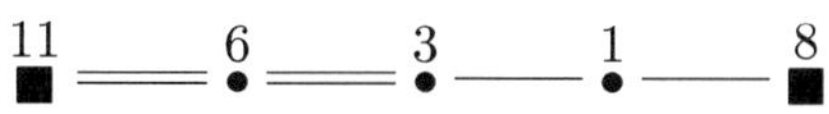

One aspect is fairly straightforward: the $\tilde{C}_2$-subdiagram (the adjacent double bonds) in the link of H_1 is just Kantor's well-known 2-local geometry for $U_4(3)$, a group which appears as a section of $H_1/U_1 \cong 3U_4(3){:}2$. This is a geometry of rank 3, defined using one maximal 2-local subgroup $2^{1+4} : (\Sigma_3 \times \Sigma_3)$, and two maximal

2-locals of the form $2^4 : \mathrm{Alt}_6$, as indicated in [**106**, p. 285]. The square node at 8 in effect extends this to a rank-4 geometry for $U_4(3){:}2$, using an outer automorphism which centralizes a subgroup $U_4(2)$.

Furthermore, the use of two filled square nodes ■ at the ends is meant to indicate the splicing together of two rank-4 geometries—each of which is a one-node extension of a more familiar rank-3 geometry (much as for $U_4(3)$ above, but probably more unusual). The subdiagram at 11 means that, in the link of H_{11}, we have a rank-4 extension of the usual rank-3 geometry for $H_{11}/U_{11} \cong M_{24}$ in Section 7.5; where now we *do* have objects in the geometry corresponding to the right-hand filled square node ■. We indicate some further details on this extension in Section 8.5. On the other hand, in the subdiagram at 8, the square nodes are being used in the spirit of the remarks for the subgroup Alt_8 of M_{23} in Section 7.4: Namely for the right end node, there is a corresponding non-local subgroup $\Omega_{10}^-(2)$ of Fi_{24}'; and the subdiagram at 8 could in principle be used to describe a rank-4 geometry for $\Omega_{10}^-(2)$.[11] But again for purposes of mod 2 cohomology, it seems important to use only the maximal parabolic subgroup $2^8{:}\Omega_8^-(2)$ from this $\Omega_{10}^-(2)$; so in this sense the use of a square node at the left end is analogous to the convention we used for M_{23}. And from the viewpoint of $H_8/U_8 \cong \Omega_8^-(2)$, the subdiagram at 8 just represents a rank-4 extension of the usual rank-3 polar space: Now the left-end filled square node ■ represents using also the maximal totally isotropic (but not singular) subspaces of dimension 4 in the natural module—in addition to the usual totally singular subspaces of dimensions 3, 2, 1 corresponding to the nodes in the C_3 subdiagram arising from twisting (in the sense of [**43**, Sec. 13.3]) the D_4 diagram for orthogonal 8-space of plus-type. Again we will indicate some further details on this extension in Section 8.5.

We apologize for the complication of these conventions in the diagram—we blame it on the apparently inherent complication of the group and the geometry. (Or alternatively, this may be evidence that the geometric approach is not really natural for treating Fi_{24}'.)

The "classical" 2-local geometry Δ of rank four for Fi_{24}' is described in Ronan and Smith [**106**, p. 285]; it is essentially the rank four 2-local complex determined by the H_i just for $i \in \{1, 3, 6, 11\}$. But see also Ivanov and Shpectorov, for example [**75**], for the viewpoint of Petersen and tilde geometries. It seems that the geometry Δ is suitable for many group-theoretical purposes—but it does not seem adequate for our present cohomological purposes. In particular, we mention that the argument in Sawabe [**111**, Sec. 6.4] unfortunately omitted for H_8 the verification of the hypothesis (1) in Theorem 2 there—and so does not in fact establish a homotopy equivalence of Δ with $\mathcal{B}_2^{cen}(G)$. (For example, the reduced Euler characteristic of Δ is divisible only by the power 2^{13}, rather than 2^{18} as for $\mathcal{B}_2^{cen}(G)$.) So here we do use H_8; that is, we choose as our 2-local geometry the rank five 2-local complex $\Delta_{\mathcal{H}_I}$—though more typically, we will prefer to use the variant of that complex given by the intersection complex $\Delta_{\mathcal{A}_I}$ developed below.

[11] Although not the usual rank-4 polar space geometry for that group, which would instead have a diagram of type C_4 (arising from twisting the diagram of type D_5). Instead the rank-4 geometry relevant here is obtained as a certain extension (by one node on the right, with a single bond) of the above diagram for $U_4(3)$. Furthermore the link of a left-end vertex in this geometry is the same as for the usual polar space of type C_4, namely a polar space of type C_3 for $\Omega_8^-(2)$—and this structure is all that we really use in the discussion here of the link of H_8 for Fi_{24}'.

We expand a little further on our earlier remark that seemingly $\Delta_{\mathcal{H}_I}$ cannot be described within the usual conventions of diagram geometries: For our group H_8 naturally corresponds to the right-hand node unfilled square node $\square$ in the diagram in Ronan and Smith [**106**, p. 285] for the rank-4 geometry which does not include H_8. This feature was perhaps first observed by Sawabe; we thank him for supplying us with some unpublished material, which established some of the details of the homotopy equivalence given below of $\Delta_{\mathcal{A}_I}$ with $\mathcal{B}_2^{cen}(G)$. But the usual convention for that square node $\square$ (as indicated in the case of M_{24} in Section 7.5) is inappropriate for our rank five geometry $\Delta_{\mathcal{H}_I}$ here: since that convention assumes that there is no corresponding subgroup in the geometry, whereas the group H_8 in Fi_{24}' does play a role in our cohomology decomposition. However, there would also be difficulties with just replacing that square node with an ordinary node $\bullet$: for while we do here need a rank five diagram, the links of H_{11} and H_8 are not "ordinary" rank four extensions of the usual rank three geometries in those links. So it seemed simpler to us to adopt the convention indicated earlier, of giving the "node" at H_8 a double identity—using $\blacksquare$ for describing the extended-M_{24} link of H_{11}, as well as the extended-$\Omega_8^-(2)$ link of H_8.

The approach via the intersection complex $\Delta_{\mathcal{A}_I}$. We are not aware of any determination of $\mathcal{A}_2(G)$. But in contrast to the cases of Fi_{22} and Fi_{23} in the previous two sections, this time we will give fuller details of the approach via an intersection complex determined by certain members of $\mathcal{A}_2(G)$.

For $i \in I$, we set $A_i := \Omega_1 Z\big(O_2(H_i)\big)$; we obtain subgroups A_i of rank i in $\mathcal{A}_2(G)$, with $N_G(A_i) = H_i$. Using the usual method of 6.2.17 (2), we get $A_1 < A_3 < A_6 < A_{11}$ and $A_1 < A_8$. We also have further intersections: $A_3 \cap A_8 =: A_{3,8}$ of rank two; $A_6 \cap A_8 =: A_{6,8}$ of rank three; and $A_8 \cap A_{11} =: A_{8,11}$ of rank four. We will use the intersection complex $\Delta_{\mathcal{A}_I}$ defined by the $A_i (i \in I)$ as in Definition 6.3.2. Again these further intersections show that $\Delta_{\mathcal{A}_I}$ is more complicated than an almost-poset complex as in Definition 6.3.9.

Note that A_1 is generated by a 2-central involution z, in the class denoted by $2B$ in the Atlas [**47**, p. 207]. Observe also that A_{11} affords the irreducible cocode module for $H_{11}/U_{11} \cong M_{24}$; indeed the incident vertices of types $i \in \{1, 3, 6, 8\}$ determine, by the intersection of A_i with A_{11}, spaces of dimensions $1, 3, 6, 4$. The first three of these exhibit the external poset complex for the cocode module for M_{24} indicated in Example 6.3.7; we will discuss the use of the 4-spaces in addition during our later treatment of flag-transitivity. Further A_6 is the natural orthogonal module for the factor $\Omega_6^+(2) \cong L_4(2)$ of H_6/U_6; with incident vertices of types $i \in \{1, 3, 8\}$ determining (via their intersections with A_6) spaces of dimensions $1, 3, 3$. Also A_8 is the natural orthogonal module for $H_8/U_8 \cong \Omega_8^-(2)$. Finally A_3 is the natural module for the factor $L_3(2)$ in H_3/U_3; with incident vertices of types $i \in \{1, 8\}$ determining (via their intersections with A_3) spaces of dimensions $1, 2$—namely the points and lines of the projective space $\mathbf{P}(A_3)$.

Some features of intersections among the A_i. In support of our definition of incidence via Definition 6.3.2, we will provide in Section 8.5 some details for showing that the intersections of G-conjugates of A_8 with G-conjugates A_i for $i \in \{3, 6, 11\}$ are as described above. From among those later details, we select a few for mention here, which will be useful as we continue our main treatment.

First $A_{8,11} = A_8 \cap A_{11}$ of rank four can be realized as the commutator subspace $[A_{11}, x]$, where $x \in O_2(H_{6,11}) \setminus A_{11}$ determines a 2-central involution in the quotient $H_{11}/A_{11} \cong M_{24}$. Further $A_{8,11}$ is a totally isotropic (but not totally singular) 4-subspace of A_8 in the orthogonal form determined by H_8: it has a 3-subspace, namely $A_{8,6}$, which is totally singular in the form—since its seven involutions lie in class $2B$ of G; while the remaining eight nonsingular vectors from $A_{8,11} \setminus A_{6,8}$ lie in the class $2A$ of G. Now the 2-central involution x has permutation type $1^8 2^8$ in M_{24}; where the even subsets of the eight points fixed by x span a 6-subspace A_6 of the module A_{11}. This leads to the coincidence of stabilizers

$$H_{8,11} = H_{6,8,11},$$

which we will derive in more detail in Section 8.5, in the latter part of our discussion there of the case "$i = 11$".

Next, the above space $A_{6,8}$ is the 3-subspace of A_6 which is given by the fixed points in A_6 (indeed in A_{11}) of $O_2(C_{M_{24}}(x)) \cong 2_+^{1+6}$. Further we saw in the previous paragraph that $A_{6,8}$ is the 3-subspace of the 4-space $A_{8,11}$ consisting of elements in $A_{8,11}$ lying in class $2B$ in G, which is totally singular in the form on A_8 determined by H_8; it is also a totally singular 3-subspace of A_6 in the form determined by H_6, but it is in the H_6-conjugacy class of such spaces other than that of A_3. In addition, the 3-subspaces $A_{6,8}$ and A_3 of A_8 are not conjugate under H_8, since A_3 is not totally singular in the orthogonal form on A_8. Here is an independent argument for this non-conjugacy, supplied by Satoshi Yoshiara: Note that A_3 is G-conjugate to the group denoted in Kitazume and Yoshiara [**80**, Sec. 2.1.5] by

$$Z(T) = \langle 0123, 0145, 0246 \rangle,$$

while $A_{6,8}$ is G-conjugate to the group denoted

$$W = \langle 0123, 0145, 0247 \rangle$$

in Lemma 7 (2) and in the last line of the proof of Lemma 8 (2) in [**80**, Sec. 2.1.5]. So the two groups are not G-conjugate, since the normalizer of W is contained in a conjugate of H_6 by [**80**, Lemma 7 (2)], but the normalizer of $Z(T)$ is a G-conjugate of H_3.

Finally, $A_{3,8}$ is a totally singular 2-space in A_8 in the orthogonal form coming from the action of H_8.

Flag-transitivity of Fi_{24}' on $\Delta_{\mathcal{A}_I}$. As usual, we wish to conclude that G is flag-transitive on the complex $\Delta_{\mathcal{A}_I}$. In our standard inductive method, it will suffice by Proposition 6.4.5 to establish flag-transitivity on links of four of the five types.

First for the link of A_6, we observed earlier that incident vertices of types $i \in \{1, 3, 8\}$ determine totally singular subspaces of dimensions $1, 3, 3$ in the orthogonal 6-space A_6; these spaces determine an external almost-poset complex, from the orthogonal module A_6, for the building for the group $\Omega_6^+(2) \cong L_4(2)$ in H_6–the spaces just correspond with the lines, points, and planes of the usual projective space for a 4-space over $\mathbb{F}_2$ under $L_4(2)$. But in this link we must also examine vertices of type 11: there are $|H_6 : H_{6,11}| = 3$ conjugates of A_{11} over A_6, which we regard as a projective line over $\mathbb{F}_2$. Thus the full link of A_6 is the join of this projective line with the building for $L_4(2)$ above; and so we may apply Remark 6.4.6 (3) to this product building, under the action of $L_2(2) \times L_4(2)$ in H_6, for flag-transitivity on the link of A_6.

Next for the link of A_3, we argue rather similarly: We observed earlier that incident vertices of types $i \in \{1, 8\}$ determine the points and lines in the projective space $\mathbf{P}(A_3)$. Calculating much as above, but now with incidence in H_3 rather than in H_6, we conclude that incident vertices of types $i \in \{6, 11\}$ determine the 15 projective points and 15 totally isotropic lines in a polar space of type $Sp_4(2)$. Again we may apply Remark 6.4.6 (3) to this product building, under the action of $Sp_4(2)' \times L_3(2)$ in H_3.

Finally we consider the links of A_{11} and A_8; but we sketch only an outline here, postponing fuller arguments to Section 8.5. We choose briefer notation C_4, C_3, and C_2 for the earlier subgroups $A_{8,11}$, $A_{6,8}$, and $A_{3,8}$ of dimension four, three, and two over $\mathbb{F}_2$; and set $C_1 := A_1$. And we describe the two links using this language:

The link of the vertex A_{11} is the rank four intersection complex on the H_{11}-conjugates of A_6, A_3, A_1, and C_4; and incidence is by inclusion, except for: conjugates of A_6 and C_4, where $A_6 \cap C_4 = C_3$ is of dimension three; and conjugates of A_3 and C_4, where $A_3 \cap C_4 = C_2$ is of dimension two. Notice that this is a rank four extension of the usual rank three 2-local geometry of M_{24} given by the subspaces A_6, A_3, A_1 of the cocode module in Example 6.3.7. We postpone consideration of flag-transitivity on this extension for a moment.

Fairly similarly, the link of the vertex A_8 is the rank four complex on the H_8-conjugates of the subspaces in the orthogonal space A_8 given by C_j of dimension $j = 1, 2, 3, 4$ over $\mathbb{F}_2$: note that the first three are totally singular in A_8, while C_4 is instead a maximal totally *isotropic* subspace. So this gives a rank four extension of the usual building geometry for $H_8/A_8 \cong \Omega_8^-(2)$—the building is given by the conjugates just of the totally *singular* subspaces C_3, C_2, and C_1.

Since both the above geometries are unusual one-node extensions of the geometries we have normally been considering, we have (as mentioned above) postponed further details of the necessary flag-transitivity arguments to Section 8.5.

Now assuming those arguments, we have verified flag-transitivity on the links of four of the five types, which suffices for flag-transitivity of G by Proposition 6.4.5. In particular, we have $\mathsf{N}(\Delta_{\mathcal{A}_I}/G) = \mathcal{H}_I$.

Finally one can as usual adapt the above proof of flag-transitivity on $\Delta_{\mathcal{A}_I}$ to also apply to $\Delta_{\mathcal{H}_I}$, in order to conclude via the standard Lemma 6.5.4 that the normalizer set-map N extends to an isomorphism of complexes from $\Delta_{\mathcal{A}_I}$ to the 2-local complex $\Delta_{\mathcal{H}_I}$.

The relationship of $\Delta_{\mathcal{A}_I}$ with $\mathcal{B}_2^{cen}(G)$. The maximal subgroups of Fi_{24}' are classified in Linton and Wilson [**84**]; and the poset $\mathcal{B}_2(G)$ of 2-radical subgroups of Fi_{24}' is determined in Kitazume and Yoshiara [**80**]. There are 33 conjugacy classes of 2-radical subgroups, of which 29 are 2-centric.

The poset $\mathcal{B}$ and its relationship to $\mathcal{B}_2^{cen}(G)$. We first change our viewpoint temporarily away from the complex $\Delta_{\mathcal{A}_I}$, to a related poset $\mathcal{B}$ which is more conveniently comparable with $\mathcal{B}_2(G)$. Our analysis closely parallels a similar but more elementary discussion for the group M_{24}, which we present toward the end of Section 8.5; the reader may wish to use that discussion for background, before continuing here.

The minimal elements of the poset $\mathcal{B}$ are the conjugates of the subgroups $U_i = O_2(H_i)$ for $i \in I$. For each nonempty subset $J \subseteq I$, we had earlier set

$B_J := \prod_{j \in J} U_j$. Then $\mathcal{B}$ is the poset determined by the conjugates of the B_J. Containments among these conjugates are fairly easy to analyze, since the U_i other than U_8 are weakly closed.

From experience with other groups in earlier sections, we might expect that the $2^{|I|} - 1 = 31$ members B_J in $\mathcal{B}$ would represent the conjugacy classes of the poset $\mathcal{B}_2^{cen}(G)$ of centric radical subgroups; but we already mentioned above that there are only 29 such classes.[12] And indeed as these numbers suggest, there are two classes in $\mathcal{B}$ which are not 2-radical in Fi_{24}'.

The situation closely parallels that for M_{24} indicated in Section 8.5; and indeed arises essentially because of the involvement of M_{24} in the stabilizer H_1 for Fi_{24}'. In particular, notice that the types $\{6, 3, 1, 8\}$ for Fi_{24}' may be regarded in the link of A_{11} via dimensions $\{6, 3, 1, 4\}$, since $A_{11} \cap A_8$ is of rank four; and these latter dimensions correspond to the type $\{O, T, S, \square\}$ for M_{24} in Section 8.5. Hence the coincidence of stabilizers $H_{8,11} = H_{6,8,11}$ for Fi_{24}' corresponds in $H_1/U_1 \cong M_{24}$ to the coincidence of stabilizers $N_{M_{24}}(B_\square) = N_{M_{24}}(B_{O\square})$. And so in Fi_{24}' we see that $B_{8,11}$ of order 2^{15} is proper in $B_{6,8,11}$ of order 2^{17}, which is normal in $N_G(B_{6,8,11}) = N_G(B_{8,11})$, so that $B_{8,11}$ is not 2-radical in Fi_{24}'. In a similar way the coincidence $N_{M_{24}}(B_{S\square}) = N_{M_{24}}(B_{OS\square})$ in M_{24} leads to the coincidence $N_G(B_{1,6,8,11}) = N_G(B_{1,8,11})$ in Fi_{24}'; so since $B_{1,8,11}$ of order 2^{18} is proper in $B_{1,6,8,11}$ of order 2^{20}, we see that $B_{1,8,11}$ is not 2-radical in Fi_{24}'.

On the other hand, the remaining 29 groups B_J are 2-radical for Fi_{24}'. In particular, we mention that for the other types J which properly contain $\{8, 11\}$ but do not contain 1, we have $N_G(B_J) \not\leq N_G(B_{6,8,11})$; and so we do not get the above coincidence of normalizers with the type $\{6\} \cup J$. (This is in contrast with the situation for stabilizers H_J of the simplices σ_J in the complex Δ_{A_I}, where via intersections we do get $H_J = H_{\{6\} \cup J}$; this distinction arises in our discussion below of homotopy equivalence.)

For convenience of comparison of the 29 classes in $\mathcal{B}_2^{cen}(G)$, we provide a guide for translating between our notation B_J and the notation of V_y^x (for suitable indices x, y) in the final 29 rows in Table 2 on page 41 of [**80**]; we are grateful to Masato Sawabe, who first adopted this approach, and provided us with his unpublished notes. Our $B_i = U_i$ for $i \in I = \{1, 3, 6, 8, 11\}$ are their five V_y, for a subscript y in their set $\{2, 6, 4, 8, 5\}$. In their V_2^x, the superscripts x involve a subset of $r, s, l, \prime$, where these in turn correspond to our indices $11, 6, 3, 8$ (and the type J in addition contains 1). There are just 14 such groups; of course they do not list any group with the notation $V_2^{r'}$, since we saw that $B_{1,8,11}$ is not radical. Observe in particular that their groups V_2^x, where the superscript x contains a prime, involve an outer automorphism of the $3U_4(3)$ in H_1/U_1; these are our B_J for J containing $\{1, 8\}$. In their seven V_6^x, the superscripts x begin with[13] p or $\emptyset$, corresponding to containing or not containing our index 8; and end with one of $0, 1, 2, 3$, corresponding to our $\{\emptyset, 6, 11, \{6, 11\}\}$ (and the type J in addition contains 3). In their three V_4^x, the superscripts x involve a subset of $p, 1$, where these in turn correspond to our $11, 8$ (and the type J in addition contains 6). Finally there is no entry in their table

[12] We are grateful to Bob Oliver for pointing out an error in the discussion in this area in the original version of our work.

[13] Here we are following Table 3 in [**80**], in using notation p for the superscript; however note that in the notation of Step 7 of the proof of Theorem 12 there, it would be more appropriate to name these groups using the notation l for that superscript.

after V_5 corresponding to the product V_5V_8, since that group is our $B_{8,11}$ which we saw is not radical. Thus we have $5 + 14 + 7 + 3 = 29$ subgroups.

The homotopy equivalence of $\Delta_{\mathcal{A}_I}$ *with* $\mathcal{B}$. Now, much as in the case of M_{24} in Section 8.5, we observe that the removal of the two non-radical classes represented by $B_{8,11}$ and $B_{1,8,11}$, to reduce from $\mathcal{B}$ to $\mathcal{B}_2^{cen}(G)$, is a homotopy equivalence via a G-map: This is essentially part of the argument of Dwyer in [**52**, 10.4]; cf. also the middle line in the tables for Theorems 1.1 and 1.2 in Grodal and Smith [**67**].

It remains to describe a homotopy equivalence between our earlier intersection complex $\Delta_{\mathcal{A}_I}$ and the more recent poset $\mathcal{B}$. As in the case of M_{24} in Section 8.5, again it is notationally convenient here to replace $\Delta_{\mathcal{A}_I}$ by the 2-local complex $\Delta_{\mathcal{H}_I}$ defined by the H_i for $i \in I$; we indicated earlier the isomorphism of these two complexes. We proceed at the level of the poset $\mathcal{R}$ of simplices of $\Delta_{\mathcal{H}_I}$; we map the standard simplices σ_J of the 31 types J to the corresponding members B_J representing the classes of $\mathcal{B}$. As usual this turns out to be a poset homotopy equivalence, obtained by making certain identifications in $\mathcal{R}$ and proceeding to the quotient poset $\bar{\mathcal{R}}$.

But this time, in comparison to the case of M_{24} in Section 8.5, there are more identifications to be made—since H_8 does not contain a full Sylow 2-subgroup S of G, but only a subgroup of index 2 in S. For simplicity of exposition, we will break up the above quotient process into two stages, beginning with those corresponding to S-orbits.

Thus we specify an initial set of identifications, for S-orbits of length 2 on simplices σ_J—where the type J properly contains 8; however, we except the type $\{8,1\}$, as well as 8 itself, from any identifications. Let $p\colon \mathcal{R} \to \bar{\mathcal{R}}'$ denote this first quotient map. In effect, we are further extending the method of Lemma 6.5.7, with distinguished index k given by 8. However, recall that in the extensions of that method for Fi_{22} and Fi_{23} in the previous sections, we could still identify a further singleton subset $J = \{j\}$ of $I \setminus \{k\}$, such that "most" of the identifications were confined to types K^+ containing k and j; by contrast this time we have identifications in almost all types containing $k = 8$.

This in turn means that there are more types K^+ for which we need a nontrivial argument, giving the contractibility of the preimages $p^{-1}(\bar{\mathcal{R}}'_{\leq \bar{r}})$ where $\bar{r} \in \bar{\mathcal{R}}'$ has type K^+; namely all K^+ properly containing 8, and other than $\{8,1\}$. But the argument here is essentially that from the proof of 6.5.7: namely for $K := K^+ \setminus \{8\}$, the preimage is the poset of simplices of the join $\sigma_K * \{H_8, H_{8'}\}$, where $H_{8'}$ is the other member of the S-orbit on the vertex $\{H_8\}$. This orbit is visible for example in $N_G(B_{11,6,8})$: indeed Fi_{23} is a subgroup of Fi'_{24}, and the configuration here is an extension of that indicated for $N_G(H_{11,7})$ in Section 7.17, where now an involution of $U_{11} \setminus H_8$ interchanges two conjugates of U_8 in $B_{11,8}$. And as always, the underlying reason for the acyclicity of these configurations is that the variation in the preimages is confined to a single vertex type, namely 8.

Then the second stage of identifications corresponds to certain exceptional types mentioned for M_{24} in Section 8.5: Namely the length-3 orbits that occurred for types $T\square$ and $ST\square$ in the case of M_{24} now imply similar length-3 orbits for types $11,3,8$ and $11,3,1,8$ in Fi'_{24}. Furthermore one checks, via the structure of the normalizers of the B_J, that these are the only length-3 orbits to arise.

So we now let $q\colon \bar{\mathcal{R}}' \to \bar{\mathcal{R}}$ denote the quotient map corresponding the the above identifications along orbits of length 3 for those two types J; and now $\bar{\mathcal{R}}$ has the poset structure of $\mathcal{B}_2^{cen}(G)$. Furthermore the preimage $q^{-1}(\bar{\mathcal{R}}_{\leq \bar{r}})$, for $\bar{r} \in \bar{\mathcal{R}}$ which has type set containing one of our possibilities for J, involves essentially one of the same set of contractible configurations that we examined for M_{24} in Section 8.5. (The simplex r to which we collapse might now have one dimension larger.) Once again we conclude that q is a poset homotopy equivalence using Quillen's homotopy fiber theorem [**104**, 1.6].

And now p followed by q gives the desired homotopy equivalence of $\mathcal{R}$ (and hence of $\Delta_{\mathcal{A}_I}$) with $\mathcal{B}_2^{cen}(G)$.

The poset $\mathcal{B}_2^{cen}(G)$ has reduced Euler characteristic

$$2^{18} \cdot 23 \cdot 79^2 \cdot 139 \cdot 23687$$

which is not divisible by $|G|_2 = 2^{21}$, and so by Remark 5.9.4 it is not in the standard ample G-homotopy type of $\mathcal{A}_2(G)$.

Main results for Fi_{24}'. The above discussion leads in the usual way to:

THEOREM 7.18.1. *For $G = Fi_{24}'$, the collection $\mathcal{B}_2^{cen}(G)$ and the intersection complex $\Delta_{\mathcal{A}_I}$ are ample and normalizer-sharp. In particular, we obtain from the* 31 *representatives in $\mathsf{N}(\Delta_{\mathcal{A}_I}/G) = \mathcal{H}_I$ a simplex cohomology decomposition as in* (5.8.6)*; and the map from the homotopy colimit of the pushout* 5*-cube diagram of Example 4.6.8 to BFi_{24}', followed by* 2*-completion, is a homotopy equivalence. (The non-simplex decomposition over $\mathcal{B}_2^{cen}(G)$ uses only* 29 *representatives.)*

PROOF. The argument for $\Delta_{\mathcal{A}_I}$ follows that in the case of HS in Theorem 7.10.2; of course now we are in a geometry of rank five rather than three. □

As in the earlier case of J_3 (and also the later case of Ly), we can alternatively obtain a decomposition over a smaller complex, by considering coincidences of spaces in the underlying homotopy colimit. To summarize: we can use the fact that $H_{6,8,11} = H_{8,11}$, which implies the further coincidences $H_{1,6,8,11} = H_{1,8,11}$ and $H_{3,6,8,11} = H_{3,8,11}$ and $H_{1,3,6,8,11} = H_{1,3,8,11}$, to make an argument similar to that for J_3 in Section 7.8, using the Cofinality Theorem 4.5.22, to show that the simplex category $\mathfrak{D}_5$ can be replaced by the subcategory obtained by removing these eight groups. The resulting category will be denoted $\mathfrak{D}_{4+4}$, and consists of two pushout hypercubes (4.6.7) identified along a three dimensional face. This is the opposite of the category associated to the poset of simplices in the simplicial complex which is obtained by identifying two copies of Δ^3 along a common face. We shall not attempt to draw this category; but the analogous case $\mathfrak{D}_{3+3}$ which is one dimension smaller is drawn later at (7.26.2). This approach gives:

THEOREM 7.18.2. *Consider the $\mathfrak{D}_{4+4}$-diagram of spaces given by omitting the eight subgroups H_J for $J \supseteq \{8, 11\}$ from the $\mathfrak{D}_5$-diagram of Theorem 7.18.1. Then the map from the homotopy colimit of this diagram to BFi_{24}', followed by* 2*-completion, is a homotopy equivalence. In particular this gives a cohomology decomposition over* 23 *of the* 31 *subgroups in Theorem 7.18.1.* □

7.19. The Harada–Norton group $HN = F_5$

For the Harada–Norton $G := HN$, we take as our index set $I := \{1, 3, 6\}$, corresponding to maximal 2-locals in the Atlas [**47**, p. 166]:

$H_1 \cong 2_+^{1+8}(\mathrm{Alt}_5 \wr 2)$,

$H_3 \cong 2^{3+2+6}(3 \times L_3(2))$, and

$H_6 \cong 2^6 \cdot \Omega_6^-(2)$.

As usual the H_i determine a subgroup simplex $\mathcal{H}_I$ as in Definition 4.6.2, and a 2-local complex $\Delta_{\mathcal{H}_I}$ as in Definition 6.3.5. Here H_6 does not contain a Sylow 2-subgroup of G; and the common intersection H_I of order 2^{12} is of index four in a Sylow group. As usual we set $U_i := O_2(H_i)$ for $i \in I$.

There does not seem to be any very natural diagram for the geometry $\Delta_{\mathcal{H}_I}$, probably due to the unusual wreath product in H_1.

We are not aware of any determination of $\mathcal{A}_2(G)$. We could approach the geometry as usual via certain members of $\mathcal{A}_2(G)$, a viewpoint we will only present in summary as we have not attempted to check all details: For $i \in I$ we set $A_i := \Omega_1 Z(O_2(H_i))$, and we obtain A_i of rank i in $\mathcal{A}_2(G)$ with $N_G(A_i) = H_i$. Via the usual method of 6.2.17 (2) we have $A_6 \cap A_3 =: A_2 > A_1$. So we see that the intersection complex $\Delta_{\mathcal{A}_I}$ defined by the $A_i (i \in I)$ is an internal almost-poset complex as in Definition 6.3.9. As usual one could adjust the proof below for flag-transitivity on $\Delta_{\mathcal{H}_I}$ to apply also to $\Delta_{\mathcal{A}_I}$; indeed that proof already investigates the links of the H_i in the language of the A_i. This would deliver the isomorphism of the complexes via the standard method of Lemma 6.5.4.

A rank two geometry for G is indicated in [**107**, p. 66], essentially as the 2-local complex using only H_1 and H_3. But as in several of our earlier cases such as Fi_{22} in Section 7.16, for our cohomological purposes it is natural to use an extension of that geometry; namely we add consideration of the further maximal 2-local subgroup H_6, obtaining the 2-local complex $\Delta_{\mathcal{H}_I}$.

Flag-transitivity of HN on $\Delta_{\mathcal{H}_I}$. We obtain flag-transitivity in the usual inductive way:

Observe that the link of H_3 is isomorphic (by intersecting A_3 with incident A_i for $i \in \{1, 6\}$) to the projective space $\mathbf{P}(A_3)$, so we may apply Remark 6.4.6 (2) to the $L_3(2)$ in H_3 to obtain flag-transitivity on this link.

And the link of H_6 is isomorphic (again by intersecting A_6 with incident A_i for $i \in \{1, 3\}$) with the projective points and totally singular lines of the usual polar space giving the building for the subgroup $\Omega_6^-(2)$ in H_6; so we may apply Remark 6.4.6 (3).

This gives flag-transitivity on the links of two of the three types, which suffices by Proposition 6.4.5 for the flag-transitivity of G. In particular, $\mathsf{N}(\Delta_{\mathcal{H}_I}/G) = \mathcal{H}_I$.

The relationship with $\mathcal{B}_2^{cen}(G)$. The poset $\mathcal{B}_2(G)$ of 2-radical subgroups of HN is determined in Yoshiara [**130**, Table 6, p. 262]; and independently in An and O'Brien [**17**, Table 2]. There are 13 conjugacy classes in $\mathcal{B}_2(G)$, of which eight are 2-centric.

Furthermore $\mathcal{B}_2^{cen}(G)$ is homotopy equivalent via a G-map to the subposet $\mathcal{B}_2^{cen-}$ determined by removing the G-class denoted by $R_5^{(1)}$ in [**130**, Table VI, p. 262]: This proceeds via the usual method of Lemma 6.5.1—but we postpone the details of this removal to Section 8.6. Consequently $\mathcal{B}_2^{cen-}$ is ample and normalizer-sharp by Proposition 5.11.2.

Next the seven remaining classes in $\mathcal{B}_2^{cen-}$ are in fact afforded by the products $B_J := \prod_{i \in J} U_i$ for nonempty $J \subseteq I$. Here is a guide for translating between our notation B_J and the notation in [**130**, Table VI, p. 262] (after the first five rows there, which correspond to non-centric classes). Our groups U_i for $i \in I = \{1, 3, 6\}$

are there denoted R_j for $j \in \{5,4,3\}$. Our B_J for $J = 16, 36, 13, 136$ are there denoted R_5^i, R_4^ℓ, $R_5^{(12)}$, $R_5^{(12i)}$. (Thus the only remaining member from the eight rows for $\mathcal{B}_2^{cen}(G)$ is $R_5^{(1)}$, which we removed above to get down to $\mathcal{B}_2^{cen-}$.)

Of course the B_J are in natural bijection with the members of the subgroup simplex $\mathcal{H}_I$ giving representatives of the G-classes in $\Delta_{\mathcal{H}_I}$. As in many previous cases, this bijection of G-orbits does not quite pull back to a poset isomorphism, from the poset $\mathcal{R}$ of simplices of $\Delta_{\mathcal{H}_I}$ to $\mathcal{B}_2^{cen-}$; instead there is a quotient relationship, determined by making a small number of identifications, which is a poset homotopy equivalence.

The procedure is again a variation on the method of Lemma 6.5.7; indeed it is slightly simpler than the variations we saw for Fi_{22} and Fi_{23} in Sections 7.16 and 7.17.

Namely we fix the distinguished vertex type $k := 6$, and distinguished subset $J := \{3\} \subseteq I \setminus \{6\}$. We consider the two possibilities for K with $J \subseteq K \subseteq I \setminus \{6\}$, namely $K = 3$ and 13; and make identifications for the two types $K^+ := K \cup \{6\}$, namely 36 and 136. For type 36, we identify the two members whose vertices of type 6 are the length-two orbit of the vertex $\{H_6\}$ under $N_G(H_I)$; this group has Sylow 2-subgroup of order 2^{13}, of index two in a Sylow 2-subgroup S of G. We can exhibit this orbit in $B_{3,6}$, which contains two conjugates of U_6 under U_3, where the conjugates intersect in a subgroup of rank 3. For type 136, we identify according to a length-four orbit under S. Here an element of $U_1 \setminus B_{3,6}$ in fact interchanges the previous $N_G(H_I)$-orbit of length two with a further conjugate. These identifications give the quotient $\bar{\mathcal{R}}$ the structure of $\mathcal{B}_2^{cen-}$.

Finally the required contractibility of the preimages $p^{-1}(\bar{\mathcal{R}}_{\leq \bar{r}})$ follows as in the earlier arguments for Fi_{22} and Fi_{23}, and indeed somewhat more simply—for this time we have no identifications outside the above types K^+ determined by J. Again it is relevant that the $N_G(H_I)$-orbits of length two are contained in the S-orbits of length four. And the contractibility then gives our poset homotopy equivalence via Quillen's homotopy fiber theorem [**104**, 1.6].

The poset $\mathcal{B}_2^{cen-}$ has reduced Euler characteristic $2^{10} \cdot 13 \cdot 677627$ which is not divisible by $|G|_2 = 2^{14}$, and so by Remark 5.9.4 it is not in the standard ample G-homotopy type represented by $\mathcal{B}_2(G)$.

Main results for HN. The above discussion leads in the usual way to the following theorem.

THEOREM 7.19.1. *For $G = HN$, the 2-local complex $\Delta_{\mathcal{H}_I}$ and the collection $\mathcal{B}_2^{cen-}$ are ample and normalizer-sharp. In particular, we obtain from the groups in* $\mathsf{N}(\Delta_{\mathcal{H}_I}/G) = \mathcal{H}_I$ *a simplex cohomology decomposition as in* (5.8.6)*; and the map from the homotopy colimit of the pushout cube diagram* (4.6.5) *to* $B(HN)$*, followed by* 2*-completion, is a homotopy equivalence.*

PROOF. The proof follows that for Co_1 in Theorem 7.15.1; except that we are in rank three rather than four. □

7.20. The Thompson group $Th = F_3$

For $G := Th$, we take as our index set $I := \{1, 5\}$, corresponding to maximal 2-locals in the Atlas [**47**, p. 177]:

$H_1 \cong 2_+^{1+8} \cdot \mathrm{Alt}_9$, and

$H_5 \cong 2^5 \cdot L_5(2)$.

As usual the H_i determine a subgroup simplex $\mathcal{H}_I$ as in Definition 4.6.2, and a 2-local complex $\Delta_{\mathcal{H}_I}$ as in Definition 6.3.5—which in this case is just a bipartite graph (and so no diagram is given for $\Delta_{\mathcal{H}_I}$). The common intersection $H_I = H_{1,5} \cong 2_+^{1+8} \cdot L_4(2)$ contains a Sylow 2-subgroup of G, and is of index nine in H_1 since $L_4(2) \cong \mathrm{Alt}_8 < \mathrm{Alt}_9$.

We are not aware of any explicit determination of $\Delta(\mathcal{A}_2(G))/G$. However, such a determination could be straightforwardly (if somewhat tediously) deduced, using a fact from [**116**, Sec. 2.14]: namely that an elementary abelian 2-subgroup of rank at least 2 is contained in a unique G-conjugate of at least one of the groups U_1 and U_5 defined below.

A rank two geometry Δ for G is given in [**107**, p. 65]; this arises essentially as $\Delta_{\mathcal{H}_I}$ in our present language. As usual we define $U_i := O_2(H_i)$ for $i \in I$, and set $U_{1,5} := U_1U_5$. It is shown in Smith and Yoshiara [**116**, Sec. 2.14] that $\Delta_{\mathcal{H}_I}$ is homotopy equivalent to $\mathcal{A}_2(G)$; and as usual that equivalence can be seen to be via a G-map.

Since $\Delta_{\mathcal{H}_I}$ is of rank two, the action of G is flag-transitive by Remark 6.4.6 (1). In particular, $\mathsf{N}(\Delta_{\mathcal{H}_I}/G) = \mathcal{H}_I$,

The poset $\mathcal{B}_2(G)$ of 2-radical subgroups of Th is determined in Yoshiara [**128**, Thm. 18]. There are 16 conjugacy classes of subgroups in $\mathcal{B}_2(G)$ (all are in fact 2-centric). However, the argument in [**116**, Sec. 2.14] shows that each contains a unique G-conjugate of a member of the subposet $\mathcal{B}_2^-$ given by the subgroups there denoted "A, E, $A^{(4)} = AE$"; in our notation, these are just the three subgroups U_J for nonempty $J \subseteq I$. This uniqueness gives the usual situation of Remark 6.5.2, and hence the hypothesis of Lemma 6.5.1, so we conclude that $\mathcal{B}_2^-$ is homotopy equivalent to $\mathcal{B}_2(G)$ via a G-map.

Now the normalizer set-map N induces an isomorphism from the vertices of $\Delta_{\mathcal{H}_I}$ to the minimal members of the poset $\mathcal{B}_2^-$. In fact we are in the trivial case of the setup of Lemma 6.5.7, in which no identifications are made (and in particular, the U_i are weakly closed). So we have a poset isomorphism from the poset of simplices of $\Delta_{\mathcal{H}_I}$ to $\mathcal{B}_2^-$. And $\mathcal{B}_2^-$ is ample and normalizer-sharp by Proposition 5.9.1.

This discussion leads as usual to the following theorem, which was first established in Benson [**26**].

THEOREM 7.20.1. *For $G = Th$, the 2-local complex $\Delta_{\mathcal{H}_I}$ and the collection $\mathcal{B}_2^-$ are ample and normalizer-sharp. In particular, we obtain from the groups in $\mathsf{N}(\Delta_{\mathcal{H}_I}/G) = \mathcal{H}_I$ a simplex cohomology decomposition as in* (5.8.6)*; and the obvious map of groups from the amalgamated free product $H_1 *_{H_{12}} H_2$ to Th induces a mod-2 cohomology isomorphism. The corresponding map from the homotopy colimit of the pushout square diagram* (4.5.6) *to BTh, followed by 2-completion, is a homotopy equivalence.*

PROOF. This follows from the above discussion and Lemma 4.5.10. □

Although the cohomology ring of Th has not been calculated, it is proved in Benson [**26**] that there is a polynomial subring $\mathbb{F}_2[x_{16}, x_{24}, x_{28}, x_{30}, x_{31}]$ (the subscripts denote the degrees of the generators) over which the cohomology ring $H^*(Th, \mathbb{F}_2)$ is finitely generated as a module, and which restricts to the Dickson invariants of rank 5 on each maximal elementary abelian 2-subgroup of Th (they all have rank 5).[14]

This should be compared with the discussion following Theorem 7.13.1 for the Conway group Co_3: in that case, the inclusion of the Dickson invariants of rank 4 is induced by a map of spaces $BCo_3 \to BDI(4)$, where the latter is the Dwyer–Wilkerson space indicated there. But there can be no full analogue of this for the case of rank 5 in this section: for it is proved in Lin and Williams [**83**] that there is no space with cohomology exactly equal to the Dickson invariants of rank 5.

7.21. The Baby Monster $B = F_2$

For $G := B$, we take as our index set $I := \{1, 2, 3, 5, 9\}$, corresponding to maximal 2-local subgroups in the Atlas [**47**, p. 217]:

$H_1 \cong 2_+^{1+22} \cdot Co_2$,
$H_2 \cong 2^{2+10+20}(M_{22}{:}2 \times \Sigma_3)$,
$H_3 \cong 2^3 2^{[32]}\big(\Sigma_5 \times L_3(2)\big)$,
$H_5 \cong 2^{5+5+10+10} L_5(2)$, and
$H_9 \cong 2^{9+16} Sp_8(2)$.

As usual the H_i determine a subgroup simplex $\mathcal{H}_I$ as in Definition 4.6.2, and a 2-local complex $\Delta_{\mathcal{H}_I}$ as in Definition 6.3.5. We mention that H_5 does not contain a Sylow 2-subgroup of G; and the common intersection H_I of order 2^{40} is of index two in a Sylow 2-group $N_G(H_I)$ of G. As always we set $U_i := O_2(H_i)$ for $i \in I$.

The usual diagram for $\Delta_{\mathcal{H}_I}$ is given in [**106**, p. 287] as

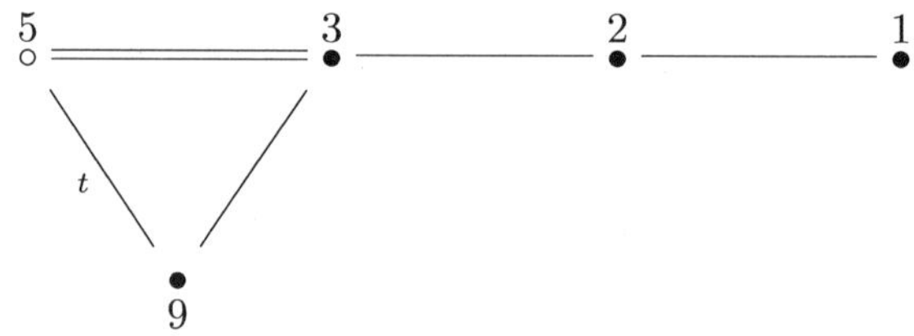

[14] The proof of Theorem 1.2 of [**26**] is incomplete as it stands, and it seems a good idea to set the record straight here. Namely, it is mistakenly stated in the proof of Lemma 3.4 and Proposition 3.5 that Th has only two conjugacy classes of maximal elementary abelian 2-subgroups. In fact it has three conjugacy classes, all of rank five, so it is necessary to deal with the third class.

A representative E of the third class still has the property that it is contained in a unique conjugate of the extraspecial group P of order 2^{1+8}. Its normaliser $N_{Th}(E)$ is of the form $2^{1+8}.2^3.L_3(2) = 2^5.2^{1+6}.L_3(2)$, and it stabilises a flag of subspaces of dimensions 1 and 4 in E. To complete the proof of Proposition 3.5, we must prove the analogue of Lemmas 3.2 and 3.3 for E. The invariant ring $H^*(E, \mathbb{F}_2)^{N_{Th}(E)}$ is a polynomial ring on generators a_1, a_8, a_{12}, a_{14} and a_{16} of degrees 1, 8, 12, 14, 16 respectively, where a_{16} is the Dickson invariant of this degree. The degree 16 terms in this ring are spanned by a_1^{16}, $a_1^8 a_8$, $a_1^4 a_{12}$, $a_1^2 a_{14}$, a_8^2 and a_{16}.

Now every isotropic 3-space in the 8-dimensional orthogonal space over $\mathbb{F}_2$ is contained in isotropic 4-spaces in both orbits of $O_8^+(2)$. We have already dealt with all the elementary abelian 2-subgroups of rank 5 in one of these orbits, in Lemmas 3.2 and 3.3. It follows that whatever the restriction of $x_{16} \in H^*(Th, \mathbb{F}_2)$ to E is, its further restriction to a subgroup of order 2^4 containing the centre of P has to be the same as the restriction of a_{16}. The element a_{16} is the only linear combination of the elements listed above that has this property. It follows that Lemma 3.4 remains true for the third class of maximal elementary abelian 2-subgroups, and this completes the proof of Proposition 3.5 and hence also of Theorem 1.2.

In particular the rank-four subdiagram in the link of 1 is that for Co_2 given in Section 7.14. Indeed our overall treatment will be parallel to that given for Co_2 in that section.

The standard 2-local geometry Δ for G in Ronan and Smith [**106**, p. 287] is essentially the 2-local complex $\Delta_{\mathcal{H}_I}$ in our present language. But see also Ivanov and Shpectorov, for example [**75**], for the viewpoint of Petersen geometries.

Some associated members of $\mathcal{A}_2(G)$. We are not aware of any determination of $\mathcal{A}_2(G)$. However, we now indicate certain particular members of $\mathcal{A}_2(G)$, which will be relevant for our further discussion of $\Delta_{\mathcal{H}_I}$.

In fact, we will even sketch a possible variant approach to a geometry for B, in terms of the corresponding subposet of $\mathcal{A}_2(G)$, for which we have not attempted to verify all details.

For $i \in I$, we set $A_i := \Omega_1 Z\big(O_2(H_i)\big)$. Then A_i is a member of $\mathcal{A}_2(G)$ of rank i, with $N_G(A_i) = H_i$. We may define the intersection complex $\Delta_{\mathcal{A}_I}$ using these A_i as in Definition 6.3.2.

Via the usual method of 6.2.17 (2) we obtain

$$A_1 < A_2 < A_3 < A_5 \cap A_9 =: A_4$$

where A_4 is of rank four; so we obtain an internal almost-poset complex as in Definition 6.3.9.

To connect this geometry with the more usual 2-local geometry for B, it should be routine to adjust the proof given below of flag-transitivity on $\Delta_{\mathcal{H}_I}$ to also apply to $\Delta_{\mathcal{A}_I}$; and then to conclude as usual by Lemma 6.5.4 that the normalizer set-map N extends to an isomorphism of complexes from the almost-poset complex $\Delta_{\mathcal{A}_I}$ to the 2-local complex $\Delta_{\mathcal{H}_I}$.

Flag-transitivity of B on $\Delta_{\mathcal{H}_I}$. The incidence in the 2-local complex $\Delta_{\mathcal{H}_I}$ can be analyzed via the method demonstrated for M_{24} in Remark 6.2.17 (1), especially as modified for M_{22} in Section 7.3. In summary, one just verifies that the links of vertices are the geometries indicated in the diagram.

We can then establish flag-transitivity of G on $\Delta_{\mathcal{H}_I}$ in the usual inductive way:

For $i = 2, 3, 5$, the link of H_i is isomorphic (as usual by intersecting A_i with incident A_j of type $j \neq i$) with a geometry for the group H_i/U_i—constructed as a product of one or more terms which are either buildings, or else flag-transitive geometries already considered (namely those for M_{22} and Σ_5). In these cases, we apply parts (3) or (4) of Remark 6.4.6 for flag-transitivity on those links.

The case $i = 9$ is covered by similar remarks—except that we must make a routine adjustment for the 2-split cover of the building of $Sp_8(2)$, just as we did for the case of $Sp_6(2)$ for Co_2 in Section 7.14.

The link of H_1 is isomorphic (by considering the images in U_1/A_1, of ranks $1, 2, 4, 7$, of the intersections with U_1 of A_i for $i \in \{2, 3, 5, 9\}$) to an external almost-poset complex, as in Definition 6.3.9, for the group Co_2 in H_1—arising from the 22-dimensional module for Co_2 which is a section of the Leech lattice modulo 2. Now flag-transitivity for Co_2 on its usual 2-local geometry was established in Section 7.14; as usual we can adjust that proof to apply also to the external almost-poset complex. So we may apply Remark 6.4.6 (4) for flag-transitivity on the link of type 1.

We have established the condition for all the types of links, which more than suffices by Proposition 6.4.5 to establish flag-transitivity for G. In particular, $\mathsf{N}(\Delta_{\mathcal{H}_I}/G) = \mathcal{H}_I$.

The relationship with $\mathcal{B}_2^{cen}(G)$. The poset $\mathcal{B}_2(G)$ of 2-radical subgroups of the Baby Monster is determined in Yoshiara [**131**, Table 1]. There are 35 conjugacy classes of subgroups in $\mathcal{B}_2(G)$, of which 31 are 2-centric.

Now the poset of simplices of $\Delta_{\mathcal{H}_I}$ has the $2^{|I|}-1 = 31$ conjugacy classes of subgroups H_J determined by the subgroup simplex $\mathcal{H}_I$. The 31 classes from $\mathcal{B}_2^{cen}(G)$ are in natural bijection with these: indeed $B_J := O_2\big(N_G(H_J)\big)$ is a member of $\mathcal{B}_2(G)$; and we may also exhibit B_J as $\prod_{i \in J} U_i$.

For convenience of comparison, we indicate the translation between our B_J for type sets J and the indices x, y in the notation V_y^x used in Yoshiara [**131**, Table 1]: Our $B_i = U_i$ for $i \in I = \{1,2,3,5,9\}$ is the V_y for y in the set $\{4,5,6,7,8\}$ there. The final eight of the subgroups following V_4 have superscript x beginning with H and followed by a subset of p, l, π there; these are for our J beginning with 19, and continuing with the corresponding subset of our $2,3,5$. The first seven subgroups following V_4 there are for our type subsets beginning with 1 and continuing with $2, 3, 32, 5, 53, 532, 52$ respectively. The seven V_5^x are for J beginning with 2, and continuing with a subset of $3,5,9$, corresponding to x from Q, H, O' there. The three subgroups V_6^x are for J beginning with 3, and continuing with our sets $5, 9, 59$, corresponding to x given by $1,2,3$ there. The subgroup $V_8^{(4)}$ is for $J = 95$.

In particular, H_I is of index 2 in a Sylow group $N_G(H_I)$, which is the 2-radical subgroup B_I corresponding to type I. As in the earlier case for Co_2 in Section 7.14, which involve the triples geometry for Σ_5 (Example 6.5.5), the above bijection of G-orbits does not quite pull back to an isomorphism of the poset $\mathcal{R}$ of simplices of $\Delta_{\mathcal{H}_I}$ with $\mathcal{B}_2^{cen}(G)$; instead there is a quotient relationship from identifying certain pairs, which gives a homotopy equivalence of posets. This comes from an application of Lemma 6.5.7, and the overall setup is much like that for Co_2; again we summarize the details fairly briefly: We take $k := 5$ as our distinguished type, and $J := \{9\}$ as our distinguished subset of $I \setminus \{5\}$. Hence we must verify (6.5.8) for the types K given by the eight over-sets of 9 in $I \setminus \{5\}$. So we check that the Sylow 2-subgroup $N_G(H_I)$ does indeed lie in each such H_K. Then we turn to considering the types $K^+ = K \cup \{5\}$; where we check that $N_G(H_I)$ does indeed supplement H_{K^+} in $N_G(H_{K^+})$.[15] As a sample of these latter verifications, note that $B_{5,9}$ contains a subgroup 2^{4+2} in which two conjugates of U_5 are interchanged by an involution from U_9. The cases for larger types K^+ are subconfigurations visible inside $N_G(B_{5,9})$. So Lemma 6.5.7 gives the poset homotopy equivalence of $\mathcal{R}$ with the quotient $\bar{\mathcal{R}}$ determined by the identification of orbits of length 2 for types K^+.

Finally we combine the poset homotopy equivalence from $\mathcal{R}$ to $\bar{\mathcal{R}}$ with the observation that $\bar{\mathcal{R}}$ just reproduces the poset $\mathcal{B}_2^{cen}(G)$.

The poset $\mathcal{B}_2^{cen}(G)$ has reduced Euler characteristic

$$2^{38} \cdot 23 \cdot 157 \cdot 389 \cdot 599$$

[15]These follow from the structure of the normalizers in [**131**, Table 1]—modulo correcting the following slight misprints there: The values 2^3 and 2^4 in the entries for $V_5^{(Q)}$ and $V_5^{(O')}$ need to be interchanged. And also—following up the misprint we had mentioned for Co_2 in Section 7.14—the values 2^2 following 2^5 in the entries for $V_4^{(H,pl)}$ and $V_4^{(H,p\pi)}$ need to be changed to 2^3.

which is not divisible by $|G|_2 = 2^{41}$, and so by Remark 5.9.4, $\mathcal{B}_2^{cen}(G)$ is not in the standard ample G-homotopy type represented by $\mathcal{B}_2(G)$.

Main results for B. This discussion leads as usual to the following theorem.

THEOREM 7.21.1. *For $G = B = F_2$, The 2-local complex $\Delta_{\mathcal{H}_I}$ and the collection $\mathcal{B}_2^{cen}(G)$ are ample and normalizer-sharp. In particular, we obtain from the groups in $\mathsf{N}(\Delta_{\mathcal{H}_I}/G) = \mathcal{H}_I$ a simplex cohomology decomposition as in (5.8.6); and the map from the homotopy colimit of the pushout 5-cube diagram of Example 4.6.8 to BF_2, followed by 2-completion, is a homotopy equivalence.*

PROOF. The argument follows that for Fi_{22} in Theorem 7.16.1; of course now we are in rank five rather than rank four. □

7.22. The Fischer–Griess Monster $M = F_1$

For $G := M$, we take as our index set $I := \{1, 2, 3, 5, 10\}$, corresponding to maximal 2-local subgroups in the Atlas [**47**, p. 234]:

$H_1 \cong 2_+^{1+24} \cdot Co_1$,
$H_2 \cong 2^{2+11+22}(M_{24} \times \Sigma_3)$,
$H_3 \cong 2^{3+6+12+18}\big(3 \cdot \Sigma_6 \times L_3(2)\big)$,
$H_5 \cong 2^{5+10+20}\big(\Sigma_3 \times L_5(2)\big)$, and
$H_{10} \cong 2^{10+16} \cdot \Omega_{10}^+(2)$.

As usual the H_i determine a subgroup simplex $\mathcal{H}_I$ as in Definition 4.6.2, and a 2-local complex $\Delta_{\mathcal{H}_I}$ as in Definition 6.3.5. The common intersection H_I of order $2^{46}3$ contains a Sylow 2-subgroup of G with index three. As always we set $U_i := O_2(H_i)$ for $i \in I$.

The usual diagram for $\Delta_{\mathcal{H}_I}$ is given in [**106**, p. 285] as

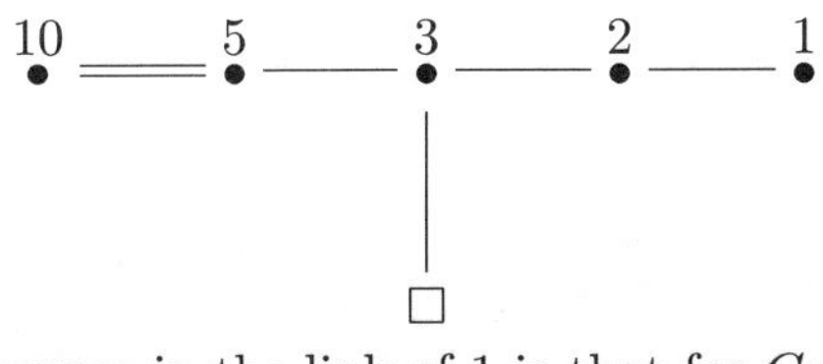

In particular the subdiagram in the link of 1 is that for Co_1 given in Section 7.15. Indeed our overall treatment will be parallel to that given for Co_1 in that section.

We are not aware of any determination of $\mathcal{A}_2(G)$. But as for Co_1, again we can approach the geometry via members of $\mathcal{A}_2(G)$: namely the intersection complex $\Delta_{\mathcal{A}_I}$ defined by the $A_i (i \in I)$ given by $A_i := \Omega_1 Z\big(O_2(H_i)\big)$, with A_i of rank i and $N_G(A_i) = H_i$. Via the usual method of 6.2.17 (2) we get

$$A_1 < A_2 < A_3 < A_5 < A_{10}.$$

We see then that the intersection complex $\Delta_{\mathcal{A}_I}$ is an internal poset complex as in Proposition 6.3.6.

Flag-transitivity of G on $\Delta_{\mathcal{H}_I}$ follows closely the case of B in Section 7.21: The incidence in the 2-local complex $\Delta_{\mathcal{H}_I}$ can be analyzed via the method demonstrated for M_{24} in Remark 6.2.17 (1). In summary, one just verifies that the links of vertices are the geometries indicated in the diagram. Then we can consider the various individual links. For $i = 2, 3, 5, 10$, the link of H_i is isomorphic (as usual by intersecting A_i with incident A_j of type $j \neq i$) with a geometry for the group H_i/U_i, which is a product of one or more terms which are either suitable truncations of buildings, or flag-transitive geometries considered earlier such as that for M_{24}. In particular, the link of H_{10} is a truncation of the usual polar space for $\Omega_{10}^+(2)$ in H_{10}, given by ignoring one class of 5-dimensional totally singular subspaces. In these cases, we apply parts (3) or (4) of Remark 6.4.6 for flag-transitivity on those links. The link of H_1 is isomorphic (by considering the images in U_1/A_1, of ranks $1, 2, 4, 8$, of the intersections with U_1 of A_i for $i \in \{2, 3, 5, 10\}$) to an external poset complex as in Proposition 6.3.6 for Co_1, arising from its 24-dimensional module in the Leech lattice modulo two. Recall that flag-transitivity for Co_1 on its usual 2-local geometry was established in Section 7.15, so adjusting that proof to also apply to the external poset complex, we may apply Remark 6.4.6 (4) for flag-transitivity on the link of type 1. We have established the condition for all the types of links, which more than suffices by Proposition 6.4.5 to give flag-transitivity for M. In particular, $\mathsf{N}(\Delta_{\mathcal{H}_I}/G) = \mathcal{H}_I$.

As usual the above argument for flag-transitivity depends essentially only on the links in the diagram, and so is easily adapted to apply to $\Delta_{\mathcal{A}_I}$ also. So by the standard Lemma 6.5.4, we see that the normalizer set-map N extends to an isomorphism of complexes from the external poset complex $\Delta_{\mathcal{A}_I}$ to the 2-local complex $\Delta_{\mathcal{H}_I}$.

The standard 2-local geometry Δ for G in Ronan and Smith [**106**, p. 285] is essentially the 2-local complex $\Delta_{\mathcal{H}_I}$ in our present language. But see also Ivanov and Shpectorov, for example [**75**], for the viewpoint of tilde geometries.

The poset $\mathcal{B}_2(G)$ of 2-radical subgroups of M is determined in Yoshiara [**131**, Table 2]. There are 63 conjugacy classes of subgroups in $\mathcal{B}_2(G)$, of which 58 are 2-centric. These remaining 58 classes all have the property that their normalizers contain a Sylow 2-subgroup. In fact these normalizers all lie in one of H_i for $i \in I$ above. Furthermore in [**112**, Sec. 7], Sawabe applies his equivalence method to M to show that $\mathcal{B}_2^{cen}(G)$ is homotopy equivalent to the subposet given by the members of $\mathcal{O}_2(\Delta_{\mathcal{H}_I})$—which has 31 conjugacy classes, represented by the $B_J := \prod_{i \in J} U_i$ for nonempty $J \subseteq I$. As usual the equivalence can be seen to be via a G-map. So following Notation 5.11.4 we will write $\mathcal{B}_2^{cen-}$ for the poset $\mathcal{O}_2(\Delta_{\mathcal{H}_I})$. Furthermore we recover the poset of simplices of $\Delta_{\mathcal{H}_I}$ as $\mathsf{N}(\mathcal{B}_2^{cen-})$. (As was the case for M_{24} in Section 7.5, here we have $N_G(H_J) = H_J$ for all types J, in effect giving the vacuous case of the hypotheses Lemma 6.5.7, from which no actual identifications are made.) So both $\mathcal{B}_2^{cen-}$ and $\Delta_{\mathcal{H}_I}$ are ample and normalizer-sharp by Proposition 5.11.2.

For completeness, we now provide a more detailed guide for translating between our notation B_J and the notation V_y^x in [**131**, Table 2]—namely we indicate which indices x, y there give members of $\mathcal{B}_2^{cen-}$. Our $B_i = U_i$ for $i \in I = \{1, 2, 3, 5, 10\}$ is the group V_y for y in the set given by $\{3, 4, 5, 6, 7\}$ there; so we count subgroups according to the value of y. For $y = 7$ corresponding to our 10, $\mathcal{B}_2^{cen-}$ includes the two subgroups V_y^x indicated there—namely it also includes $V_7^{(5.1)}$; these correspond to B_J for our J given by $\{10\}$ and $\{10, 5\}$. For $y = 6$ corresponding to our 5, we include the two V_6^x for x not involving 1; the further superscript 4 there corresponds to our 10. For $y = 5$ corresponding to our 3, we include the four groups V_5^x there for for x not containing $*$; those x are based on O, T corresponding to our $10, 5$. For $y = 4$ corresponding to our 2, we similarly include the eight groups V_4^x there for x

not containing $*$; those x are based on O, T, S corresponding to our $10, 5, 3$. Finally for $y = 3$ corresponding to our 1, we consider the subgroups denoted by V_3^x: The superscript x there uses notation for the 2-radical subgroups of Co_1, defined by Sawabe in [**109**, Table I, p. 132], as we mentioned in Section 7.15; though in [**131**, Table 2], the $Q_1.N_i$ and $R.P_i$ of Sawabe appear instead in the form Q_1^w and R^w for specific superscripts w. Translating our discussion in Section 7.15 for Co_1 into the present indexing set $I = \{1, 2, 3, 5, 10\}$: We use first five groups beginning with V_3, and also the three Q_1^w for w not involving 3—hence corresponding to our present indices $2, 3, 5$. And we use the seven R^w for w not involving 3—hence corresponding to our present indices $3, 5$. (We also exclude V_3V, V_3V2 and V_3F.) This process then reproduces the $5 + 3 + 7 = 15$ members of $\mathcal{B}_2^{cen-}$ for Co_1 in Section 7.15. So now considering all y, we have accounted for the $2 + 2 + 4 + 8 + 15 = 31$ subgroups of $\mathcal{B}_2^{cen-}$.

The poset $\mathcal{B}_2^{cen-}$ has reduced Euler characteristic

$$2^{42} \cdot 23 \cdot 5179 \cdot 318245027 \cdot 1936981015559,$$

which is not divisible by $|G|_2 = 2^{46}$, and so by Remark 5.9.4 it is not in the standard ample G-homotopy type represented by $\mathcal{B}_2(G)$.

This discussion leads as usual to the following theorem.

THEOREM 7.22.1. *For the Monster $G = F_1 = M$, the 2-local complex $\Delta_{\mathcal{H}_I}$ and the collection $\mathcal{B}_2^{cen-}$ are ample and normalizer-sharp. In particular, we obtain from the groups in $\mathsf{N}(\Delta_{\mathcal{H}_I}/G) = \mathcal{H}_I$ a simplex cohomology decomposition as in (5.8.6); and the map from the homotopy colimit of the pushout 5-cube diagram of Example 4.6.8 to BF_1, followed by 2-completion, is a homotopy equivalence.*

PROOF. The argument follows that for Co_1 in Theorem 7.15.1; of course now we are in rank five rather than rank four. $\square$

We mention that the complex $\mathcal{B}_2^{cen-}$ of dimension four provides a smaller decomposition than $\mathcal{B}_2^{cen}(G)$, which would require dimension five.

7.23. The Held group He

For $G := He$, we take as our index set $I := \{1, 6, 6'\}$, corresponding to maximal 2-locals in the Atlas [**47**, p. 104]:

$H_1 \cong 2_+^{1+6} L_3(2)$,

$H_6 \cong 2^6 : 3^{\cdot}\Sigma_6$, and

$H_{6'} \cong 2^6 : 3^{\cdot}\Sigma_6$.

As usual the H_i determine a subgroup simplex $\mathcal{H}_I$ as in Definition 4.6.2, and a 2-local complex $\Delta_{\mathcal{H}_I}$ as in Definition 6.3.5. The subgroups H_6 and $H_{6'}$ are conjugate by an outer automorphism of G. The common intersection $H_I = N_G(H_I)$ of order 2^{10} is a Sylow 2-subgroup of G. As always we set $U_i := O_2(H_i)$ for $i \in I$.

The standard diagram for $\Delta_{\mathcal{H}_I}$ is given in [**106**, p. 288] as

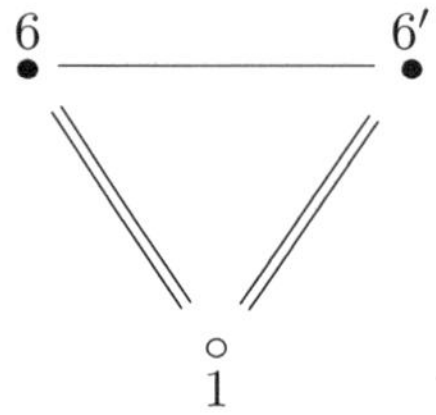

In this case, the hollow node $\circ$ at 1 indicates "a 3-split cover" in the sense of that paper.

We are not aware of any determination of $\mathcal{A}_2(G)$. However, we can approach the geometry via members of $\mathcal{A}_2(G)$: For $i \in I$, as usual we set $A_i := \Omega_1 Z\big(O_2(H_i)\big)$, obtaining A_i of rank i in $\mathcal{A}_2(G)$, with $N_G(A_i) = H_i$. Via the usual method of Remark 6.2.17 (2) we have $A_1 < A_6 \cap A_{6'} =: A_4$, where A_4 is of rank four. Thus we have an internal almost-poset complex $\Delta_{\mathcal{A}_I}$ as in Definition 6.3.9.

The standard 2-local geometry Δ for G in Ronan and Smith [**106**, p. 288] is essentially the 2-local complex $\Delta_{\mathcal{H}_I}$ in our present language.

We obtain flag-transitivity of G on $\Delta_{\mathcal{H}_I}$ in the usual inductive way: The links of H_i for $i = 6, 6'$ are isomorphic (via the intersection of A_i with incident vertices of type A_j for $j \neq i$) with the geometry of 45 vectors and 15 special $\mathbb{F}_4$ 2-spaces in A_6 regarded as a 3-dimensional faithful $\mathbb{F}_4$-module for $H_6/U_6 \cong 3\Sigma_6$. This geometry is a 3-split cover of the polar space of 15 points and 15 isotropic lines for $Sp_4(2) \cong \Sigma_6$. And as usual (cf. the 2-split cover appearing for M_{22} in Section 7.3), we simply extend from flag-transitivity of $Sp_4(2)$ on that polar space, by observing that the normal subgroup of order three in H_6/U_6 is in the stabilizer of a point-line flag, and acts transitively on the three vectors corresponding to that point. We have established the condition on links of the two of the three types, which suffices by Proposition 6.4.5 for the flag-transitivity of G. In particular, $\mathsf{N}(\Delta_{\mathcal{H}_I}/G) = \mathcal{H}_I$.

As usual the above argument for flag-transitivity on on $\Delta_{\mathcal{H}_I}$ involves essentially just the geometry of the links indicated by the diagram; and so we can adjust to obtain flag-transitivity also on $\Delta_{\mathcal{A}_I}$ (indeed the proof above is already partly phrased in that language). So by the standard Lemma 6.5.4 we see that the normalizer set-map N extends to an isomorphism of complexes from the almost-poset complex $\Delta_{\mathcal{A}_I}$ to the 2-local complex $\Delta_{\mathcal{H}_I}$.

The poset $\mathcal{B}_2(G)$ of 2-radical subgroups of He is determined in An [**11**, 2D] There are 14 conjugacy classes of subgroups in $\mathcal{B}_2(G)$, of which 11 are 2-centric. However, Sawabe in [**111**, 6.6] shows that $\mathcal{B}_2^{cen}(G)$ is homotopy equivalent with the proper subposet $\mathcal{B}_2^{cen-}$ given by the conjugates of the seven subgroups $B_J := \prod_{j \in J} U_i$ for nonempty $J \subseteq I$.

Furthermore we recover the poset of simplices of $\Delta_{\mathcal{H}_I}$ as $\mathsf{N}(\mathcal{B}_2^{cen-})$. (As was the case for M_{24} in Section 7.5, here we have $N_G(H_J) = H_J$ for all types J, in effect giving the vacuous case of the hypotheses Lemma 6.5.7, from which no actual identifications are made.) So both $\mathcal{B}_2^{cen-}$ and $\Delta_{\mathcal{H}_I}$ are ample and normalizer-sharp by Proposition 5.11.2.

Here is a guide for translating between our notation B_J and the subgroups in [**11**, 2D] which lie in the subposet $\mathcal{B}_2^{cen-}$: Our H_i for $i = 1, 6, 6'$ correspond to the subgroups denoted 2_+^{1+6}, 2^6, $(2^6)^*$ there. Then the B_J for our $J = 16, 16', 66', 166'$ correspond to the subgroups $2_+^{1+6}.2^2$,$(2_+^{1+6}.2^2)^*$, $2^4.2^4$, and S.

The poset $\mathcal{B}_2^{cen-}$ has reduced Euler characteristic $2^7 \cdot 8753$ which is not divisible by $|G|_2 = 2^{10}$, and so by Remark 5.9.4 it is not in the standard ample G-homotopy type represented by $\mathcal{B}_2(G)$.

This discussion leads as usual to the following theorem.

THEOREM 7.23.1. *For $G = He$, the 2-local complex $\Delta_{\mathcal{H}_I}$ and the collection $\mathcal{B}_2^{cen-}$ are ample and normalizer-sharp. In particular, we obtain from the groups in $\mathsf{N}(\Delta_{\mathcal{H}_I}/G) = \mathcal{H}_I$ a simplex cohomology decomposition as in* (5.8.6)*; and the map from the homotopy colimit of the pushout cube diagram* (4.6.5) *to BHe, followed by 2-completion, is a homotopy equivalence.*

PROOF. The proof follows that for HN in Theorem 7.19.1. □

7.24. The Rudvalis group Ru

For $G := Ru$, we take as our index set $I := \{1, 3, 6\}$, corresponding to maximal 2-locals in the Atlas [**47**, p. 126]:

$H_1 \cong (2 \cdot 2^{4+6}) : \Sigma_5$,

$H_3 \cong 2^{3+8} : L_3(2)$, and

$H_6 \cong 2^6 : G_2(2)$.

As usual the H_i determine a subgroup simplex $\mathcal{H}_I$ as in Definition 4.6.2, and a 2-local complex $\Delta_{\mathcal{H}_I}$ as in Definition 6.3.5. Note that H_6 does not contain a Sylow 2-subgroup of G; the common intersection H_I of order 2^{12} is of index four in a Sylow group. As always we set $U_i := O_2(H_i)$.

The usual diagram for $\Delta_{\mathcal{H}_I}$ is given in [**106**, p. 288] as

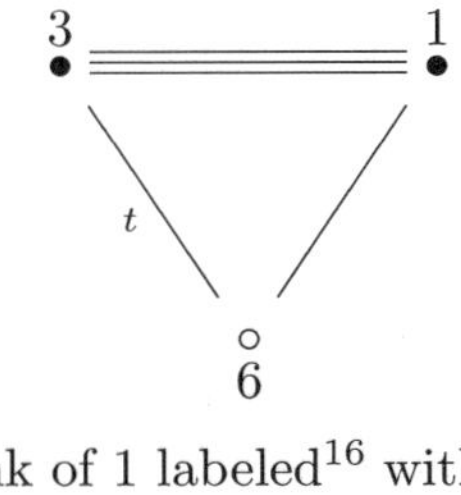

Here the subdiagram in the link of 1 labeled[16] with t is a 2-split cover in the sense of that paper of the triples geometry for the Σ_5 in H_1 (see Example 6.5.5); we saw such a cover in the geometry for $U_4(2) : 2$, involved in the geometry for Fi_{22} as discussed in Section 8.4.

The standard 2-local geometry for G as in Ronan and Smith [**106**, p. 288] arises essentially as the 2-local complex $\Delta_{\mathcal{H}_I}$ in our present language.

Our treatment in this section will be somewhat similar to that for Suz in Section 7.12, in that we indicate two approaches: First we will derive a further subposet $\mathcal{E}_2^-$ of the subposet $\mathcal{E}_2(G)$ of $\mathcal{A}_2(G)$, which is homotopy equivalent to $\Delta_{\mathcal{A}_I}$ above via a G-map; however the corresponding decomposition will not be over the subgroup simplex $\mathcal{H}_I$, but over a closely related poset. After that derivation, we

[16]Here we deviate from the use of $2t$ as the label in [**106**, p. 288]; our use of t is for consistency with some other occurrences, such as in the larger subdiagram for M_{22} in the geometry for Fi_{22} in Section 7.16.

will indicate an easier and perhaps more natural approach via $\mathcal{B}_2^{cen}(G)$, which does give a decomposition over $\mathcal{H}_I$.

An approach via $\mathcal{A}_2(G)$. We are not aware of any treatment of $\mathcal{A}_2(G)$. However, we can obtain one decomposition using a geometry based on members of $\mathcal{A}_2(G)$.

Namely we use the intersection complex $\Delta_{\mathcal{A}_I}$ defined by the $A_i (i \in I)$ given by $A_i := \Omega_1 Z\big(O_2(H_i)\big)$, with A_i of rank i and $N_G(A_i) = H_i$, Via the usual method of 6.2.17 (2) we have $A_1 < A_3 \cap A_6 =: A_2$, where A_2 is of rank 2. So $\Delta_{\mathcal{A}_I}$ is an internal almost-poset complex as in Definition 6.3.9.

The poset $\mathcal{E}_2^-$. We temporarily set aside our complex $\Delta_{\mathcal{A}_I}$, in order to derive a certain related poset $\mathcal{E}_2^-$. This time the treatment is sufficiently complicated that we will only outline it here; we will postpone somewhat fuller details to Section 8.7.

There are two conjugacy classes of involutions in G, denoted by $2A$ and $2B$ in the Atlas [**47**]; of these, the involutions in the center of a Sylow 2-subgroup lie in class $2A$. As usual from the character table of Ru (for example using the ClassMultiplicationCoefficient function in GAP), we can compute the value

$$\xi(2A, 2A, 2B) = 0$$

for the structure constant defined in (5.10.5). Hence the collection $\mathcal{E}_2(G)$ in Definition 5.10.1 is given by the elementary abelian 2-subgroups whose nontrivial elements are in class $2A$; and the collection is ample and normalizer-sharp by Proposition 5.10.2.

We will indicate at least some details here rather than later in Section 8.7; one reference for many of the following properties of Ru is Chapter J of Aschbacher and Smith [**20**].

Since our chosen A_1 is in the center of a Sylow group, it is generated by a member of class $2A$, and hence A_1 is in the subposet $\mathcal{E}_2(G)$; and we have

$$H_1 = N_G(A_1) = C_G(A_1) \cong 2^{1+4+2+4}{:}\Sigma_5$$

using the ascending composition series for H_1 on $O_2(H_1)$ given in [**20**, J.2.2]. We have also chosen subgroups A_i above for $i = 3, 6$; and since $A_1 < A_3 \cap A_6$, from the transitivity on involutions of A_i of a subgroup $L_3(2)$ or $G_2(2)$ in H_i, we see that those involutions also lie in class $2A$, so that A_3 and A_6 are in the subposet $\mathcal{E}_2(G)$. Now we may choose $A_3 \leq U_1$. Then the normal closure V_1 of A_3 in H_1 is of rank five, and is a subgroup of U_1 lying in $\mathcal{E}_2(G)$; that is, V_1 is also a pure-$2A$ subgroup. Next for any subgroup X (such as A_2) of rank two in A_3, we find that A_3 is the only G-conjugate of itself above X. However, we cannot choose A_3 to lie in A_6; instead we have chosen A_2 of rank two as the particular intersection $A_3 \cap A_6$. The conjugates of A_2 in A_6 are totally isotropic 2-subspaces of A_6 in the natural symplectic form for $G_2(2) < Sp_6(2)$; indeed they are lines of the generalized hexagon which gives the building for $G_2(2)$.

We let $\mathcal{E}_2^-$ denote the subposet of $\mathcal{E}_2(G)$ given by the G-conjugates of A_1, A_2, A_3, and A_6. Now $\mathcal{E}_2(G)$ is ample and normalizer-sharp; and we postpone to Section 8.7 the details of a deduction that $\mathcal{E}_2^-$ is also ample and normalizer-sharp, using a variant of our usual methods for homotopy equivalence via a G-map.

The poset $\mathcal{E}_2^-$ has reduced Euler characteristic $2^{12} \cdot 3 \cdot 823$ which is not divisible by $|G|_2 = 2^{14}$, and so by Remark 5.9.4, $\mathcal{E}_2^-$ is not in the standard ample G-homotopy type represented by $\mathcal{A}_2(G)$.

Now A_3 is contained in no G-conjugate of A_6; indeed we saw that conjugates of A_3 and A_6 can intersect at most in a conjugate of A_2. As a result, we obtain two G-classes of maximal chains in $\mathcal{E}_2^-$, represented by $A_1 < A_2 < A_3$ and $A_1 < A_2 < A_6$. Consequently the homotopy decomposition obtained from the ample collection $\mathcal{E}_2^-$ is not indexed by a pushout hypercube diagram; that is, it is not naturally indexed by the simplex $\mathcal{H}_I$ of subgroups.

Relating $\mathcal{E}_2^-$ to $\Delta_{\mathcal{A}_I}$. But we will see that the decomposition over $\Delta(\mathcal{E}_2^-)$ is equivalent to a decomposition which *is* naturally indexed by the subgroup simplex $\mathcal{H}_I$: namely we will show that the collection $\mathcal{E}_2^-$ is homotopy equivalent via a G-map to the almost-poset complex $\Delta_{\mathcal{A}_I}$ indicated above. This will follow using the method of Lemma 6.5.3, with $\mathcal{E}_2^-$ in the role of "$\mathcal{E}$": In particular, note that $\mathcal{E}_2^-$ contains the intersection $A_2 = A_3 \cap A_6$. Indeed in the notation of that result, the intersections "E_σ" for σ of types 13 or 16 or 136 are conjugates of A_1, while those of type 36 are conjugates of A_2. We must show for each $E \in \mathcal{E}_2^-$ that the subcomplex

$$L_E := \{\sigma \in \Delta_{\mathcal{A}_I} : E \leq E_\sigma\}$$

is contractible via a G_E-map. For $E = A_3$ or A_6, we see that L_E is the single vertex E, so we have the usual situation of Remark 6.5.2 giving contractibility via a G_E-map. For $E = A_1$, the subcomplex L_{A_1} is the cone given by the join of the vertex A_1 with the subcomplex of vertices of types A_3 and A_6 above A_1, and again we invoke Remark 6.5.2. For $E = A_2$, the subcomplex L_{A_2} is the cone given by the join of A_3 (the unique vertex of that type above A_2) with the vertices of type A_6 above A_2, so again Remark 6.5.2 applies for contractibility. via a G_{A_2}-map. This establishes the hypotheses needed for Lemma 6.5.3 to give the homotopy equivalence via a G-map. Thus $\Delta_{\mathcal{A}_I}$ is also ample and normalizer-sharp by Proposition 5.10.2.

We obtain flag-transitivity of G on $\Delta_{\mathcal{A}_I}$ by the usual inductive method: The link of H_6 is isomorphic, by intersection with A_6 of incident conjugates of A_j for $j \neq 6$, to the usual building for the subgroup $G_2(2)$ in H_6, namely the generalized hexagon for $G_2(2)$. So we can apply part (3) of Remark 6.4.6 to obtain flag-transitivity on those links. Similar remarks apply to the link of H_3—though we must extend from flag-transitivity of the $L_3(2)$ in H_3 or the projective space $\mathbf{P}(A_3)$, to account for the 4-split cover that occurs here; this proceeds much as for the 2-split cover in the geometry for M_{22} in Section 7.3 (or the 4-split covers in the geometry for $U_4(2):2$ in Section 8.4). So we have established the condition for links of two of the three types, which suffices by Proposition 6.4.5 for flag-transitivity of G. In particular, $\mathsf{N}(\Delta_{\mathcal{A}_I}/G) = \mathcal{H}_I$.

As usual, it is straightforward to adjust the above flag-transitivity proof on $\Delta_{\mathcal{A}_I}$ to also apply to $\Delta_{\mathcal{H}_I}$; and so as by the standard Lemma 6.5.4 we see that the normalizer set-map N extends to an isomorphism of complexes from $\Delta_{\mathcal{A}_I}$ to the 2-local complex $\Delta_{\mathcal{H}_I}$.

An approach via $\mathcal{B}_2(G)$. As indicated in our earlier remark, we now turn to an easier approach to the 2-local geometry of Ru, working via $\mathcal{B}_2(G)$.

The poset $\mathcal{B}_2(G)$ of nontrivial 2-radical subgroups for Ru is determined in An and O'Brien [**16**, Table 6], and also in Sawabe [**110**, Table 1, p. 414]. There are eight conjugacy classes, of which seven are 2-centric.

Indeed representatives of the classes in $\mathcal{B}_2^{cen}(G)$ are given by the seven subgroups $B_J := \prod_{i \in J} U_i$ for nonempty $J \subseteq I$. Here is a guide for translating between our notation B_J and that for the seven subgroups in [**110**, Table 1, p. 414] which lie in $\mathcal{B}_2^{cen}(G)$: Our $B_i = U_i$ for $i \in I = \{1,3,6\}$ are given by J, Q, X there. Then our $J = 13, 16, 36, 136$ correspond to $J : 2^2$, $J : 2$, $Q : 2^2$, $J : D_8$ there.

As in earlier cases, the above bijection between G-orbits of vertices of $\Delta_{\mathcal{H}_I}$ and minimal members of $\mathcal{B}_2^{cen}(G)$ does not quite pull back to a poset isomorphism between that poset $\mathcal{R}$ of simplices of $\Delta_{\mathcal{H}_I}$ and $\mathcal{B}_2^{cen}(G)$; instead there is a quotient relationship determined by making a small number of identifications in $\mathcal{R}$, which is a poset homotopy equivalence.

Again we have a variation on the method of Lemma 6.5.7; this is very similar to that appearing in the geometry for $U_4(2){:}2$ in Section 8.4.

As there, we have H_I of index two in $N_G(H_I)$, which in turn is of index two in a Sylow 2-subgroup S of G; where S gives the class B_I in $\mathcal{B}_2^{cen}(G)$ corresponding to type I.

So we parallel the setup for 6.5.7: we specify a distinguished type $k = 6$ and a distinguished subset $J = \{3\}$ of $I \setminus \{6\}$. Then for $K = \{3\}$ or its unique over-set $\{3,1\}$ in $I \setminus \{6\}$, we have that S lies in H_K. For $K^+ = \{6,3\}$ and $\{6,3,1\}$ (that is, I), we make identifications corresponding to a full S-orbit of length 4. We then continue beyond the setup of 6.5.7: for the further type $\{6,1\}$ which is not of the form K^+, we make identifications corresponding to an $N_G(H_I)$-orbit of length 2. These are the identifications which give the quotient $\bar{\mathcal{R}}$ the poset structure of $\mathcal{B}_2(G)$. (The orbit of length 2 can be seen in $B_{6,1} = U_6U_1$, in which U_1 interchanges two conjugates of U_6; the full S-orbit of length 4 is visible in $B_{6,1,3} = S$, in which an involution in U_3 interchanges two of the above orbits of length 2.)

The verification of contractibility of the preimage $p^{-1}(\bar{\mathcal{R}}_{\leq \bar{r}})$, by considering $\bar{r}$ of each possible type in turn, essentially just follows as in the case of $U_4(2) : 2$ in Section 8.4. Again it is relevant that the $N_G(H_I)$-orbit is contained in the S-orbit on the vertex $\{H_6\}$.

These extensions to the method of proof of 6.5.7 complete the proof of the poset homotopy equivalence of $\mathcal{R}$ with $\bar{\mathcal{R}}$, namely with $\mathcal{B}_2^{cen}(G)$. In particular $\Delta_{\mathcal{H}_I}$ is ample and normalizer-sharp by Proposition 5.11.2.

Main results for ***Ru*****.** This discussion via our usual approach leads to the following theorem.

THEOREM 7.24.1. *For $G = Ru$, the 2-local complex $\Delta_{\mathcal{H}_I}$ and the collections $\mathcal{E}_2^-$ and $\mathcal{B}_2^{cen}(G)$ are ample and normalizer-sharp. In particular, we obtain from the groups in $\mathsf{N}(\Delta_{\mathcal{H}_I}/G) = \mathcal{H}_I$ a simplex cohomology decomposition as in* (5.8.6)*; and the map from the homotopy colimit of the pushout cube diagram* (4.6.5) *to BRu, followed by 2-completion, is a homotopy equivalence.*

PROOF. The arguments follow those for Suz in Theorem 7.12.1. □

REMARK 7.24.2. We had mentioned that the poset $\mathcal{E}_2^-$ has maximal chains $A_1 < A_2 < A_3$ and $A_1 < A_2 < A_6$; and indeed the orbit complex $\Delta(\mathcal{E}_2^-)/G$ is given by that pair of 2-simplices, which share the face $A_1 < A_2$. Hence the corresponding decomposition in the above Theorem could be in fact be given in the language of the category $\mathfrak{D}_{3+3}$. However, this is somewhat different in spirit from the use of $\mathfrak{D}_{3+3}$ in the case of the Lyons group Ly in the discussion after Theorem 7.26.1: for in that latter case (like those for earlier J_3, and Fi_{24}'), the reduction to a non-simplex category proceeds via the Cofinality Theorem 4.5.22, using a coincidence of stabilizers. However, one could presumably produce a somewhat analogous derivation in the present case of Ru: namely by adjusting the earlier arguments giving the homotopy equivalence of $\mathcal{E}_2^-$ with the intersection complex $\Delta_{\mathcal{A}_I}$ based on A_1, A_3, and A_6, to actually show equivalence of $\mathcal{E}_2^-$ with a larger intersection complex based instead on the subgroups A_1, A_2, A_3, and A_6—in order to reduce from the category $\mathfrak{D}_4$ down to $\mathfrak{D}_{3+3}$, essentially by removing simplices of types containing $\{3, 6\}$.

Comment. We mention that we have saved the remaining two sporadic groups G given by $O'N$ or Ly until last, since they exhibit some of the strangest behavior (along with earlier Fi'_{24}) with regard to our usual viewpoint of decompositions over a subgroup simplex $\mathcal{H}_I$. We observe that each of these two final groups has just one G-conjugacy class of involutions, so that $\mathcal{E}_2(G)$ does not differ from $\mathcal{A}_2(G)$. Furthermore each group has comparatively "few" 2-local subgroups; so that there seems to be no very convincingly natural 2-local geometry for these groups in the geometric literature. On the other hand, the mod 2 cohomology of both groups has been determined in the topological literature; and along the way, a considerable amount of structural information on the groups has been established. Thus our analysis of decompositions for these two groups is at least partly *a posteriori*.

7.25. The O'Nan group $O'N$

For $G := O'N$, we take as our index set $I := \{1, 3\}$, corresponding to G-classes of maximal 2-local subgroups in the Atlas [**47**, p. 132]:

$H_1 \cong 4{}^{\cdot}L_3(4){:}2$, and

$H_3 \cong 4^3{}^{\cdot}L_3(2)$.

As usual the H_i determine a subgroup simplex $\mathcal{H}_I$ as in Definition 4.6.2, and a 2-local complex $\Delta_{\mathcal{H}_I}$ as in Definition 6.3.5, which in this case is just a bipartite graph. The common intersection $H_I = H_{1,3}$ of order $2^9 3$ contains a Sylow 2-subgroup of G with index three.

The geometry $\Delta_{\mathcal{H}_I}$ is of rank two, and so is not customarily described via a diagram.

The orbit complex $\Delta(\mathcal{A}_2(G))/G$ is determined in Adem and Milgram [**8**, p. 289]; we will shortly indicate some details leading up to their description of this orbit complex, and we will refer also to Yoshiara [**130**]. Later, we will indicate a different cohomology decomposition, which can be approached from the viewpoint of $\mathcal{B}_2(G)$.

But before entering into those details, we mention that the cohomology decomposition for $O'N$ eventually obtained in [**8**, Thm. 4.2] is indexed by a complex smaller than $\Delta(\mathcal{A}_2(G))/G$: namely they use the subgroup simplex $\mathcal{H}_I$, which we have indicated above; this is the orbit 1-simplex for the usual 2-local geometry for $O'N$ as in Ronan and Stroth [**107**, p. 65], which arises essentially from the 2-local complex $\Delta_{\mathcal{H}_I}$ in our present language. Adem and Milgram reduce from the larger complex $\Delta(\mathcal{A}_2(G))/G$ to the 1-simplex $\Delta_{\mathcal{H}_I}/G = \mathcal{H}_I$ by cancellation primarily at the level of cohomology—cf. our discussion in Section 7.6 of their application of this method to J_1 in their work in [**6**, V.2]. We emphasize that in the case of $O'N$, we do *not* know of any proof of an analogous reduction at the level of homotopy equivalence—say using removal methods such as Lemma 6.5.1, which we have been able to use for the other sporadic groups. (We mention this as a possible open problem for study.) So in our discussion below, we indicate some of the structural complications which seem (at least so far) to prevent any deeper proof of the relationship of $\mathcal{A}_2(G)$ with $\Delta_{\mathcal{H}_I}$.

The approach via $\mathcal{A}_2(G)$. We now proceed with our overview of $\mathcal{A}_2(G)$ and related structures.

As usual for $i \in I$ we set $A_i := \Omega_1 Z\big(O_2(H_i)\big)$; we obtain A_i of rank i with $N_G(A_i) = H_i$. Via the usual method of 6.2.17 (2) we have $A_1 < A_3$. So the intersection complex $\Delta_{\mathcal{A}_I}$ determined by A_1 and A_2 is an internal poset complex as in Proposition 6.3.6. We obtain flag-transitivity of G on $\Delta_{\mathcal{A}_I}$ using Remark 6.4.6 (1); and similarly on $\Delta_{\mathcal{H}_I}$, so that the two complexes are isomorphic by the standard Lemma 6.5.4. However we do *not* here establish equivalence of $\Delta_{\mathcal{A}_I}$ with $\mathcal{A}_2(G)$, and so do not ourselves establish the cohomology decomposition based on $\mathcal{H}_I$—thus we are instead quoting [**8**] for ampleness and normalizer sharpness of $\Delta_{\mathcal{H}_I}$.

There is one G-conjugacy class of involutions in G, denoted $2A$ in the Atlas. Our group A_1 is a member of $\mathcal{A}_2(G)$ of rank one, say with generator z; then

$$H_1 = N_G(A_1) = C_G(z) \cong 4^{\cdot}L_3(4){:}2$$

with $O_2(H_1) = Z(H_1') \cong \mathbb{Z}/4$, and $A_1 = \Omega_1\big(Z(H_1')\big)$; where the indicated outer automorphism of $L_3(4)$ (denoted 2_1 in the Atlas and in [**6**]) is unitary, and hence can be represented by the inverse-transpose map. Set $\bar{H}_1 := H_1/Z(H_1')$. As usual from the character table of $O'N$ (for example using the ClassMultiplicationCoefficient function in GAP), we can compute the value of $\xi(2A, 2A, 2A)$ defined in (5.10.5); we obtain the class-algebra multiplication coefficient 1750. Of the corresponding 1750 G-conjugates of z in $H_1 \setminus Z(H_1')$, we find that 630 arise as the two involutions in the preimages of the three involutions in each of the 105 root groups in the commutator subgroup $\bar{H}_1' \cong L_3(4)$. Let A_3 denote the member of $\mathcal{A}_2(G)$ of rank three spanned by z and the six further involutions in such a root group preimage—this group is denoted by I in [**8**, Thm. A]; then $H_3 = N_G(A_3) \cong 4^3 L_3(2)$. Next if A_2 is any subgroup of A_3 of rank two containing A_1, we find that A_3 is the unique G-conjugate of A_3 over A_2; in particular, $H_2 := N_G(A_2) \leq N_G(A_3)$, so that in our usual notation of $H_{2,3} := H_2 \cap H_3$, we have

$$H_2 = H_{2,3} \cong 4^{3}{}^{\cdot}\big(2^2 : L_2(2)\big) \cong 4^{3}{}^{\cdot}\Sigma_4.$$

The remaining $1750 - 630 = 1120$ conjugates of z in H_1 lie outside H_1': there are 4 such conjugates in each of the H_1'-cosets of the 280 conjugates of a unitary outer automorphism of $\bar{H}_1$. We let t denote such a conjugate, with $B_2 := \langle z, t\rangle$. Since we have now accounted for all 1750 conjugates mentioned above, we conclude that A_2 and B_2 represent the only G-classes of members of $\mathcal{A}_2(G)$ of rank 2. As t is a unitary outer automorphism, we find (cf. [**130**, p. 256]) that[17]

$$C_G(B_2) \cong B_2 \times (3^2{:}4),$$

and then

$$N_G(B_2) \cong \big((3^2{:}4) \times \mathrm{Alt}_4\big){:}2;$$

and $N_G(B_2)$, H_1, and H_3 give the G-classes of maximal 2-local subgroup of G. Letting a denote the unique involution in a Sylow 2-subgroup $\mathbb{Z}/4$ of $3^2{:}4$ in $C_G(B_2)$, the group $B_3 := B_2\langle z\rangle = \langle z, a, t\rangle$ of rank three represents the final remaining G-class in $\mathcal{A}_2(G)$—and is denoted by II in [**8**, Thm. A]. We have $N_G(B_3) \leq C_G(a) \cong H_1$;

[17] We mention that in the description of $N_G(B_2)$ in the diagram on [**8**, p. 289], the 3^4 indicated there is a misprint for 3^2.

then the image in $C_G(a)/Z(C_G(a)')$ of B_3 intersects two distinct root groups in the unipotent radical of a rank one parabolic of $L_3(4)$, and we find that $N_G(B_3)$ is an extension of

$$Z\big(C_G(a)'\big)B_3 \cong \mathbb{Z}/4 \times (\mathbb{Z}/2)^2$$

by Σ_4. Thus our A_1, A_2, A_3, B_2, and B_3 correspond to the five vertices of $\Delta\big(\mathcal{A}_2(G)\big)/G$ in the diagram of that orbit complex in [**8**, p. 289]. That diagram further indicates that the 4 G-classes of maximal simplices in $\Delta\big(\mathcal{A}_2(G)\big)/G$ are represented at the level of vertices by the 3 inclusions $A_1 < A_2 < A_3$, $A_1 < A_2 < B_3$, and $A_1 < B_2 < B_3$; and indeed for the inclusion $A_1 < A_2 < B_3$, there are in fact two G-classes of edges $A_1 < B_3$ and $A_1 < B_3'$ using G-conjugates of B_3, and correspondingly two G-classes of such maximal chains.

In particular, observe that G is not flag-transitive on $\mathcal{A}_2(G)$; hence the homotopy decomposition over $\Delta\big(\mathcal{A}_2(G)\big)/G$ from the normalizer decomposition for $\mathcal{A}_2(G)$ in (5.6.9) is not indexed by a pushout n-cube diagram. As mentioned earlier, Adem and Milgram use cancellations in cohomology to proceed from $\Delta\big(\mathcal{A}_2(G)\big)/G$ to the pushout square diagram determined by the orbit 1-simplex given by $\Delta_{\mathcal{H}_I}/G = \mathcal{H}_I$.

Some possible alternative approaches via $\mathcal{B}_2(G)$. We turn next to a discussion of the poset $\mathcal{B}_2(G)$. In summary: we will see that $\mathcal{B}_2(G)$ is not as naturally related with the normalizers of the subgroups in $\mathcal{A}_2(G)$ as was the case in our earlier discussions of a number of the other sporadic groups; however, we will sketch some details, in order to indicate some possible approaches beginning with $\mathcal{B}_2(G)$—to lead to cohomology decompositions indexed by suitable orbit graphs, in the spirit of the Adem-Milgram decomposition over $\Delta_{\mathcal{H}_I}/G$.

The poset $\mathcal{B}_2(G)$ of 2-radical subgroups of $O'N$ is determined in An and O'Brien [**16**, Table 3] and in Yoshiara [**130**, Table V, p. 257]; see also Uno and Yoshiara [**121**]. There are eight conjugacy classes, of which five are 2-centric; we will give an initial description of all eight classes, though later we will restrict attention to the five centric classes. We note first that our subgroups denoted above by H_1, H_3, and $N_G(B_2)$ are the normalizers of the 2-groups denoted C, V, and V_2 in [**130**, Table V, p. 257]; where $C = Z(H_1') \cong \mathbb{Z}/4$ and $V_2 := B_2$ are not 2-centric, while $V = O_2(H_3) \cong (\mathbb{Z}/4)^3$ is 2-centric. Further $H_{1,3} = H_1 \cap H_3$ is the normalizer of the 2-centric 2-group there denoted $C^{(pl)} := O_2(H_{1,3})$; and the non-maximal 2-local H_2 is the normalizer of the 2-centric 2-group there denoted $V^{(l)} := O_2(H_2)$, so it follows that

$$H_{1,2,3} := H_1 \cap H_2 \cap H_3 = H_1 \cap H_2 =: H_{1,2}$$

is the (self-normalizing) Sylow 2-subgroup there denoted $C^{(\iota pl)} := O_2(H_{1,2,3})$, which of course is also 2-centric. Next $H_1 \cap N_G(B_2)$ realizes the normalizer of the non-2-centric 2-group there denoted

$$C^{(\iota)} := O_2\big(H_1 \cap N_G(B_2)\big).$$

However, the final 2-centric 2-group denoted $C^{(p)}$ there does not arise naturally from the normalizer of a subgroup in $\mathcal{A}_2(G)$: $C^{(p)}$ has order 2^6, and is the preimage of a unipotent radical of rank four in $\bar{H}_1 \cong L_3(4)$; with normalizer containing an $L_2(4)$ in a Levi complement of that parabolic. For brevity we will write $H_p := N_G(C^{(p)})$.

We now review some aspects of the above discussion, from the viewpoint of possible cohomology decompositions:

Note that the five members of $\mathcal{B}_2^{cen}(G)$ are the 2-groups denoted in [**130**, Table V, p. 257] by V, $V^{(l)}$, $C^{(p)}$, $C^{(pl)}$, and $C^{(ipl)}$; and their normalizers are given respectively by H_3, $H_2 = H_{2,3}$, H_p, $H_{1,3}$, and $H_{1,2,3}$. We have: $V^{(l)} \cong V \cdot 2^2$, and $C^{(p)} \cong C \cdot 2^4$, and $C^{(pl)} \cong C^{(p)} : 2^2 \cong C \cdot 2^{2+4}$, while $C^{(\iota pl)} \cong C^{(pl)} \cdot 2$ is a Sylow 2-group.

Unfortunately this structure of $\mathcal{B}_2^{cen}(G)$ itself, even taking cancellations into account, does not seem to lead to a decomposition over a simplex category such as $\mathcal{H}_{I^+}$ for $I^+ := \{1, 2, 3\}$: For observe that the group H_1 would not be available, since it is the normalizer of the non-2-centric subgroup C. Furthermore it does not seem clear that results in the literature would guarantee that the process of adding the G-class of C, corresponding to H_1, to $\mathcal{B}_2^{cen}(G)$ should be a homotopy equivalence; and even if it were, we are not aware of any equivalence arguments that would then allow removing $C^{(p)}$ from the poset, so that its normalizer H_p could be removed from the decomposition.

So instead, we consider some possible approaches to decompositions via reducing $\mathcal{B}_2^{cen}(G)$ using our usual methods. We first indicate the relevant inclusions in the orbit poset $\mathcal{B}_2^{cen}(G)/G$: we observe that V is contained in $V^{(l)}$ and $C^{(pl)}$; while $C^{(p)}$ is contained in $C^{(pl)}$, but in no conjugate of $V^{(l)}$; and of course all these subgroups lie in the Sylow group $C^{(\iota pl)}$.

Now V is weakly closed in the Sylow 2-subgroup $C^{(\iota pl)}$; we postpone to Section 8.8 further details of the easy argument for this fact. Then it follows, using the restrictions on inclusions in the previous paragraph, that $\mathcal{B}_2^{cen}(G)_{<V^{(l)}}$ consists just of V; thus we have the standard situation of Remark 6.5.2, so that the link of $V^{(l)}$ in $\mathcal{B}_2^{cen}(G)$ is contractible via a $G_{V^{(l)}}$-map. Hence by Lemma 6.5.1, $\mathcal{B}_2^{cen}(G)$ is homotopy equivalent via a G-map to the subposet $\mathcal{B}_1$ with the G-class of $V^{(l)}$ removed. Next we claim that $C^{(pl)}$ is weakly closed in the Sylow 2-group $C^{(\iota pl)}$; again we postpone to Section 8.8 the details of an argument for this claim. Since we removed $V^{(l)}$ to get $\mathcal{B}_1$, it now follows using the above restrictions on inclusions that $(\mathcal{B}_1)_{<C^{(\iota pl)}}$ has unique maximal element $C^{(pl)}$. Again it follows using Remark 6.5.2 for the hypothesis of Lemma 6.5.1 that $\mathcal{B}_1$ is homotopy equivalent via a G-map to the subposet $\mathcal{B}_2$ with the G-class of $C^{(\iota pl)}$ removed. Notice that $\mathcal{B}_2$ is a graph, with edges of types $C^{(p)} < C^{(pl)}$ and $V < C^{(pl)}$. As usual, $\mathcal{B}_2$ is ample and normalizer-sharp by Proposition 5.11.2.

Indeed we could reduce one more step beyond $\mathcal{B}_2$, using cancellation techniques as in the proof of Adem and Milgram [**8**, Thm. 4.2]: For we can now use once again the fact that V is weakly closed, to see that the stabilizer $H_{1,3}$ of the vertex $C^{(pl)}$ is the same as the stabilizer of the edge $V < C^{(pl)}$, so that the corresponding terms in the homology decomposition cancel. Note however that removing these terms is not really natural from a geometric viewpoint, since the configuration which remains is no longer a graph: the remaining terms include vertices of types V and $C^{(p)}$; but there are edges only given by conjugates of $C^{(p)} < C^{(pl)}$, where the former vertex type is not in the configuration. We will discuss how to remedy this deficiency in a moment. Now observe that the normalizers indexing the decomposition over the remaining terms are H_3, H_p, and $H_p \cap H_{1,3}$; indeed since $H_p \leq H_1$, we have

$$H_p \cap H_{1,3} = H_p \cap H_3.$$

This is not the same decomposition as that obtained by Adem and Milgram in [**8**, Thm. 4.2]—the normalizers indexing their decomposition are the members of H_3, H_1, and $H_{1,3}$ corresponding to the complex $\Delta_{\mathcal{H}_I}$.

Finally we indicate how one can obtain the decomposition of the previous paragraph based on the poset related to H_p, using an orbit graph from a suitable simplicial complex of dimension one. We let $\mathcal{B}^-$ denote the set of G-conjugates of V and $C^{(p)}$; thus in effect we could use a type set $I' := \{3, p\}$. We define the corresponding intersection complex $\Delta_{\mathcal{A}_{I'}}$ as in Definition 6.3.2, by letting conjugates of V and $C^{(p)}$ be incident when their intersection is a conjugate of $V \cap C^{(p)}$; this turns out to be equivalent to their product being a conjugate of $VC^{(p)} = C^{(pl)}$. We obtain flag-transitivity on the intersection complex using Remark 6.4.6 (1); in particular, $\mathsf{N}(\Delta_{\mathcal{A}_{I'}}/G) = \mathcal{H}_{I'}$. Furthermore one can show that $\Delta_{\mathcal{A}_{I'}}$ is homotopy equivalent via a G-map to $\mathcal{B}_2^{cen}(G)$; again we postpone to Section 8.8 a sketch of the details. Hence $\Delta_{\mathcal{A}_{I'}}$ is ample and normalizer-sharp by by Proposition 5.11.2. And we see that the corresponding decomposition is indexed by the subgroup 1-simplex $\mathcal{H}_{I'}$ defined by H_3, H_p, and $H_p \cap H_3$ as above.

Main results for $O'N$. In conclusion, we summarize the above discussion on decompositions.

We saw that the poset $\mathcal{A}_2(G)$ leads via cancellation in the work of Adem and Milgram [**8**, Thm. 4.2] to a cohomology decomposition indexed by the pushout square for $\Delta_{\mathcal{H}_I}/G = \mathcal{H}_I$. But we are not aware of any homotopy equivalence of $\Delta_{\mathcal{H}_I}$ with any of the standard ample collections in the literature; indeed computation of the reduced Euler characteristic (as in Remark 5.9.4) seems to rule out any equivalence with those usual suspects. Thus it appears that we can only obtain the conclusion that the 2-local complex $\Delta_{\mathcal{H}_I}$ is ample and normalizer-sharp, by using some "direct" argument such as Lemma 4.5.10 based on an amalgamated product of stabilizers; we have not obtained it not using our usual equivalence methods.

In contrast, our argument above giving homotopy equivalence via a G-map with $\mathcal{B}_2^{cen}(G)$ does show in our usual way that the complex $\Delta_{\mathcal{A}_{I'}}$ is ample and normalizer-sharp. However, its decomposition indexed by H_3, H_p and $H_3 \cap H_p$ is not as "strong" as that for $\Delta_{\mathcal{H}_I}$, in view of the proper inclusion $H_p < H_1$.

The complex $\Delta_{\mathcal{A}_{I'}}$ has reduced Euler characteristic $2^8 \cdot 993337$, which is not divisible by $|G|_2 = 2^9$, and so by Remark 5.9.4 $\Delta_{\mathcal{A}_{I'}}$ is not in the standard ample G-homotopy type represented by $\mathcal{B}_2(G)$.

We combine this discussion of the two results to state the following theorem.

THEOREM 7.25.1. *Let $G := O'N$, with 2-local complex $\Delta_{\mathcal{H}_I}$, and intersection complex $\Delta_{\mathcal{A}_{I'}}$ related to $\mathcal{B}_2^{cen}(G)$ as described above. Then $\Delta_{\mathcal{H}_I}$ and $\Delta_{\mathcal{A}_{I'}}$ are ample and normalizer-sharp. Indeed from the groups in $\mathsf{N}(\Delta_{\mathcal{H}_I}/G) = \mathcal{H}_I$, we obtain a simplex cohomology decomposition as in* (5.8.6)*; in particular, the obvious map of groups from the amalgamated free product $H_1 *_{H_{13}} H_3$ to $O'N$ induces a mod-2 cohomology isomorphism. The analogous statement holds for the groups in $\mathsf{N}(\Delta_{\mathcal{A}_{I'}}/G) = \mathcal{H}_{I'}$, and the amalgamated free product $H_p *_{H_p \cap H_3} H_3$. The corresponding maps from the homotopy colimit of the pushout square diagram* (4.5.6) *to $B(O'N)$, followed by 2-completion, are homotopy equivalences.*

The cohomology of the O'Nan group was computed by Adem and Milgram [**8**], at least as a module over a large subring, by obtaining the cohomology decomposition above in the case of $\Delta_{\mathcal{H}_I}$ and then doing a computation. It is a Cohen–Macaulay ring of Krull dimension three, with Poincaré series

$$\sum_{i=0}^{\infty} \dim_{\mathbb{F}_2} H^i(O'N, \mathbb{F}_2)\, t^i$$
$$= (1 + t^3 + t^4 + t^5 + 2t^6 + t^7 + t^8 + 3t^9 + 2t^{10} + 2t^{11} + 2t^{12} + 2t^{13} + 2t^{14} + 3t^{15} + t^{16} + t^{17} + 2t^{18} + t^{19} + t^{20} + t^{21} + t^{24}) \;/\; (1 - t^7)(1 - t^8)(1 - t^{12}).$$

7.26. The Lyons group Ly

For $G := Ly$, we take as our index set $I := \{1, 2, 3, 4\}$, corresponding to 2-local subgroups which can be determined using the Atlas [**47**, p. 174]:

$H_1 \cong 2^{\cdot}\mathrm{Alt}_{11}$,
$H_2 \cong \big((2^{2+4}{:}(3 \times 3){:}2) \times 3\big){:}2$,
$H_3 \cong 2^3{}^{\cdot}L_3(2)$, and
$H_4 \cong (2^4 \times 3)^{\cdot}\mathrm{Alt}_7$.

As usual the H_i determine a subgroup simplex $\mathcal{H}_I$ as in Definition 4.6.2, and a 2-local complex $\Delta_{\mathcal{H}_I}$ as in Definition 6.3.5. The maximal 2-locals H_2 and H_4 are not maximal subgroups of G, since they are proper in a subgroup $M := 3^{\cdot}McL{:}2$. The maximal 2-local H_3 is also not a maximal subgroup of G, since it is proper in a subgroup $G_2(5)$. Furthermore H_3 does not contain a Sylow 2-subgroup of G; the common intersection H_I of order 2^6 is of index four in a Sylow group.

There seems to be no natural diagram associated with this geometry; presumably since none is known for the Alt_{11} involved in H_1.

An approach via $\mathcal{A}_2(G)$. The G-classes in $\mathcal{A}_2(G)$ are a fundamental ingredient in the determination of the mod 2 cohomology of Ly by Adem, Karagueuzian, Milgram and Umland in [**2**]—although the full orbit complex $\Delta\big(\mathcal{A}_2(G)\big)/G$ is not explicitly discussed there. We will recall certain relevant details below; in particular, we will indicate an approach to a decomposition via a suitable intersection complex $\Delta_{\mathcal{A}_I}$ as in Definition 6.3.2.

Some members of $\mathcal{A}_2(G)$ and the complex $\Delta_{\mathcal{A}_I}$. For $i \in I$, as usual we define $A_i := \Omega_1 Z\big(O_2(H_i)\big)$; we obtain A_i of rank i with $N_G(A_i) = H_i$. We use the A_i to define our intersection complex $\Delta_{\mathcal{A}_I}$. Via the usual method of 6.2.17 (2) we have $A_1 < A_2 = A_3 \cap A_4$, where A_2 is of rank two; so $\Delta_{\mathcal{A}_I}$ affords an internal almost-poset complex as in Definition 6.3.9.

We now derive some relevant properties of the subgroups in $\mathcal{A}_2(G)$.

There is one G-conjugacy class of involutions in G, denoted by $2A$ in the Atlas [**47**, p. 174]. Let z denote such an involution, generating A_1 of rank one; then $H_1 = C_G(z) = N_G(A_1) \cong 2^{\cdot}\mathrm{Alt}_{11}$, the nonsplit double cover of the alternating group Alt_{11}. Now involutions of $\bar{H}_1 := H_1/A_1 \cong A_{11}$ with cycle type $1^7 2^2$ correspond to elements of G of order four, since they arise in subgroups $2A_5 \cong SL_2(5)$ of H_1 in which z is the unique involution. Thus involutions in G arise only from the other conjugacy class of involutions of A_{11}, namely those with cycle type $1^3 2^4$;

let a denote such an involution from H_1. Then $\bar{a}$ is regular involution in a subgroup Alt_8 of $\bar{H}_1$ on the 8 points in its four 2-cycles; and this Alt_8 lies in a maximal subgroup $(\mathrm{Alt}_8 \times \mathrm{Alt}_3){:}2$ of $\mathrm{Alt}_{11} = \bar{H}_1$. The centralizer in $\bar{H}_1$ of $\bar{a}$ lies inside the latter group, and its centralizer in Alt_8 is a parabolic subgroup of $L_4(2) \cong \mathrm{Alt}_8$ of structure $2_+^{1+4}{:}\Sigma_3$.

Choose $\langle z, a\rangle$ as our group A_2, so that A_2 represents the unique conjugacy class of 4-subgroups of G. Further $H_2 = N_G(A_2)$ is an extension of $C_G(A_2)$ (which from the above discussion has structure $((2^{2+4}{:}3) \times 3){:}2)$ by a group Σ_3 acting as $L_2(2)$ on A_2. The central 3-element in $C_G(A_2)'$ shows that H_2 lies in $M \cong 3{}^{\cdot}McL{:}2$.

Next we recall from our discussion of McL in Section 7.11 that McL has an elementary abelian 2-subgroup A_4 of rank four, normalized by a group Alt_7. Now inside the maximal subgroup M of Ly, we find that the normalizer in M of A_4 is given by $N_M(A_4) \cong (2^4 \times 3)\mathrm{Alt}_7$; and this in fact equals the full normalizer $H_4 = N_G(A_4)$ in G.

Further under the flag-transitive action of Alt_7 on the projective space $\mathbf{P}(A_4)$, there is just one conjugacy class of subgroups of rank three, represented say by B_3 containing A_2. The image $\bar{A}_4$ of rank three lies in the extraspecial group 2_+^{1+4} of $C_{\bar{H}_1}(\bar{a})$ above; indeed since all elements of $\bar{A}_4$ are regular involutions, $\bar{A}_4$ of rank three is a full transvection subgroup in Alt_8 regarded as $L_4(2)$ acting on its natural module. There are only two choices for $\bar{A}_4$ over $\bar{A}_2$ in 2_+^{1+4}, and they intersect exactly in $\bar{A}_2$ which generated by $\bar{a}$; these two choices are in fact conjugate in $H_1 \backslash H_1'$. In particular, $\bar{B}_3$ lies in $\bar{A}_4$, but not in the other choice for a transvection subgroup over $\bar{A}_2$. We conclude that A_4 is the unique member of its conjugacy class above B_3.

Finally (cf. [**2**, Lemma 5.1]) there is another class of members of $\mathcal{A}_2(G)$ of rank three; we will denote a representative by A_3: Then A_4 and A_3 represent the G-classes of maximal members of $\mathcal{A}_2(G)$. In particular, A_3 is contained in no conjugate of A_4; and we may choose A_3 to satisfy the restriction $A_3 \cap A_4 = A_2$ indicated earlier. Furthermore in $\bar{H}_1 \cong \mathrm{Alt}_{11}$, $\bar{A}_3$ is a rank two group of involutions of cycle type $2^4 1^3$; since A_3 does not lie in a conjugate of A_4, the three involutions of $\bar{B}_3$ determine a 4-subset of 11 points on which all are regular, and a further 6-subset on which each has cycle type $2^2 1^2$. We can check that there are 36 choices of A_3 over a fixed A_2. We can also locate $H_3 = N_G(A_3) \cong 2^3{}^{\cdot}L_3(2)$ inside a maximal subgroup $G_2(5)$ of Ly. Since $A_3 \cap A_4 = A_2$, we see that $H_{3,4} = H_3 \cap H_4 \leq H_2$; so that we have a coincidence of stabilizers given by

$$H_{3,4} = H_{2,3,4} = H_2 \cap H_3 \cap H_4.$$

Equivalence of $\Delta_{\mathcal{A}_I}$ with $\mathcal{A}_2(G)$ via the subposet $\mathcal{A}_2^-$. Now observe that by the above uniqueness of A_4 over B_3, we have the situation of Remark 6.5.2; so we can apply Lemma 6.5.1 to see that $\mathcal{A}_2(G)$ is homotopy equivalent via a G-map to the subposet $\mathcal{A}_2^-$ defined by removing the G-class of B_3. Thus $\mathcal{A}_2^-$ is ample and normalizer-sharp by Proposition 5.9.1. Now for $i = 3, 4$, by the transitivity on flags in A_i of the subgroup $L_3(2)$ or Alt_7 in H_i, we see that G has two orbits on maximal chains in $\mathcal{A}_2^-$, represented by $A_1 < A_2 < A_4$ and $A_1 < A_2 < A_3$. In particular, the homotopy decomposition arising from $\Delta(\mathcal{A}_2^-)/G$ in the normalizer decomposition for $\mathcal{A}_2^-$ is not indexed by a simplex corresponding to a pushout 4-cube.

However, much as for McL in Section 7.11, we can fairly naturally recover the decomposition from $\mathcal{A}_2^-$ as a reduction via cancellations from a different decomposition over a suitable pushout 4-cube diagram: namely we can use the almost-poset complex $\Delta_{\mathcal{A}_I}$ defined by the A_i for $i \in I$.

However, we first indicate why we cannot work instead with a pushout 3-cube; that is, why we do *not* get a homotopy equivalence, via the method of Lemma 6.5.3, of $\mathcal{A}_2^-$ with the smaller intersection complex $\Delta_{\mathcal{A}_{I^-}}$ defined by using only the types in $I^- := \{1, 3, 4\}$: For in attempting to implement such a proof, we would find that the subcomplex "L_{A_2}" would be the join of the two subcomplexes given by the 36 vertices of type A_3 over A_2, and by the two vertices of type A_4 over A_2. Since both these subcomplexes are discrete and of size greater than 1, the corresponding join has 1-cycles and so is not contractible.

We now return to considering the almost-poset complex $\Delta_{\mathcal{A}_I}$ mentioned above. Here we do get a homotopy equivalence via a G-map of $\mathcal{A}_2^-$ with $\Delta_{\mathcal{A}_I}$ using the method of Lemma 6.5.3; for now each subcomplex L_E is just the join of the vertex E with the subcomplex given by vertices of types corresponding to proper subgroups of E. Thus $\Delta_{\mathcal{A}_I}$ is ample and normalizer-sharp by Proposition 5.9.1.

Flag-transitivity of Ly on $\Delta_{\mathcal{A}_I}$. Establishing flag-transitivity on G on $\Delta_{\mathcal{A}_I}$ is more complicated than usual—primarily as a consequence of the stabilizer coincidence $H_{3,4} = H_{2,3,4}$ noted above.

We consider first the link of the vertex A_3. Transitivity on flags involving only conjugates of A_1 and A_2 in A_3 is clear from the natural action of $H_3/A_3 \cong L_3(2)$ on the projective space $\mathbf{P}(A_3)$. It remains to consider flags in the link of A_3 involving A_4; and it will in fact suffice to consider such flags which are over $A_1 < A_2$. Now the stabilizer of $A_1 < A_2$ in $L_3(2)$ is a subgroup D_8, which is in fact transitive on the two conjugates of A_4 above A_2, as required.

Next we consider the link of A_4. Again transitivity on flags involving only conjugates of A_1 and A_2 in A_4 is clear, from the natural action of $H_4/C_G(A_4) \cong \mathrm{Alt}_7$ on the projective space $\mathbf{P}(A_4)$ as a flag-transitive subgroup of $L_4(2) \cong \mathrm{Alt}_8$. It remains to consider flags in the link of A_4 which involve A_3; and again it will suffice to consider further such flags over $A_1 < A_2$. Now the stabilizer in Alt_7 of $A_1 < A_2$ is a subgroup Σ_4, which (using also the central 3-element from M) is in fact transitive on the 36 choices of conjugates of A_3 over A_2, as required.

Finally we consider the link of A_1. Again transitivity on flags involving only conjugates of A_2 and A_4 over A_1 is clear from the action $H_1/A_1 \cong A_{11}$ on the corresponding parabolics of the subgroup Alt_8 regarded as $L_4(2)$, from our earlier discussion. It remains to consider flags in the link of A_1 which involve A_3; and again it suffices to consider further such flags incident with $A_2 < A_4$. The stabilizer in A_{11} of $A_2 < A_4$ is a subgroup $(2_+^{1+4}{:}\Sigma_3) \times 3$, which again is transitive on the 36 conjugates of A_3 over A_2, as required.

We have now checked flag-transitivity in links of three of the four types; this suffices for flag-transitivity of G by Proposition 6.4.5. In particular, we have $\mathsf{N}(\Delta_{\mathcal{A}_I}/G) = \mathcal{H}_I$.

As usual it is straightforward to adopt the above proof of flag-transitivity to apply also to $\Delta_{\mathcal{H}_I}$; so by the standard Lemma 6.5.4 we obtain the isomorphism of the almost-poset complex $\Delta_{\mathcal{A}_I}$ with the 2-local complex $\Delta_{\mathcal{H}_I}$.

Finally we examine the implications of the stabilizer coincidence $H_{3,4} = H_{2,3,4}$, from the viewpoint of our complex $\Delta_{\mathcal{A}_I}$: Observe that a 1-simplex of type A_4—A_3

is incident to a unique vertex of type A_2; so arguing as usual with Remark 6.5.2 and then Lemma 6.5.1, we may remove the G-class of the star of such 1-simplices from our complex $\Delta_{\mathcal{A}_I}$. Notice then that from the 15 stabilizers H_J for nonempty $J \subseteq I$, this removes those types containing $\{3,4\}$; and so reduces to the 11 stabilizers for subsets J corresponding to subchains of the maximal chains $A_1 < A_2 < A_3$ and $A_1 < A_2 < A_4$ in our original poset $\mathcal{A}_2^-$.

Remarks on other possible geometries. The poset $\mathcal{B}_2(G)$ of 2-radical subgroups for Ly is determined in Yoshiara [**130**, Table II, p. 248]. There are 17 conjugacy classes, of which 13 are 2-centric. In particular, A_1 is 2-radical, but is not 2-centric; thus $\mathcal{B}_2^{cen}(G)$ does not seem to be naturally related to the above posets arising in connection with $\mathcal{A}_2(G)$. However there might well be some homotopy equivalence with the above poset determined by the 11 members of $\Delta(\mathcal{A}_2^-)/G$.

Two possible 2-local geometries for Ly were given in Ronan and Stroth [**107**, p. 65]; however, these do not seem to use the full normalizers H_i above, but only certain proper subgroups of H_1 and H_2. We are not aware of any homotopy equivalence of either of those geometries with the above configurations related to $\mathcal{A}_2(G)$ and $\mathcal{B}_2(G)$. Thus from our viewpoint of homotopy decompositions in this work, the most natural choice for a 2-local geometry for Ly would seem to be the complex $\Delta_{\mathcal{A}_I}$ above, including its equivalent subcomplex with simplices of types containing $\{3,4\}$ removed.

Main results for Ly. This discussion leads (more or less) as usual to the following theorem.

THEOREM 7.26.1. *For $G = Ly$, the collection $\mathcal{A}_2^-$ and the almost-poset complex $\Delta_{\mathcal{A}_I}$ are ample and normalizer-sharp. In particular, we obtain from the groups in $\mathsf{N}(\Delta_{\mathcal{A}_I}/G) = \mathcal{H}_I$ a simplex cohomology decomposition as in* (5.8.6)*; and the map from the homotopy colimit of the pushout* 4*-cube diagram* (4.6.7) *to BLy, followed by* 2*-completion, is a homotopy equivalence.*

PROOF. The proof follows the argument for J_3 in Theorem 7.8.1; but of course now for rank four rather than three. □

As in the cases of J_3 and Fi_{24}', we can use the fact that $H_{3,4} = H_{2,3,4}$, and hence also $H_{1,3,4} = H_{1,2,3,4}$, to make an argument similar to that for J_3 in Section 7.8—using the Cofinality Theorem 4.5.22, to show that the simplex category $\mathfrak{D}_4$ can be replaced by the subcategory $\mathfrak{D}_{3+3}$ obtained by removing these four groups. The resulting category consists of two pushout cubes identified along a face (determined by B_1, B_2, and B_{12} in the diagram below). This is the opposite of the category associated to the poset of simplices in the simplicial complex obtained by identifying two copies of the simplex Δ^2 along a common edge. Here is the corresponding diagram of classifying spaces:

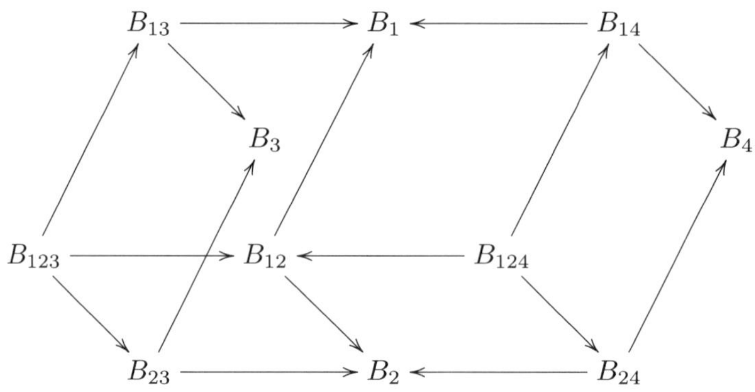

(7.26.2)

THEOREM 7.26.3. *The map from the homotopy colimit of the diagram above of type* $\mathfrak{D}_{3+3}$ *to* BLy*, followed by* 2*-completion, is a homotopy equivalence.* □

In particular the cohomology decomposition from this theorem is over the subset of 11 stabilizers for types not containing $\{3, 4\}$ which was indicated earlier.

The cohomology ring of Ly was computed by Adem, Karagueuzian, Milgram and Umland [**2**]. Their main theorem shows that there is a short exact sequence, obtained using the 2-local subgroups H_4 and H_3:

$$0 \to H^*(Ly, \mathbb{F}_2) \to H^*(2^4, \mathbb{F}_2)^{\mathrm{Alt}_7} \oplus H^*(2^3, \mathbb{F}_2)^{L_3(2)} \to \mathbb{F}_2[d_4, d_6] \to 0.$$

Here, the components of the maps are ring homomorphisms. The components of the left map are given by restrictions, and the right map is given by $d_8 \to d_4^2$ and $d_{12} \to d_6^2$. Another way of writing this information is

$$H^*(Ly, \mathbb{F}_2) = H^*(2^4, \mathbb{F}_2)^{\mathrm{Alt}_7} \oplus d_7.\mathbb{F}_2[d_4, d_6, d_7].$$

This gives the entire ring structure, as a direct sum of a subring and a two sided ideal. It follows that the Quillen map [**101, 102**]

$$H^*(Ly, \mathbb{F}_2) \to \lim_{\substack{\longleftarrow \\ E \in \mathcal{A}_2(Ly)}} H^*(E, \mathbb{F}_2)$$

is an isomorphism.

The invariant ring $H^*(2^4, \mathbb{F}_2)^{\mathrm{Alt}_7}$ is calculated in Adem and Milgram [**5**]. The answer is that it is a free module over the rank four Dickson invariants, on eight generators in degrees 0, 18, 20, 21, 24, 25, 27 and 45. So the Poincaré series for this invariant ring and for Ly itself are

$$\sum_{i=0}^{\infty} \dim_{\mathbb{F}_2} H^i(2^4, \mathbb{F}_2)^{\mathrm{Alt}_7}\, t^i = \frac{1 + t^{18} + t^{20} + t^{21} + t^{24} + t^{25} + t^{27} + t^{45}}{(1 - t^8)(1 - t^{12})(1 - t^{14})(1 - t^{15})}$$

$$\sum_{i=0}^{\infty} \dim_{\mathbb{F}_2} H^i(Ly, \mathbb{F}_2)\, t^i = \sum_{i=0}^{\infty} \dim_{\mathbb{F}_2} H^i(2^4, \mathbb{F}_2)^{\mathrm{Alt}_7}\, t^i + \frac{t^7}{(1 - t^4)(1 - t^6)(1 - t^7)}.$$

CHAPTER 8

Details of proofs for individual groups

In this final chapter, we provide details (or at least sketches) for some of the lengthier arguments, which we had postponed from the sections on the individual sporadic groups in Chapter 7.

Some of these arguments develop more detailed geometric properties of the particular 2-local geometry Δ for the relevant sporadic group G.

Other arguments are devoted to exhibiting a homotopy equivalence of Δ via a G-map with one of the ample collections of Chapter 5.

8.1. Details for M_{12}

We sketch the proof of the claim in Section 7.2 that $\mathcal{E}_2(G)$ is homotopy equivalent to the subposet $\mathcal{E}_2^-$; we will show using Lemma 6.5.1 that we can remove from $\Delta(\mathcal{E}_2(G))$ the G-conjugates of the groups which we had denoted in Section 7.2 by R_1 of rank two, and K of rank three—this latter we group we will now denote by A_3.

Geometric background. The proof is perhaps most easily described in a geometric context: in which we regard each G-conjugate of A_2 as a projective line over $\mathbb{F}_2$, whose three projective points are the G-conjugates of A_1 which it contains. Thus points A_1 and $g^{-1}A_1g$ are in the line relation if and only if the set product $A_1(g^{-1}A_1g)$ is a G-conjugate of A_2; we had mentioned in Section 7.2 that this is equivalent to the (symmetric) relation $g^{-1}A_1g \leq O_2(C_G(A_1))$.

We describe certain aspects of this geometry by quoting some computations: From the character table of M_{12}, for example ClassMultiplicationCoefficient function in GAP, we can compute the value $\xi(2B, 2B, 2B)$ of the structure constant defined in (5.10.5); we obtain the class-algebra multiplication coefficient 30, which says that A_1 commutes with 30 distinct G-conjugates of itself. We can describe the interrelation of these using the *suborbit diagram*, which gives a picture of the orbits of $H_1 = C_G(A_1)$ on points and lines; this diagram is straightforward to compute, and is given for example in Buekenhout [**38**, Sec. 7]. The diagram shows that just six of these conjugates lie in $O_2(H_1)$; and these six conjugates are partitioned into three pairs, each pair defining the remaining two points, other than A_1, of a line; in particular we obtain:

(a) there are exactly three lines containing the point A_1.

Furthermore the diagram shows that there are no further lines between points on distinct members of these three lines on A_1: that is, there are no triangles in the geometry (other than the degenerate "triangles" from pairs among the three points within a single line); so that these six points collinear with A_1 are mutually at distance two in the geometry. Hence on each of these six points, the two lines not through A_1 go to the points among the $30 - 6 = 24$ points at distance two from A_1. It follows using the suborbit diagram (indeed, just from the calculation that $6 \cdot 2 \cdot 2 = 24$) that each of those 24 points is collinear with exactly one of the six points at distance one from A_1. Thus

(b) there are no (nondegenerate) 4-gons in the geometry;

and furthermore

(c) two points at distance two determine a unique point collinear with each,

where in fact this unique point c is also collinear with the point ab given by the product of the involutions a, b defining the original two points at distance two.

The homotopy equivalence argument. We will first describe the link in $\mathcal{E}_2(G)$ of R_1. We begin by choosing two points A_1 and $g^{-1}A_1g$ at distance two. Our calculation above showed that two such points commute when regarded as groups; so the set product of the rank one groups for these two points is a 4-group in $C_G(A_1)$ which contains A_1. However $g^{-1}A_1g \not\leq O_2\big(C_G(A_1)\big)$, since that containment would instead determine collinear points—for which the set product would lie in the G-class of A_2. Thus we can take our product $A_1(g^{-1}A_1g)$ to be R_1, since we saw that R_1 represents the only other class of 4-groups in $\mathcal{E}_2(G)$. Now by (c), there is a unique conjugate $h^{-1}A_1h$ at common distance one from both A_1 and $g^{-1}A_1g$, and indeed from the point for the product involution determined by this pair; hence the stabilizer G_{R_1} must also normalize this unique point $h^{-1}A_1h$ collinear with all three points in R_1. Furthermore the points A_1, $g^{-1}A_1g$, and $h^{-1}A_1h$ mutually commute, and do not lie on any one single line; so their set product

$$X := (h^{-1}A_1h)R_1 = (h^{-1}A_1h)A_1(g^{-1}A_1g)$$

is an elementary abelian subgroup of rank three. We may abbreviate these three points by a, b, c; and their collinearity relations by $a - c - b$. Next recall that we began our discussion of $\mathcal{E}_2(G)$ in Section 7.2 with the condition $\xi(2B, 2B, 2A) = 0$; so all involutions in X are of class $2B$—and hence X must lie in the unique G-class in $\mathcal{E}_2(G)$ of rank three, represented by A_3. Indeed we claim that

(d) X is the unique conjugate of A_3 over R_1.

To obtain this, assume that R_1 is contained in some further conjugate X' of A_3. Then $X \cap X' = R_1$ by rank considerations, and in particular $h^{-1}A_1h \not\leq X'$. Now the points A_1 and $g^{-1}A_1g$ are at distance two in the graph on $\mathcal{E}_2(G)$, and so cannot be collinear in X'. So these two non-collinear points in the conjugate X' of A_3 must be then be at distance two via collinearity with an appropriate conjugate in X' of $h^{-1}A_1h$—which cannot lie in X, since $X \cap X' = R_1$ contains no lines. We abbreviate the point X' by d, with collinearity relations $a - d - b$. Now we have the nondegenerate 4-gon $a - c - b - d - a$, contradicting (b); so we have established (d).

Since $\mathcal{E}_2(G)$ has no members of rank 4, (d) shows that

$$\mathcal{E}_2(G)_{>R_1} \text{ has the unique member } X;$$

and we saw that the stabilizer G_{R_1} normalizes $h^{-1}A_1h$, and so it also normalizes $R_1(h^{-1}A_1h) = X$. So we have the usual special case in Remark 6.5.2: for we see that the link of R_1 in $\mathcal{E}_2(G)$ is the cone given by the join of the unique rank-3 vertex X of $\mathcal{E}_2(G)_{>R_1}$ with the three points contained in $\mathcal{E}_2(G)_{<R_1}$; this is contractible via the obvious G_{R_1}-map. So we conclude by Lemma 6.5.1 that $\mathcal{E}_2(G)$ is homotopy equivalent via a G-map to the poset $\mathcal{E}_2(G)'$ defined by removing the G-class of R_1.

Finally we calculate the link L in $\mathcal{E}_2(G)'$ of A_3: This link contains the seven points in A_3; but since in the previous paragraph we have excluded the rank two subgroups conjugate to R_1 by passing to the subposet $\mathcal{E}_2(G)'$, the rank two subgroups in our link L are just the three which are conjugate to A_2—namely the three lines on A_1 as in (a). In particular, G_{A_3} stabilizes the unique point A_1 common to these lines. Now L is contractible, indeed via a G_{A_3}-map, as we can see via a sequence of two applications of Remark 6.5.2: First, the link in L of each of the six points P other than A_1 is just a single vertex—given by the unique line on P through A_1; and of course this unique line is fixed by $G_{A_3,P}$. Thus the link is already contracted to a single vertex, using just the identity map which is of course $G_{A_3,P}$-equivariant. So applying Remark 6.5.2 with G_{A_3} in the role of "G", we see using Lemma 6.5.1 that the removal from L of the G_{A_3}-orbit of P given by these six points is a homotopy equivalence via a G_{A_3}-map. The subcomplex L' of L remaining after this removal is just the join of the single point A_1, which is fixed by G_{A_3}, with the three lines on A_1. In particular this is a cone on A_1, which contractible by a G_{A_3}-map as in Remark 6.5.2. So using Lemma 6.5.1 now with G itself in the role of "G", we remove the G-conjugates of A_3 from $\mathcal{E}_2(G)'$; thus completing our desired reduction via a G-map homotopy equivalence to $\mathcal{E}_2^-$.

We mention that this final argument for the contractibility of L is also easily made using the "conical contractibility" method of Brown (cf. Quillen [**104**, 1.5]): for given any member Y of the link L, we have the group inclusions $Y \leq YA_1 \geq A_1$ among members of L, giving a zig-zag of equivalences from L to the point A_1.

8.2. Details for J_3

We provide details for two claims made in Section 7.8.

Uniqueness of the class of rank four. We first sketch the proof of the claim that A_4 of rank four represents the unique G-conjugacy class of maximal elements of $\mathcal{A}_2(G)$.

We write B for a representative of some initially-unspecified conjugacy class of members of $\mathcal{A}_2(G)$. Since G has just one conjugacy class of involutions, all members of $\mathcal{A}_2(G)$ of rank one are conjugate to A_1. Thus we may as well assume that B contains A_1, and hence that $B \leq C_G(A_1) = H_1$.

Now assume that B is maximal in $\mathcal{A}_2(G)$; in particular, $B > A_1$ since for example A_2 represents a class with $A_2 > A_1$. In the action of $H_1/U_1 = \mathrm{Alt}_5$ as $\Omega_4^-(2)$ on the natural orthogonal 4-space $\tilde{U}_1 := U_1/A_1$ over $\mathbb{F}_2$, an involution centralizes a degenerate 2-subspace with a unique singular vector; and a Sylow 2-subgroup centralizes just the 1-space of a singular vector. Since $B > A_1$, it follows that $C_{U_1}(B)$ contains an involution not in A_1; hence by the maximality of B, $B \cap U_1 > A_1$. However $U_1 \cong 2_-^{1+4}$ has 2-rank two; so $B \cap U_1$ is of rank two and hence conjugate in H_1 to A_2, so we may as well assume that $A_2 \leq B$. Since A_4 represents a class with $A_4 > A_2$, by maximality of B we have $A_2 < B$. In particular $B \leq C_{H_1}(A_2)$; but $C_{H_1}(A_2)$ is the group $A_4 C_{U_1}(A_2)$ of order 2^6 indicated in Section 7.8, namely $C_{H_1}(A_2) = O_2\big(C_G(A_2)\big) = U_2 \cong 2^{2+4}$. Now we also indicated there that U_2 is isomorphic to a Sylow 2-subgroup of $L_3(4)$, and so in particular has just two maximal elementary abelian subgroups, each of rank four; and A_4 is conjugate to the other rank four subgroup of U_2 by an element of A_4U_1, which normalizes A_2; so we may as well assume that $B \leq A_4$, and hence by maximality of B that $B = A_4$. This completes the proof that A_4 represents the unique G-conjugacy class of maximal elements of $\mathcal{A}_2(G)$.

The homotopy equivalence. We now turn to the proof of the claim that $\mathcal{A}_2^-$ is homotopy equivalent to $\mathcal{A}_2(G)$ via a G-map. (The argument here can be considered a variation, for $\mathcal{A}_2^-$, of the equivalence proof for the 2-local geometry of J_3 given in [**116**, 2.11]).

The overall argument here will be similar in outline to that for HS in Section 8.3, in that we will remove from $\mathcal{A}_2(G)$ members B of rank three, as well as members C of rank two not conjugate to A_2. We will apply Lemma 6.5.1, showing containment of B or C in a unique conjugate of A_4, to see that we have the situation of Remark 6.5.2.

So suppose first that B denotes a member of of rank three in $\mathcal{A}_2(G)$. We just showed that A_4 represents the unique class of maximal members of $\mathcal{A}_2(G)$, so we may as well assume that $B \leq A_4$. We saw in Section 7.8 that the conjugates of A_2 in A_4 are the five 1-subspaces of A_4 regarded as a 2-space over $\mathbb{F}_4$, each containing three conjugates of A_1. Since B contains seven conjugates of A_1, it must contain at least two involutions from one of the five 1-spaces over $\mathbb{F}_4$, and hence it contains that entire 1-space; so by conjugacy under H_4 we may assume that B contains A_2, and hence $B > A_2$. Now let D denote any member of rank four over B in $\mathcal{A}_2(G)$. Then in particular, $D \leq C_G(B) \leq C_G(A_2)$; and so D lies in U_2, since this is in fact the unique Sylow 2-subgroup of $C_G(A_2)$. But we saw in our earlier arguments that U_2 has the structure of a Sylow 2-subgroup of $L_3(4)$, so that B of rank 3 lies in a unique subgroup of U_2 of rank four. Since we chose B in A_4, we conclude that $D = A_4$; so we have shown that B of rank 3 lies in a unique conjugate of A_4. Thus we may apply Lemma 6.5.1—via Remark 6.5.2, in which we use the uniqueness of A_4 over B to see that the stabilizer G_B normalizes A_4. We obtain a homotopy equivalence of $\mathcal{A}_2(G)$ via a G-map with the subposet $\mathcal{A}^1$ defined by removing members B of rank three.

Finally assume that C is of rank two in $\mathcal{A}_2(G)$, but not conjugate to A_2. As usual we may assume that $C > A_1$, so that $C \leq H_1$. Since the 4-subgroups of U_1 are conjugate to A_2, we have $C \not\leq U_1$. Again since A_4 represents the unique class of maximal members of $\mathcal{A}_2(G)$, C is contained in some conjugate of A_4; and since that conjugate lies in $C_G(C) \leq C_G(A_1) = H_1$, while rank-four subgroups of H_1 are conjugate in H_1, we may in fact assume that $C \leq A_4$, and hence C centralizes A_2. Then C lies in the unique Sylow 2-group U_2 of $C_G(A_2)$. Next using our description of the action of H_1/U_1 at the start of the section, the image A_2/A_1 is the unique singular vector of U_1/A_1 centralized by the image of C in H_1/U_1; hence any further subgroup D of rank four over C normalizes A_2. Indeed D lies in U_2: for involutions of H_2 interchanging the two groups of rank 4 in U_2 can only lie in subgroups of rank up to three, formed from their product with A_2. Again from the structure of U_2 as a Sylow 2-group of $L_3(4)$, we see since $C > A_1$ but $C \not\leq U_1$ that C is contained in only one subgroup of rank 4 in U_2; since we chose $C \leq A_4$, we conclude that $D = A_4$. Thus we have shown that C lies in a unique conjugate of A_4. So as in the previous paragraph, we may apply Lemma 6.5.1 via Remark 6.5.2, to obtain a homotopy equivalence of $\mathcal{A}^1$ via a G-map with the subposet $\mathcal{A}_2^-$ defined by removing members of rank two which are not conjugate to A_2. This completes the proof.

8.3. Details for HS

We provide details for the two claims whose proofs we postponed from Section 7.10.

The reduction from $\mathcal{E}_2(G)$ to $\mathcal{E}_2^-$. We first sketch a proof of the claim that $\mathcal{E}_2^-$ is homotopy equivalent to $\mathcal{E}_2(G)$ via a G-map; as usual we will use Lemma 6.5.1, typically in the situation of Remark 6.5.2.

We will use notation B_i for certain further subgroups (so that we are *not* in this area using the earlier convention $B_i = U_i$ for the subgroups B_J in Section 7.10).

We begin by characterizing the members $B_2 \in \mathcal{E}_2(G)$ of rank 2 which are not G-conjugate to A_2. Since B_2 consists of involutions of type $2A$, up to G-conjugacy we may assume that $A_1 \leq B_2$; and it will be convenient to make the remainder of our preliminary arguments under this further assumption. In particular we have $B_2 \leq C_G(A_1) = H_1$. Now in Section 7.10, we saw that $A_2 \leq O_2(H_1) = U_1$, and further by H_1-conjugacy that all involutions of $U_1 \setminus A_1$ generate with A_1 a G-conjugate of A_2. Indeed there is a converse statement: since $N_G(A_2)$ is transitive on the three involutions of A_2, we see that the G-conjugates of A_2 containing A_1 are just the H_1-conjugates of A_2; these all lie in the normal subgroup U_1 of H_1. Thus

(a) For B_2 a member of $\mathcal{E}_2(G)$ of rank 2 containing A_1,
B_2 is not G-conjugate to A_2 if and only if $B_2 \not\leq U_1$.

We continue to study properties of B_2 in the situation of (a). In particular note that $B_2 \cap U_1 = A_1$. Also B_2 does not lie in any conjugate of A_3—since we saw in Section 7.10 that all rank-two subgroups of A_3 are conjugate to A_2. Next from the structure of H_3 in Section 7.10 we see that $A_3 = \Omega_1\big(C_G(A_3)\big)$. So since all involutions of the commutator subgroup H_1' are H_1-conjugate to involutions in $H_3 = N_G(A_3)$, we see since B_2 is in no conjugate of A_3 that

(b) $B_2 \not\leq H_1'$, so that $B_2 \cap H_1' = U_1$.

Now an involution $b \in B_2 \backslash H_1'$ induces a field automorphism on $H_1'/U_1 \cong L_2(4)$; and then b centralizes just three nonzero vectors of $\tilde{U}_1$, one from each of three distinct 1-spaces over $\mathbb{F}_4$. The involutions in the subgroup generated by b and the preimages of these three vectors then define a conjugate of A_4, which we may as well call A_4. In particular, this establishes the characterization of A_4 via $b \in H_1 - H_1'$ which we had indicated in Section 7.10. In addition we will show that

(c) B_2 lies in a unique G-conjugate of A_4.

For a further conjugate C of A_4 over B_2 would require some further involution c commuting with b, but lying outside U_1—since the involutions in $C_{U_1}(b)$ are in A_4. Then c with A_1 would generate a 4-group B_2' not lying in U_1, and hence not conjugate to A_2 by (a). So applying to B_2' the arguments following (b), we see that B_2' is in fact defined by A_1 and a c given by some further field automorphism of $L_2(4)$. But now the product bc would define an involution in $H_1' - U_1$, namely an inner involution of H_1/U_1, with $\langle bc \rangle A_1$ a 4-group not lying in U_1, and hence not conjugate to A_2 by (a); and this is contrary to (b). Thus (c) holds.

We are now ready to proceed with our equivalence argument.

We consider first any member B_3 of $\mathcal{E}_2(G)$ which is of rank three, but is not conjugate to A_3; as usual we may assume by conjugacy that $A_1 \leq B_3$, so that $B_3 \leq H_1$. We will reduce to the case where $B_3 = A_2 B_2$, where B_2 is described in (a).

Suppose first that $B_3 \leq U_1$. Then since B_3 is not conjugate to A_3, we see in the quotient 4-space $\tilde{U}_1$ that the three nonzero vectors $\tilde{v}_1$, $\tilde{v}_2$, $\tilde{v}_3$ in the 4-group $\tilde{B}_3$ lie in different 1-subspaces over $\mathbb{F}_4$. We may as well choose A_2 as being given by the involutions in the preimage of $\langle \tilde{v}_1 \rangle$: say by an involution v_1, along with the generator a_1 of A_1. Now by the earlier characterization (7.10.1), A_3 is determined as the intersection $U_1 \cap U_{v_1}$ of conjugates of U_1; and this intersection does not contain v_2, by the distinctness of the 1-spaces just indicated. In particular, $v_2 \notin U_{v_1}$. We may write $B_3 = A_2 B_2$, where B_2 is generated by v_1 and a suitable involution v_2 from the preimage of $\langle \tilde{v}_2 \rangle$. Furthermore we also make a change of notation, to regard v_1 rather than a_1 as the generator of A_1; this amounts to conjugating by an element of $N_G(A_2)$ which interchanges v_1 and a_1. Hence we now have $A_1 \leq B_2$; and the earlier statement $v_2 \notin U_{v_1}$ now takes the form $v_2 \notin U_1$. Thus $B_3 = A_2 B_2$ with B_2 as described in (a), completing our reduction under the present assumption. (Of course that original assumption was $B_3 \leq U_1$, which is changed under our new notation to a case where $B_3 \not\leq U_1$, since $B_2 \not\leq U_1$ by (a).)

Now suppose instead that $B_3 \not\leq U_1$. Then B_3 of rank 3 contains a subgroup B_2 of rank 2 containing A_1, with $B_2 \not\leq U_1$; hence B_2 is described in (a). If we had $B_3 \cap U_1 = A_1$, then since B_3 is of rank 3, it would contain a second 4-subgroup B_2' of rank 2 and containing A_1, but not lying in U_1—and this leads to the same contradiction via "bc" as in the proof of (c). Hence $B_3 \cap U_1$ is of rank 2 and contains A_1, and we saw in proving (a) that such a 4-group is a conjugate of A_2. So letting A_2 denote that conjugate, we have $B_3 = A_2 B_2$, and the reduction is complete in this remaining case as well.

We now proceed on the basis of the reduction to $B_3 = A_2 B_2$, with B_2 as in (a): then by (c), B_2 lies in a unique conjugate of A_4, which again we may as well call A_4. It follows that the upper poset $\mathcal{E}_2(G)_{>B_3}$ has the unique element A_4. Hence we are in the usual situation of Remark 6.5.2, so that by Lemma 6.5.1, the process of removing all G-conjugates of B_3, reducing $\mathcal{E}_2(G)$ to a subposet $\mathcal{E}'$, is a homotopy equivalence via a G-map.

The previous paragraph removed members of $\mathcal{E}_2(G)$ of rank three other than the conjugates of A_3. It remains to remove members of $\mathcal{E}'$ of rank two which are not conjugates of A_2—namely members B_2 described in (a). But since we saw just after (a) that B_2 is contained in no conjugate of A_3, it follows again using (c) that the upper poset $\mathcal{E}'_{>B_2}$ has the unique element A_4; so again we apply Remark 6.5.2 and Lemma 6.5.1, so that the reduction of $\mathcal{E}'$ to $\mathcal{E}_2^-$ is a homotopy equivalence via a G-map, as required.

The reduction to $\Delta_{\mathcal{A}_I}$. Next we turn to our sketch of the proof that $\Delta(\mathcal{E}_2^-)$ is homotopy equivalent via a G-map to the almost-poset complex $\Delta_{\mathcal{A}_I}$ with type set $I = \{1, 3, 4\}$. Observe that $\mathcal{E}_2^-$ plays the role of "$\mathcal{E}$" in Lemma 6.5.3: in particular, $\mathcal{E}_2^-$ contains the intersection $A_2 = A_3 \cap A_4$. Indeed in the notation that result, the intersections "E_σ" for σ of types 13 or 14 or 124 are conjugates of A_1, while those of type 34 are conjugates of A_2.

So to apply Lemma 6.5.3, it remains to verify the hypothesis of contractibility of the subcomplex "L_E" for each $E \in \mathcal{E}_2^-$: namely for each $i \in \{1, 2, 3, 4\}$, we consider the subcomplex of $\Delta_{\mathcal{A}_I}$ given by the subcomplex

$$L_{A_i} = \{\sigma \in \Delta_{\mathcal{A}_I} : A_i \leq E_\sigma\}.$$

Observe for A_1 of rank one that A_1 itself is the only vertex v of type 1 with E_v (a conjugate of A_1) containing A_1. Indeed we see that the subcomplex L_{A_1} is the join of the vertex A_1 with the complex determined by the conjugates of A_3 and A_4 containing A_1. This is a cone on A_1, and hence contractible via an H_1-map as in

Remark 6.5.2. Next for A_2 of rank two, there is no vertex v of type 1 with $A_2 \leq E_v$ since E_v is a conjugate of A_1 of rank one. We saw in Section 7.10 that A_3 is the unique conjugate of A_3 over A_2; so A_3 is the unique vertex v of type 3 with $A_2 \leq E_v$. Indeed we see that the subcomplex L_{A_2} is the graph given by the join of A_3 with the two conjugates of A_4 over A_2 observed earlier; again this is a cone on A_2, and hence contractible via an $N_G(A_2)$-map as in Remark 6.5.2. Finally recall that A_3 and A_4 are maximal under inclusion in $\mathcal{E}_2(G)$, and hence in $\mathcal{E}_2^-$. So for $i = 3$ or 4, the subcomplex L_{A_i} consists of the single vertex A_i, trivially contractible via an H_i-map. So by Lemma 6.5.3, we obtain the desired homotopy equivalence of $\mathcal{E}_2^-$ with $\Delta_{\mathcal{A}_I}$ via a G-map.

8.4. Details for Fi_{22}

In this section we provide some further details in support of our treatment of Fi_{22} in Section 7.16.

These details primarily concern a certain rank 3 geometry for $U_4(2){:}2$—which arises in the link, in the rank 4 geometry for Fi_{22} in Section 7.16, of a vertex of the type there denoted by $H_1 \cong \left(2 \times 2_+^{1+8}{:}U_4(2)\right){:}2$.

A rank 3 geometry for $U_4(2) : 2$**.** This particular geometry for $U_4(2) : 2$ has seemingly not been explicitly studied in the earlier literature on group geometries. However, one could regard it as being implicitly treated by the corresponding discussion of the 2-local structure of $U_4(2) : 2$ in the work of Kitazume and Yoshiara [**80**], particularly for their Lemma 14. (That analysis was in turn part of their determination of the poset of 2-radical subgroups for Fi_{22}.)

It arises as a one-node extension of the triples geometry for Σ_5 in Example 6.5.5. And indeed—just as was the case there—our construction of a geometry of rank 3 from *local* subgroups (in contrast to rank 2 for the building of the commutator subgroup $U_4(2)$) requires the use of an outer automorphism of $U_4(2)$, whose centralizer appears as the group denoted H_T below.

We now introduce the geometry from our usual viewpoint of a 2-local complex. We take index set $I := \{T, P, \Sigma\}$; this notation reflects the remark made just above, in that the link of a vertex of type Σ will be determined from such a triples geometry, with points P and triples T.

As always these vertices are determined by 2-local subgroups, from the Atlas [**47**, p. 163]:

$H_T \cong 2 \times \Sigma_6$,

$H_P \cong 2^{1+4}3^2 2^2$, and

$H_\Sigma \cong 2^4{:}\Sigma_5$.

As usual the H_i determine a subgroup simplex $\mathcal{H}_I$ as in Definition 4.6.2, and a 2-local complex $\Delta_{\mathcal{H}_I}$ as in Definition 6.3.5. The group H_P is the centralizer of a 2-central involution. Here H_T does not contain a Sylow 2-subgroup of G; and the common intersection H_I of order 2^5 is of index four in a Sylow group. As usual we set $U_i := O_2(H_i)$ for $i \in I$.

We propose the following diagram for this geometry:

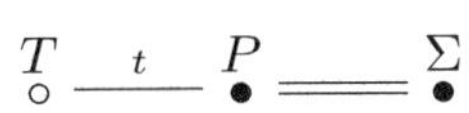

The subdiagram at T indicates the usual polar space for $H_T/U_T \cong \Sigma_6$ regarded as $Sp_4(2)$.

The subdiagram at Σ indicates that the link of H_Σ corresponds to the triples geometry for $H_\Sigma/U_\Sigma \cong \Sigma_5$ in Example 6.5.5; but since now H_I is of index 4 in a Sylow 2-subgroup S of G, there will now be 4 rather than 2 conjugates of H_I in S. In the complex $\Delta_{\mathcal{H}_I}$, these correspond to 2-simplices determined by the join of the edge $\{H_\Sigma\} * \{H_P\}$ with the 4 S-conjugates of the vertex $\{H_T\}$—and we now describe these conjugates. Set $\bar{H}_\Sigma := H_\Sigma/U_\Sigma \cong \Sigma_5$. If t is the central involution of H_T, then $\bar{t}$ determines a transposition of Σ_5; and we saw in Example 6.5.5 that $\bar{S}$ conjugates $\bar{t}$ to some commuting transposition $\bar{t}'$. Thus t has an S-conjugate t' in the preimage of $\bar{t}'$. Indeed there are two conjugates in each preimage: For $H_{\Sigma,T}$ contains only the rank-3 subspace $C_{U_\Sigma}(t)$ of U_Σ of rank 4; and the full group U_Σ conjugates t to $t^* := tu$, where u generates the commutator space $[U_\Sigma, t]$ of rank 1. Thus the S-conjugates of t are $t, t^*, t', (t')^*$; the corresponding vertices of our complex are given by their centralizers.

We need the following property in our treatment of Fi_{22} in Section 7.16.

Flag-transitivity of $U_4(2):2$ on $\Delta_{\mathcal{H}_I}$. The incidence in the 2-local complex $\Delta_{\mathcal{H}_I}$ can be analyzed via the method demonstrated for M_{24} in Remark 6.2.17 (1), especially as modified for M_{22} in Section 7.3. In summary, one just verifies that the links of vertices are the geometries indicated in the diagram.

We can then establish flag-transitivity of G on $\Delta_{\mathcal{H}_I}$ in the usual inductive way:

The link of H_T is just the polar space for $H_T/U_T \cong Sp_4(2) \cong \Sigma_6$; so we can apply Remark 6.4.6 (3).

The link of $H_\Sigma \cong 2^4\Sigma_5$ is the variant described above of the triples geometry of Example 6.5.5. We had showed at 6.5.5 that Σ_5 is flag-transitive on the triples geometry itself; and then the argument above that U_Σ conjugates t to t^* in fact extends that to flag-transitivity on our actual link.

We have now verified the condition in two of the three types of links in $\Delta_{\mathcal{H}_I}$, which suffices by Proposition 6.4.5 for flag-transitivity of $U_4(2){:}2$ on $\Delta_{\mathcal{H}_I}$.

Digression: the relationship with $\mathcal{B}_2(G)$. In fact we can show that $\Delta_{\mathcal{H}_I}$ is homotopy equivalent via a G-map with $\mathcal{B}_2(G)$—yielding a simplex cohomology decomposition for $U_4(2){:}2$, though that is not really required for our main results on sporadic groups. Nonetheless we will outline a proof, since it gives a comparatively elementary demonstration of some ways of extending the method of Lemma 6.5.7, which we will use in our treatment of Fi_{22} and the other Fischer groups.

The poset $\mathcal{B}_2(G)$ of 2-radical subgroups of $U_4(2){:}2$ was determined by Kitazume and Yoshiara in [**80**, Lemma 14]. There are 7 classes, and these are in natural bijection with the members of the subgroup simplex $\mathcal{H}_I$—more precisely we have for each type set $J \subseteq I$ that $B_J := \prod_{j \in J} U_j$ is a member of $\mathcal{B}_2(G)$. For convenience of comparison, we mention that the seven rows in [**80**, Table 3] correspond to our type sets

$$P, \Sigma, P\Sigma, T, TP, T\Sigma, TP\Sigma = I.$$

This bijection of G-orbits does not quite pull back to an isomorphism of the the poset of simplices of $\mathcal{H}_I$ with $\mathcal{B}_2(G)$. Instead we get a quotient relationship via certain identifications—a process which this time is slightly more complicated than the method of Lemma 6.5.7.

One aspect of the failure of that method is that now H_I is of index 2 in $N_G(H_I)$, which in turn is of index 2 in a Sylow 2-subgroup S of G—where S gives the class B_I in $\mathcal{B}_2(G)$ corresponding to H_I.

This means that, in proceeding from the poset $\mathcal{R}$ of simplices of $\mathcal{H}_I$ to a quotient $\bar{\mathcal{R}}$ which we can recognize as $\mathcal{B}_2(G)$, we will need to give a more ad-hoc description of the identifications that we want to make: usually the role of $N_G(H_I)$-orbits in 6.5.7 will now be played by S-orbits. Furthermore we will need to distinguish a little more carefully among certain particular possible types for the simplex $\bar{r}$, in mimicking the part of the proof of 6.5.7 establishing the contractibility of the preimage $p^{-1}(\bar{\mathcal{R}}_{\leq \bar{r}})$.

Thus we begin by paralleling the setup for 6.5.7: we specify a distinguished type $k = T$ and a distinguished subset $J = \{P\}$ of $I \setminus \{T\}$. Then for $K = \{P\}$ or its unique over-set $\{P, \Sigma\}$ in $I \setminus \{T\}$, we have that S lies in H_K. For $K^+ = \{P, T\}$ and $\{P, \Sigma, T\}$ (that is, I), we make identifications corresponding to a full S-orbit of length 4. We then continue beyond the setup of 6.5.7: for the further type $\{\Sigma, T\}$ which is not of the form K^+, we make identifications corresponding to an $N_G(H_I)$-orbit of length 2; this corresponds to pairs denoted t, t^* in our earlier discussion. These are the identifications which give the quotient $\bar{\mathcal{R}}$ the poset structure of $\mathcal{B}_2(G)$.

We can now verify contractibility of the preimage $p^{-1}(\bar{\mathcal{R}}_{\leq \bar{r}})$, by considering $\bar{r}$ of each possible type in turn. In fact this analysis proceeds essentially in parallel with the treatment of types as organized via K in the proof of 6.5.7—where for the types K^+, the preimage arises from the join of a simplex with the S-orbit of length 4 on the vertex $\{H_T\}$. The only nontrivial new case corresponds to type $\{\Sigma, T\}$—where the preimage arises from the join of the vertex $\{H_\Sigma\}$ with the $N_G(H_I)$-orbit of length two on $\{H_T\}$, which would also contain $\{H_{T^*}\}$ in notation corresponding to our earlier discussion. In particular, since this $N_G(H_I)$-orbit is contained in the S-orbit on $\{H_T\}$, these preimages are included in the above description of the full preimage for the strictly larger type I indicated above.

These extensions to the method of proof of 6.5.7 complete the proof of the poset homotopy equivalence of $\mathcal{R}$ with $\bar{\mathcal{R}}$, namely with $\mathcal{B}_2(G)$.

8.5. Details for Fi'_{24}

We provide details supporting a number of statements in Section 7.18.

The intersections defining the incidences in $\Delta_{\mathcal{A}_I}$. We first provide a more detailed derivation of certain intersections of the groups $A_i (i \in I)$, which define the incidences in the intersection complex $\Delta_{\mathcal{A}_I}$; primarily we need to show for $i \in \{11, 6, 3\}$ that $A_{8,i}$ (that is, $A_8 \cap A_i$) is a certain subgroup of rank $4, 3, 2$, respectively.

From the Atlas [**47**, p. 207], the 2-central involutions of $G := Fi'_{24}$ are in class $2B$ (including z generating A_1); while the non-2-central involutions are in class $2A$, and have centralizer $2^{\cdot}Fi_{22}{:}2$ with Sylow group of order 2^{19}.

We will discuss each $H_i (i \in I)$ in turn. Since A_1 lies in each A_i, it is useful to begin with H_1.

Now $A_1 = Z(U_1)$ for $U_1 \cong 2_+^{1+12}$. Set

$$\bar{H}_1 := H_1/U_1 \cong 3^{\cdot}U_4(3){:}2$$

and

$$\tilde{U}_1 := U_1/A_1 \cong 2^{12}.$$

Then $\tilde{U}_1$ affords a 6-dimensional $\mathbb{F}_4$-module for $\bar{H}_1' \cong 3U_4(3)$; furthermore suitable elements of $\bar{H}_1 \setminus \bar{H}_1'$ induce outer automorphisms of order two of $\bar{H}_1'$, which act semilinearly on $\tilde{U}_1$—that is, they act as field automorphisms of $\mathbb{F}_4$ over $\mathbb{F}_2$, which act freely on a suitable $\mathbb{F}_2$-basis of $\tilde{U}_1$. That module structure is determined by the inclusion of $3^{\cdot}U_4(3)$ in $SU_6(2)$, where $SU_6(2)$ acts on $\tilde{U}_1$ as its natural unitary module of $\mathbb{F}_4$-dimension six. In order to distinguish this unitary form on $\tilde{U}_1$ from other forms on other spaces which will appear in our further analysis, we will use expressions involving $\sim$, such as "$\sim$-isotropic", to specify "isotropic in the form $\sim$ on $\tilde{U}_1$", and so on. In particular, for each $i = 11, 6, 3$, $\bar{H}_{i,1}$ is the intersection of $\bar{H}_1$ with the parabolic subgroup of $SU_6(2)$ stabilizing a totally $\sim$-isotropic $\mathbb{F}_4$-subspace $\tilde{V}_i$ of $\tilde{U}_1$, with $\tilde{V}_{11}$, $\tilde{V}_6$, $\tilde{V}_3$ of $\mathbb{F}_4$-dimension 3,2,1 (respectively).

We next discuss H_8—from the viewpoint of the situation of $H_{8,1}$ in H_1. Set

$$H_8^* := H_8/A_8 \cong \Omega_8^-(2).$$

Then A_8 is the natural orthogonal module for $\Omega_8^-(2)$; for this form, we will use expressions involving $*$, such as "$*$-singular".

Since the centralizer in H_8 of a $*$-singular vector in A_8 has order divisible by $|H_8|_2 = 2^{20} > 2^{19}$, a $*$-singular vector lies in the 2-central class $2B$ of G rather than class $2A$; thus A_1 generated by z in class $2B$ of G is $*$-singular. However the centralizer in H_8 of a $*$-nonsingular vector involves $Sp_6(2)$, a group which is not involved in

$$C_G(z) = H_1 \cong 2^{1+12} \cdot 3U_4(3)2;$$

so the $*$-nonsingular vectors in A_8 lie in class $2A$ of G.

In particular, $H_{8,1}^*$ is the stabilizer in H_8^* of the $*$-singular vector z generating $A_1 \le A_8$; hence $H_{8,1}^* \cong 2^6 : \Omega_6^-(2)$. Let

$$W_1 := A_1^{\perp}$$

denote the $*$-perpendicular space of A_1. Then the quotient $\Omega_6^-(2)$ in $H_{8,1}^*$ acts irreducibly on the rank six section W_1/A_1 of A_8; this section of course also appears as the 6-subspace $\tilde{W}_1$ of $\tilde{U}_1$. Since the 2-rank of $\bar{H}_1 \cong 3^{\cdot}U_4(3){:}2$ is less than 6, it follows by the above irreducibility that $A_8 \cap U_1$ contains W_1 of rank 7. So since $U_1 \cong 2_+^{1+12}$ has 2-rank 7, we conclude that $A_8 \cap U_1$ is of rank exactly 7, with $A_8 \cap U_1 = W_1$. We let t denote an element of $A_8 \setminus U_1 = A_8 \setminus W_1$; then $\langle \bar{t} \rangle = \bar{A}_8 = A_8U_1/U_1$ is of rank one, and hence $\bar{t}$ is central in $\bar{H}_{8,1}$. Since $H_{8,1}^*/O_2(H_{8,1}^*) \cong \Omega_6^-(2)$, we see that $C_{\bar{H}_{8,1}}(\bar{t})$ involves $\Omega_6^-(2)$. Further as $\bar{H}_{8,1}$ does not contain the central 3-element of $3U_4(3)$, we see that $\bar{t}$ induces an outer automorphism of $3U_4(3)$ inverting that center. It follows that $\bar{t}$ induces an automorphism in the class which is denoted by 2_2 in the Atlas [**47**, p. 52], and that $\bar{H}_{8,1} \cong \Omega_6^-(2) \times \langle \bar{t} \rangle$. We conclude further that $\tilde{U}_1$ affords the sum of two 6-dimensional orthogonal $\mathbb{F}_2$-modules for $\Omega_6^-(2)$ which are interchanged by t; so $\tilde{W}_1 = C_{\tilde{U}_1}(t) = [\tilde{U}_1, t]$ is a natural orthogonal module for the $\Omega_6^-(2)$ in $\bar{H}_{8,1}$. For this latest form defined now on $\tilde{W}_1$, we may use expressions such as $\approx$-isotropic.

For $i \in \{11, 6, 3\}$, we can now begin to translate the above information about the situation of $H_{8,1}$ in H_1, via investigation of $H_{8,1,i}$ and $H_{1,i}$, into information about the situation of $H_{8,i}$ in H_i. (Indeed the case $i = 11$ will almost suffice for the other i.)

So for each $i = 11, 6, 3$, we recall our earlier definition of the totally $\sim$-isotropic subspace $\tilde{V}_i$ of $\mathbb{F}_4$-dimension $3, 2, 1$ in $\tilde{U}_1$—and now use the viewpoint of the $\approx$-form over $\mathbb{F}_2$ introduced just above. We see that $\bar{H}_{8,1,i}$ is the product of $\langle \bar{t} \rangle$ with the stabilizer in $\Omega_6^-(2)$ of the totally $\approx$-isotropic $\mathbb{F}_2$-subspace

$$\tilde{W}_i := C_{\tilde{V}_i}(t)$$

of $C_{\tilde{U}_1}(t) = \tilde{W}_1$, with $\tilde{W}_i$ of $\mathbb{F}_2$-dimension $3, 2, 1$ (respectively). Of these, the spaces $\tilde{W}_6$ and $\tilde{W}_3$ are totally $\approx$-singular. Further $\tilde{W}_{11}$ uniquely determines $\tilde{W}_6$ as the subspace consisting of all the $\approx$-singular vectors in $\tilde{W}_{11}$. In particular, we see that $H_{8,1,11} = H_{8,1,6,11}$; this is a first step toward our eventual verification of the desired coincidence of stabilizers $H_{8,11} = H_{8,6,11}$. The preimage W_i of $\tilde{W}_i$ is a totally $*$-isotropic subspace (of dimension $4, 3, 2$) of the 7-subspace W_1 of A_8. Eventually we will see that W_i also lies in A_i; and we will conclude that $W_i = A_8 \cap A_i = A_{8,i}$, which is the primary goal of this part of the proof.

We consider first $i = 11$; we begin by discussing H_{11} and $H_{11,1}$. Set

$$H_{11}^+ := H_{11}/A_{11} \cong M_{24}.$$

Recall that A_{11} is the cocode module for H_{11}^+ as in our discussion in Example 6.3.7; that is, for Ω a 24-set permuted by M_{24}, we can identify A_{11} with the quotient of the 23-subspace of even subsets of Ω, modulo its 12-subspace given by the Golay code on the Steiner system $\mathcal{S}(5, 8, 24)$. Now a 4-subset of Ω determines a unique sextet of the Steiner system; the corresponding sextet vector of A_{11} is centralized by the sextet stabilizer of M_{24}, which contains a Sylow 2-subgroup of M_{24}. Thus this sextet vector is 2-central in G, so we may take it to be the 2-central element z, in the class $2B$ of G, which generates A_1. Thus $H_{11,1}^+$ is the stabilizer in M_{24} of the sextet vector A_1; this stabilizer has structure $2^6 : 3{\cdot}\Sigma_6$. A nonzero vector in the other M_{24}-class in A_{11} is determined by a 2-subset of Ω, so that its centralizer involves M_{22}, which is not involved in

$$C_G(z) = H_1 \cong 2_+^{1+12}{\cdot}3U_4(3)2;$$

hence these vectors lie in the non-2-central class $2A$ of G. Next viewing the image $\bar{H}_{11,1}$ in $\bar{H}_1 \cong 3{\cdot}U_4(3){:}2$, we obtain a subgroup of structure $2^4{:}3{\cdot}\Sigma_6$. In particular, our earlier involution $t \in A_8 \setminus U_1$, which induces an outer automorphism of $3{\cdot}U_4(3)$ of type 2_2, lies outside the subgroup $3{\cdot}\mathrm{Alt}_6$ of the quotient $\bar{H}_{11,1}/O_2(\bar{H}_{11,1}) \cong 3{\cdot}\Sigma_6$.

Now it is standard (see for example [**20**, H.16.1.1]) that the composition series on A_{11} of $H_{11,1}^+ \cong 2^6{:}3{\cdot}\Sigma_6$ has descending irreducible sections of dimensions 4/6/1. Arguing on 2-ranks much as in the earlier case of A_8, we conclude that $A_{11} \cap U_1$ is of rank seven, coming from the factors 6/1; indeed we see that $A_{11} \cap U_1$ is the preimage V_{11} of the 3-dimensional $\mathbb{F}_4$-space $\tilde{V}_{11}$ described earlier. Thus $C_{\tilde{V}_{11}}(t) = \tilde{W}_{11}$ is of $\mathbb{F}_2$-dimension three. We also saw earlier that $C_{\tilde{U}_1}(t) = \tilde{W}_1$, where $W_1 = A_1^\perp$ of $\mathbb{F}_2$-dimension seven was defined in the $*$-form on A_8; so we conclude that $A_{11,8} = A_{11} \cap A_8$ contains at least the preimage W_{11} of $\tilde{W}_{11}$, where W_{11} is of rank four. On the other hand, t centralizes at most a 2-subspace of the 4-dimensional composition factor above, so that $A_{11,8}$ is of rank at most six.

We may also consider the situation of $A_{11,8} \geq W_{11}$ in H_8: It is standard that the overgroup $O_{10}^-(2)$ of H_8 has 2-rank 10 (see for example [**64**, Table 3.3.1]); hence A_{11} of rank 11 does not lie in $H_8 = N_G(A_8)$. However, a Sylow 2-subgroup of H_8 is of index two in a Sylow of G, so we conclude that $A_{11} \cap H_8 = N_{A_{11}}(A_8)$

is of rank 10. Since $A_{11,8}$ is of rank at most six, $\overline{A_{11} \cap H_8}$ is of rank at least four. Furthermore $\overline{A_{11} \cap H_8}$ lies in the centralizer in $\bar{H}_8$ of the totally $*$-isotropic 4-subspace W_{11} of $A_{11,8} \leq A_8$; the image of this centralizer lies in the unipotent radical 2^{3+6} of a parabolic subgroup $2^{3+6}L_3(2)$ of $\Omega_8^-(2)$. And now the action of an $L_3(2)$ in $H_{11,6,8}$ on that radical forces $\overline{A_{11} \cap H_8}$ of rank at least four to in fact have rank at least six—and hence rank exactly six, since that is the 2-rank of the radical. It follows that $A_{11,8}$ has rank exactly four, and so from the containment in the previous paragraph must equal W_{11}.

The above determination of $A_{11,8}$ as W_{11} of rank four accomplishes the primary purpose of our analysis of the case $i = 11$. However we will continue our study of this case, in order to also provide details supporting the above mentioned coincidence of stabilizers $H_{11,8} = H_{11,6,8}$, which is relevant to the structure of the indexing sets for the decompositions in our main results for Fi_{24}' in Section 7.18. We are grateful to Satoshi Yoshiara for suggestions in this area of the proof.

So we return to considering the situation of $A_{11,8}$ in H_{11}. Since $A_{11,8}$ is of rank four, the quotient

$$A_8/(A_8 \cap A_{11,8}) \cong A_8^+$$

is of rank four; and of course A_8^+ lies in the centralizer C^+ in H_{11}^+ of the 4-space $A_{11,8}$. Since we identified $A_{11,8}$ above as W_{11}, we see that C^+ lies in the centralizer D^+ in H_{11}^+ of the earlier 3-subspace W_6 of W_{11}. Now W_6 is in fact a 3-subspace of A_6, with W_6 totally singular in the orthogonal form on A_6 defined by the subgroup $\Omega_6^+(2) \cong L_4(2)$ in $H_{11,6}^+ \cong 2^4{:}L_4(2)$; but W_6 is in the $H_{11,6}$-conjugacy class different from that of A_3. We see then that D^+ contains the centralizer E^+ in $H_{11,6}^+$ of W_6, namely a subgroup 2_+^{1+6}, which is the extension of $O_2(H_{11,6}^+)$ of rank four by the unipotent radical of the corresponding parabolic $2^3 : L_3(2)$ of $H_{11,6}^+/O_2(H_{11,6}^+) \cong L_4(2)$. We let x^+ denote the generator of $Z(E^+)$; then x^+ is a 2-central involution of $H_{11}^+ \cong M_{24}$. It follows that $H_{11,6}^+$ is the unique maximal subgroup of H_{11}^+ containing the centralizer F^+ in H_{11}^+ of x^+; and $H_{11,6}^+$ does not normalize W_6. We conclude then that F^+ is the full normalizer in H_{11}^+ of W_6, and hence that $D^+ = E^+ \cong 2_+^{1+6}$ is the full centralizer in H_{11}^+ of W_6. Since A_8^+ is of rank four and lies in $C^+ \leq D^+$, we see that $x^+ \in A_8^+$. Now $H_{11,8}$ stabilizes the 4-space $A_{11,8} = W_{11}$ and its singular 3-subspace W_6, and hence normalizes $C_{H_{11}^+}(W_6) = E^+$; so $H_{11,8}^+$ centralizes $Z(E^+) = \langle x^+ \rangle$, and then

$$H_{11,8}^+ \leq C_{H_{11}^+}(x^+) = F^+ \leq H_{11,6}^+.$$

This finally gives the desired coincidence of stabilizers $H_{11,8} = H_{11,6,8}$.

We consider next the case $i = 6$, briefly reviewing the structure of H_6. We have $H_6/U_6 \cong \Sigma_3 \times L_4(2)$; we let $H_6^!$ denote the subgroup of H_6/U_6 isomorphic to $L_4(2) \cong \Omega_6^+(2)$. Then A_6 is a natural 6-dimensional orthogonal module for $\Omega_6^+(2)$; for this form we may write such expressions as !-singular. Since $A_6 \leq A_{11}$, we see that

$$A_{8,6} = A_8 \cap A_6 \leq A_8 \cap A_{11} = A_{8,11},$$

where $A_{8,11}$ is described above. Indeed our discussion above showed that $A_{8,11} \cap A_6$ is the subgroup W_6 of rank three in $W_{11} = A_{8,11}$ which contains all the $*$-nonsingular vectors of W_{11}. We conclude that $A_{8,6} = W_6$ is of rank three; and that W_6 is a totally !-singular 3-subspace of A_6, in the $H_6^!$-class distinct from that of A_3. From our earlier discussion we obtain that $A_{8,6}$ is the subspace of A_{11} fixed by $O_2(C_{H^+_{11,6}}(x^+)) \cong 2^{1+6}_+$. These were the details we needed concerning $A_{8,6}$.

Similar arguments apply to the case $i = 3$. We recall that

$$H_3/U_3 \cong 3\,\Sigma_6 \times L_3(2),$$

where A_3 is a natural module for the quotient $L_3(2)$ of H_3. Again since $A_3 \leq A_{11}$, we have

$$A_8 \cap A_3 \leq A_8 \cap A_{11} = A_{8,11} = W_{11}.$$

This time the earlier discussion shows in fact that $A_{8,11} \cap A_3$ is the 2-subspace W_3 of $W_6 \leq W_{11}$, so that $A_{8,3} = W_3$ is of rank two, as required.

This completes our derivation of the intersections of the subgroups A_i which determine the incidences in our intersection complex $\Delta_{\mathcal{A}_I}$.

Flag-transitivity of rank four extension geometries for M_{24} and $\Omega_8^-(2)$. We now give details for the flag-transitivity indicated in Section 7.18, of the groups M_{24} and $\Omega_8^-(2)$ on the rank four extensions of their usual geometries.

Orthogonal groups using isotropic subspaces. In the case of $\Omega_8^-(2)$, we may as well indicate a general inductive proof for the flag-transitivity of $X := \Omega_{2n}^-(2)$ ($n \geq 2$) on the rank n geometry given by adjoining, to the usual building of totally singular subspaces of rank up to $n-1$, the class of maximal totally *isotropic* subspaces of dimension n. The argument can be regarded as an application of Witt's lemma, and so is surely well known at least in principle; however the use of the maximal totally isotropic subspaces of dimension n is seemingly not customary in the geometric literature.

The base case of the induction is for $n = 2$, namely for $\Omega_4^-(2)$. Here we just observe that the stabilizer Σ_4 of a singular point has quotient Σ_3 which is transitive on the three totally isotropic 2-subspaces containing that point. (This geometry is essentially that of $\mathcal{B}_2(\Omega_4^-(2))$; by contrast the triples geometry for Σ_5 in Example 6.5.5 has six rather than three triples on each of the five points.) For later use, we also mention that the case $n = 1$ would correspond just to the transitivity of $\Omega_2^-(2) \cong L_2(2)$ on the three isotropic (but nonsingular) points of orthogonal 2-space of minus type.

For the inductive case with general n, in our usual method it suffices by Proposition 6.4.5 to establish flag-transitivity on links of all but one type. That is, we take a space Y of each dimension in the geometry (we are allowed to omit one possible dimension, but need not); and we check flag-transitivity on the link of Y by the stabilizer of Y. For example, for a totally singular subspace A of dimension 1, the link can be exhibited in $A^\perp/A$ as the corresponding geometry for the case $n-1$: so that we have flag-transitivity by the inductive assertion. More generally for a totally singular subspace Y of dimension i with $1 < i \leq n-1$, the link is the join of the projective space from Y, of rank $i-1$, with the corresponding geometry of isotropic subspaces in $Y^\perp/Y$, of rank $n-i$. So since the stabilizer contains the product of the linear group on Y with the orthogonal group on $Y^\perp/Y$, and induction applies to the latter factor (using our earlier observation following the base case when $i = n-1$), we can apply Remark 6.4.6 (2) to obtain flag-transitivity on the link. Finally for a totally isotropic subspace W of dimension n, since W has a unique totally singular subspace V of dimension $n-1$ consisting of all the singular vectors of W, the link of W is just the join of the vertex V with the projective space $\mathbf{P}(V)$. Flag-transitivity on this link then comes just from the flag-transitivity in Remark 6.4.6 (2) of the subgroup X_W, of index three in the stabilizer X_V, on that projective space.

We comment finally that though we have flag-transitivity, some of the stabilizers are not entirely independent: for the above stabilizer X_W is of index three in X_V, with similar relations following for stabilizers of larger flags.

The rank four geometry of M_{24}. It remains to consider $G := M_{24}$ on the rank four extension of its usual rank three 2-local geometry, where the extension is given using the type set $\{6, 3, 1, 4\}$ described in Section 7.18, for an intersection complex on the module A_{11}. We will refer to this complex by the name Δ. Again by the usual inductive method, we consider flag-transitivity on their links in Δ of the stabilizers of vertices A_6, A_3, and C_4, which suffices by Proposition 6.4.5 for the flag-transitivity of G.

We consider first the link in Δ of A_6; let S denote the stabilizer

$$S := N_{M_{24}}(A_6) \cong 2^4\Omega_6^+(2).$$

The incidence of conjugates of C_4 with A_6 is determined by containment of a conjugate of $C_3 = C_4 \cap A_6$. From our earlier discussion of intersections, C_3 is a totally !-singular 3-subspace of the orthogonal space A_6, determined by C_4 as its set of involutions in class $2B$ of G. The 15 M_{24}-conjugates of C_3 in A_6 are in turn determined by the S-conjugates of the involutions in $O_2(S)$ (say x^+ in our earlier notation), as the subgroup of all $2B$-involutions in the commutator space $[A_{11}, x^+]$. Hence conversely, C_3 is contained in a unique M_{24}-conjugate of C_4, determined via x^+ as $C_4 = [A_{11}, x^+]$. We conclude that the link of A_6 is isomorphic to the complex of the particular totally !-singular subspaces of A_6 given by the S-classes of A_3, A_1, and C_3: this is just the external almost-poset complex view of the usual building for the $\Omega_6^+(2) \cong L_4(2)$ in S. So flag-transitivity holds here via Remark 6.4.6 (3) using that property for buildings.

The link of C_4 is touchier to analyze: Here we are not dealing with a poset complex, and the geometry is not naturally described by a diagram—indeed, diagrammatic methods might lead us to expect this link to be a projective space for $L_4(2)$—whereas the actual link is smaller. We begin by observing that our earlier discussion of C_4 as $A_{11,8}$ essentially showed that A_6 is the unique member of its M_{24}-conjugacy class containing C_4; here is a more detailed argument for this fact: Let A_6' be an M_{24}-conjugate of A_6 containing C_4. Then $C_{M_{24}}(A_6')$ lies in $C_{M_{24}}(C_3) \cong 2_+^{1+6}$, where the latter subgroup has center x^+ in our earlier notation. Now $C_{M_{24}}(A_6) \cong 2^4$; so by conjugacy, also $C_{M_{24}}(A_6') \cong 2^4$. So this group is of rank four in 2_+^{1+6}, and hence contains x^+. Next x^+ centralizes just a 7-subspace of A_{11}, including A_6, and also including A_6' in view of the previous sentence; it follows that $A_6' \cap A_6$ is of rank at least 5. But the centralizer in $\Omega_6^+(2)$ of a hyperplane of A_6 is trivial; hence $C_{M_{24}}(A_6') \leq C_{M_{24}}(A_6)$, so that by conjugacy, in fact equality holds. Since $A_6 = C_{A_{11}}(C_{M_{24}}(A_6))$, we conclude that $A_6' = A_6$. This completes the proof that C_4 is incident with a unique conjugate of A_6, namely A_6 itself. Thus the link of C_4 contains only one vertex A_6 of that class. It follows that the full flags in the link of C_4 are formed with conjugates of A_3 and A_1 contained in A_6. Of course, we use only those conjugates of A_1 lying in C_4, and hence lying in the subgroup C_3 of elements of class $2B$ in C_4. Similarly, we use only those conjugates of A_3 intersecting C_4 in a 2-subspace of C_3. We conclude that the link of C_4 is isomorphic to the join of the vertex A_6 with the link in A_6 of C_3; where the latter link has the structure of the projective space $\mathbf{P}(C_3)$. Hence again we have flag-transitivity using Remark 6.4.6 (3) in view of that property for buildings.

Finally we consider the link of A_3. Again the viewpoint of diagram geometries might suggest that this link should be the join of a projective line (given by the three vertices of type A_6 over A_3) with the projective space $\mathbf{P}(A_3)$. But the actual structure of the link is not a join: for we saw in the previous paragraph that each conjugate of C_4 is incident with a unique conjugate of A_6, rather than all three indicated above. Thus there are indeed seven conjugates of A_1 contained in A_3. But there are 21 conjugates of C_4 incident to A_3: one for each of the 21 2-central involutions in $O_2(S)$ for the stabilizer $S := N_{M_{24}}(A_3)$; these occur in sets of three which intersect A_3 in the same choice of one of the seven conjugates of the 2-subspace $C_2 = A_3 \cap C_4$ of A_3. And each of the three conjugates of C_4 for such a triple of involutions is incident with just one of the three conjugates of A_6 over A_3. Although this structure is more complicated than usual, the links are still not difficult to treat: The link of A_6 is isomorphic to the projective space $\mathbf{P}(A_3)$ via the intersections of form $C_2 = C_4 \cap A_3$, so we may apply Remark 6.4.6 (2). And the link of C_4 is the join of the unique vertex A_6 of its type with a projective line given by the three conjugates of A_1 in C_2, so we may apply Remark 6.4.6 (3) to this product building. Thus verifying the condition in these sublinks of two of the three types suffices by Proposition 6.4.5 to give flag-transitivity on the link of A_3 of rank three.

Hence our verification of the condition on the above rank three links, of three of the four types, suffices by Proposition 6.4.5 to establish the flag-transitivity for M_{24} on Δ of rank four.

Finally, as in the orthogonal case, we remark that although we have flag-transitivity, we also have coincidences of stabilizers: since the stabilizer in M_{24} of C_4 is in fact the stabilizer of A_6 and C_4.

Digression: Further remarks on rank four geometries for M_{24}. We conclude the section with some discussion which is not in fact required for our main results on Fi_{24}' in Section 7.18; but which may provide useful background, by demonstrating a fairly elementary example of some of the homotopy equivalence arguments presented there.

Namely we will sketch for $G := M_{24}$ the close relationship of the rank four geometry Δ described above with the poset $\mathcal{B}_2(G)$ of 2-radical subgroups. (And so one consequence is a cohomology decomposition for M_{24} using a larger indexing category than that presented in our main results for M_{24} in Section 7.5.)

The poset $\mathcal{B}$ and its relationship with $\mathcal{B}_2(G)$. In fact it will be convenient to change our viewpoint for a while: Above we have been considering the rank four extension Δ of the usual 2-local geometry for M_{24}—realized as an external intersection complex on the cocode module A_{11} via certain subspaces having dimensions $\{6, 3, 1, 4\}$. This is not an actual poset complex; that is, it is not the collection of chains from some poset. However, we will now turn our attention to a related construction which is a poset complex—namely an internal poset complex $\Delta(\mathcal{B})$ determined by a certain poset $\mathcal{B}$ of 2-subgroups of G; and we will comment on the specific relationship of Δ to $\mathcal{B}$ during the course of our development.

The members of $\mathcal{B}$ will be represented up to conjugacy by subgroups B_J, having types J given by subsets of the type set $I := \{O, T, S, \square\}$. This set just extends, by the use of $\square$, the notation we had used for our rank three type set for M_{24} in Section 7.5.

In overview, we will use the fact that M_{24} has the same Sylow 2-subgroup as the Lie type group $L_5(2)$ of BN-rank four. The 15 representatives B_J for M_{24} will then be isomorphic to the 15 classes in $\mathcal{B}_2(L_5(2))$; but only 13 of the B_J will turn out to be 2-radical for M_{24}.

Now the 2-radical subgroups for $L_5(2)$ are easily exhibited in matrix form; in particular they arise as the products of the unipotent radicals of the maximal parabolic subgroups. In M_{24}, we have already seen three of these radicals, namely the subgroups $U_i = O_2(H_i)$ for $i \in \{O, T, S\}$ in the notation of Section 7.5. So it only remains to describe the remaining group $U_\square$.

We fix an involution in the center of our Sylow 2-group, say denoted by x^+ as above; in particular we take $x^+ \in U_O$. Then $C_G(x^+) \cong 2^{1+6} : L_3(2)$; and we can identify $U_\square$ as the rank four preimage of the 3-subspace of $O_2(C_G(x^+))/\langle x^+\rangle$ on which $L_3(2)$ has module action dual to that on $U_O/\langle x^+\rangle$.

Now one feature of M_{24} which does not hold for $L_5(2)$ is that we have $C_G(x^+) < H_O \cong 2^4 : L_4(2)$ for the above octad O. And we might regard the poset $\mathcal{B}$ as roughly adding, to the original 2-local geometry of M_{24}, a further stabilizer subgroup corresponding to the centralizer $C_G(x^+)$.

Furthermore, to begin to relate $\mathcal{B}$ to the previous viewpoint of Δ defined on the module A_{11}, we could also realize $U_\square$ as the centralizer of one of the defining 4-subspaces of Δ (indeed the one given earlier as "C_4", that is, $[A_{11}, x^+]$). This property of $U_\square$ parallels that of the U_i for $i \in \{O, T, S\}$, since they arise as the centralizers of the subspaces in Δ of dimensions $\{6, 3, 1\}$ in A_{11}.

Returning to the above approach via the language of unipotent radicals of $L_5(2)$, we can now summarize the construction of the poset $\mathcal{B}$: For each nonempty subset $J \subseteq I$, we set $B_J := \prod_{j \in J} U_i$. And then $\mathcal{B}$ is the poset of subgroups of G determined by the G-conjugates of the B_J.

As mentioned earlier, there are $2^{|I|} - 1 = 15$ nonempty types J; whereas we saw in Section 7.5 that there are 13 conjugacy classes in $\mathcal{B}_2(G)$ for M_{24} (tabulated in An and Conder [**13**, 2J]).

And as these numbers suggest, there are two classes in $\mathcal{B}$ which are not 2-radical for M_{24}—namely $B_\square = U_\square$ and $B_{S\square}$. One checks that both their normalizers lie in $C_G(x^+)$; and then it follows using the "unexpected" containment $C_G(x^+) < H_O$ mentioned above that we get coincidences of stabilizers:

$$N_G(B_\square) = N_G(B_{O\square}) = C_G(x^+) < H_O \text{ and}$$
$$N_G(B_{S\square}) = N_G(B_{OS\square}) = N_{C_G(x^+)}(B_S).$$

So since $B_\square$ of order 2^4 is proper in $B_{O\square}$ of order 2^7, and $B_{S\square}$ of order 2^7 is proper in $B_{OS\square}$ of order 2^9, we see that $B_\square$ and $B_{OS\square}$ are indeed not 2-radical.

On the other hand, the remaining 13 groups B_J are 2-radical for M_{24}. In particular, we mention that for the other types J which properly contain $\square$ but do not contain O, we have $N_G(B_J) \nleq C_G(x^+)$, and so we do not get the above coincidence of normalizers with the type $\{O\} \cup J$. (This is in contrast with the situation for stabilizers H_J of the simplices σ_J in the complex Δ, where via intersections we do get $H_J = H_{\{O\} \cup J}$; this distinction arises in our discussion below of homotopy equivalence.) For convenience of comparison, we mention that the 13 rows in [**13**, 2J] correspond respectively to our type sets:

$$O, T, S, O\square, OT, OS, T\square, TS, OTS, TS\square, OT\square, OS\square, OTS\square = I.$$

The homotopy equivalence of Δ *with* $\mathcal{B}$. We are now in a position to discuss homotopy equivalence arguments.

The removal of the two non-radical classes represented by $B_\square$ and $B_{OS\square}$, to reduce from $\mathcal{B}$ to $\mathcal{B}_2(G)$, is a homotopy equivalence via a G-map: This is essentially part of the argument of Dwyer in [**52**, 10.4]; cf. also the middle line in the tables for Theorems 1.1 and 1.2 in Grodal and Smith [**67**].

Furthermore we had indicated in Section 7.5 that $\mathcal{B}_2(G)$ is in turn equivalent with the usual rank three 2-local geometry (call it Δ_3 for the moment) of M_{24}. So composing with the map in the previous paragraph, $\mathcal{B}$ is also equivalent with Δ_3.

It remains to describe a homotopy equivalence between $\mathcal{B}$ and the rank-four complex Δ determined on the module A_{11}. We mention first an indirect route: The equivalence in the previous paragraph can be composed with an equivalence of Δ with Δ_3 obtained via [**116**, 1.5]—we can remove the vertices corresponding to the 4-spaces "C_4" of A_{11}, since each is incident with a unique vertex of type A_6. (More precisely, the link of C_4 is contractible: it is the join of A_6 with seven vertices of each of the types A_3 and A_1, in the configuration of a projective plane over $\mathbb{F}_2$.)

But for comparison with arguments made for Fi'_{24} in Section 7.18, we will describe a more direct relationship, comparing the poset of simplices of Δ with the poset $\mathcal{B}$. The argument will involve yet another variation on the ideas in Lemma 6.5.7

First it will be notationally convention to replace Δ by the corresponding rank four 2-local complex $\Delta_{\mathcal{H}_I}$: That is, for $i \in I$ (where we continue to use $I = \{O, T, S, \square\}$), we start with the 2-local subgroup $N_G(A_i) = H_i$—where now $H_\square$ is the subgroup $N_G(U_\square) = C_G(x^+) \cong 2^{1+6} : L_3(2)$. Here, contrary to our usual practice, $H_\square$ is not a maximal 2-local, since it is proper in H_O. Indeed this gives us a coincidence of stabilizers $H_\square = H_{O\square}$; but that coincidence is not an obstacle[1] to the usual axiomatic construction of $\Delta_{\mathcal{H}_I}$. It is then straightforward to adjust the above flag-transitivity proof on Δ to also apply to $\Delta_{\mathcal{H}_I}$; and so as usual by Lemma 6.5.4 we see that $\Delta_{\mathcal{H}_I}$ really is a legitimate replacement for Δ.

We now let $\mathcal{R}$ denote the poset of simplices of $\Delta_{\mathcal{H}_I}$; and we will make certain identifications, so that the quotient poset $\bar{\mathcal{R}}$ realizes $\mathcal{B}$.

However we will not have not the description of identifications in terms of H_J and $N_G(H_J)$ that we had in 6.5.4: this time the image of H_J in the quotient will be stabilized by $N_G(B_J)$. (Furthermore we could regard a subgroup of order 3 in $O_{2,3}(H_S)$ as in effect playing the role of "$N_G(H_I)$" in 6.5.4, since it will be transitive on the $N_G(B_J)$-orbits of simplices being identified.)

Now there will be only two type sets J for which we make identifications, namely $T\square$ and $ST\square$. In particular, this is a departure from previous situations—in which identifications for a type set J would always be mirrored by similar identifications for all larger type sets containing J. Nonetheless we will still be able to make

[1]Indeed this coincidence implies the further coincidences $H_{J\square} = H_{JO\square}$ for each $J \subseteq I \setminus \{O, \square\} = \{T, S\}$. But these coincidences will also not be an obstacle to the construction; nor will they be relevant in the later equivalence arguments.

homotopy equivalence arguments using Quillen's homotopy fiber theorem [**104**, 1.6].

Before doing so, we briefly examine what is involved in the identifications. We consider the case of type set $T\square$; the configuration for the case of type $ST\square$ will then be just the join of the $T\square$ configuration with a vertex of type S. We consider first the $\mathcal{B}$-side, in order to see which identifications should be made on the $\Delta_{\mathcal{H}_I}$-side. One checks that the subgroup $B_{T\square} = U_T U_\square$ of order 2^8 in fact contains three conjugates of $U_\square$ under $N_G(B_{T\square})$; we will denote these by $U_\square$, $U_{\square'}$, and $U_{\square''}$. In particular we have a coincidence of members of $\mathcal{B}$ given by $B_{T\square} = B_{T\square'} = B_{T\square''}$, which might not have been expected on the basis of the notation alone. By contrast on the $\Delta_{\mathcal{H}_I}$-side, our standard simplex $\sigma_{T\square}$ is by definition the pair $\{H_T, H_\square\}$, whose stabilizer is the intersection $H_T \cap H_\square$, which has index three in $N_G(B_{T\square})$. Correspondingly we get an $N_G(B_{T\square})$-orbit of length 3, consisting of distinct simplices $\{H_T, H_\square\}$, $\{H_T, H_{\square'}\}$, and $\{H_T, H_{\square''}\}$. So we will identify these members (with a similar analysis for type $ST\square$), in passing from $\mathcal{R}$ to $\bar{\mathcal{R}}$. Finally we investigate the effect of the fact that we are *not* making identifications in $\mathcal{R}$ for larger type sets—notably $OT\square$. On the $\Delta_{\mathcal{H}_I}$-side, each vertex of type $\square$ is incident with a unique vertex of type O; and it follows that our three distinct simplices above are each contained in a unique simplex of the higher type, which we may denote by $\{H_T, H_\square, H_O\}$, $\{H_T, H_{\square'}, H_{O'}\}$, and $\{H_T, H_{\square''}, H_{O''}\}$. On the $\mathcal{B}$-side, we now also obtain three different members of order 2^9 when we take the product of $B_{T\square}$ with U_O, $U_{O'}$, and $U_{O''}$. Using our earlier equalities, we can denote these groups by $B_{T\square O}$, $B_{T\square' O'}$, and $B_{T\square'' O''}$. So we see that at this higher level in the posets, we have returned to a 1:1 correspondence, rather than the 3:1 correspondence we had at the $T\square$ level.

Now we indicate the argument that the projection p from $\mathcal{R}$ onto $\bar{\mathcal{R}}$ is a poset homotopy equivalence via [**104**, 1.6]; for $\bar{r} \in \bar{\mathcal{R}}$, we must show contractibility of the preimage $p^{-1}(\bar{\mathcal{R}}_{\leq \bar{r}})$. As usual there is nothing to prove unless $\bar{r}$ has a type involving identifications; and the most basic case is type $T\square$. Here the preimage is just the poset of simplices from the cone (contractible!) given by the join X of $\{H_T\}$ with the three isolated vertices $H_\square$, $H_{\square'}$, and $H_{\square''}$. For the type $ST\square$, we make a similar argument in the join of $\{H_S\}$ with this configuration X. For $\bar{r}$ of type strictly larger than $T\square$, the preimage is poset of simplices of the complex given by the direct limit of a simplex r with the configuration X, over their intersection given by the edge $\{H_T, H_\square\}$—and this complex is collapsible back to the simplex r. Similar remarks apply for types larger than $ST\square$; namely for the full type I. These considerations give the necessary contractibility conditions, and so establish the poset homotopy equivalence.

8.6. Details for HN

Let $G := HN$. We sketch the proof of the claim in Section 7.19 that $\mathcal{B}_2^{cen}(G)$ is homotopy equivalent via a G-map to the subposet $\mathcal{B}_2^{cen-}$ obtained via removing the G-class of $R := R_5^{(1)}$: We will show that $\mathcal{B}_2^{cen}(G)_{<R}$ consists of the single vertex $B_1 = U_1$; so that we are in the usual situation of Remark 6.5.2, and hence the link in $\mathcal{B}_2^{cen}(G)$ of R is contractible via an $N_G(R)$-map to U_1, giving the hypothesis of Lemma 6.5.1.

Set $\bar{H}_1 := H_1/U_1$ and $\tilde{U}_1 := U_1/Z(U_1)$.

First, $R := R_5^{(1)}$ is of order 2^{11}; and it cannot contain any G-conjugate of U_3 of order 2^{11}: since containment would force equality, whereas the two groups have normalizers of different structures in [**130**, Table 6, p. 262].

Next, R contains no conjugate X of U_6: For otherwise, since $Z_1 := Z(U_1) = Z(H_1)$ with $R \leq H_1'$, we have $Z_1 \leq C_G(X)$, and then $Z_1 \leq X$ since $U_6 = C_G(U_6)$. But now $Z_1 \leq X \leq R \leq H_1'$; and using transitivity of the $\Omega_6^-(2)$ in $N_G(X)$ on the G-conjugates of Z_1 in X, we see using [**130**, Lemma 16 (2)] that X contains four-subgroups over Z_1 which have exactly one involution in the class denoted $2A$ in the Atlas [**47**, p. 166]. And now we see that $X \leq H_1'$ contradicts the equivalence of (ii) and (iii) on page 259 of [**130**].

Any remaining members of $\mathcal{B}_2^{cen}(G)$, namely those not containing a G-conjugate of U_3 or U_6, are in G-classes of subgroups containing U_1. We next show by a standard argument that U_1 is weakly closed in R: For suppose that R contains a G-conjugate of U_1, other than U_1; we may denote that conjugate by U_X, where $X := Z(U_X)$ is G-conjugate to Z_1; but $X \neq Z_1$ since $U_1 = O_2(H_1)$ where $H_1 = C_G(Z_1)$. Since U_1 is of index 2^2 in R, the intersection $I := U_1 \cap U_X$ is of order at least 2^7. Further I is abelian, since

$$[I, I] \leq [U_1, U_1] \cap [U_X, U_X] = Z_1 \cap X = 1.$$

But this contradicts the order 2^5 of the maximal abelian subgroups of U_1. Thus U_1 is the unique member of its G-conjugacy class in R. It also follows using this uniqueness that R contains no G-conjugate of any other members of $\mathcal{B}_2^{cen}(G)$ (such as $U_{1,6}$ of order 2^{10} but not lying in H_1') which properly contain U_1.

Hence $\mathcal{B}_2^{cen}(G)_{<R}$ consists just of U_1, as claimed; and we saw that this suffices to give the needed homotopy equivalence.

8.7. Details for Ru

We provide at least some details for the claim in Section 7.24 that the subposet $\mathcal{E}_2^-$ of $\mathcal{E}_2(G)$, defined by the G-conjugates just of A_1, A_2, A_3, and A_6, is ample and normalizer-sharp. The argument proceeds by a variant of our usual removal methods giving a homotopy equivalence via a G-map; where we start with the full poset $\mathcal{E}_2(G)$, which is ample and normalizer sharp by Proposition 5.10.2.

Since Section 7.24 indicates a substantially easier approach to a homotopy decomposition via $\mathcal{B}_2^{cen}(G)$, we could have omitted the discussion of $\mathcal{E}_2^-$ in this work. However since the decomposition from $\mathcal{E}_2^-$ is somewhat different, it seemed worthwhile to include it, with suitable supporting details. But since Ru is a rather large group, the needed arguments are fairly elaborate. For this reason, we have chosen to provide here a treatment which is perhaps closer to an outline than to a complete proof. For example, we will frequently write expressions such as "we find" in lieu of a full proof for a statement that is fairly straightforward, but might require several lines to check in detail.

Since the conjugates of A_1 are points of the generalized hexagon on A_6 for $H_6/U_6 \cong G_2(2)$, we will often refer to conjugates of a generator z of A_1 (in the 2-central class $2A$) as "points".

As indicated in Section 7.24, many of the facts about Ru that we quote can be obtained from Chapter J of Aschbacher-Smith [**20**]. However, we will use notation adapted for this work rather than the notation used there.

Some preliminary results. Before beginning the process of removing subgroups from $\mathcal{E}_2(G)$, we sketch some general background and properties. and establish some preliminary results.

Recall that in Section 7.24, we defined V_1 as the normal closure of A_3 in H_1. Indeed from [**20**, J.2.2] we have a series under the action of H_1:

$$A_1 < V_1 < C_{U_1}(V_1) < U_1 < H_1' < H_1,$$

where V_1 of rank five is generated by the conjugates of A_2 over A_1 in U_1. Then V_1/A_1 and U_1/C_{U_1} of rank four are natural modules for $H_1'/U_1 \cong SL_2(4)$, and $C_{U_1}(V_1)/V_1$ is of rank two. Further V_1 consists of involutions of class $2A$ (cf. J.2.6.3 in [**20**]) and so lies in $\mathcal{E}_2(G)$. In fact in reducing $\mathcal{E}_2(G)$ to $\mathcal{E}_2^-$, we will first reduce to the subposet $\mathcal{E}'$ defined by the G-conjugates of A_1, A_2, A_3, V_1, and A_6.

To discuss $\mathcal{E}_2(G)$, we first wish to classify its members of rank two; up to conjugacy these can be assumed to contain A_1, and we will describe them with respect to the series in the previous paragraph. As usual from the character table of Ru (for example using the ClassMultiplicationCoefficient function in GAP), we can compute the value $\xi(2A, 2A, 2A)$ for the structure constant defined in (5.10.5); we obtain the class-algebra multiplication coefficient 1390. Then (cf. J.3.4 in [**20**]) the 1390 distinct conjugates of a generator z of A_1 lying in $C_G(A_1) = H_1$ fall into H_1-classes of sizes

$$30,\ 240,\ 480,\ 640; \tag{8.7.1}$$

these lie in $V_1 \setminus A_1$, $U_1 \setminus C_{U_1}(V_1)$, $H_1' \setminus U_1$, and $H_1 \setminus H_1'$, respectively. Further $C_{U_1}(V_1) \setminus V_1$ contains no involutions; cf. J.2.6 and J.3.4 of [**20**]. For x in each of these classes, we denote a representative of the class of the 4-group $\langle z, x\rangle$ by A_2, B_2, D_2, and C_2. Of course eventually we will need to remove B_2, C_2, and D_2 from $\mathcal{E}_2(G)$, as well as suitable overgroups (which we will not need to classify in full detail).

The representatives A_2, B_2, and C_2 can be chosen to lie in A_6; and each is of form $\langle z, x\rangle$ for a pair of points z, x at respective distance $1, 2, 3$ (cf. [**119**, 3.34]) in the generalized hexagon on A_6 under $H_6/A_6 \cong G_2(2)$. We will typically write a, b, c for a point x at distance $1, 2, 3$ from z in this sense.

We will be exhibiting homotopy equivalences via G-maps: given by removing G-classes of subgroups X in $\mathcal{E}_2(G)$, in the spirit of Lemma 6.5.1; where for suitable X, contractibility of the link of X via a G_X-map will be guaranteed, as in Remark 6.5.2, by certain uniqueness statements—such as (a), (e) and (g) below.

For the first of these, we note that the conjugates of A_3 over A_1 in V_1 afford 1-subspaces over $\mathbb{F}_4$ under the action of $H_1'/U_1 \cong SL_2(4)$, and we obtain:

(a) A_2 is contained in a unique G-conjugate of A_3.

We mention that the uniqueness in (a) also essentially provides the reason for the statement established in Section 7.24, that the poset $\mathcal{E}_2^-$ is homotopy equivalent via a G-map to the almost-poset complex $\Delta_{\mathcal{A}_I}$ defined by A_1, A_3, and A_6 (but not A_2). In fact, (a) is fundamental throughout the course of our reductions to $\mathcal{E}_2^-$. For example, suppose that A_2 and a G-conjugate A_2', both containing A_1, arise from different $\mathbb{F}_4$ 1-subspaces of V_1; then we get

(b) $V_1 = A_3A_3'$

using the corresponding G-conjugates A_3 and A_3' of A_3 determined in (a).

Furthermore, we can now extend the notion of distance one in A_6 (corresponding to lines of the generalized hexagon) in a natural way to more general commuting conjugates of z; and using this more general notion of distance one, we define a graph and the corresponding distance function on the entire G-conjugacy class $2A$ represented by z.

First, we saw that $A_2 = \langle z, a\rangle$ represents a pair at distance one in the hexagon A_6; and also that the H_1-conjugates of A_2 cover V_1. Thus distance one in the class $2A$ is determined by the relation $a \in V_1 \setminus A_1$. Such conjugate pairs are indeed at distance one in some conjugate of A_6; and similarly they also lie in some conjugate of A_3. Furthermore we find that the relation is symmetric: we have $a \in V_1 = V_z$ if and only if $z \in V_a$.

Next consider a point b at distance two from z in the graph we are now defining on the class $2A$: this means that there is some suitable a at distance one from both z and b. One shows, using the action of the conjugate H_a of H_1, that $\langle z, b\rangle$ is G-conjugate to either B_2 or D_2, the subgroups determined by the orbits of sizes 240 and 480 in our earlier count (8.7.1). Indeed one obtains

(c) $V_z \cap V_b = \langle a\rangle$, so that there are no nontrivial 4-gons in the class $2A$;

that is, there are no 4-gons outside a conjugate of A_3. This is of course an extension to the class $2A$ of the corresponding property in the generalized hexagon on A_6; in effect, we are looking at the property in the intersection of $2A$ with H_1'. In fact (arguing with the sizes 240 and 480 much as for statement (b) for M_{12} in Section 8.1), one finds that:

(d) The pair z, b determines a unique a at distance one from each of z, b.

We had indicated that conjugates of B_2 arise in conjugates of A_6; and the relation of distance two in A_6 determines conjugates of B_2. Now $X := \langle b, z\rangle$ is of type B_2 or D_2; and using the symmetry of the V-relation for distance one, from (c) and (d) we obtain that

(e) $\langle b, z\rangle$ lies in a unique V_a—described as in (b) for V_a in the role of "V_1".

We also want to further describe $C_2 := \langle z, c\rangle$ for c at distance three from z in the hexagon A_6. It is standard that in that hexagon over $\mathbb{F}_2$, there are exactly 3 points a_1, a_2, a_3 which are at distance one from z in A_6, but at distance two from c in A_6. We find that these three points lie in different (but linearly dependent) $\mathbb{F}_4$ 1-subspaces of V_1/U_1; and that with z they span $A_6 \cap V_1$, where

(f) $A_6 \cap V_1$ is of rank three; while $A_6 \cap U_1$ is of rank five.

Arguing with (f) and further generators b_1, b_2 for $(A_6 \cap U_1)/(A_6 \cap V_1)$, we find that $A_6 = \langle z\rangle \times (U_z \cap U_c) \times \langle c\rangle$, where $U_z = U_1$, and U_c denotes the obvious G-conjugate of U_1. In particular, we obtain

(g) C_2 lies in a unique G-conjugate of A_6, but in no G-conjugate of V_1,

since hexagon-distances (hence distances in the class $2A$) of pairs in V_1 are at most 2.

From this discussion, we see that B_2 corresponds to z, b at distance two in A_6; but D_2 is contained in no G-conjugate of A_6. Indeed for z, b generating B_2, with path z—a—b determined as in (e), we can compute further with (8.7.1) and other properties above to show that

(h) $\langle z, a, b\rangle$ of rank three is contained in exactly two G-conjugates of A_6.

This concludes our collection of preliminary results.

The reduction to $\mathcal{E}_2^-$. Before starting our reductions from $\mathcal{E}_2(G)$, we mention some technical features which make the case of Ru more difficult than (for example) the case of HS in Section 7.10. Here V_1 and A_6 are fairly obvious G-classes of maximal elements of $\mathcal{E}_2(G)$—but there is another class of maximal elements, with representative C defined below. We will need to remove this C, as essentially the first order of business—and that process seems to require a refinement of our usual methods. Furthermore the existence of four G-classes of members of $\mathcal{E}_2(G)$ of rank two demonstrates that the poset is more complicated than in our earlier, smaller examples.

The argument via removals which now follows is fairly delicate; and in particular, it seems to depend rather crucially on the order in which we remove members of $\mathcal{E}_2(G)$. Very likely this argument can be improved on!

1) Removal of overgroups of C_2 other than A_6.

We take $C_2 := \langle z, c\rangle$ for z, c at distance three in the hexagon A_6, which is the unique G-conjugate of A_6 over C_2 by (g).

Step 1.a) We will first discuss how to remove from $\mathcal{E}_2(G)$ all proper overgroups $E > C_2$ which also satisfy $E \not\leq A_6$.

We begin by describing some properties of such an E: Set $C := C_G(E)$, so that $E \leq C$. As E contains C_2, C normalizes A_6 using (g) and so lies in H_6. Arguing with properties of the generalized hexagon A_6, we conclude that

(i) $C = \langle z, a, b, c, d\rangle$,

where z—a—b—c is a path of length three in the hexagon A_6, and d is an involution in $H_6 \setminus A_6$; and as $E \not\leq A_6$, we may assume that $E \geq \langle z, c, d\rangle$. (In particular, C is of rank five, and is maximal under inclusion in $\mathcal{E}_2(G)$, as mentioned earlier.)

We first remove such the G-classes of such E other than C itself: For we have just seen in (i) above that for such an E, $\mathcal{E}_2(G)_{>E}$ has the unique maximal member C; and then using Remark 6.5.2 for contractibility via a G_E-map of the link of E, we obtain the hypothesis of Lemma 6.5.1: so $\mathcal{E}_2(G)$ is homotopy equivalent via a G-map to the subposet $\mathcal{E}_1$ of $\mathcal{E}_2(G)$ obtained by removing all such E. Observe that if E is a member of $\mathcal{E}_1$ with $E > C_2$ and $E < C$, then we have $E \leq A_6$.

In particular, $\mathcal{E}_1$ is also ample and normalizer sharp by Proposition 5.10.2.

The removal of the G-class of C itself is somewhat touchier: We will argue on $(\mathcal{E}_1)_{<C}$; so consider a general member $X \in (\mathcal{E}_1)_{<C}$. In particular, X has rank at most 4.

We sketch fairly briefly the following technical point. Suppose Y is a complement to $\langle a, b\rangle$ in C. Then Y is of rank 3 and $Y \not\leq A_6$. Furthermore the complement property implies that the intersection of Y with $\langle z, a, b\rangle$ and $\langle a, b, c\rangle$ (respectively) is also a complement (of rank 1) to $\langle a, b\rangle$ in each of those groups, generated say by z' and c', which are collinear with a, b. And then $d(z', c') = 3$, since a hexagon contains no circuits of length ≤ 5. Thus Y properly contains the G-conjugate $\langle z', c'\rangle$ of C_2.

Suppose now that $X\langle a, b\rangle = C$. Then X contains a complement Y to $\langle a, b\rangle$ in C; and X inherits from Y the properties in the previous paragraph: namely X properly contains a G-conjugate W of C_2 lying in A_6, and $X \not\leq A_6$—which is the unique G-conjugate of A_6 above W by (g). Further since X has rank at most 4, X

is proper in every G-conjugate of C over W. But now the remark at the end of the above reduction to $\mathcal{E}_1$ shows that $X \leq A_6$, which we just saw is not the case.

Thus in fact we have $X\langle a,b\rangle < C$. It follows that $(\mathcal{E}_1)_{<C}$ is conically contractible (the standard method of Brown, see Quillen [**104**, 1.5]) via the inclusions in the following relation, defined for each X as above:

$$X \leq X\langle a,b\rangle \geq \langle a,b\rangle.$$

We let $\mathcal{E}_2$ denote the result of this removal. Thus $\mathcal{E}_2$ contains only those proper overgroups E of C_2 satisfying $E \leq A_6$.

But here a complication for our usual method arises: it is not clear that the equivalences from the above inclusions are G_C-equivariant; so that $\mathcal{E}_2$ might conceivably not be homotopy equivalent to $\mathcal{E}_1$ via a G-map. Indeed for G_C-equivariance above, we would need to know (say) that G_C preserves $\langle a,b\rangle$ (which would follow if G_C preserves C_2 and hence A_6); but it is not necessarily clear even that C_2 should be the unique conjugate of itself in C.

So we make a variation on our usual method: instead of arguing on the normalizer decomposition, we will argue initially on the centralizer decomposition of Section 5.4 for $\mathcal{E}_2(G)$.

First, we claim that by construction, our subposet $\mathcal{E}_2$ is closed under taking nontrivial subgroups: This property holds for the original poset $\mathcal{E}_2(G)$; and the reduction to $\mathcal{E}_2$ considered only overgroups of C_2. So it suffices to consider some $E > C_2$ which lies in $\mathcal{E}_2$, and show that each nontrivial $F < E$ also lies in $\mathcal{E}_2$. Now since E was not removed in the reduction, we have $E \leq A_6$, and hence $F \leq A_6$. Further F could have been removed in the reduction only if F properly contains some G-conjugate Y of C_2. But then that conjugate Y is also in A_6, and so by (g), A_6 is also the unique conjugate of itself over Y. And now since $F \leq A_6$, we see that F was not in fact removed in the reduction to $\mathcal{E}_2$.

Next, we re-phrase our equivariance arguments, now using the action of the centralizer $C_G(C) = C$, rather than $N_G(C) = G_C$: Certainly the elementary abelian group C centralizes subgroups of itself; so the inclusions in the conical contractibility relations above are C-equivariant; and hence the removal of C itself during the reduction from $\mathcal{E}_1$ to $\mathcal{E}_2$ is equivariant under $C_G(C)$. Furthermore the removal of E, during the earlier reduction from $\mathcal{E}_2(G)$ to $\mathcal{E}_1$, was equivariant under $N_G(C)$ where $C = C_G(E)$; and so in particular is also $C_G(E)$-equivariant.

Now we recall the space X for the centralizer decomposition in Definition 5.2.4—namely that corresponding to the functor $\tilde{\alpha}_{\mathcal{C}}$ in Section 5.4 (and in the discussion of sharpness in Section 5.8), for the collection $\mathcal{C} := \mathcal{E}_2(G)$. In particular in the action of G on the underlying conjugation category $\mathfrak{A}_{\mathcal{C}}$ there, the natural stabilizer "G_F" in G of a member $F \in \mathcal{E}_2(G)$ is given by $C_G(F)$ rather than $N_G(F)$. So we can use centralizers as the appropriate stabilizers in the analogue of Lemma 6.5.1 (2) for the G-space X. Thus the observations in the previous paragraph show that the reduction from $\mathcal{E}_2(G)$ to $\mathcal{E}_2$ is a homotopy equivalence via a G-map of the spaces for the corresponding centralizer decompositions. In particular, the centralizer decomposition for $\mathcal{E}_2$ is ample and centralizer sharp.

Finally since we saw that the subposet $\mathcal{E}_2$ is closed under taking nontrivial subgroups, we see from Grodal and Smith [**67**, 2.7 (2)] that $\mathcal{E}_2$ must also be normalizer-sharp (and hence ample).

Step 1.b) We can now remove subgroups E with $C_2 \leq E < A_6$, including C_2 in particular: For again $C_G(E) \leq H_6$ by (g); but the previous step eliminated the possibility that $E \not\leq A_6$, so that $(\mathcal{E}_2)_{>E}$ has unique maximal element A_6. So we can argue with Remark 6.5.2: since G_E must normalize A_6 by its uniqueness over C_2 and hence over E, we obtain contractibility via a G_E-map, as needed for Lemma 6.5.1. We let $\mathcal{E}_3$ denote the result of this reduction via a G-equivariant homotopy equivalence.

In particular, $\mathcal{E}_3$ is also ample and normalizer sharp: these properties are inherited from $\mathcal{E}_2$ in view of Lemma 5.6.8 and Theorem 5.8.11. Indeed from this point onwards, in contrast to our argument above using the centralizer decomposition, our further reductions will propagate the conditions of ampleness and normalizer-sharpness in this more usual way; namely via removals which are homotopy equivalences via G-maps, in which the stabilizers of subgroups are given by their normalizers. So we will not explicitly repeat these references at each such reduction.

2) Removal of B_2 and D_2 as well as suitable overgroups.

Now assume that $E \in \mathcal{E}_3$, with E of rank at least 2, but that E is not conjugate to A_6. The previous removals imply that any 4-subgroup of E is of type A_2, B_2, or D_2; in particular,

(j) involutions in E are at distance at most two in the class $2A$.

Furthermore using previous facts, it is not difficult to establish another extension to the class $2A$ (compare with (c) above) of a standard fact in the hexagon on the points in A_6:

(k) there are no nontrivial 5-gons on points of the class $2A$.

Step 2.a) We will first remove overgroups of conjugates of D_2 (including D_2 itself) other than those of type V_1.

Let $D_2 = \langle z, b \rangle$ for z, b at distance two, but not of type B_2 as described earlier. By (d), the pair z, b determines a unique a at mutual distance 1 from z, b. Then for $E \in (\mathcal{E}_3)_{\geq D_2}$ one can show, using the nonexistence of 4-gons and 5-gons in (c) and (k), that $E \leq V_a$. In fact, since we do not wish to remove the conjugates of V_1, we may assume that $E < V_a$. Hence working inside $C_G(E)$, we may apply the containment of such overgroups in V_a to conclude that $(\mathcal{E}_3)_{<E}$ has unique maximal element V_a. In particular, G_E normalizes V_a and hence lies in the conjugate H_a of H_1 centralizing a. So we argue as usual with Remark 6.5.2 and Lemma 6.5.1; and we let $\mathcal{E}_4$ denote the result of removing all such E from $\mathcal{E}_3$ (as usual a homotopy equivalence via a G-map).

Step 2.b) We will next remove conjugates of overgroups of B_2 (including B_2 itself) other than those of types V_1 and A_6.

Observe first that the previous step in particular removed D_2; so the analogue of (j) for $\mathcal{E}_4$ has 4-subgroups only of types A_2 and B_2. Again we let $B_2 = \langle z, b \rangle$, with a uniquely determined as in (d).

We begin by considering $E \in (\mathcal{E}_4)_{>B_2}$.

We will first remove those E containing no G-conjugate of A_2; that is, such that all 4-subgroups of E are of type B_2.

We claim that E is contained in one of the two G-conjugates A_6 or A_6^g above $\langle z, a, b \rangle$ in (h). We sketch fairly briefly the proof of this technical point, which proceeds via considerations much like those in Step 1.a. So assume there is $c \in E$ with c not lying in either A_6 or A_6^g. Certainly c lies in $C_G(E) \leq C_G(B_2)$. Indeed analyzing more closely the structures underlying (h) shows that $C_G(B_2) = C_G(\langle z,a,b\rangle) = A_6A_6^g$ with $A_6 \cap A_6^g = \langle z,a,b\rangle$, and $a \notin E$ since $\langle z,a\rangle = A_2$ is not of type B_2. If c determines the coset of a short-root involution of H_6/A_6, then c must be the product of $e \in A_6^g$ with a member of $C_{A_6}(e) = \langle z,a,b\rangle \leq A_6^g$—which forces $c \in A_6^g$, contrary to our assumption. So c is the product of e determining a long-root involution coset of H_6/A_6 with a member d of $C_{A_6}(e) = \langle z,a,b,d\rangle$; where z—a—b—d is a path of length 3 in the hexagon A_6, and z—a—b—e is a path in A_6^g. Since c is not in A_6 or A_6^g, in fact c lies in $de\langle z,a,b\rangle$. But in the group $\Sigma_5 \cong H_1/U_1$, d and e determine commuting transpositions, so that de represents a root involution of $SL_2(4) \cong H_1'/U_1$. Further $\langle z,a,b,d,e\rangle \cap U_1 = \langle z,a,b\rangle$. The analysis via (8.7.1) shows that this forces $\langle z,c\rangle$ to be a conjugate of D_2; whereas E contains no such subgroups by the previous reduction to $\mathcal{E}_4$.

Thus we may assume that $E \leq A_6$. Then any pair of points in E is at mutual distance 2; and since $E > \langle z,b\rangle$, one checks in the hexagon A_6 that E is of rank 3, and is normalized by a subgroup $L_3(2)$ of H_6/A_6. For this configuration one obtains $C_G(E) = A_6$, by applying the above discussion of $C_G(B_2)$ to two conjugates of under $L_3(2)$ of B_2 over A_1. This shows as usual that $(\mathcal{E}_4)_{>E}$ has unique maximal member A_6, so that $G_E \leq H_6$. We argue as always with Remark 6.5.2 and Lemma 6.5.1; and let $\mathcal{E}_5$ denote the result of the corresponding removal.

Thus next we will consider E lying in $(\mathcal{E}_5)_{>B_2}$, where now E does contain some G-conjugate of A_2. As indicated in the statement of Step 2.b, we now also assume that E is not of type V_1 or A_6. Then using the nonexistence of 4-gons and 5-gons in (c) and (k), one can show that E is uniquely determined as $\langle z,a,b\rangle$. In particular, this uniqueness shows that G_{B_2} normalizes $E = \langle z,a,b\rangle$; so arguing as usual with Remark 6.5.2 and Lemma 6.5.1, we may now remove the G-class of B_2 from $\mathcal{E}_5$; let $\mathcal{E}_6$ denote the result.

Notice that we have not yet removed the above overgroup $E = \langle z,a,b\rangle$ of B_2. However we can now do so: for since we have removed the class of B_2, the 4-subgroups of E lying in $\mathcal{E}_6$ are all of type A_2. Then letting zb denote a point at distance 1 from a but at distance two from z and b, we find that $(\mathcal{E}_6)_{<E}$ consists of the seven points of E and the three conjugates of A_2 in E, each of which is generated by a and one of z,b,zb. In particular, this poset is conically contractible via the inclusions defined for each member X of $(\mathcal{E}_6)_{<E}$ in:

$$X \leq X\langle a\rangle \geq \langle a\rangle.$$

Now G_E fixes the unique point a in E at distance at most one from all points in E; so the equivalences from these inclusions are G_E-equivariant. So we let $\mathcal{E}'$ denote the result of this removal.

3) Removal of the G-class of V_1.

We have now removed all conjugates of B_2, C_2, and D_2. Furthermore we have removed all conjugates of their overgroups, except for those of types V_1 and A_6; so for any $E \in \mathcal{E}'$ of rank greater than 2, all subgroups of rank two are G-conjugates of A_2—and A_3 represents the unique G-class of subgroups satisfying the latter condition. Thus $\mathcal{E}'$ consists of the G-conjugates of A_1, A_2, A_3, V_1, and A_6; so we have completed the primary reduction mentioned at the start of our argument.

It remains to remove the G-class of V_1: Observe that $\mathcal{E}'_{<V_1}$ consists of the 31 points of V_1 (conjugates of A_1); together with the five conjugates of A_3 in V_1 (which contain A_1) and the various conjugates of A_2 containing A_1 in those conjugates of A_3. In particular, for $X \in \mathcal{E}'_{<V_1}$ we have

$$X \leq XA_1 \geq A_1,$$

so that $\mathcal{E}'_{<V_1}$ is conically contractible. Since G_{V_1} centralizes the unique point A_1 of V_1 at distance at most one from all points of V_1, the equivalences from the above inclusions are G_{V_1}-equivariant.

Notice that the result of this final removal is the poset $\mathcal{E}_2^-$ defined by the G-conjugates of A_1, A_2, A_3, and A_6—as desired.

8.8. Details for $O'N$

In this section, we give arguments supporting several claims made in Section 7.25.

First we show that the group $V \cong (\mathbb{Z}/4)^3$ is weakly closed. We may assume for some $g \in G$ that the conjugate $g^{-1}Vg$ lies in $N_G(V) = H_3 \cong 4^3L_3(2)$, and we must show that $g^{-1}Vg = V$. Set $H_3^+ := H_3/V \cong L_3(2)$, and for elements and subgroups of H_3, we use a superscript $+$ to denote image in $L_3(2)$. Note that the conjugation action of H_3 on V passes down to an action of H_3^+. The conjugate $(g^{-1}Vg)^+$ is an abelian 2-subgroup of H_3^+, so that $|(g^{-1}Vg)^+| \leq 2^2$; hence $g^{-1}Vg \cap V$ is of order at least 2^4, and is centralized by $(g^{-1}Vg)^+$. Thus $(g^{-1}Vg)^+$ is not cyclic of order four, since an element of order four centralizes a subgroup of rank only one in each of the natural $L_3(2)$-modules given by the quotients in the series $V > \Omega_1(V) > 1$ where of course $\Omega_1(V) = A_3$. Thus $(g^{-1}Vg)^+$ is elementary abelian, so that $\Omega_1(g^{-1}Vg) \leq V$. So $\Omega_1(g^{-1}Vg) \leq \Omega_1(V)$, and hence $\Omega_1(g^{-1}Vg) = \Omega_1(V)$ since both have order 2^3. It follows that $g^{-1}Vg$ centralizes $\Omega_1(V) = A_3$, and hence $g^{-1}Vg \leq C_G(A_3) = V$; and now since both have order 2^6, we obtain $g^{-1}Vg = V$, as desired.

Next we show that $C^{(pl)}$ is weakly closed: this time we assume that a conjugate $g^{-1}(C^{(pl)})g$ lies in the Sylow 2-group $C^{(\iota pl)}$ with center $\langle z \rangle$, and so in particular lies in $C_G(z) = H_1 \cong 4{}^\cdot L_3(4) : 2$. Now $g^{-1}C^{(pl)}g$ has a quotient of rank four corresponding to a unipotent radical in $C^{(pl)}$; while no subgroup of index two in $C^{(\iota pl)}$ not lying in H_1' has such a quotient. Then $g^{-1}C^{(pl)}g \leq H_1' \cap C^{(\iota pl)} = C^{(pl)}$, so that $g^{-1}C^{(pl)}g = C^{(pl)}$ since both are of index two in $C^{(\iota pl)}$.

Finally we sketch the homotopy equivalence via a G-map of $\Delta_{\mathcal{A}_{I'}}$ with $\mathcal{B}_2^{cen}(G)$.

We will work via the collection $\mathcal{B}_2$ defined earlier in Section 7.25, which we saw there is homotopy equivalent via a G-map with $\mathcal{B}_2^{cen}(G)$.

Indeed first we will work toward a variant $\mathcal{B}_3$ of $\mathcal{B}_2$, as follows: Set $I := V \cap C^{(p)}$, and let $\mathcal{B}^+$ denote the poset $\mathcal{B}_2^{cen}(G)$ with the G-class of I added; that class now affords the unique class of minimal elements of $\mathcal{B}^+$. Since I contains $\Omega_1(V) = A_3$, using arguments like those in our proof above of weak closure of V, we can check that V is the only G-conjugate of itself containing I. It follows that if I lies in a conjugate $g^{-1}C^{(p)}g$, then $V \,.\, g^{-1}C^{(p)}g = g^{-1}C^{(pl)}g$. Then the link in $\mathcal{B}^+$ of I is conically contractible to V, using the map

$$X \leq VX \geq V$$

on elements X of that link. Since G_I preserves V by its uniqueness as in Remark 6.5.2, the equivalences from these inclusions are G_I-equivariant. And then by Lemma 6.5.1, $\mathcal{B}^+$ is homotopy equivalent via a G-map to $\mathcal{B}_2^{cen}(G)$. Next, we may remove from $\mathcal{B}^+$ the G-conjugates of $V^{(l)}$ and $C^{(\iota pl)}$, by essentially the same arguments used in Section 7.25 to remove them from $\mathcal{B}_2^{cen}(G)$, obtaining a subposet $\mathcal{B}_1^+$. But subsequently we can also remove the conjugates of $C^{(pl)}$: for the link in $\mathcal{B}_1^+$ of $C^{(pl)}$ is a cone on the vertex I—which is joined to V, $C^{(p)}$, and $t(C^{(p)})$, where t is the outer automorphism of the $L_3(4)$ in H_1 described earlier. Note that $G_{C^{(pl)}}$ stabilizes I since it is the unique vertex of its type in this link. As usual we may apply Remark 6.5.2 and then Lemma 6.5.1; let $\mathcal{B}_3$ denote the result of this last removal.

We now finish by sketching the homotopy equivalence via a G-map of the poset $\mathcal{B}_3$ with the graph $\Delta_{\mathcal{A}_{I'}}$; we will use the method of Lemma 6.5.3, with $\mathcal{B}_3$, $\Delta_{\mathcal{A}_{I'}}$ in the roles of "$\mathcal{E}$, $\Delta_{\mathcal{A}_I}$". Thus the type set "I" in that result is represented by the G-classes V and $C^{(p)}$ (that is, by $I' = \{3, p\}$). Indeed now we are using the notation I instead to denote the group $V \cap C^{(p)}$. So this group I now plays the role of "E_σ" for the edge σ defined by V and $C^{(p)}$; and by construction, I is in the poset $\mathcal{B}_3$. So it remains to establish, for each $E \in \mathcal{B}_3$, the contractibility via a G_E-map of the subcomplex $L_E := \{\sigma \in \Delta_{\mathcal{A}_{I'}} : E \leq E_\sigma\}$. For E conjugate to V, it suffices to consider V itself: then L_V is the cone given by the join of the vertex V with all the G-conjugates of I in V (corresponding to the conjugates of $C^{(p)}$ incident with V); as usual we may apply Remark 6.5.2. Similarly for E given by $C^{(p)}$, we find that the subcomplex $L_{C^{(p)}}$ is the join of the vertex $\{C^{(p)}\}$ with all the G-conjugates of the subgroup I in $C^{(p)}$ (corresponding to the conjugates of V incident with $C^{(p)}$); again we may apply Remark 6.5.2. Finally for E conjugate to I, the subcomplex L_I consists just of the vertices V, $C^{(p)}$, and $t(C^{(p)})$, together with the two corresponding edges V—$C^{(p)}$ and V—$t(C^{(p)})$; we see directly that since G_I normalizes the unique vertex V of its type in this subcomplex, the subcomplex is contractible to V via a G_I-map. This completes the verification of the contractibility hypotheses needed for Lemma 6.5.3; and that result then completes the proof.

Bibliography

1. A. Adem, J. Carlson, D. Karagueuzian, and R. J. Milgram, *The cohomology of the Sylow 2-subgroup of the Higman–Sims group*, J. Pure & Applied Algebra **164** (2001), 275–305.
2. A. Adem, D. Karagueuzian, R. J. Milgram, and K. Umland, *The cohomology of the Lyons group and the double covers of the alternating groups*, J. Algebra **208** (1998), 452–479.
3. A. Adem, J. Maginnis, and R. J. Milgram, *Symmetric invariants and cohomology of groups*, Math. Ann. **287** (1990), 391–411.
4. ———, *The geometry and cohomology of the Mathieu group M_{12}*, J. Algebra **139** (1991), 90–133.
5. A. Adem and R. J. Milgram, *Invariants and cohomology of groups*, Bol. Soc. Mat. Mexicana **37** (1992), 1–26.
6. ———, *Cohomology of finite groups*, Grundlehren der mathematischen Wissenschaften, vol. 309, Springer-Verlag, Berlin/New York, 1994.
7. ———, *The cohomology of the Mathieu group M_{22}*, Topology **34** (1995), 389–410.
8. ———, *The subgroup structure and mod-2 cohomology of O'Nan's sporadic simple group*, J. Algebra **176** (1995), 288–315.
9. ———, *The cohomology of the McLaughlin group and some associated groups*, Math. Zeit. **224** (1997), 495–517.
10. C. Allday and V. Puppe, *Cohomological methods in transformation groups*, Cambridge Studies in Advanced Mathematics, vol. 32, Cambridge University Press, 1993.
11. J. An, *The Alperin and Dade conjectures for the simple Held group*, J. Algebra **189** (1997), 34–57.
12. ———, *The Alperin and Dade conjectures for the simple Conway's third group*, Israel J. Math. **112** (1999), 109–134.
13. J. An and M. Conder, *The Alperin and Dade conjectures for the simple Mathieu groups*, Commun. in Algebra **23** (1995), 2797–2823.
14. J. An and E. A. O'Brien, *A local strategy to decide the Alperin and Dade conjectures*, J. Algebra **206** (1998), 183–207.
15. ———, *The Alperin and Dade conjectures for the Fischer simple group* Fi_{23}, Internat. J. Algebra Comput. **9** (1999), 621–670.
16. ———, *The Alperin and Dade conjectures for the O'Nan and Rudvalis simple groups*, Commun. in Algebra **30** (2002), 1305–1348.
17. ———, *Conjectures on the character degrees of the Harada–Norton simple group HN*, Israel J. Math. **137** (2003), 157–181.
18. ———, *The Alperin and Uno conjectures for the Fischer simple group Fi_{22}*, Commun. in Algebra (2005), 1529–1557.
19. M. Aschbacher, *Flag structures on Tits geometries*, Geom. Dedicata **14** (1983), no. 1, 21–32.
20. M. Aschbacher and S. D. Smith, *The classification of quasithin groups: I, Structure of strongly quasithin $\mathcal{K}$-groups*, Surveys and Monographs of the Amer. Math. Soc., vol. 111, American Math. Society, Providence RI, 2004.
21. D. J. Benson, *Representations and Cohomology I: Basic representation theory of finite groups and associative algebras*, Cambridge Studies in Advanced Mathematics, vol. 30, Cambridge University Press, 1991, reprinted in paperback, 1998.
22. ———, *Representations and Cohomology II: Cohomology of groups and modules*, Cambridge Studies in Advanced Mathematics, vol. 31, Cambridge University Press, 1991, reprinted in paperback, 1998.
23. ———, *Polynomial invariants of finite groups*, London Math. Soc. Lecture Note Series, vol. 190, Cambridge University Press, 1993.

24. ———, *Conway's group Co_3 and the Dickson invariants*, Manuscripta Math. **85** (1994), 177–193.
25. ———, *Cohomology of sporadic groups, finite loop spaces, and the Dickson invariants*, Proceedings of the 1994 Durham conference on Geometry and Cohomology in Group Theory (P. Kropholler, G. Niblo, and R. Stöhr, eds.), London Math. Soc. Lecture Note Series, vol. 252, 1998, pp. 10–23.
26. ———, *On the classifying space and cohomology of Thompson's sporadic simple group*, Finite Groups 2003, de Gruyter, Berlin/New York, 2004, pp. 47–57.
27. D. J. Benson and J. F. Carlson, *Diagrammatic methods for modular representations and cohomology*, Commun. in Algebra **15** (1987), 53–121.
28. ———, *Cohomology of the double cover of the Mathieu group M_{12}*, J. Algebra (2000), 547–576.
29. D. J. Benson and C. W. Wilkerson, *Finite simple groups and Dickson invariants*, Homotopy theory and its applications, Cocoyoc, Mexico, 1993, Contemp. Math., vol. 188, 1995, pp. 39–50.
30. S. Bouc, *Homologie de certains ensembles ordonnés*, Comptes Rendus Acad. Sci. Paris, Série I **299** (1984), 49–52.
31. A. K. Bousfield and D. M. Kan, *Homotopy limits, completions and localizations*, Lecture Notes in Mathematics, vol. 304, Springer-Verlag, Berlin/New York, 1972.
32. G. E. Bredon, *Equivariant cohomology theories*, Lecture Notes in Mathematics, vol. 34, Springer-Verlag, Berlin/New York, 1967.
33. K. S. Brown, *Euler characteristics of groups: the p-fractional part*, Invent. Math. **29** (1975), 1–5.
34. ———, *Groups of virtually finite dimension*, Homological group theory (Proc. Sympos., Durham, 1977), London Math. Soc. Lecture Note Series, vol. 36, Cambridge University Press, 1979, pp. 27–70.
35. ———, *Cohomology of groups*, Graduate Texts in Mathematics, vol. 87, Springer-Verlag, Berlin/New York, 1982.
36. ———, *Buildings*, Springer-Verlag, Berlin/New York, 1989.
37. F. Buekenhout, *Diagrams for geometries and groups*, J. Combin. Theory Ser. A **27** (1979), no. 2, 121–151.
38. ———, *Geometries for the Mathieu group M_{12}*, Combinatorial theory (Schloss Rauischholzhausen, 1982), Lecture Notes in Mathematics, vol. 969, Springer-Verlag, Berlin/New York, 1982, pp. 74–85.
39. ———, *Diagram geometries for sporadic groups*, Finite groups—coming of age (Montreal, Que., 1982), Contemp. Math., vol. 45, Amer. Math. Soc., Providence, RI, 1985, pp. 1–32.
40. ———, *The geometry of the finite simple groups*, Buildings and the geometry of diagrams (Como, 1984), Lecture Notes in Math., vol. 1181, Springer, Berlin, 1986, pp. 1–78.
41. J. F. Carlson, J. Maginnis, and R. J. Milgram, *The cohomology of the sporadic groups J_2 and J_3*, J. Algebra **214** (1999), 143–173.
42. H. Cartan and S. Eilenberg, *Homological algebra*, Princeton Mathematical Series, no. 19, Princeton Univ. Press, 1956.
43. R. W. Carter, *Simple groups of Lie type*, Pure and Applied Mathematics, vol. 28, John Wiley & Sons, London-New York-Sydney, 1972.
44. G. R. Chapman, *Generators and relations for the cohomology ring of Janko's first group in the first twenty one dimensions*, Groups—St. Andrews 1981, London Math. Soc. Lecture Note Series, vol. 71, Cambridge University Press, 1982, pp. 201–206.
45. E. Cline, B. Parshall, and L. Scott, *Cohomology of finite groups of Lie type, I*, Publ. Math. Inst. Hautes Études Sci. **45** (1975), 169–191.
46. F. R. Cohen and R. Levi, *On the homotopy theory of p-completed classifying spaces*, Group representations: cohomology, group actions and topology (Seattle, WA, 1996), Proc. Symp. Pure Math., vol. 63, American Math. Society, 1998, pp. 157–182.
47. J. H. Conway, R. T. Curtis, S. P. Norton, R. A. Parker, and R. A. Wilson, *Atlas of Finite Groups*, Oxford University Press, 1985.
48. C. W. Curtis and I. Reiner, *Representation theory of finite groups and associative algebras*, Wiley Classics Library, John Wiley & Sons Inc., New York, 1988, Reprint of the 1962 original, A Wiley-Interscience Publication.
49. E. C. Dade, *Counting characters in blocks, I*, Invent. Math. **109** (1992), 187–210.

50. E. Dror Farjoun, *Fundamental group of homotopy colimits*, Adv. in Math. **182** (2004), 1–27.
51. W. G. Dwyer, *Homology decompositions for classifying spaces of finite groups*, Topology **36** (1997), 783–804.
52. ———, *Sharp homology decompositions for classifying spaces of finite groups*, Group representations: cohomology, group actions and topology (Seattle, WA, 1996), Proc. Symp. Pure Math., vol. 63, American Math. Society, 1998, pp. 197–220.
53. ———, *Classifying spaces and homology decompositions*, Homotopy theoretic methods in group cohomology (W. G. Dwyer and H.-W. Henn, eds.), Advanced Courses in Mathematics—CRM Barcelona, Birkhäuser Verlag, Basel, 2001, pp. 3–53.
54. W. G. Dwyer and D. M. Kan, *A classification theorem for diagrams of simplicial sets*, Topology **23** (1984), 139–155.
55. W. G. Dwyer and J. Spaliński, *Homotopy theories and model categories*, Handbook of algebraic topology, North Holland, Amsterdam, 1995, pp. 73–126.
56. W. G. Dwyer and C. W. Wilkerson, *A new finite loop space at the prime two*, J. Amer. Math. Soc. **6** (1993), 37–64.
57. G. Entz and H. Pahlings, *The Dade conjecture for the McLaughlin simple group*, Groups St. Andrews 1997 in Bath, I, London Math. Soc. Lecture Note Series, vol. 260, Cambridge University Press, 1999, pp. 253–266.
58. L. Evens, *Cohomology of groups*, Oxford University Press, 1991.
59. W. Feit, *The representation theory of finite groups*, North Holland, Amsterdam, 1982.
60. Z. Fiedorowicz and S. Priddy, *Homology of classical groups over finite fields and their associated infinite loop spaces*, Lecture Notes in Mathematics, vol. 674, Springer-Verlag, Berlin/New York, 1978.
61. D. G. Flaass, *2-local subgroups of Fischer groups*, Mat. Zametki **35** (1984), 333–342, Translation: Math. Notes **35** (1984), 173–178.
62. P. Gabriel and M. Zisman, *Calculus of fractions and homotopy theory*, Ergebnisse der Mathematik und ihrer Grenzgebiete, vol. 35, Springer-Verlag, Berlin/New York, 1967.
63. P. G. Goerss and J. F. Jardine, *Simplicial homotopy theory*, Progress in Mathematics, vol. 174, Birkhäuser Verlag, Basel, 1999.
64. D. Gorenstein, R. Lyons, and R. M. Solomon, *The Classification of the Finite Simple Groups. Number 3: Part I, Chapter A, Almost Simple $\mathcal{K}$-groups*, Surveys and Monographs, vol. 40.3, American Math. Society, 1998.
65. M. J. Greenberg and J. R. Harper, *Algebraic topology*, Mathematics Lecture Note Series, vol. 58, Benjamin/Cummings Publishing Co. Inc. Advanced Book Program, 1981, A first course.
66. J. Grodal, *Higher limits via subgroup complexes*, Ann. of Math. **155** (2002), 405–457.
67. J. Grodal and S. D. Smith, *Propagating sharpness of group homology decompositions*, Adv. in Math. **200** (2006), 525–538.
68. The GAP group, *GAP—Groups, Algorithms, and programming*, http://www-gap.idcs.st-and.ac.uk/~gap, 2000.
69. K. Gruenberg, *Cohomological topics in group theory*, Lecture Notes in Mathematics, vol. 143, Springer-Verlag, Berlin/New York, 1970.
70. J. Gunawardena, J. Lannes, and S. Zarati, *Cohomologie des groupes symétriques et application de Quillen*, Advances in homotopy theory (Cortona, 1988), Lecture Notes in Mathematics, vol. 139, Cambridge University Press, 1989, pp. 61–68.
71. Le Minh Ha and K. Lesh, *The cohomology of symmetric groups and the Quillen map at odd primes*, J. Pure & Applied Algebra **190** (2004), 137–153.
72. N. M. Hassan and E. Horváth, *Dade's conjecture for the simple Higman–Sims group*, Groups St. Andrews 1997 in Bath, I, London Math. Soc. Lecture Note Series, Cambridge University Press, 1999, pp. 329–345.
73. A. Hatcher, *Algebraic topology*, Cambridge University Press, 2002.
74. P. Hilton and U. Stammbach, *A course in homological algebra*, Lecture Notes in Mathematics, vol. 4, Springer-Verlag, Berlin/New York, 1971.
75. A. A. Ivanov, *Geometry of sporadic groups. I*, Encyclopedia of Mathematics and its Applications, vol. 76, Cambridge University Press, 1999, Petersen and tilde geometries.
76. S. Jackowski and J. E. McClure, *Homotopy decomposition of classifying spaces via elementary abelian subgroups*, Topology **31** (1992), 113–132.

77. S. Jackowski, J. E. McClure, and R. Oliver, *Homotopy classification of self-maps of BG via G-actions, I*, Ann. of Math. **135** (1992), 183–226.
78. I. M. James and G. B. Segal, *On equivariant homotopy type*, Topology **17** (1978), 267–272.
79. D. M. Kan, *A combinatorial definition of homotopy groups*, Ann. of Math. **67** (1958), 282–312.
80. M. Kitazume and S. Yoshiara, *The radical subgroups of the Fischer simple groups*, J. Algebra **255** (2002), 22–58.
81. S. N. Kleinerman, *The cohomology of Chevalley groups of exceptional Lie type*, Mem. AMS, vol. 268, American Math. Society, 1982.
82. S. Kotlica, *Verification of Dade's conjecture for the Janko group J_3*, J. Algebra **187** (1997), 579–619.
83. J. P. Lin and F. Williams, *On 14-connected finite H-spaces*, Israel J. Math. **66** (1989), 274–288.
84. S. A. Linton and R. A. Wilson, *The maximal subgroups of the Fischer group Fi_{24} and Fi'_{24}*, Proc. London Math. Soc. **63** (1991), 113–164.
85. W. Lück, *Transformation groups and algebraic K-theory*, Lecture Notes in Mathematics, vol. 1408, Springer-Verlag, Berlin/New York, 1989.
86. J. Maginnis, *Local control of group cohomology*, J. Pure & Applied Algebra **103** (1995), 81–90.
87. J. Maginnis and S. Onofrei, *On a homotopy relation between the 2-local geometry and the Bouc complex for the sporadic group Co_3*, J. Algebra **315** (2007), 1–17.
88. ———, *On a homotopy relation between the 2-local geometry and the Bouc complex for the sporadic group McL*, to appear in Arch. Math., 2008.
89. B. Mann, *The cohomology of the symmetric groups*, Trans. Amer. Math. Soc. **242** (1978), 157–184.
90. ———, *The cohomology of the alternating groups*, Michigan Math. J. **32** (1985), 267–277.
91. J. P. May, *Simplicial objects in algebraic topology*, Van Nostrand, 1967, Reprinted in the Chicago Lectures in Mathematics series, University of Chicago Press, 1992.
92. R. J. Milgram, *The 14-dimensional Kervaire invariant and the sporadic group M_{12}*, Asian J. Math. **1** (1997), 314–329.
93. ———, *On the relation between simple groups and homotopy theory*, Group representations: cohomology, group actions and topology (Seattle, WA, 1996), Proc. Symp. Pure Math., vol. 63, American Math. Society, 1998, pp. 419–440.
94. ———, *The cohomology of the Mathieu group M_{23}*, J. Group Theory **3** (2000), 7–26.
95. J. Milnor, *On spaces with the homotopy type of a CW-complex*, Trans. Amer. Math. Soc. **90** (1959), 272–280.
96. J. C. Moore, *Homotopie des complexes monoidaux, I*, Séminaire Henri Cartan (1954–5).
97. H. Mùi, *Modular invariant theory and cohomology algebras of symmetric groups*, J. Fac. Sci. Univ. Tokyo Sect. 1A Math. **22** (1975), 319–369.
98. J. Murray, *Dade's conjecture for the McLaughlin Simple Group*, Thesis, University of Illinois at Urbana-Champaign, 1998.
99. M. Nakaoka, *Homology of the infinite symmetric group*, Ann. of Math. **73** (1961), 229–257.
100. D. G. Quillen, *Homotopical algebra*, Lecture Notes in Mathematics, vol. 43, Springer-Verlag, Berlin/New York, 1967.
101. ———, *The spectrum of an equivariant cohomology ring, I*, Ann. of Math. **94** (1971), 549–572.
102. ———, *The spectrum of an equivariant cohomology ring, II*, Ann. of Math. **94** (1971), 573–602.
103. ———, *On the cohomology and K-theory of the general linear groups over a finite field*, Ann. of Math. **96** (1972), 552–586.
104. ———, *Homotopy properties of the poset of nontrivial p-subgroups of a finite group*, Adv. in Math. **28** (1978), 101–128.
105. M. A. Ronan, *Lectures on buildings*, Perspectives in Mathematics, vol. 7, Academic Press, New York/London, 1989.
106. M. A. Ronan and S. D. Smith, *2-local geometries for some sporadic groups*, The Santa Cruz Conference on Finite Groups (Univ. California, Santa Cruz, Calif., 1979) (Providence, R.I.), Proc. Symp. Pure Math., vol. 37, American Math. Society, 1980, pp. 283–289.

107. M. A. Ronan and G. Stroth, *Minimal parabolic geometries for the sporadic groups*, European J. Combin. **5** (1984), no. 1, 59–91.

108. A. J. E. Ryba, S. D. Smith, and S. Yoshiara, *Some projective modules determined by sporadic geometries*, J. Algebra **129** (1990), 279–311.

109. M. Sawabe, *2-radical subgroups of the Conway simple group* Co_1, J. Algebra **211** (1999), 115–133.

110. ———, *The 3-radicals of* Co_1 *and the 2-radicals of Rud*, Arch. Math. (Basel) **74** (2000), 414–422.

111. ———, *The centric p-radical complex and a related p-local geometry*, Math. Proc. Camb. Phil. Soc. **133** (2002), 383–398.

112. ———, *On the reduced Lefschetz module and the centric p-radical subgroups, II*, J. London Math. Soc. **73** (2006), 126–140.

113. S. D. Smith, *Residual geometries for sporadic and classical groups—a survey*, Finite groups—coming of age (Montreal, Que., 1982) (Providence, RI), Contemp. Math., vol. 45, American Math. Society, 1985, pp. 315–334.

114. ———, *Geometric methods in group theory*, Algebraic, extremal and metric combinatorics, 1986 (Montreal, PQ, 1986) (Cambridge), London Math. Soc. Lecture Note Series, vol. 131, Cambridge University Press, 1988, pp. 219–237.

115. ———, *Subgroup Complexes*, Surveys and Monographs, American Math. Society, to appear (2008?), preprint at http://www.math.uic.edu/~smiths/book/master.ps.

116. S. D. Smith and S. Yoshiara, *Some homotopy equivalences for sporadic geometries*, J. Algebra **192** (1997), 326–379.

117. E. Spanier, *Algebraic Topology*, McGraw-Hill, New York, 1966.

118. J. Thévenaz and P. Webb, *Homotopy equivalence of posets with a group action*, J. Comb. Th. A **56** (1991), 173–181.

119. J. Tits, *Buildings of Spherical Type and Finite BN-Pairs*, Lecture Notes in Mathematics, vol. 386, Springer-Verlag, Berlin/New York, 1974.

120. ———, *A local approach to buildings*, The geometric vein, Springer-Verlag, Berlin/New York, 1981, pp. 519–547.

121. K. Uno and S. Yoshiara, *Dade's conjecture for the simple O'Nan group*, J. Algebra **249** (2002), 147–185.

122. P. J. Webb, *A local method in group cohomology*, Comment. Math. Helvetici **62** (1987), 135–167.

123. ———, *Subgroup complexes*, The Arcata Conference on Representations of Finite Groups (Arcata, Calif., 1986), Proc. Symp. Pure Math., vol. 47, American Math. Society, 1987, pp. 349–365.

124. ———, *A split exact sequence of Mackey functors*, Comment. Math. Helvetici **66** (1991), 34–69.

125. C. A. Weibel, *Introduction to homological algebra*, Cambridge Studies in Advanced Mathematics, vol. 38, Cambridge University Press, 1994.

126. R. A. Wilson, *The complex Leech lattice and maximal subgroups of the Suzuki group*, J. Algebra **84** (1983), 151–188.

127. S. Yoshiara, *Some geometries for* J_3 *and* $O'N$, European J. Combin. **10** (1989), 499–506.

128. ———, *The radical 2-subgroups of the sporadic simple groups* J_4, Co_2 *and* Th, J. Algebra **233** (2000), 309–341.

129. ———, *Radical subgroups of the sporadic simple group of Suzuki*, Groups and combinatorics—in memory of Michio Suzuki (Eiichi Bannai, Hiroshi Suzuki, Hiroyoshi Yamaki, and Tomoyuki Yoshida, eds.), Advanced Studies in Pure Mathematics, vol. 32, Mathematical Society of Japan, Tokyo, 2001, pp. 453–464.

130. ———, *The radical 2-subgroups of some sporadic simple groups*, J. Algebra **248** (2002), 237–264.

131. ———, *Radical 2-subgroups of the Monster and Baby Monster*, J. Algebra **287** (2005), 123–139.

Index

Titles in This Series

147 **David J. Benson and Stephen D. Smith,** Classifying Spaces of Sporadic Groups, 2008

146 **Murray Marshall,** Positive Polynomials and Sums of Squares, 2008

145 **Tuna Altinel, Alexandre V. Borovik, and Gregory Cherlin,** Simple Groups of Finite Morley Rank, 2008

144 **Bennett Chow, Sun-Chin Chu, David Glickenstein, Christine Guenther, James Isenberg, Tom Ivey, Dan Knopf, Peng Lu, Feng Luo, and Lei Ni,** The Ricci flow: Techniques and applications, Part II: Analytic aspects, 2008

143 **Alexander Molev,** Yangians and classical Lie algebras, 2007

142 **Joseph A. Wolf,** Harmonic analysis on commutative spaces, 2007

141 **Vladimir Maz′ya and Gunther Schmidt,** Approximate approximations, 2007

140 **Elisabetta Barletta, Sorin Dragomir, and Krishan L. Duggal,** Foliations in Cauchy-Riemann geometry, 2007

139 **Michael Tsfasman, Serge Vlăduţ, and Dmitry Nogin,** Algebraic geometric codes: Basic notions, 2007

138 **Kehe Zhu,** Operator theory in function spaces, 2007

137 **Mikhail G. Katz,** Systolic geometry and topology, 2007

136 **Jean-Michel Coron,** Control and nonlinearity, 2007

135 **Bennett Chow, Sun-Chin Chu, David Glickenstein, Christine Guenther, James Isenberg, Tom Ivey, Dan Knopf, Peng Lu, Feng Luo, and Lei Ni,** The Ricci flow: Techniques and applications, Part I: Geometric aspects, 2007

134 **Dana P. Williams,** Crossed products of C^*-algebras, 2007

133 **Andrew Knightly and Charles Li,** Traces of Hecke operators, 2006

132 **J. P. May and J. Sigurdsson,** Parametrized homotopy theory, 2006

131 **Jin Feng and Thomas G. Kurtz,** Large deviations for stochastic processes, 2006

130 **Qing Han and Jia-Xing Hong,** Isometric embedding of Riemannian manifolds in Euclidean spaces, 2006

129 **William M. Singer,** Steenrod squares in spectral sequences, 2006

128 **Athanassios S. Fokas, Alexander R. Its, Andrei A. Kapaev, and Victor Yu. Novokshenov,** Painlevé transcendents, 2006

127 **Nikolai Chernov and Roberto Markarian,** Chaotic billiards, 2006

126 **Sen-Zhong Huang,** Gradient inequalities, 2006

125 **Joseph A. Cima, Alec L. Matheson, and William T. Ross,** The Cauchy Transform, 2006

124 **Ido Efrat, Editor,** Valuations, orderings, and Milnor K-Theory, 2006

123 **Barbara Fantechi, Lothar Göttsche, Luc Illusie, Steven L. Kleiman, Nitin Nitsure, and Angelo Vistoli,** Fundamental algebraic geometry: Grothendieck's FGA explained, 2005

122 **Antonio Giambruno and Mikhail Zaicev, Editors,** Polynomial identities and asymptotic methods, 2005

121 **Anton Zettl,** Sturm-Liouville theory, 2005

120 **Barry Simon,** Trace ideals and their applications, 2005

119 **Tian Ma and Shouhong Wang,** Geometric theory of incompressible flows with applications to fluid dynamics, 2005

118 **Alexandru Buium,** Arithmetic differential equations, 2005

117 **Volodymyr Nekrashevych,** Self-similar groups, 2005

116 **Alexander Koldobsky,** Fourier analysis in convex geometry, 2005

115 **Carlos Julio Moreno,** Advanced analytic number theory: L-functions, 2005

114 **Gregory F. Lawler,** Conformally invariant processes in the plane, 2005

TITLES IN THIS SERIES

113 **William G. Dwyer, Philip S. Hirschhorn, Daniel M. Kan, and Jeffrey H. Smith,** Homotopy limit functors on model categories and homotopical categories, 2004

112 **Michael Aschbacher and Stephen D. Smith,** The classification of quasithin groups II. Main theorems: The classification of simple QTKE-groups, 2004

111 **Michael Aschbacher and Stephen D. Smith,** The classification of quasithin groups I. Structure of strongly quasithin K-groups, 2004

110 **Bennett Chow and Dan Knopf,** The Ricci flow: An introduction, 2004

109 **Goro Shimura,** Arithmetic and analytic theories of quadratic forms and Clifford groups, 2004

108 **Michael Farber,** Topology of closed one-forms, 2004

107 **Jens Carsten Jantzen,** Representations of algebraic groups, 2003

106 **Hiroyuki Yoshida,** Absolute CM-periods, 2003

105 **Charalambos D. Aliprantis and Owen Burkinshaw,** Locally solid Riesz spaces with applications to economics, second edition, 2003

104 **Graham Everest, Alf van der Poorten, Igor Shparlinski, and Thomas Ward,** Recurrence sequences, 2003

103 **Octav Cornea, Gregory Lupton, John Oprea, and Daniel Tanré,** Lusternik-Schnirelmann category, 2003

102 **Linda Rass and John Radcliffe,** Spatial deterministic epidemics, 2003

101 **Eli Glasner,** Ergodic theory via joinings, 2003

100 **Peter Duren and Alexander Schuster,** Bergman spaces, 2004

99 **Philip S. Hirschhorn,** Model categories and their localizations, 2003

98 **Victor Guillemin, Viktor Ginzburg, and Yael Karshon,** Moment maps, cobordisms, and Hamiltonian group actions, 2002

97 **V. A. Vassiliev,** Applied Picard-Lefschetz theory, 2002

96 **Martin Markl, Steve Shnider, and Jim Stasheff,** Operads in algebra, topology and physics, 2002

95 **Seiichi Kamada,** Braid and knot theory in dimension four, 2002

94 **Mara D. Neusel and Larry Smith,** Invariant theory of finite groups, 2002

93 **Nikolai K. Nikolski,** Operators, functions, and systems: An easy reading. Volume 2: Model operators and systems, 2002

92 **Nikolai K. Nikolski,** Operators, functions, and systems: An easy reading. Volume 1: Hardy, Hankel, and Toeplitz, 2002

91 **Richard Montgomery,** A tour of subriemannian geometries, their geodesics and applications, 2002

90 **Christian Gérard and Izabella Łaba,** Multiparticle quantum scattering in constant magnetic fields, 2002

89 **Michel Ledoux,** The concentration of measure phenomenon, 2001

88 **Edward Frenkel and David Ben-Zvi,** Vertex algebras and algebraic curves, second edition, 2004

87 **Bruno Poizat,** Stable groups, 2001

86 **Stanley N. Burris,** Number theoretic density and logical limit laws, 2001

85 **V. A. Kozlov, V. G. Maz′ya, and J. Rossmann,** Spectral problems associated with corner singularities of solutions to elliptic equations, 2001

84 **László Fuchs and Luigi Salce,** Modules over non-Noetherian domains, 2001

For a complete list of titles in this series, visit the
AMS Bookstore at **www.ams.org/bookstore/**.